Surface and Interfacial Defects in Nanomaterials for Sustainable Energy Production and Storage

Surface and Interfacial Defects in Nanomaterials for Sustainable Energy Production and Storage

Edited by

Noé Arjona
Lorena Álvarez Contreras
Minerva Guerra Balcázar

WILEY-VCH

Editors

Dr. Noé Arjona
Centro de Investigación en Materiales Avanzados
Miguel de Cervantes 120
Complejo Industrial Chihuahua
Chihuahua, Chih. C.P. 31136
México

Prof. Lorena Álvarez Contreras
Centro de Investigación en Materiales Avanzados
Miguel de Cervantes 120
Complejo Industrial Chihuahua
Chihuahua, Chih. C.P. 31136
México

Prof. Minerva Guerra Balcázar
División de Investigación y Posgrado
Facultad de Ingeniería
Universidad Autónoma de Querétaro
76010 Querétaro
México

Cover Image: © berCheck/Shutterstock

Typesetting: Straive, Chennai, India

Printing and Binding: CPI Group (UK) Ltd, Croydon, CR0 4YY

Library of Congress Card No.: applied for

British Library Cataloguing-in-Publication Data A catalogue record for this book is available from the British Library.

Bibliographic information published by the Deutsche Nationalbibliothek
The Deutsche Nationalbibliothek lists this publication in the Deutsche Nationalbibliografie; detailed bibliographic data are available on the Internet at <http://dnb.d-nb.de>.

Print ISBN: 978-3-527-35464-1
ePDF ISBN: 978-3-527-85026-6
ePub ISBN: 978-3-527-85025-9
oBook ISBN: 978-3-527-85027-3

C9783527354641_111025

Contents

Preface

Currently, the world is facing challenges in energy production, storage, and distribution. As the global energy demand continues to rise, environmental pollution intensifies the need for innovative, low-carbon energy solutions. These may arise from either the development of novel technologies or optimization of existing processes. In this context, electrochemical energy conversion and storage systems have emerged as promising alternatives, and are increasingly being integrated into stationary and mobile applications, including electromobility. Despite significant advances, the electrochemical processes that govern these technologies remain far from fully optimized. Thus, electrocatalysis is relevant in electrochemical energy systems. Since the 1990s, the rapid expansion of nanomaterial synthesis has led to remarkable progress in the development of electrocatalysts. Researchers have explored the influence of particle size, morphology, and composition, as well as electronic factors such as metal-metal (M–M) and metal-support (M–S) interactions, in the quest for improved catalytic efficiency. Recently, the emergence of defect engineering as a scientific frontier has redefined material design through intentional manipulation of matter at the sub-nanometric scale. This approach not only targets bulk and surface defects, but also encompasses interface engineering, focusing on M–M and M–S interactions at the atomic level. Although defect science is widely recognized in the field of semiconductors, its application in electrochemical energy systems, particularly energy conversion and storage, remains relatively unexplored. The scarcity of dedicated literature on this topic motivated the creation of this book. We consider this volume as the first deliberate step toward establishing defect and interface engineering as a strategic approach in the rational design of electrochemical systems. More than a collection of current advances, this work articulates a scientific vision in which atomic-level control of matter is no longer optional but essential to achieving significant improvements in efficiency, durability, and sustainability. By integrating emerging knowledge across disciplines, this volume sets the stage for a new paradigm in materials science, where structural imperfections become a tool, and the interface, a platform for innovation. We hope that it will serve not only as a reference but also to catalyze future research and technological development at the intersection of matter, energy, and function.

April 2025

Noé Arjona
Centro de Investigación en
Materiales Avanzados
Miguel de Cervantes 120
Complejo Industrial Chihuahua
Chihuahua, Chih. C.P. 31136
México

1

Fundamentals of Nanomaterials in Energy Systems

Ricardo Antonio Escalona-Villalpando[1], Fabiola Ilian Espinosa-Lagunes[1], Luis Gerardo Arriaga Hurtado[2], and Janet Ledesma-García[1]

[1] *Universidad Autónoma de Querétaro, Facultad de Ingeniería, División de Investigación y Posgrado, Cerro de las Campanas s/n, Santiago de Querétaro, 76010, México*
[2] *Tecnológico de Monterrey, Institute of Advanced Materials for Sustainable Manufacturing, Epigmenio González 500, Santiago de Querétaro, 76130, México*

1.1 Introduction

Nanomaterials, such as nanoparticles, nanotubes, nanowires, and quantum dots, are of interest in various application areas of technology. They are revolutionizing energy systems and improving the efficiency of energy conversion, storage, and utilization processes due to their properties, such as surface with excellent electron transport capacity, physicochemical properties, mechanical strength, and unique dimensions, which offer advantages and advances in new technologies [1–5]. These properties of nanomaterials have become interesting for supercapacitors, fuel cells, solid-state batteries, photocatalysis, light-emitting diodes, hydrogen storage systems, etc. [1, 2, 6–10].

However, the integration of nanomaterials into energy systems is associated with challenges, such as sustainability and renewable energy, improving energy storage, reducing environmental impact, and conditioning materials with specific system properties [1, 11]. For example, suppose that the size of nanomaterials in semiconductors is reduced (Figure 1.1). In this case, the band gap increases due to the quantum confinement effect, and the optical properties such as absorption, emission, and excitation can be adjusted [12–14]. In solar energy applications, nanomaterials can significantly improve the light absorption and conversion efficiency of photovoltaic cells. In battery technology, these materials achieve higher energy densities, fast charging times, and a long service life [10, 15, 16]. For the development of fuel cells, supercapacitors, and hydrogen storage, nanomaterials facilitate efficient electrochemical reactions in the storage of large amounts of energy in a compact form, making them indispensable in the search for next-generation energy solutions [6, 12, 17–21]. Compared to carbon-based nanomaterials, which have minimal surface defects, this composition is critical as an alternative to stainless

Surface and Interfacial Defects in Nanomaterials for Sustainable Energy Production and Storage, First Edition. Edited by Noé Arjona, Lorena Álvarez Contreras, and Minerva Guerra Balcázar.

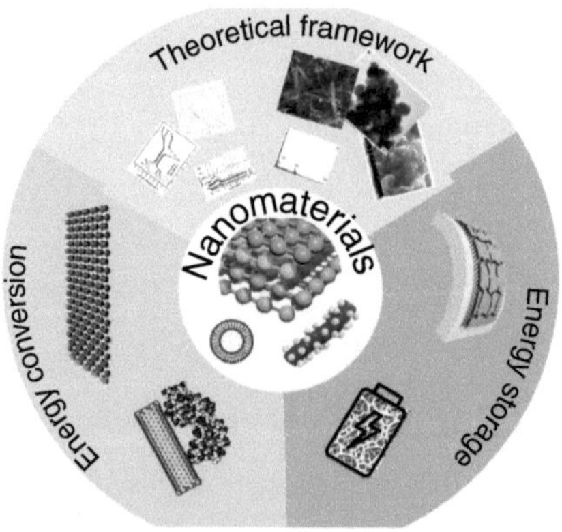

Figure 1.1 Schematic overview of the applications of nanomaterials in various fields.

steel or iron for mechanical strength and stability. Further studies are required to confirm its feasibility as a replacement.

Understanding the fundamentals of nanomaterials and their applications in energy systems is critical for research groups seeking to innovate and improve energy technologies [4, 13, 22, 23]. This introduction provides an overview of the progress and potential future directions in the field of nanomaterials for energy storage and conversion of energy systems.

This chapter highlights future research opportunities and proposes solutions to bridge the gap between fundamental research and real-world applications in next-generation energy technologies.

1.1.1 Nanomaterials in Battery Systems

Nanomaterials have significantly improved energy conversion and storage in battery systems, especially in lithium-ion, Li-S, metal-air, and solid-state batteries. They improve electrochemical reaction kinetics; reduce ion diffusion pathways; and improve stability, surface area, cycle life, and miniaturization. The main components of battery systems in which energy conversion takes place are the cathode, the anode, the separator, and the electrolyte. Together they ensure the cell voltage (E), the energy and potential density, the stability, and the service life.

Lithium-ion batteries (LIBs), as shown in Figure 1.2, typically use graphite for the anode and $LiCoO_2$ for the cathode, resulting in an energy density of around 150 Wh kg^{-1}. This energy density is achieved by the movement of lithium ions through the internal passageway and electrons through the external circuit, which eventually return to the positive electrode. The following equation illustrates the ideal electrochemical reactions that take place during the charge/discharge process, with the cathode reaction represented by given Eqs. (1.1) and (1.2) showing what happens in the anode [24].

$$LiCoO_2 \leftrightarrow xLi^+ + xe^- + Li_{1-x}CoO_2 \tag{1.1}$$

$$6C + xLi^+ + xe^- \leftrightarrow Li_xC_6 \tag{1.2}$$

Figure 1.2 Schematic of a lithium-ion battery (LIB).

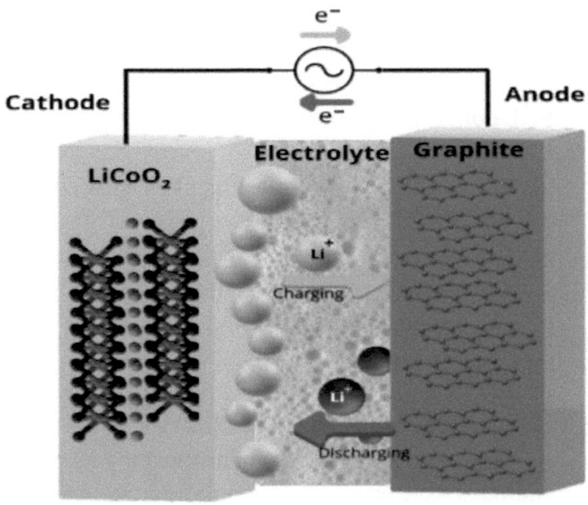

Research into nanomaterials in LIBs has significantly improved their efficiency in meeting new energy needs, e.g. in electric vehicles and trains, and even in meeting lower energy needs, e.g. in biomedical devices. For example, Si-based nanomaterials have attracted interest as anode electrode materials due to their theoretical capacity of 4200 mAh g^{-1} compared to that of graphite (372 mAh g^{-1}).

i) Nanomaterials for the Anode
Conventionally, graphite is the most widely used material in LIBs, although this material tends to degrade, limiting its useful life. Carbon-based nanomaterials, such as carbon nanotubes (CNTs), carbon nanofibers, porous carbon, and graphene, have been investigated to reduce this drawback. In addition, other nanostructured materials that are not carbon-based have been reported. These include Ti/TiO_2-based materials, Si-based materials, spinel $Li_4Ti_5O_{12}$, germanium, and metal oxides as iron and cobalt, which have the potential to improve the performance of the anode [25].
ii) Nanomaterials for the Cathode
Lithium nickel manganese cobalt oxide (NMC) and lithium manganese oxide (LMO) usually have a long lifetime with about 500 charge/discharge cycles. NMC offers specific capacities of 150–220 Wh kg^{-1}, while LMO delivers 100–150 Wh kg^{-1}. Other Li-based nanomaterials have been reported, such as $LiCoO_2$, $LiMn_2O_4$, $LiNiO_2$, $LiFePO_4$, $LiMnPO_4$, $LiFeSO_4F$, $LiCoNiMnO_2$, $LiNiCoMnO_2O$, and lithium–nickel–cobalt–manganese (Li–Ni–Co–Mn), in which there are several monoanionic and polyanionic compounds whose chemistry is favored for their high-energy storage capacity [24].
iii) Separator
The separator is usually a porous material that separates the anode and cathode to prevent short circuits, allowing the battery to charge/discharge. Polymers such as polyethylene and polypropylene are usually used, but they have limitations such as low stability in mechanical strength, low melting points, and

lower chemical resistance, which reduce the efficiency of the battery's overall performance. Nanomaterials such as SiO_2, TiO_2, Al_2O_3, and SiO_2 are alternatives to these inconvenient materials, which mainly bring advantages such as high porosity and chemical and mechanical resistance [25].

iv) Electrolyte
Electrolytes enable the movement of positive lithium ions between the electrodes in LIBs. There are two types of electrolyte materials: solid polymer electrolytes (SPEs) and liquid electrolytes (LEs). An SPE consists of a solid polymer matrix infused with lithium salts. Ceramic nanopowders investigated in this context include Al_2O_3, SiO_2, TiO_2, and $Li_{1.4}Al_{0.4}Ti_{1.6}(PO_4)_3$, which can be incorporated into polyethylene-based electrolytes to improve stability. In addition, SiO_2 (MA-SiO_2)-nanostructured materials functionalized with methacrylate are used for electrolytes in LIBs. In recent years, solid-state and gel electrolytes have been investigated and nanomaterials or nanostructured additives have been integrated to improve ion transport in batteries [25].

v) Lithium–Sulfur Batteries (LSBs)
Another type of battery used in energy systems are lithium–sulfur batteries (LSBs), which have a theoretical capacity of 1675 mA hg^{-1} and a gravimetric and volumetric energy density of 2500 Wh kg^{-1} and 2800 W h L^{-1}, respectively, compared to LIBs. The components of this type of battery are the same as those of LIBs: cathode, anode, separator, and electrolyte. Nanomaterials such as $LiCoO_2$, $LiFePO_4$, $LiMnO_2$, tetratitanium heptoxide (Ti_4O_7), and Ti–O_2–TiN heterostructure have been described for use as cathodes for LSBs, with a discharge capacity of 704 mAh g^{-1} at the 2000th cycle. In addition, perovskite-based materials such as $La_{0.6}Sr_{0.4}CoO_3$-δ and $Ba_{0.5}Sr_{0.5}Co_{0.8}Fe_{0.2}O_3 - \delta$ can immobilize polysulfides through chemical bonding between lithium and oxygen or sulphidic and sulfur, resulting in improved performance [26].

CNTs were used for the anode. Graphene, a 2D nanomaterial, forms nanoparticles together with 3D hybrid nanomaterials and a metal oxide or sulfide clusters, which are added to the sulfur–carbon nanostructures to achieve high performance. Examples are CeO_2, Al_2O_3, La_2O_3, MgO, and CaO.

vi) Zinc–Air Batteries
Zinc–air batteries (ZABs) use Zn as the reaction electrode due to their cost advantage, reversibility, and a promising theoretical specific energy of 1350 Wh kg^{-1}. However, their practical specific energy is only about 200 Wh kg^{-1}. Their disadvantage is their limited performance and application. This type of battery is divided into alkaline and non-alkaline systems. In the mechanism of ZABs shown in Figure 1.3, zinc and oxygen react to complete the redox process and generate electricity, using Zn at the anode and air at the cathode in the presence of an electrolyte [27, 28]. In charging, the overall reaction involves ZnO to generate Zn and O_2, whereas in discharging the process is reversible and cyclic.

Carbon nanomaterials such as CNTs and CNFs were used as substrates and even mixed with nanomaterials. Among the nanomaterials for Zn electrodes,

Figure 1.3 The mechanism of the zinc–air batteries (ZABs).

Charge

Cathode: $ZnO + H_2O + 2OH^- \rightarrow Zn(OH)_4^{2-}$

$Zn(OH)_4^{2-} + 2e^- \rightarrow Zn + 4OH^-$

Anode: $2OH^- \rightarrow \frac{1}{2} O2 + H_2O + 2e^-$

Overall reaction: $2ZnO \rightarrow 2Zn + O_2$

Discharge

Negative electrode: $Zn + 4OH^- \rightarrow Zn(OH)_4^{2-} + 2e^-$

$Zn(OH)_4^{2-} \rightarrow ZnO + H_2O + 2OH^-$

Positive electrode: $\frac{1}{2} O2 + H_2O + 2e^- \rightarrow 2OH^-$

Overall reaction: $2Zn + O_2 \rightarrow 2ZnO$

the use of ZnZnO/C, Zn/CNT, ZnZu/Zn, Zn-ZnO@/C, ZnO/C, Zn@ZSO, C/Zn, Zn/ZnO–N–C, Ti_3C_2Tx@Zn, and Zn/CNT has been reported, and it should be noted that core–shell materials and bimetallic materials have been reported. While the electrolyte is an important component of this type of battery, they have been improved by the addition of nanomaterials to increase their efficiency in ion transport. 3D Nanoporous composites, Zn nanoarrays, nanoshells/nanoparticles, Cu nanowire Zn–Si–O nanosheets, Zn nanosheets, Zn–N–C microsheets, and ZnO nanorod-GO have been reported to show improvements in efficiency, energy, and cycling [29].

1.1.2 Nanomaterials in Fuel Cells

A fuel cell (FC) is a type of electrochemical system for generating electricity through chemical reactions using a constant flow of fuel and oxidizing agent. The components of an FC are the anode, the cathode, the electrolyte, the fuel, the oxidizing agent, and the membrane (except for microfluidic FC membranes) [30]. There are different types of fuels such as proton exchange membrane fuel cells (PEMFCs), phosphoric acid fuel cells (PAFCs), molten carbonate fuel cells (MCFCs), solid oxide fuel cells (SOFCs), alkaline fuel cells (AFCs), and microbial fuel cells (MFCs). It can even include biofuel cells and microfluidic fuel cells. Depending on the type of FC, they produce different power outputs at different operating temperatures; AFCs can operate at room temperature and are reported to have a power output of 100 kW, followed by PEMFCs between 100 and 250 kW, MCFCs at 1000 kW, PAFCs between 1000 and 2000 kW, and SOFCs up to 3000 kW but at a temperature of up to 1000 °C [31].

The diverse and continuous research of FCs is mainly due to their applications and prospects, such as fuel storage, green applications, automotive, aerospace, stationary power, portable power, small electronics, and military applications, which is why they continue to be investigated [31]. A search in Scopus® using the keywords nanomaterials and fuel cell shows an increase in publications since 2007 to a maximum of 600 articles per year. In this context, nanomaterials have played an important role in improving performance, lifetime, output power, and current, opening up the possibility of commercializing this device (Figure 1.4).

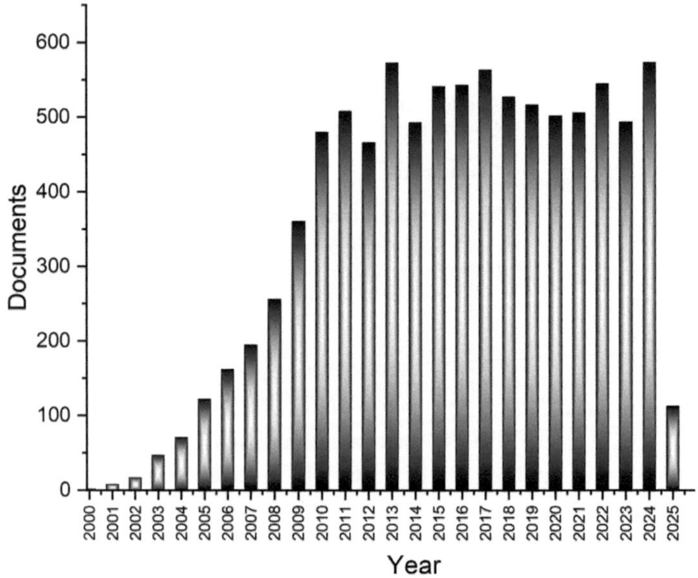

Figure 1.4 Technological advances in nanomaterials for fuel cells. Source: Adapted from Scopus®.

1.1.3 Nanocatalysts for Fuel Oxidation

In FCs, the oxidation of the fuel, which can be mainly methanol, ethanol, or hydrogen, takes place at the anode, where electrons are generated that travel in an external circuit, while ions in the electrolyte migrate to the cathode. The first nanomaterials used came from noble metals such as Au, Ag, Pd, and Pt as well as from bimetals in the form of alloys, layers, and core–shells such as Au@Pd, Au@Pt, Pt@Pd, and Pd@Pt, but also from other combinations with metals that help to reduce the amount of noble metal, such as Pt–Co–ZrO_2/MWCNTs, Pt–Fe in a nitrogen-doped carbon carrier, and Pt–Ga nanowires made of an ultra-thin alloy. It is worth mentioning that several of these nanocatalysts are supported on carbonaceous nanoparticles such as MWCNTs, rGO, GO, nanofibers, or some non-carbonaceous ones such as ZrO_2, TiO_2, indium oxides, alumina, silica and tungsten oxide, ceria, and conducting polymer nanomaterials [32].

While, in the cathode, for the oxygen reduction reaction (ORR), the chemical reaction in which oxygen is reduced to water, Pt is usually preferred or mixed with other metals to reduce the amount used but with similar efficiencies such as Pt–Fe, Pt–Cr and Pt–Cr–Co, FeNCS, Ni/C, FeNi–NC, $AgAg_2WO_4$, AgCoCu-rGO, Fe–N–C, Zn–N–C, FeN_4-O-NCR, Cu/CuONC-900, Mn, Ni/NCNTs, Bi_2SiO_5NG, N-doped porous CNFs, N, F co-doped CNFs, Fe–N, No–N–CNFs, FeNi–N, NiCo–N–NFs, NiCo, ZnCo, $LaCoO_3$, and $CaMnO_3$ [33].

In the FCs using proton exchange membranes (PEM) (Figure 1.5), whose function is ion transport, which can be protonic or anionic, Nafion® was mainly used as the perfluorosulfonate resin. However, other membranes with similar properties were also investigated, such as poly(ether ether ketone) (PEEK),

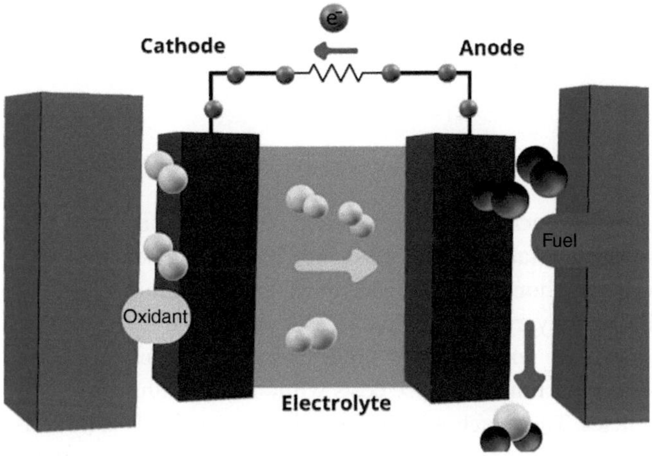

Figure 1.5 FCs that use PEM membranes.

[4, 13, 22–24] polybenzimidazole (PBI), [29–32] polystyrene sulfonate (PSS), [33, 34] polyvinylidene fluoride (PVDF) and its copolymers, and 41–44 polysulfone (PSU), which has its advantages and disadvantages, such as solubility, low ionic conductivity, low mechanical, or temperature stability, which is why its composition with nanomaterials as silica titania MOFs, MoS_2 two-dimensional nanomaterials, graphene oxide, boron nitride, zeolite, ZrP, and silica/bentonite was investigated to improve proton conductivity, membrane stability, permeability, and terminal and mechanical stability [34–36].

1.1.4 Nanomaterials for Photovoltaic Cells

Solar or photovoltaic (PV) cells are technological energy systems that convert solar energy into sustainable and environmentally friendly electrical energy. This promising energy system is currently used for powering homes, water heating, solar power plants, solar-powered transportation, space applications, portable electronics, and charging stations. The amount of energy produced can be calculated by taking into account the theoretical energy efficiency of the cell, which can produce in a given time given a percentage of its stated output. This means that a cell with an energy efficiency of 300 W for five hours and 80% of its power could produce 1200 Wh or 1.2 kW day^{-1}, which could theoretically cover the energy consumption of an average household. On the other hand, it has some disadvantages, such as intermittent energy, the installation is expensive, the technology has not yet reached a high efficiency to meet the daily demand in medium/large cities, and constant maintenance or replacement of the cells is necessary [37].

The power conversion efficiency (Z) of these devices is given by Eq. (1.3). Pinc is the incident solar power, V_{OC} is the open circuit voltage, J_{SC} is the short-circuit current density, and FF is the fill factor. J_{SC} is the current density at zero bias voltage and V_{OC} is the maximum voltage that a solar cell can generate. This conversion efficiency can vary depending on the material used and physical aspects such as light irradiation, light absorption, separation of electrons and holes, transport of charge carriers generated by light to the electrodes, and energy dissipation in the circuit [38].

$$n = \frac{V_{OC} J_{SC} FF}{P_{inc}} \tag{1.3}$$

The development of PV cells includes the first generation of silicon solar cells, which were characterized by a relatively high efficiency but were associated with high manufacturing and installation costs. The second generation of thin-film solar cells used amorphous silicon, polycrystalline silicon, CdTe, and Cu–Ir–Ga–Se; these technologies were less expensive and increased production throughput. The third generation of solar technologies aims to reduce costs while maintaining high solar energy conversion efficiencies. These technologies include dye-sensitized solar cells (DSSCs), organic photovoltaics, inorganic photovoltaics in the solution process, and colloidal quantum dot solar cells. Currently, solar cells made from crystalline, polycrystalline, and amorphous silicon account for more than 90% of global production [38].

In this sense, nanomaterials have emerged as a new alternative to achieve higher efficiency in the conversion of solar energy into electrical energy, reduce costs, find more affordable materials, extend useful life, increase energy efficiency in portable devices, and miniaturize them. In addition to Si nanoparticles and Si-based materials, quantum dots (QDs) and quantum rods (QRs), nanomaterials that have the property of being semiconductors and utilize the absorption of light across the spectrum, depending on the choice of a semiconductor material due to its intrinsic bandgap tuning properties, have also been presented. Lead sulfide (PbS) and group IV–VI QDs have already been documented, each absorbing in different spectral regions. Similarly, the quantum confinement effect allows the absolute positions of quantized energy levels to be adjusted. For example, modifications to cadmium selenide (CdSe) and cadmium telluride (CdTe) QDs have shown the potential to manipulate the charge separation at interfaces. In addition, consideration of the plasmonic properties of nanostructures can improve light concentration and energy efficiency. Various materials, including Si, InP/InGaAsP, and GaAs, and group IVB transition metal nitrides (TMNs), such as titanium nitride (TiN), zirconium nitride (ZrN), and hafnium nitride (HfN), have already been described [39].

One strategy to increase the performance of PV cells is to use inorganic semiconductor/polymer nanocomposites with hole-conducting polymers and semiconductor quantum materials such as CdSe, PbS, and PbSe QDs or QRs. This approach combines the advantages of both materials: low-cost, easy processing in solution, lightweight, flexible substrates, and high electron mobility, which is influenced by size, optical bandgap, and plasmonic properties. Some quantum-confined nanocrystals such as PbS, PbSe, PbTe, CdSe, InAs, InP, and Si-modified materials exhibit multiple exciton generation with a quantum yield of over 100%, improving the short-circuit current of solar cells [40].

1.1.5 Nanomaterials for the Application of Capacitors

Electrochemical capacitors enable energy storage by ion adsorption (electrochemical double-layer capacitors) or in the non-Faradaic part of the charge/discharge at the interface between a material and an electrolyte (pseudocapacitors). We can also mention ultracapacitors or supercapacitors, which can be charged/discharged quickly and have a lower energy density (5 Wh kg^{-1}) but a high power (10 kW kg^{-1}). Hybrid capacitors are a combination of a capacitor electrode and a battery electrode, combining both the advantages and disadvantages of both systems. They have gained attention due to their applications, especially in transportation, such as hybrid electric vehicles, subways, and streetcars [41].

Electrochemical double-layer capacitors (EDLCs) store energy by electrostatic charging using the reversible absorption of electrolyte ions within the electroactive material; charge separation occurs upon polarization at the electrode–electrolyte interface and is described by Eq. (1.4).

$$C/A = \frac{\varepsilon_r \varepsilon_0}{d} \tag{1.4}$$

where ε_r is the electrolytic dielectric constant, ε_0 is the dielectric constant, d is the double layer thickness, and A is the electrode surface area. The capacitance model of Goury–Chapman–Stern–Geary, in which the stored energy (E) is proportional to the square of the voltage, is also accepted, as shown in Eq. (1.5):

$$E = \tfrac{1}{2}\,\mathrm{CV}^2 \tag{1.5}$$

Capacitance is investigated using electrochemical techniques such as cyclic voltammetry (CV), galvanostatic charge/discharge, and electrochemical impedance spectroscopy. These methods enable the calculation of the specific capacitance (C_{sp}). The key variables include the charge (Q), electrode mass (m), potential window (ΔV), discharge current (I), and discharge time (Δt).

$$C_{sp} = \frac{Q}{m\Delta V}\ \text{by cyclic voltammetry} \tag{1.6}$$

$$C_{sp} = \frac{I\Delta t}{m\Delta V}\ \text{by galvanostatic charge/discharge} \tag{1.7}$$

A C_{sp} shows a more efficient capacitor for storing energy and the energy distribution with the charging/discharging times, Figure 1.6 [42]. As can be observed in both systems, the efficiency of the capacitance is directly related to the type of material used, making nanomaterials a widely used resource due to their versatility, large surface area, and charge storage. These nanomaterials can be made from carbon, metal oxides, and hybrids with 0, 1, 2, and 3 dimensions [43].

1.1.6 Carbon Nanomaterials

CNTs are used as supports, hydrogen storage devices, batteries, sensors/biosensors, fuel cells, semiconductors, and supercapacitors. Due to their high porosity, larger surface area, and higher charge transport capacity, they can be doped with hydrogen atoms that improve their capacitive properties. This type of material has been

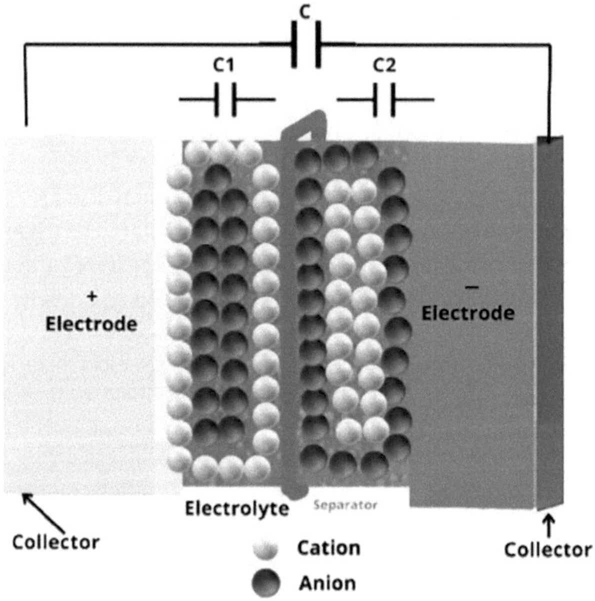

Figure 1.6 Schematic of an electrochemical double-layer capacitor.

reported to have a higher capacitance, close to the estimated theoretical capacitance (71–178 Fg^{-1}) [44].

Graphene foam is a two-dimensional material with sp^2 hybridization of polycyclic aromatic hydrocarbons. It has demonstrated an energy density of 26 Wh kg^{-1} using an organic electrolyte (1 M $LiPF_6$). Other features include a high theoretical specific surface area of 2630 $m^2\ g^{-1}$ and the ability to facilitate electron transfer along its surface, with a very low charge transfer resistance [44].

1.1.6.1 Activated Carbon

Activated carbon is characterized by a large surface area, favorable electrical properties, low cost, and capacitance properties compared to those of graphene and CNTs [44].

Mesoporous carbon (MC) is ideal for supercapacitor electrodes due to its large specific surface area with different pore sizes. In the synthesis of ordered mesoporous carbon, a mixture containing P-doped MC, which has a better porous structure than undoped MC, is pyrolyzed. This structure has a specific surface area of 403.5 $m^2\ g^{-1}$, a pore volume of 0.545 $m^3\ g^{-1}$, and an average pore size of 4.64 nm. P-doped MC has a higher specific capacitance of 288 F g^{-1} and a robust cycle stability of 97.6% [44].

1.1.7 Metal Oxides/hydroxides

The capacitance values of metal oxides/hydroxides are significantly lower than those of carbon nanomaterials, as their specific pseudocapacitance exceeds that of carbon materials. In addition, they are very versatile and can be used in different types of structures. They support corrosion, are friendly to the supporting electrolyte, and improve their specific capacities [42].

Ruthenium oxide (RuO_2) is well studied because of its high pseudocapacitance, estimated at 1450 Fg^{-1} and 1360 Fg^{-1}. Various nanostructures such as nanotubes, nanorods, nanoflowers, and nanosheets have been investigated. Mesoporous RuO_2 with a high surface area of 140 m^2 g^{-1} has a capacity of 202 Fg^{-1} compared to 146 Fg^{-1} with a lower surface area of 39 m^2 g^{-1}, emphasizing the importance of surface area.

Another much-studied metal oxide is manganese oxide (MnO_2), which is inexpensive and less toxic than RuO_2. MnO_2 has a high theoretical specific capacity of 1370 Fg^{-1}. Nickel-based compounds such as $NiO/Ni(OH)_2$ have also been investigated in various structures including porous films, hollow nanospheres, nanoribbon nanoplatelets, nanorod arrays, nanowires, nanoflakes, nanosheets, nanowhiskers, flower-like microspheres, 3D dendrites, and nanoflowers due to their diverse surfaces for Faradaic reactions and high specific capacitance.

Other metal oxides with pseudocapacitance properties have also been reported, including $Co_3O_4/Co(OH)_2$, Fe_2O_3/Fe_3O_4, Cu_2O/CuO, V_2O_5, SnO_2, Bi_2O_3, and titanium oxide-based nanowires. However, individual metal oxides are often subject to various limitations and generally have a lower energy storage capacity. To solve these problems, researchers have explored different types of core–shell and bimetallic metal oxides. Examples include various spinel metal oxides such as $MnCo_2O_4$, $CoMn_2O_4$, $MnNi_2O_4$, $NiCo_2O_4$, $NiMn_2O_4$, and $ZnCo_2O_4$. For example, porous nanowires made of $MnCo_2O_4$ have been reported to have a surface area of 106.6 m^2g^{-1} and a high specific capacitance of 2108 Fg^{-1} [42].

1.1.8 Conductive Polymers

Conductive polymers (CPs) provide pseudocapacitance through fast reversible redox reactions of p-conjugated polymer chains but have limited nanostructural morphology and lower capacitive performance than metal oxides. Their advantages lie in their flexibility, higher conductivity, and ability to form thin films. Examples of these CPs are polyaniline, polypyrrole, and polythiophene, whose pseudocapacitance performance largely depends on phase crystallinity, surface area, morphology, conductivity, and mass loading over the electrode substrates [44].

Hybrid materials that combine metal oxides and conductive polymers have problems with cyclicity due to volume changes. To solve this problem, nanostructural composites with stable carbon materials such as carbon nanotubes, graphene, carbon dots, and activated carbon are preferred. For example, CuO@MnO2/N-MWCNT composite electrodes achieve a specific capacitance of 184 $F\,g^{-1}$ at 0.5 $A\,g^{-1}$. Similarly, SiC@PANI electrodes offer a specific capacitance of 352 $mF\,cm^{-2}$ at 1 $mA\,cm^{-2}$, outperforming SiC or PANI (polyaniline) alone. In addition, 20 wt% PANI/MoOx nanowires exhibit a specific capacitance of 450 Fg^{-1} [45].

Hybrid supercapacitors showed improved electrochemical performance when two different electrodes with different materials were used, such as $ZnSnO_3@In_2O_3$ core–shell microfibers as well as $Fe_2O_{3\text{-}x}@rGO/MnO_2/rGO$, CS@ZnO core–shell

NCs, and RGO/MnOx@HCNs exhibit a high specific surface area ranging from 493 to 668 m^2 as significant MnO, which has a specific capacitance of 355 Fg^{-1} at a current density of 1 Ag^{-1} and 270 Fg^{-1} at 1.0 Ag^{-1} in a two-electrode system using 6 M KOH as the electrolyte medium. These technologies improve stability and cycle life, and, in some cases, lead to a higher capacity by reducing the pseudocapacitance [45].

1.2 Conclusions

In summary, nanomaterials have fundamentally advanced energy systems by improving the performance, stability, and longevity of battery components. Their diverse applications in anodes, cathodes, separators, and electrolytes underline their central role in the development of high-performance energy storage solutions. Continued research and development of nanomaterials promises further improvements in the efficiency and performance of future energy systems.

References

1. Durgalakshmi, D., Balakumar, S., Rajendran, S., and Naushad, M. (2020). Functional nanomaterial in energy and environmental science. In: *Nanomaterials for Sustainable Energy and Environmental Remediation*, 1–23. Elsevier.
2. Gupta, C. and Bhardwaj, A. (2020). Summary and future perspectives of nanomaterials and technologies. In: *Nanomaterials for Sustainable Energy and Environmental Remediation*, 333–353. Elsevier.
3. Nanomaterials for Su stainable Energy Applications [Internet]. Available from: www.routledge.com/.
4. Jaffer, J.M. and Mohamedd, S. *Modern Nanotechnology Volume 1: Environmental Sustainability and Remediation*. Springer Nature.
5. Naik, R., Nagaswarupa, H.P., Ananda Murthy, H.C., and MET, S. (ed.) (2024). *Advanced Nanomaterials for Energy Storage Devices [Internet]*. Cham, Switzerland: Springer Nature. (Nanostructure Science and Technology). Available from: https://link.springer.com/10.1007/978-3-031-74730-4.
6. Xin, S., Gao, H., and Guo, Y.G. (2019). Charge transfer and storage of an electrochemical cell and its nano effects. In: *Nanostructures and Nanomaterials for Batteries: Principles and Applications*, 29–87. Springer Singapore.
7. Xin, S., Gao, H., Li, Y., and Guo, Y.G. (2019). Introduction to electrochemical energy storage. In: *Nanostructures and Nanomaterials for Batteries: Principles and Applications*, 1–28. Springer Singapore.
8. ALOthman, Z.A., Rodriguez-Padron, D., Luque, R., and Balu, A.M. (2021). Innovative nanomaterials for energy storage: moving toward nature-inspired systems. In: *Current Opinion in Green and Sustainable Chemistry*, vol. 32. Elsevier B.V.

9 Wu, X.L., Guo, J.Z., and Guo, Y.G. (2019). Traditional nanostructures and nanomaterials in batteries. In: *Nanostructures and Nanomaterials for Batteries: Principles and Applications*, 313–357. Springer Singapore.

10 İnada, A.A., Arman, S., and Safaei, B. (2022). A novel review on the efficiency of nanomaterials for solar energy storage systems. *J. Energy Storage*. Elsevier Ltd 55: 105661.

11 Nanomaterials for Sustainable Energy Applications [Internet]. Available from: www.routledge.com/.

12 Yu, R., Feng, S., Sun, Q. et al. (2024). Ambient energy harvesters in wearable electronics: fundamentals, methodologies, and applications. *J. Nanobiotechnol.* 22: 497.

13 Mohammed, H., Mia, M.F., Wiggins, J., and Desai, S. (2025). Nanomaterials for energy storage systems—A review. *Molecules* 30 (4): 883. Available from: https://www.mdpi.com/1420-3049/30/4/883.

14 Mishra, K., Devi, N., Siwal, S.S. et al. (2023). Hybrid semiconductor photocatalyst nanomaterials for energy and environmental applications: fundamentals, designing, and prospects. *Adv. Sustain. Syst.*John Wiley and Sons Inc; 7: 2300095.

15 Tyagi, P.K., Kumar, R., and Said, Z. (2022). Recent advances on the role of nanomaterials for improving the performance of photovoltaic thermal systems: trends, challenges and prospective. *Nano Energy*. Elsevier Ltd; 93: 106834.

16 Cheng, P., Wang, D., and Schaaf, P. (2022). A review on photothermal conversion of solar energy with nanomaterials and nanostructures: from fundamentals to applications. *Adv. Sustain. Syst.*John Wiley and Sons Inc; 6: 2200115.

17 Amardeep, A., Freschi, D.J., Wang, J., and Liu, J. (2023). Fundamentals, preparation, and mechanism understanding of Li/Na/Mg-Sn alloy anodes for liquid and solid-state lithium batteries and beyond. *Nano Res.*; 16: 8191–8218.

18 Xu, Y., Li, Y., Gao, L. et al. (2024). Advances and prospects of nanomaterials for solid-state hydrogen storage. *Nanomaterials*. Multidisciplinary Digital Publishing Institute (MDPI); 14: 1036.

19 Cavaliere, P. (2023). *Water Electrolysis for Hydrogen Production*, 1–836. Springer International Publishing.

20 Modupeola, D. and Popoola, P. (2023). High entropy nanomaterials for energy storage and catalysis applications. *Front. Energy Res.*; 11.

21 You, Y. and Guo, Y.G. (2019). Conclusions and perspectives on new opportunities of nanostrucutresnanostructures and nanomaterials in batteries. In: *Nanostructures and Nanomaterials for Batteries: Principles and Applications*, 359–379. Springer Singapore.

22 Liu, Y., Zhou, G., Liu, K., and Cui, Y. (2017). Design of complex nanomaterials for energy storage: past success and future opportunity. *Acc. Chem. Res.* 50 (12): 2895–2905.

23 Hasan, N., Muthu, M., Hakami, O., and Gopal, J. (2025). Assessing the sustainability of energy-related nanomaterial synthesis: emphasizing the need for energy-efficient nanomaterial preparation techniques. *Energies*. Multidisciplinary Digital Publishing Institute (MDPI); 18: 523.

24 Bruce, P.G., Scrosati, B., and Tarascon, J.M. (2008). Nanomaterials for rechargeable lithium batteries. *Angew. Chem. Int. Ed.* 47: 2930–2946.

25 Hur, J. (2022). Nanomaterials for ion battery applications. *Nanomaterials*. MDPI; 12: 2293.

26 Liu, H.K. (2013). An overview - Functional nanomaterials for lithium rechargeable batteries, supercapacitors, hydrogen storage, and fuel cells. *Mater. Res. Bull.* 48 (12): 4968–4973.

27 Contreras-Martínez, M.V., Arredondo-Espínola, A., Guerra-Balcázar, M. et al. (2025). Development of electrospun and casted poly(vinyl alcohol)/poly(acrylic acid) membranes and their effect on the performance of rechargeable zinc-air batteries. *J. Mater. Sci.* 60 (9): 4502–4518. Available from: https://doi.org/10.1007/s10853-025-10717-9.

28 Béjar, J., Álvarez–Contreras, L., Espinosa–Magaña, F. et al. (2021). Zn–air battery operated with a 3DOM trimetallic spinel ($Mn_{0.5}Ni_{0.5}Co_2O_4$) as the oxygen electrode. *Electrochim. Acta* 391: 138900.

29 Zhang, P., Wei, M., Wang, K. et al. (2025). Performance optimization of zinc-air batteries via nanomaterials. *Energy Stor. Mater.*; 75: 104109.

30 Estrada-Osorio, D.V., Escalona-Villalpando, R.A., Gurrola, M.P. et al. (2023). Abiotic, hybrid, and biological electrocatalytic materials applied in microfluidic fuel cells: a comprehensive review. *ACS Meas. Sci. Au.* 4 (1): 25–41; Available from: https://doi.org/10.1021/acsmeasuresciau.3c00044.

31 Manzo, D., Thai, R., Le, H.T., and Venayagamoorthy, G.K. (2025). Fuel cell technology review: types, economy, applications, and vehicle-to-grid scheme. *Sustain Energy Technol Assess.* Elsevier Ltd; 75: 104229.

32 Raduwan, N.F., Shaari, N., Kamarudin, S.K. et al. (2022). An overview of nanomaterials in fuel cells: synthesis method and application. *Int. J. Hydrog. Energy*. Elsevier Ltd; 47: 18468–18495.

33 Guo, S. and Wang, E. (2011). Noble metal nanomaterials: controllable synthesis and application in fuel cells and analytical sensors. *Nano Today* 6: 240–264.

34 Rico-Zavala, A., Pineda-Delgado, J.L., Carbone, A. et al. (2020). Composite sulfonated polyether-ether ketone membranes with SBA-15 for electrochemical energy systems. *Materials* 13 (7): 1570.

35 Rico-Zavala, A., Contreras-Martínez, M.V., Murillo-Borbonio, I. et al. (2020). Organic composite membrane for hydrogen electrochemical conversion devices. *Int. J. Hydrog. Energy* 45 (56): 32493–32507.

36 Qasem, N.A.A. (2024). A recent overview of proton exchange membrane fuel cells: fundamentals, applications, and advances. *Appl. Therm. Eng.*Elsevier Ltd; 252: 123746.

37 Chen, G., Seo, J., Yang, C., and Prasad, P.N. (2013). Nanochemistry and nanomaterials for photovoltaics. *Chem. Soc. Rev.*; 42: 8304–8338.

38 Yu, R., Lin, Q., Leung, S.F., and Fan, Z. (2012). Nanomaterials and nanostructures for efficient light absorption and photovoltaics. *Nano Energy*. Elsevier Ltd; 1: 57–72.

39 Günaydın, B.N., Çetinkaya, A.O., Torabfam, M. et al. (2024). Plasmonic group IVB transition metal nitrides: fabrication methods and applications in biosensing, photovoltaics and photocatalysis. *Adv. Colloid Interf. Sci.* Elsevier B.V.; 333: 103298.

40 Pusch, A., Bremner, S.P., Tayebjee, M.J.Y., and Daukes, N.J.E. (2021). Microscopic reversibility demands lower open circuit voltage in multiple exciton generation solar cells. *Appl. Phys. Lett.* 118 (15).

41 Simon, P. and Gogotsi, Y. (2008). Materials for electrochemical capacitors. *Nat. Mater.* 7: 845–854.

42 Bhoyate, S., Kahol, P.K., and Gupta, R.K. (2019). Nanostructured materials for supercapacitor applications. *SPR Nanoscience* 5: 1–29.

43 Girirajan, M., Bojarajan, A.K., Pulidindi, I.N. et al. (2024). An insight into the nanoarchitecture of electrode materials on the performance of supercapacitors. *Coord. Chem. Rev.*; 518: 216080.

44 Choi, H. and Yoon, H. (2015). Nanostructured electrode materials for electrochemical capacitor applications. *Nanomaterials.* MDPI AG; 5: 906–936.

45 Gupta, R. (2024). A review of functionalized nanomaterials for supercapacitor and hybrid capacitor technologies. *Discov. Electron.* 1 (1): 24. Available from: https://link.springer.com/10.1007/s44291-024-00029-8.

2

Basics of Surface Defects: Types, Formation, and Impact

H. Rojas-Chávez[1], *M.A. Valdés-Madrigal*[2], *and J. M. Juárez-García*[3]

[1] *Tecnológico Nacional de México, Instituto Tecnológico de Tláhuac II, Camino Real 625, Col. Jardines del Llano, San Juan Ixtayopan, Alcaldía Tláhuac, CDMX 13550, Mexico*
[2] *Tecnológico Nacional de México, Instituto Tecnológico Superior de Ciudad Hidalgo, Av. Ing. Carlos Rojas Gutiérrez 2120, Fracc. Valle de la Herradura, Ciudad Hidalgo 61100, Mexico*
[3] *Universidad Tecnológica de Querétaro, Industrial Division, Av. Pie de la Cuesta 2501, Santiago de Querétaro, Qro., Nacional, 76148, Mexico*

2.1 Introduction

The properties of the materials depend on the size and shape imparted to them by processes. However, material properties are also affected by deviations from crystalline perfection. At all working temperatures, crystals are imperfect; only at absolute zero, hypothetically, can a perfect crystal be obtained. It is well-known that the properties of nanomaterials strongly differ from those of their bulk counterparts. For instance, larger surface areas of materials are associated with a decrease in the size. At the nanoscale, the surface/volume ratio of materials becomes exceptionally large. Furthermore, the shape of the nanomaterials depends on the crystalline structure of their surface facets. Consequently, the properties of nanomaterials are expected to depend on their size and shape. However, since crystalline solids contain large numbers of defects, their properties also depend on these defects (Figure 2.1). In addition, it is well-known that defects not only are inevitable features of crystalline solids but also affect the properties of crystalline solids, and even control some of them. Thus, as the properties of crystalline materials are closely sensitive to deviations from crystalline perfection, they are the subject of intensive investigation.

This chapter covers a convenient scheme – providing a general picture – for the classification of surface defects, i.e. two-dimensional (2D) imperfections. Special emphasis is given to addressing the defect formation. More importantly, it is discussed how surface defects can interact with each other. After a brief discussion on defect formation, this chapter deals with the relevance of interfacial defects. Then, the impact of 2D imperfections on nanomaterials has been considered, which, in turn, is related to sustainable energy production and storage. Therefore, the aim of this chapter is to provide an overview of how surface defects in nanomaterials

Surface and Interfacial Defects in Nanomaterials for Sustainable Energy Production and Storage, First Edition. Edited by Noé Arjona, Lorena Álvarez Contreras, and Minerva Guerra Balcázar.

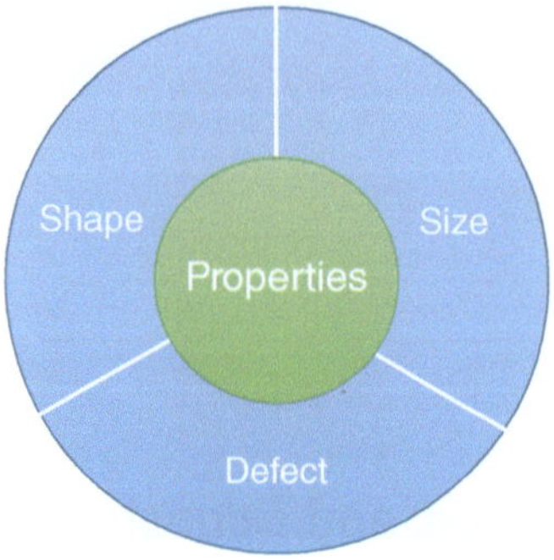

Figure 2.1 Properties of nanomaterials are influenced by shape, size, and defects.

may contribute to the progress of more efficient and more sustainable energy production and energy storage.

2.2 Surface Defect Typology

Matter can exist in three states, namely, gaseous, liquid, and solid. A key feature of solids is that they can be classified as crystalline, noncrystalline, or a mixture of both. In this context, theoretical explorations and experimental findings have confirmed the occurrence of several kinds of defects in crystalline nanomaterials. Point, line, surface, and volume defects constitute the main classification of imperfections or defects in crystalline solids (Figure 2.2). Furthermore, the classification of crystalline imperfections is based on the dimensionality of the defect (Figure 2.2). For instance, they are considered as the zero- (0D), one- (1D), two- (2D), and three-dimensional (3D) defects [1].

In general, imperfections can exist at several scales, i.e. from the macro down to the atomic scale. Of the various crystalline defects in nanomaterials, this chapter focuses on 2D defects. Surface defects, with a thickness of only several atomic layers [2], are considered as boundaries that differentiate zones of those materials that have distinct crystal structures or even crystallographic orientations, i.e. such defects break some symmetry of the ordered reference state. Among the 2D defects, grain boundaries, twins, stacking faults, phase boundaries, and external or free surfaces are the most common.

2.2.1 Grain Boundaries

Grain boundaries (GBs) confine differently oriented crystals or single grains in polycrystalline materials. The structure of GBs in nanomaterials depends on the nature

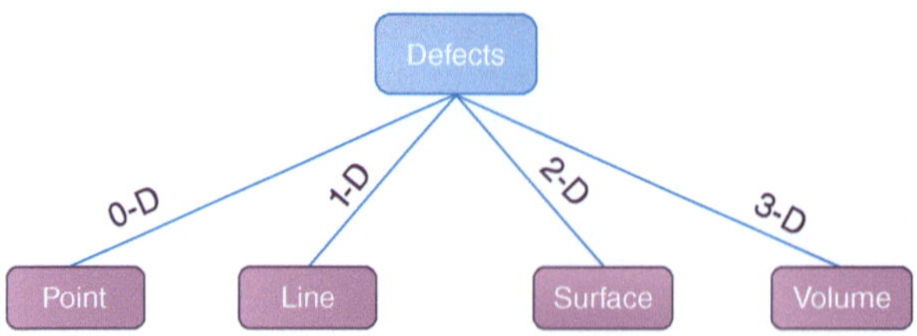

Figure 2.2 Typology of common defects in nanocrystalline solids.

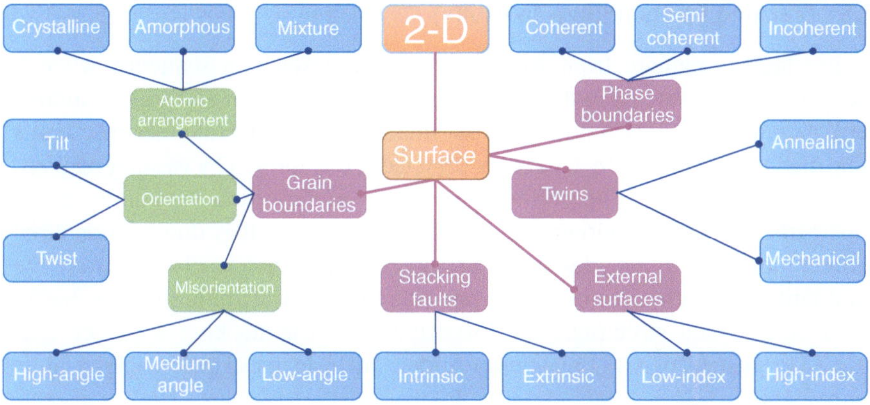

Figure 2.3 Typology of common surface defects in nanocrystalline solids.

of the material and the synthesis/processing techniques used [3]. For instance, amorphous, crystalline, or a mixture of both constitutes the main classification of GBs in nanomaterials (Figure 2.3). Furthermore, in the transition of two differently oriented adjacent crystals or single grains, there is some atomic mismatch inside the boundary region. GBs are also classified according to the crystallographic misalignment between adjacent crystals or grains, which are commonly known as low- (misorientation angle <15°) and high-angle grain boundaries (misorientation angle >15°) [4]. GBs can also be differentiated by the type of orientation distinction, e.g. tilt and twist boundaries (Figure 2.3). In the first case, the type of difference in orientation can be depicted as a relative tilting between two lattices, while, in the second case, the type of difference in orientation can be characterized as a relative rotation between two lattices [5]. It is worth noting that the atoms, along a grain boundary, are bonded less regularly, which gives way to an interfacial or grain boundary energy. According to data reported in the literature, the density of GBs in nanocrystals is too large ($\sim 10^{19}$ cm^{-3}) [4]. Therefore, GBs are more chemically reactive than the grains themselves.

2.2.2 Twins

It has been documented that a twin boundary is a special class of GB across which there is a specific symmetry rule, namely, mirror lattice symmetry. Hence, when a crystal is composed of portions that are oriented in reference to one another according to the mirror lattice symmetry, the crystal is considered as twinned, i.e. a planar fault structure in which one part of the crystal, separated via a twinning plane, is the mirror image of the other part. It has been reported that twins depend on the nature of the energetic stimuli. Thus, twins result from atomic displacements, which can be produced by thermal energy or mechanical energy (Figure 2.3). In the first case, annealing twins are produced, while in the second case, mechanical twins are obtained [6]. In both cases, they depend on the crystal structure. Moreover, according to experimental evidence, the formation of growth twins is quite sensitive to the growth conditions.

2.2.3 Stacking Faults (Piling-Up Faults)

Stacking faults originate from irregularities in the stacking sequence when a structure, mainly close-packed, is built up from the uniform stacking of atomic layers. According to data reported in the literature, intrinsic and extrinsic faults can be distinguished (Figure 2.3). In the first case, it is formed by vacancy agglomeration, i.e. the crystal is unperturbed up to the boundary surface on approaching the perturbed zone from either side. In the second case, it is due to interstitial agglomeration, i.e. it includes an intermediate layer that is distinct from the crystal lattice on both sides of the piling-up fault. It is worth mentioning that in close-packed crystal structures, the interruption in the stacking sequence (e.g. ABCABABC instead of ABC) gives way to either intrinsic or extrinsic stacking faults (Figure 2.3) [7].

2.2.4 Phase Boundaries

Many nanomaterials are found to possess more than one phase. Conventionally, a phase boundary is defined as a transition region or interface between distinct solid phases, which have different crystal structures (atom arrangement) and/or chemical compositions in a multiphase material [5]. Phase boundaries are generally differentiated as coherent (commensurate), semicoherent (partially coherent or discommensurate), and incoherent (incommensurate) [8] (Figure 2.3) based on the lattice mismatch among the corresponding planes of both lattices as well as the bonding strength at the interface. In the first case, it is present when there is the same atomic configuration of the lattice planes across the boundary, having a border in common due to the closely related crystal structures. Then, a one-to-one matching of lattices is expected. The second type of phase boundary is characterized by intervals of continuity. It occurs via the appearance of linear, misfit compensating irregularities at the interphase boundary, i.e. a coherent boundary is associated with an elastic strain, where the atomic structure of the coherent portions at the interface equals the structure of a totally coherent interface, while the misfit compensating defects accommodate the lattice mismatch partly or fully. With respect to incoherent phase boundary, it takes place when two adjacent phases have no regularity of lattice plane matching across the boundary, which is considered as an interface between two rigidly connected lattices. It has been reported that, in this kind of phase boundary, the atomic structure has many features in common with high-angle GBs [9].

2.2.5 External Surfaces (Free Surfaces)

This type of planar defect is considered the most common type of 2D defect [10]. Surface atoms are not fully comparable to those of the bulk due to the broken bonds, which, in turn, break the symmetry as the surface atoms are bonded to other atoms just on one side; that is to say, they are not bonded to the maximum number of nearest neighbors, so surface atoms have a lower number of neighbors. Then, the external surface is considered as a free surface along which the crystal structure terminates, i.e. the order of the crystal is interrupted. Therefore, free surfaces are considered as transition regions (interfaces) between solids and gases. External surfaces can be classified as low- and high-index surfaces (Figure 2.3). The first ones are referred to crystal surfaces with low Miller index values, e.g. (100), (010), and (001), which are not only normally denser but also more stable. As a result, low-index surfaces are commonly the predominant facets observed in crystals due to their lower surface energy. By contrast, high-index surfaces have higher Miller index values, e.g. (311), (511), and (711), which not only are less stable but also have a lower atomic density. Although such surfaces are not usually the dominant facets observed in crystals, they are interesting for various applications due to their unique properties compared to their low-index counterparts, thus highlighting the importance of free surfaces.

2.3 Surface Defect Formation

Nanomaterials usually contain structural irregularities and localized regions of disorder. Such imperfections, in general, are affected by the manipulation and fabrication processes of the nanomaterials as these processes do not occur slow enough, hence, giving way to a variety of defects. In a state of thermal equilibrium, surface defects are not present; therefore, such defects must be produced by other means. It is worth noting that some defects are formed more easily than others, i.e. the occurrence of lower-energy defects will be the first, followed by other defects that require more energy, i.e. the work required for their formation.

This section starts by briefly considering a correlation between surface defects and other types of defects. Many 2D defects occur in nanomaterials ranging from large-angle GBs (30–40°) to atomic planes in crystals, which contain mis-stacking of the atoms. Generally, the extended form of 1D defects (dislocations) is associated with surface defects. GBs are considered as incoherent interfaces with a relatively high energy, which give way to lower-energy defects such as twin interfaces and stacking faults. Another case of the correlation of defects is that of small-angle tilt GBs, which have been described as vertical walls of dislocations. Moreover, it has been reported that a crossed grid of two sets of screw dislocations (1D defect) gives way to twist GBs. On the other hand, in hexagonal close-packed crystal structures (HCP), the refinement process occurs through dynamic recrystallization at elevated temperatures, e.g. the nucleation of ultrafine grains over preexisting GBs or twins due to the progress of stress concentrations and often the activation of basal slip and non-basal one.

Annealing twins are expected when some general requirements are fulfilled, specifically, plastic deformation and heat treatment, i.e. annealing of cold-worked nanomaterials usually generates annealing twins. This type of 2D defect commonly occurs in face-centered cubic (FCC) nanostructures. It is worth highlighting that the surfaces of a nanomaterial inside and outside an annealing twin have distinct surface energies, such a difference is due to their different lattice orientations. Moreover, the lack of coherency is associated with a higher-energy content per unit surface. Consequently, it is expected that both surfaces, within and outside an annealing twin, respond chemically quite differently. Based on energy considerations, stacking faults are low in energy content; consequently, it is expected that nanocrystalline materials that twin readily are also likely to contain many stacking faults due to their similarity in character. Twins also occur due to liquid-to-solid or vapor-to-solid phase changes [6], and even in polycrystalline metals, twins also occur when a grain grows at the expense of another. In such cases, twins formed are also named annealing twins.

Plastic deformation is often accompanied by the formation of another type of twin, the so-called deformation twins. During phase transformations, if the atom movements in a crystal result in a new crystal of identical structure, but different orientation, the process is called mechanical twinning. Furthermore, mechanical twins are the resultant of atomic displacements that are caused when mechanical shear forces are applied. Nanosized thick mechanical twins are constantly formed in FCC structures prepared via high-strain rate processes; for instance, shock loading [11], ball milling [12], and dynamic plastic deformation [13]. It has been reported that mechanical twins are found in body-centered cubic (BCC) and HCP structures, while annealing twins are normally observed in FCC structures.

When atoms lie between two partial dislocations, they have stacking positions that contrast from those atoms that are present on the far side of either of the partial dislocations. Then, a stacking fault is expected due to the discontinuities that occur in the stacking order of the close-packed planes. This type of defect is also associated with dislocations, particularly those with FCC and HCP structures, because stacking faults – boundaries or coherent interfaces – have energies significantly lower than noncoherent interfaces, e.g. free surfaces. Stacking faults are often found in deformed nanomaterials than in annealed ones. For instance, intrinsic stacking faults can be produced via a sliding process [7, 10]. Furthermore, stacking faults play a significant role in controlling the particle shape and size through defect design.

Phase boundaries, which also sometimes are termed as phase interfaces, have a prominent level of migration since they follow the development and transformation of relevant phases. For instance, it has been reported that the movement of phase boundary between two compounds is tuned via thermal treatment in vacuum, distributing the grains within the matrix [14].

External surfaces can also be named surfaces or free surfaces, and these defects constitute the most top atomic arrangement. Even though free surfaces have a similar crystal structure, they have different lattice parameters compared to the atoms located deeper. It has been documented that, under specific parameters, the first and second topmost surface layers in the (111) plane of Pt exhibited an expansion and contraction [15].

The coexistence of different surface defects not only is feasible, but also the correlation and/or interaction among distinct types of dimensional imperfections does. The surface defects occur principally but are not limited to the following stimuli: thermal, mechanical, or a mixture of both. For instance, the configuration of such defects can be produced during crystal growth (or recrystallization) and via plastic deformation. The last one not only significantly influences the structure and properties of nanomaterials but also leads to the formation of distinct types of defects depending on the severe plastic deformation processing route.

2.4 Impact of 2D Defects

Up to now, surface defects have been considered independently of the application of nanomaterials. Now, these applications will be considered in the form of nanomaterials for sustainable energy production and storage. To a first approximation, irregularities in the crystal arrangement and periodicity are related to a certain energy associated with them, the so-called defect energy, as discussed in the preceding section. Consequently, 2D defects can be considered as active sites for many kinds of reactions, e.g. photocatalytic and electrocatalytic, among others. In fact, the intentionally introduced defects are necessary features in nanomaterials. Therefore, it is key to achieve precise control over the nature, concentration, transformation, distribution, and behavior of the defects, which determine the efficacy of the nanomaterials. These considerations show the advantage of the occurrence of defects in nanomaterials for sustainable energy technologies. In this respect, it is necessary to develop a suite of novel functional nanomaterials, with well-controlled defect incorporation, for energy production and storage. In addition, improving the efficiency of energy use is also substantial. Following this trend, recent studies provide evidence to confirm the grounds for these considerations. For instance, some oil companies now use nanocrystals in their diesel fuel production processes. Although any new sustainable energy solution comprises three major sets of technologies, namely, production, storage, and transportation, in this section, only the first two will be addressed.

2.4.1 Energy Production

To mitigate the reliance on fossil fuels, electrocatalytic energy conversion technologies have been widely considered. Such technologies have attracted considerable interest because they are clean and sustainable. Generally, solar cells are a hopeful energy conversion system that converts solar energy into electricity. It has been documented that there are two main routes for sustainable electricity and heat production, which depend on the solar energy exploitation, namely, photovoltaic (PV) and solar thermal systems. At the present technological stage, the production of electricity by third-generation PV solar cells is based on nanostructured materials for reducing or overcoming existing limitations of conventional solar cells. For instance, the performance is strongly dependent on the material characteristics. In the case of dye-sensitized solar cells, defects such as grain boundaries and other defects at the surfaces are required to reduce the charge recombination, addressing the issue of slow electron diffusion across the TiO_2 nanoparticle boundaries [16]. In this sense, it has been proved that surface defects in perovskites noticeably influence the performance and stability of perovskite solar cell films, which is closely related to light soaking behaviors [17], thus confirming a positive correlation between surface defects and the light soaking effect. However, defect modulation is a key factor in determining the performance of solar cells [18].

Fuel cells used for the direct conversion of chemical fuels to electricity have been widely investigated. Among such cells, solid oxide fuel cells (SOFCs) are considered the cleanest and the most effective ones with awesome fuel flexibility. In SOFCs, the leading role of an electrolyte is to permit the conduction of the oxide ions (O^{2-}) from the cathode to the anode, without permitting the transport of electrons, in order to prevent any risk of current leakage. It implies that defects are present in the crystalline structure to allow oxide ion diffusion. It has been reported that the surface modification, via in-situ exsolution, enhances the potential of methane-fueled SOFCs as feasible energy conversion devices, due to the interaction between cation dopants and the surrounding lattice [19]. It is achieved through the atomic rearrangement on the material surface [20], forming a metal oxide heterostructure, namely, a pinning structure. This strategy, surface modification, effectively mitigates particle coarsening at elevated temperatures, reducing the carbon accumulation on the material surface due to the strong interaction between the matrix and nanoparticles [21]. In this context, the surface reconstruction treatment has been documented to enhance the electrolyte functionality in SOFCs, resulting in enhanced ionic conduction in the material [22].

In the case of a photoelectrochemical cell (PEC), it is a device that transforms light energy into electrical energy and is based on photoelectrochemical reactions. The performance of PEC water splitting can be enhanced via the rational defect regulation, which allows the modulation of the crystal structure of CdS through the internal and surface defects [23], coupled with pyroelectric effects, giving way to the optimization of the PEC properties of CdS. By altering the surface of the semiconductor, the photoelectric performance of the photoelectrode can be intensified [24]. Furthermore, surface modification through the deposition of Au nanoparticles has been documented to enhance the photocurrent response of the $MoSe_2$ electrode [25]. Moreover, the surface of the photoanodes has been decorated – via electroless substitution – to enhance the efficiency of the PEC water splitting, thus increasing the surface area as well as the porosity [26]. Recently, it has been reported that an S-scheme photoanode formed, synthesized via a facile hydrothermal-pyrolysis, exhibited excellent PEC performance [27]. The enhanced PEC performance is attributed to the synergy of unique heterojunction formation and the photoanode hierarchical surface structure. It is documented that the unique structure of the heterojunction could provide more surface-active sites for enhancing photocatalytic activity.

Thermoelectric (TE) materials have the capacity to transform waste heat into useful electrical power. The coexistence of multidimensional defects, from 0D to 3D (twinned structure among them), in the AgCuTe has been reported. Due to the combined effects of abundant defects and twinning in the AgCuTe system, low thermal conductivity and high thermoelectric performance are enhanced, achieving high-performance thermoelectric materials [28]. The design of TEs is based on the synergistic combination of the energy filtering effect and hierarchical defects evolution via defect engineering, i.e. the development from 0-D defects to higher-dimensional ones that scatter more phonons in the GeTe, leading to greatly enhanced thermoelectric performance [29].

2.4.2 Energy Storage

Lithium-ion batteries (LIBs), supercapacitors (SCs), and fuel cells are considered the main electrochemical energy storage devices. Thermoelectric, photocatalytic, and photovoltaic applicability of materials are numerous ways of storing energy. Due to the ability to effectively store electrical energy in chemical form, chemical reactions in batteries provide power that can be converted into useful work. It has been reported that thin films manufactured utilizing nanomaterials augment the transport properties of the electronic conduction, which assists ion storage due to the increase in the surface area of the electrode materials. Nanomaterials with large surface areas are desirable because they contribute to more efficient lithium-ion transport, thus shortening ion diffusion pathways in LIBs [30]. In addition, nanomaterial coating has large surface areas for storing Li ions; it dismisses the specific current density of the active material, which, in turn, adequately enhances the stability of the electrode at high charge/discharge current densities [31]. Consequently, the main area of research is to find nanomaterials and nanostructured materials [32], with high surface areas, which can be used in electrodes to increase the energy density and capacity of the batteries, including an augment in the lifespan and efficiency of the batteries [33]. For that reason, it has been reported that functional nano-sizing surface coating on the cathodes for solid-state batteries is the mainstream application of nanotechnology from the laboratory to the industrial scale [34].

SCs are recognized as sustainable energy storage devices. Among the electrode materials, transition metal oxides (TMOs) are receiving significant attention. According to empirical and theoretical research, the design of TMOs is key to enhance specific surface areas and boost electron transport pathways, improving energy storage. The improvement in surface areas can enhance the rate performance of TMOs, including cycle stability and specific capacity [35]. Recently, a novel approach, in the field of energy storage, was proposed via the integration of $NiSe_2$ crystals onto the surface of nickel-cobalt-layered double hydroxides (NiCo-LDHs). This strategy not only significantly enhances exposure to numerous electrochemically active sites due to the surface area increment but also synergistically improves capacitance performance [36].

Hydrogen energy is an alternative source of energy that is sustainable, environmentally friendly, and highly effective. However, the utilization of hydrogen for commercial purposes is presently limited by drawbacks related to its storage, among others. Perovskite hydrides (e.g. Be_2FeH_6 and Mg_2FeH_6) are an alternative, as a novel class of materials for hydrogen storage, due to their capability to appropriately capture and preserve hydrogen either on their surface or within their structures [37].

2.5 Concluding Remarks

Nanomaterials are distinguished from their bulk counterparts primarily due to their properties, which are not only significantly affected by their size and shape, but it has been recognized that they are also defect-dependent. Thus, ever since one thought about properties of nanomaterials, defects play a considerable role in their considerations. The major types of surface defects in crystalline nanomaterials are grain boundaries, twins (boundaries), stacking faults (piling-up faults), phase boundaries, and external surfaces (free surfaces or surfaces). The properties of nanomaterials and nanostructures can be influenced by a variety of defects whose content depends on the specific synthetic process used as well as on the subsequent thermal, mechanical, chemical, or electrochemical treatment.

At the present technological stage, the potential of nanomaterials and nanostructured materials has been explored for advanced energy production and storage applications. By doing so, researchers, via experimental methods and theoretical ones, have studied the effect of surface defects in nanomaterials to improve the performance of nanomaterials for energy production and storage, regarding both the coexistence and the relation among multiple defects.

Nowadays, researchers face a challenge in finding efficient and cost-effective processes to enhance nanomaterials properties in an environmentally friendly manner, considering that climate change has established new parameters for the search of such nanomaterials. In the design of nanomaterials, it is often easier to control processing conditions (temperature, pressure, composition, processing time, etc.) than to precisely control the exact number and dimensionality of defects addition. Therefore, the ongoing research requires the precise manipulation of atoms and molecules to both design and control the planar defects to minimize or maximize them for tuning the properties of the nanomaterials, signifying the role of new nanomaterials in defect science in sustainable energy production and storage.

References

1 Murty, B.S., Shankar, P., Raj, B. et al. (2013). Unique properties of nanomaterials. In: *Textbook of Nanoscience and Nanotechnology*, 29–65. Berlin: Springer.

2 Xiong, J., Di, J., Xia, J. et al. (2018). Surface defect engineering in 2D nanomaterials for photocatalysis. *Adv. Funct. Mater.* 28: 1801983.

3 Valiev, R.Z. (2015). Grain boundary design of bulk nanomaterials for advanced properties. *Diffus. Found.* 5: 43–54.

4 Xu, X., Zhong, Y., Wajrak, M. et al. (2024). Grain boundary engineering: an emerging pathway toward efficient electrocatalysis. *InfoMat* 6: e12608.

5 Fang, T.T. (2018). Elements of structures and defects of crystalline materials. In: *Elements of Structures and Defects of Crystalline Materials*, 169–197. Elsevier.

6 Cahn, R.W. (1954). Twinned crystals. *Adv. Phys.* 3: 363–445.

7 Han, L.B., An, Q., Fu, R.S. et al. (2009). Melting of defective Cu with stacking faults. *J. Chem. Phys.* 130: 024508. https://doi.org/10.1063/1.3049799.

8 Howe, J.M., Aaronson, H.I., and Hirth, J.P. (2000). Aspects of interphase boundary structure in diffusional phase transformations. *Acta Mater.* 48: 3977–3984.

9 Bollmann, W. (1970). *Crystal Defects and Crystalline Interfaces*, XI, 254. Heidelberg: Springer Berlin https://doi.org/10.1007/978-3-642-49173-3.
10 Smallman, R.E. and Bishop, R.J. (1999). Defects in solids. In: *Modern Physical Metallurgy and Materials Engineering*, 84–124. Butterworth-Heinemann.
11 Murr, L.E., Niou, C.S., Garcia, E.P. et al. (1997). Comparison of jetting-related microstructures associated with hypervelocity impact crater formation in copper targets and copper shaped charges. *Mater. Sci. Eng. A* 222: 118–132.
12 Huang, J.Y., Wu, Y.K., and Ye, H.Q. (1996). Deformation structures in ball milled copper. *Acta Mater.* 44: 1211–1221.
13 Zhao, W.S., Tao, N.R., Guo, J.Y. et al. (2005). High density nano-scale twins in Cu induced by dynamic plastic deformation. *Scr. Mater.* 53: 745–749.
14 Li, J., Wang, S.Q., Mayer, J.W. et al. (1989). Oxygen-diffusion-induced phase boundary migration in copper oxide thin films. *Phys. Rev. B* 39: 12367.
15 Gallego, S., Ocal, C., and Soria, F. (1997). Surface and bulk reconstruction of Pt(111) 1×1. *Surf. Sci.* 377–379: 18–22.
16 Wu, C., Wang, K., Batmunkh, M. et al. (2020). Multifunctional nanostructured materials for next generation photovoltaics. *Nano Energy* 70: 104480.
17 Zhang, G., Wei, Q., Liu, G. et al. (2024). Regulating surface defects to achieve more positive light soaking effect in perovskite solar cells. *ACS Appl. Mater. Interfaces* 16: 14263–14274.
18 Wang, Y., Guo, J., Siqin, L. et al. (2024). Enhancing the photovoltaic efficiency of CZTSSe thin-film solar cells via Ag-doping induced defect modulation. *Sol. Energy Mater. Sol. Cells* 277: 113138.
19 Zhang, W., Wei, J., Yin, F. et al. (2023). Recent advances in carbon-resistant anodes for solid oxide fuel cells. *Mater. Chem. Front.* 7: 1943–1991.
20 Lee, W., Han, J.W., Chen, Y. et al. (2013). Cation size mismatch and charge interactions drive dopant segregation at the surfaces of manganite perovskites. *J. Am. Chem. Soc.* 135: 7909–7925.
21 Zhang, W., Wei, J., Zhou, Y. et al. (2024). Co-Ru bimetallic nanoparticles/oxygen deficient perovskite oxides as a highly efficient anode catalyst layer for direct-methane solid oxide fuel cells. *Chem. Eng. J.* 498: 155502.
22 Tao, W., Ye, Y., Gao, J. et al. (2024). Surface reconstruction of β-Ga_2O_3 nanorod electrolyte for LT-SOFCs. *Chem. Eng. J.* 493: 152722.
23 Chen, X., Liu, Z., Ruan, M. et al. (2024). Synergistic mechanism of 0D internal and surface defects regulation coupled with pyroelectric effects for optimizing the photoelectrocatalytic properties of CdS. *ChemCatChem* 16: e202301030.
24 Yan, X., Cao, H., Shi, B. et al. (2024). Enhanced photoresponse in Au-decorated $MoSe_2$ photoanodes for dual-wavelength photoelectrochemical applications. *J. Power Sources* 622: 235373.
25 Yan, X., Cao, H., Shi, B. et al. (2024). Enhanced photoresponse in Au-decorated $MoSe_2$ photoanodes for dual-wavelength photoelectrochemical applications. *J. Power Sources* 622: 235373.
26 Mostafazadeh, N. and Faraji, M. (2024). Anodized commercial galvanized grids decorated with Ag nanospheres and Cu worm-like nanorods for photoelectrochemical water splitting. *Int. J. Hydrog. Energy* 81: 942–949.

27 Zhou, Y., Sun, R., Li, H. et al. (2024). Design and application of hierarchical α-Fe_2O_3/In2O_3 heterojunction photoanode for enhanced photoelectrochemical water oxidation. *Int. J. Hydrog. Energy* 83: 236–245.

28 Luan, X., Li, J., Wu, S. et al. (2024). A nanoscale perspective of the coexistence of multidimensional defects in the AgCuTe system. *Nano Energy* 124: 109505.

29 Fu, G., Lin, S., Liu, Y. et al. (2024). Synergistic strategy via modulated energy filtering and hierarchical defects evolution for high performance GeTe thermoelectrics. *Chem. Eng. J.* 498: 155652.

30 Bhosale, S.S., Sun, Z., and Hong, R. (2024). Nanomaterial coating for layered lithium rich transition metal oxide cathode for lithium-ion battery. *Nanotechnol. Rev.* 13: https://doi.org/10.1515/NTREV-2024-0034.

31 Wang, Y., Li, H., He, P. et al. (2010). Nano active materials for lithium-ion batteries. *Nanoscale* 2: 1294–1305.

32 Jiang, C., Hosono, E., and Zhou, H. (2006). Nanomaterials for lithium ion batteries. *Nano Today* 1: 28–33.

33 Wang, K.X., Li, X.H., and Chen, J.S. (2015). Surface and interface engineering of electrode materials for lithium-ion batteries. *Adv. Mater.* 27: 527–545.

34 Zhang, Y.-H., Wu, L.-R., Ma, J. et al. (2023). Nanotechnology in solid state batteries, what's next? *Next Nanotechnol.* 2: 100011.

35 Tatrari, G., Ahmed, M., and Shah, F.U. (2024). Synthesis, thermoelectric and energy storage performance of transition metal oxides composites. *Coord. Chem. Rev.* 498: 215470.

36 Zhai, H., He, J., Liu, G. et al. (2024). $NiSe_2$ crystals integrated with nickel cobalt hydroxide for high-performance aqueous supercapacitors. *J. Energy Storage* 99: 113472.

37 Hakami, O. and Alathlawi, H.J. (2024). Study of mechanical, optoelectronic, and thermoelectric characteristics of Be/Mg ions Based double perovskites A_2FeH_6 (A=Be, Mg) for hydrogen storage applications. *Int. J. Hydrog. Energy* 83: 307–316.

3

Fundamentals of Interfacial Defects in Materials Science: Types, Formation, and Classification

J. Moroni Mora Muñoz, I. Olvera Rodríguez*, L. J. Salazar-Gastélum, and R. Castellanos-Espinoza*

Universidad Autónoma de Querétaro, Facultad de Ingeniería, División de Investigación y Posgrado, Carr. a Chichimequillas S/N, Ejido Bolaños, Querétaro, C.P. 76010, México

3.1 Interfacial Defects

Interfacial defects are two-dimensional (2D) atomic discontinuities formed at phase boundaries or in regions with differing crystallographic orientations within a material [1]. The alterations are attributed to a variety of factors, including but not limited to atom packing faults, discrepancies in the structure of the lattice, compositional outliers, and deviations in the crystalline orientation between phases [2]. Interfacial defects formed during processing can be classified into different types, and their formation mechanisms vary depending on the material temperature, the applied stress, and the material's crystallography [3–5]. The properties mentioned above render their study crucial for applications that demand precise control over material properties [6, 7]. In this chapter, the main types of interfacial defects, their structural characteristics, formation mechanisms, and the impact they have on the properties of various materials are examined [8–10].

3.2 Grain Boundaries: Low-Angle and High-Angle Grain Boundaries

Grain boundaries are interfacial defects between grains or single crystals in a polycrystalline material [11]. From the above, these defects found extensive study due to their significant effect on the most important properties of solids – such as mechanical strength, electrical conductivity, and diffusion mechanism [6, 12]. The type of grain boundary is determined by the misorientation of the neighboring crystals about the plane of the interface. The classification is low-angle grain boundary and high-angle grain boundary (Figure 3.1) [2, 5].

* These authors contributed equally to this work

Surface and Interfacial Defects in Nanomaterials for Sustainable Energy Production and Storage, First Edition. Edited by Noé Arjona, Lorena Álvarez Contreras, and Minerva Guerra Balcázar.

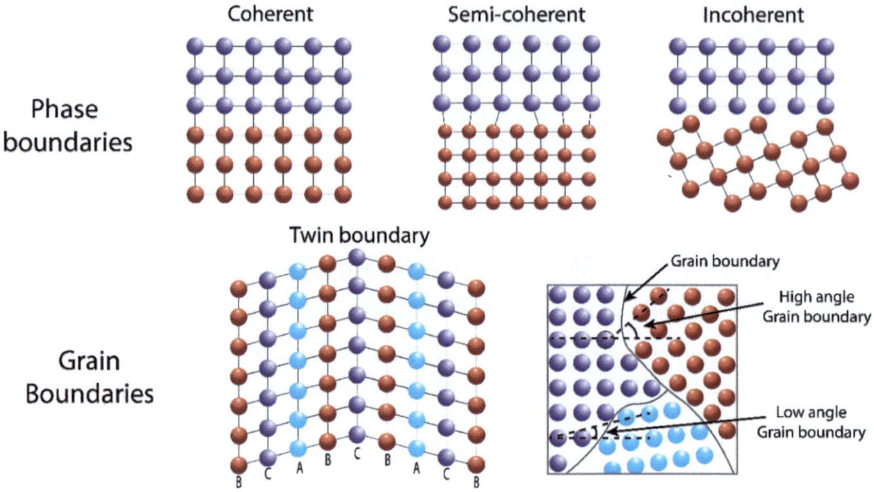

Figure 3.1 Representation of phase and grain boundaries, showing their respective subclassifications.

3.2.1 Low-Angle Grain Boundaries (LAGBs)

Low-angle grain boundaries (LAGBs) have a small misalignment angle between two neighboring grains (typically less than 15°) [2]. This small misorientation leads to the generation of dislocation arrays instead of a full tilt of the crystal lattice [5]. LAGBs have a dislocated wall structure that accounts for the misorientation; therefore, they have a lower interfacial energy than high-angle grain boundaries [12]. Because the atomic arrangement in LAGBs is relatively aligned with the neighboring phase, these boundaries act as moderate barriers to dislocation motion [6]. However, they are less effective in preventing dislocation sliding than high-angle boundaries, which give materials with a high proportion of LAGBs greater ductility. These boundaries are particularly helpful in applications that need higher creep resistance. They can handle and spread stress at high temperatures [8, 11]. Also, the way dislocations are arranged within LAGBs can create paths for diffusion across grain boundaries. This affects processes such as sintering and grain growth in ceramic and metallic materials.

3.2.2 High-Angle Grain Boundaries (HAGBs)

High-angle grain boundaries (HAGBs) are characterized by a significant misalignment angle, greater than 15°, between adjacent crystalline grains [2]. This marked misorientation generates a considerable misalignment in the atomic alignment, resulting in an interfacial region with a high degree of atomic disorder [5]. As a result, HAGBs exhibit a higher interfacial energy, a factor that plays an important role on the mechanical and thermal properties of the materials [4].

From a mechanical point of view, high-angle grain boundaries provide a barrier to dislocation propagation, which increases yield strength via the grain boundary strengthening or the Hall–Petch effect [13]. However, such interfaces also constitute preferred sites for fracture nucleation and propagation owing to stress concentration in the interfacial region that can compromise the material's fracture toughness [14].

The second group of interfacial defects is phase boundaries, which are formed between two places of various phases of matter (Figure 3.1) [15–17]. By contrast, grain boundaries delimit the boundaries between the regions of identical phase with different crystallographic directions, and phase boundaries [17] divide regions with different chemical compositions, crystal structures, or physical components. These interfaces are ubiquitous in multiphase materials, including alloys and composites, in which phase coexistence plays a crucial role in their overall properties [6, 11, 15]. The classification of phase boundary is determined by the degree of alignment between the crystalline planes of adjacent phases.

3.2.3 Coherent Phase Boundaries

The coherent phase boundaries take place when the atomic planes of one phase are aligned with those of the adjacent phase, generating a continuous crystal lattice across the interface [17]. This coherence typically appears in systems where there is a low lattice mismatch between the two phases [15]. Due to the continuity in the atomic arrangement, coherent phase boundaries exhibit reduced interfacial energy, which translates into increased mechanical and thermal stability [4]. Nevertheless, when a small mismatch in the crystal lattice occurs, elastic deformation can develop at the interface, which influences properties such as phase stability and resistance to dislocation motion [5]. The coherent phase boundaries play a key role in precipitation hardening alloys, in which finely dispersed precipitates with coherent interfaces contribute to the material strengthening by hindering the movement of dislocations [13]. This type of phase boundaries is observed in aluminum alloys, where the presence of coherent precipitates provides significant strengthening through the Orowan mechanism [15]. This phenomenon highlights the relevance of coherent phase boundaries in the optimization of the mechanical properties of advanced materials.

3.2.4 Semi-coherent Phase Boundaries

Semi-coherent boundaries, also called partially coherent, are formed when there is a major lattice mismatch between phases and are made up of coherent regions intercalated with zones of mismatch that are compensated by the presence of dislocations and/or steps [17]. These dislocations provide relief from the stress caused by lattice mismatch and generate regions with larger interfacial energy compared to coherent phase boundaries [5]. It can be argued that periodic dislocations enhance the mechanical strength of the material by impeding dislocation creep [13]. Semi-coherent boundaries are typical for multiphase composites and layered structures and are the main factor controlling many properties, such as ductility and strength [6]. Moreover, the presence of dislocations at these boundaries can favor the localized plasticity, a mechanism that allows materials to sustain higher levels of stress before failure [15]. At the level of solid-state batteries, ion transport is enhanced by semi-coherent boundaries by offering pathways for ionic mobility while not jeopardizing the mechanical stability of the material [18]. Hence, these boundaries are an important aspect in the development of high-performance materials for structural and energy technologies.

3.2.5 Incoherent Phase Boundaries

Incoherent phase boundaries with a full atomic rearrangement and lattice parameter mismatch between adjacent phases are present [17]. These misalignments lead to a large amount of atomic disorder at the interface and hence markedly increase interfacial energy [15]. Since incoherent boundaries do not impede the motion of dislocations in the same way as coherent or semi-coherent boundaries, they make materials more prone to crack propagation and interfacial failure under certain stress conditions [14]. Nevertheless, incoherent boundaries can offer benefits in applications that require high surface reactivity because their disordered atomic structure favors diffusion processes and promotes interfacial activity [10]. A notable example of this is multiphase catalysts, where incoherent boundaries provide active sites for the adsorption of reagents and the enhancement of chemical reactions [19]. In systems where incoherent boundaries cannot be eliminated, their stabilization is obtained by controlled chemical composition or introduction into matrices with a mitigated negative impact of high interfacial energy. These boundaries thus constitute a significant aspect within the design of materials for specific applications since their presence needs to be carefully taken into account concerning mechanical and chemical properties [6].

3.3 Twin Boundaries: Symmetrical Interfaces Within a Crystal

Twin boundaries are a kind of interface defect whose symmetry is described by mirror symmetry with respect to a given crystallographic plane [20]. This type of boundary has a low energy interface that is produced by two regions of crystal that are like mirrors of each other and thus have a marked effect on the mechanical and electrical properties of the material [2, 13]. These boundaries are usually in face-centered cubic (FCC) metals, such as copper and aluminum, and also in hexagonal close-packed (HCP) metals, such as magnesium and titanium [15, 21].

3.3.1 Growth Twins and Annealing Twins

The twin boundaries are organized according to their mechanism of formation. Growth twins originate during the initial crystallization or deposition processes, while annealing twins form during post-deposition heat treatments, where recrystallization favors twin formation [20]. The two types of twins contribute significantly to the toughness and ductility of the material [15]. In FCC metals, annealing twins act as dislocation migration barriers, offering an additional reinforcement mechanism by strengthening the twin boundaries [13]. In addition, twin boundaries improve fracture resistance by deflecting propagating cracks, which helps to prevent brittle materials from undergoing failure [14]. This behavior is particularly beneficial in high-strength alloys and ceramics, where the presence of twin boundaries increases the material's ability to support mechanical loads [19, 21].

3.3.2 Deformation Twins

Deformation twins form as a consequence of the application of mechanical stress and are particularly common in metals with an HCP structure [21]. Deformation twins accommodate stress by generating regions with mirror symmetry, allowing HCP metals to undergo plastic deformation even at low temperatures [15]. This quality is particularly beneficial in structural applications, where a low-weight material is needed, such as in automotive and aerospace parts [6]. Besides, deformation twins facilitate strain hardening due to the accumulation of twinned phases during deformation, which blocks dislocation movements, thus elevating the strength of the material [13]. However, excessive twinning can lead to stress localization, which increases the risk of premature material failure [5].

3.4 Free Surface Defects: Influence on Solid Interface Interactions with Other Phases

Free surface defects represent a class of interfacial disruptions that occur at the outer boundaries of materials, where the atomic structure lacks the neighboring atoms necessary to complete its bonding network [22]. These defects are inherent to any surface and play a fundamental role in determining the interactions between the solid material and its environment [23]. The absence of coordinated atoms on the surface generates unbalanced forces, which increases surface energy and gives these sites a high reactivity [24]. In general, free surface defects influence a wide range of material properties, such as adhesion, wettability, catalytic activity, and surface reactivity [19, 25]. Understanding the nature and behavior of these defects is essential for optimizing the performance of materials in various interface interactions, including solid–solid, solid–liquid, and solid–gas interfaces [10].

3.4.1 Nature and Formation of Free Surface Defects

The boundaries between these two regions create surface defects since there is no well-defined crystal lattice at the outermost surface of a material, and the atoms at the edge have an asymmetric coordination in their bonds [22]. However, in the case of surface atoms, they are not fully surrounded by neighbors as in the bulk of material, leading to the presence of dangling bonds [24]. Unlike coatings, which are passivated by their chemical composition, the development of surface defects is a product of the formation of any new surface and is greatly influenced by the crystallographic orientation of the material as well as the processing conditions and surface treatment techniques utilized [23, 25]. For instance, high-energy synthesis processes (specific examples include chemical vapor deposition [CVD] or mechanical polishing) may create a high density of surface defects due to fast surface evolution or abrasion [10, 19].

The degree of surface disorder depends on factors such as surface crystallography and atomic packing [22]. Low-index planes, which have dense atomic arrangements, tend to exhibit lower surface defect densities and hence lower surface energy [24]. On the other hand, high-index planes, characterized by less dense atomic packing, have a higher density of dangling bonds, resulting in higher surface reactivity [25]. These properties make high-index surfaces particularly attractive for applications where greater reactivity or adsorption capacity is required, such as in catalysis and sensor technologies [10, 19]. Therefore, understanding and controlling surface crystallographic orientation is essential in order to optimize the performance of materials in specific applications.

3.4.2 Influence of Free Surface Defects at Solid–Liquid Interfaces

At solid–liquid interfaces, free surface defects greatly change the wettability of material and its ability to adsorb interfacial chemical reactions [23, 24]. Because of dangling bonds and undercoordinated surface atoms, the material can have more or less affinity for the liquid explaining, for example, hydrophilicity or hydrophobicity properties. Materials with more free surface defects are more hydrophilic because the free surface energy increases, which thus lowers the effective surface tension of the liquid [10]. This improvement in hydrophilicity is useful in applications such as adhesives, coatings, and biomedical devices, where strong interfacial interactions with liquids are required.

Furthermore, in electrochemical systems, free surface defects can improve ion adsorption and charge transfer, improving the performance of electrodes in applications such as batteries, fuel cells, and supercapacitors [18, 19]. In addition, the reactive nature of these defects generates additional active sites for ionic interaction, resulting in faster charge transfer and increased storage capacity. Nonetheless, the high reactivity of free surface defects can also give rise to undesired secondary reactions, such as the formation of layer passivation that reduces performance throughout time [25]. Incorporating functional groups such as hydroxyl and carboxyl is a strategy to mitigate these effects.

When nanocomposites are used, control of the free surface defects in the nanoparticles improves their dispersion across the liquid polymer matrix [6, 8]. Surface treatments that modulate the defect density can reduce nanoparticle aggregation, which in turn improves the mechanical, electrical, and thermal properties of the composite material. These improvements are essential in applications where uniform dispersion and strong interactions between the matrix and filler are required.

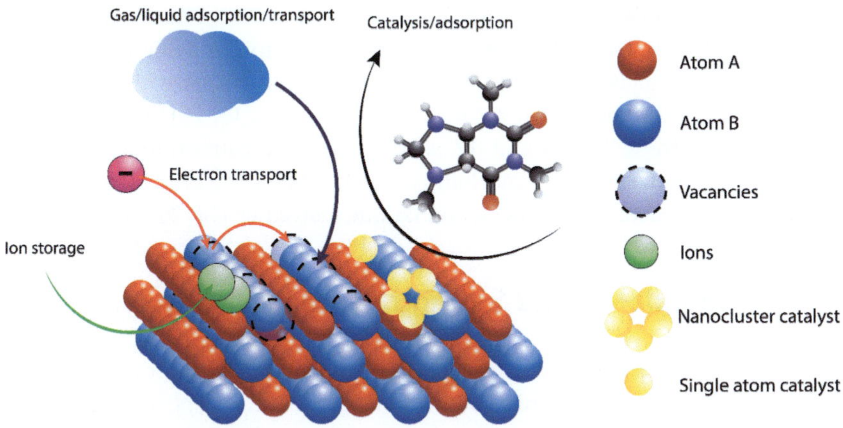

Figure 3.2 Illustration of free surface defects, highlighting atomic vacancies, dangling bonds, and adsorbed molecules/atoms on the surface. The schematic represents the interacting free surface with external molecules, atoms, or charge carriers.

3.4.3 Engineering-Free Surface Defects for Optimized Material Performance

A powerful tool in material design is the ability to manipulate the density and distribution of free surface defects, which allows for the optimization of surface properties for specific applications [10, 23]. For this purpose, advanced techniques such as chemical etching, plasma treatment, and atomic layer deposition (ALD) are employed. In particular, plasma treatments can selectively increase defect density, enhancing the catalytic activity of nanostructured materials without compromising their stability [9, 19].

In the field of separation and filtration technologies, free surface defects in membranes improve permeability and selectivity by providing preferential pathways for molecular transport [26, 27]. The functionalization of these defects with specific functional groups increases selectivity, which optimizes critical processes such as water desalination or gas separation. In the field of catalysis, defect engineering using techniques such as doping or surface oxidation allows the creation of high-density active sites, which increases the catalytic performance and selectivity of the materials used in chemical synthesis and environmental remediation [17, 19] (Figure 3.2).

3.5 Impact of Interfacial Defects on the Material Properties

3.5.1 Impact of Interfacial Defects on Mechanical Properties: Enhancing Strength, Toughness, and Durability

Due to the interactions of interfacial defects, such as grain boundaries, voids, dislocations, and twin boundaries, they have a significant impact on mechanical property, such as strength, toughness, and stress transfer ability of materials during deformation [13, 22, 28]. Depending on their type, distribution, and microstructural environment, such defects can either act as barriers that increase the strength of the material or weak points that result in its fracture [5, 15, 29]. The complex interaction between these defects and the surrounding microstructure determines the response of the material under mechanical load, influencing critical properties such as elastic limit, fracture toughness, and fatigue resistance [14, 20, 21].

3.6 Grain Boundaries and Strengthening Mechanisms

In polycrystalline materials, grain boundaries are a common kind of interfacial defects [22]. Such defects cause phonon scattering, which lowers the material's thermal conductivity by reducing the phonon mean free path [17, 30].

The effect of grain boundaries on thermal characteristics is particularly significant in high-performance composites and thermoelectric materials [12, 31]. Controlled phonon scattering at grain boundaries enhances the figure of merit (ZT) of thermoelectric materials by lowering heat conductivity without notably modifying electron mobility. However, extremely dense grain boundaries might increase thermal resistance, which could block heat transfer and lower the material's efficiency. It has been shown that phonon transport and thermal conductivity in nanostructured materials can be further enhanced by methods such as grain boundary engineering, which optimizes grain size and boundary nature [10, 15].

To mitigate the phonon scattering caused by grain boundaries in carbon nanotube (CNT) composites, CNT alignment plays an important role [3]. In applications that need efficient thermal management, such as heat transmission sinks and portable electronics, aligned carbon nanotubes (CNTs) reduce grain boundary density and provide more effective heat transfer.

3.6.1 Dislocations: Phonon Interactions and Thermal Conductivity Reduction

As linear defects in the crystal lattice, dislocations are important phonon scattering sites [22, 30]. When dislocations are present, the crystal lattice is distorted, creating regions where phonons are more likely to scatter [15, 17]. The average free path of the phonons is reduced by this scattering, which lowers the thermal conductivity of the material.

Although controlled dislocation insertion increases phonon scattering and decreases heat conductivity while maintaining high electrical conductivity, it may be beneficial in thermoelectric materials [31, 32]. Optimizing energy conversion efficiency in thermoelectric devices is crucial during equilibrium. Dislocations, however, can seriously obstruct heat flow in materials where excellent thermal conductivity is necessary, such as heat sinks and thermal interface materials. Techniques like annealing and stress relaxation, which lower the density of dislocations, align the crystal lattice, and enhance thermal transport, are used to lessen the negative consequences of dislocations [20, 24].

3.6.2 Voids: Interruptions in Thermal Pathways

Because the lack of material in the voids creates an insulating barrier that increases thermal resistance, these flaws result in areas of low thermal conductivity [30]. Thermal or mechanical stresses can exacerbate the voids, which are created during material manufacturing procedures including casting, sintering, or depositing [24].

Vapors at the interface between the solid electrolyte and the electrode material seriously impair the thermal stability of solid-state batteries [12, 18]. These holes raise the possibility of thermal runaway by creating hot spots during cycles of charging and discharging. Advanced synthesis methods including ALD and hot isostatic pressing (HIP) are used to reduce the creation of voids in order to lessen this effect [10].

3.6.3 Free Surface Defects: Enhancing or Impeding Thermal Conductivity

High-energy sites generated by free surface defects can enhance surface interactions to boost heat transport, reduce thermal conductivity, or scatter phonons [27, 30]. Surface imperfections can significantly affect the thermal properties of thin films and nanoscale materials [24].

In particular, surface vacancies in graphene and other 2D materials can create localized states, which increase heat conductivity through phonon transit facilitation [7]. In contrast, an excessive density of free surface imperfections can lead to large phonon scattering at free surfaces, which would decrease the efficiency of heat transfer [22, 23]. Surface passivation methods are routinely used to modulate the density of free surface defects [3, 10] in metallurgy for tuning thermal properties for use in microelectronics and thermal interface materials.

3.7 Optical and Photocatalytic Properties: Enhancing Light Absorption and Catalysis

In devices such as solar cells, light-emitting diodes (LEDs), photodetectors, and lasers, interfacial defects play an essential purpose when it comes to the optoelectronic characteristics of materials [33]. Key optoelectronic characteristics such as carrier mobility [9, 24], recombination rates, and light absorption efficiency [17, 22] can all be impacted by these defects. By understanding these defects and their effects, we can tailor the material properties to optimize the performance of our gadgets.

3.7.1 Recombination Centers and Lower Charge Mobility: Grain Boundaries

Grain boundaries act as a contact between grains with different crystal orientations and widely exist in thin-film materials and polycrystalline semiconductors [33, 34].

After all, these defects are under-utilized sites for charge carrier recombination as they typically have a greater density of dangling bonds and impurities [17, 22]. Grain boundaries, for instance, have been demonstrated to function as nonradiative recombination sites in perovskite solar cells, lowering the lifespan of photogenerated carriers, and hence, the energy conversion efficiency [31, 32].

Grain boundaries' capacity to trap charge carriers has a direct negative impact on optoelectronic characteristics. In photovoltaic applications, the trapped electrons and holes recombine without adding to the photocurrent, which is harmful. Techniques like grain boundary passivation, which use surface treatments to neutralize faulty states, have been shown to be successful in lowering recombination losses and increasing device efficiency in order to address this issue [23, 32].

3.7.2 Voids: Scattering Centers Affecting Light Absorption

Voids are volumetric defects introduced during thin film deposition or film growth that interrupt the continuity of the material. They are also incident light scattering centers [22]. This scattering phenomenon can be detrimental for devices such as solar cells and photodetectors in optoelectronic applications since it reduces the amount of light that gets absorbed by the active layer [33, 34]. Gaps at the absorber layer interface, for instance, deflect light away from the active region of thin-film solar cells, lowering optical absorption efficiency and jeopardizing device performance.

Advanced deposition methods such as ALD and CVD are employed to reduce gap generation in order to solve this issue [10]. By reducing porosity and enabling uniform film development, these techniques maximize photon conversion into charge carriers and improve light absorption [24, 32].

3.7.3 Phase Mismatches: Challenges in Band Alignment and Charge Transfer

The semiconductor bandgap possesses energy levels determined by semi-coherent and incoherent phase differences. These inconsistencies serve well as sites for nonradiative recombination [17, 22]. This problem relates to optoelectronic devices, such as lasers and LEDs, with the objective of achieving radiative recombination or photon emission. In gallium nitride (GaN) LEDs, dislocations significantly reduce external quantum efficiency by establishing non-radiative recombination pathways for charge carriers.

A popular method for overcoming phase mismatches is to employ epitaxial growth on substrates with comparable crystal lattices [10, 20]. Decreasing lattice mismatch during crystal formation decreases dislocation density, hence improving luminescence efficiency. Thermal annealing efficiently removes dislocations and reduces stresses in the crystal lattice, consequently improving the optoelectronic performance of semiconductor devices.

Phase mismatches cause the misalignment of energy bands at the contact [17, 22]. This misalignment inhibits charge movement, adversely affecting the performance of heterojunction devices such as solar cells and photodetectors [33, 34]. In heterojunction solar cells, phase discrepancies result in band misalignment, complicating the separation of photogenerated charge carriers. This diminishes the open-circuit voltage and power conversion efficiency.

Moreover, the addition of interfacial or interlayers is used as a method to mitigate problems caused by phase mismatches [10]. These layers allow the gradual change in electronic characteristics, reducing barriers to charge transfer while improving carrier extraction efficiency. Thus, selecting materials with suitable band structures and crystallographic characteristics is crucial for minimizing phase mismatches and improving device efficacy [24, 31].

3.8 Free Surface Defects: Impact on Surface States and Carrier Dynamics

Free surface defects induce dangling bonds and unsaturated atomic states that might act as cages for charge carriers [22]. These defects result in the emergence of surface states within the bandgap, which can entrap photogenerated electrons and holes, hence elevating recombination rates and diminishing the efficiency of optoelectronic devices [33, 34]. In photodetectors, surface imperfections can diminish responsivity by capturing carriers prior to their contribution to the photocurrent.

Surface passivation is a commonly employed method to mitigate the detrimental impacts of free surface imperfections [10, 23]. Employing self-assembled monolayers or passivation with halide ions, along with other chemical treatments, can eliminate dangling bonds and reduce surface recombination velocities [32]. This leads to prolonged carrier lifetimes and superior device performance, especially in applications necessitating high sensitivity and little noise, such as photodetectors and solar cells.

3.8.1 Barrier and Chemical Properties: Enhancing Permeability and Selectivity

The permeability and selectivity of materials are significantly influenced by the presence and characteristics of interfacial defects [10, 32]. The presence of these defects plays a crucial role in influencing the transport properties of materials utilized in gas separation membranes, barrier coatings, and filtration systems. Grasping these defects and their impacts on molecular transport mechanisms is crucial for enhancing material performance in selective permeation applications [23, 26, 31].

3.8.2 Grain Boundaries: Facilitating Enhanced Permeability

Grain boundaries serve as interfaces that indicate adjacent grains showing different crystallographic orientations. In gas and polymeric composite separation membranes, grain boundaries serve as favorable pathways for molecular transport due to their disordered structure [22, 26]. This effect typically enhances permeability;

however, it can negatively impact selectivity by letting the nonselective passage of molecules [10, 33].

In mixed matrix membranes (MMMs), interfacial voids can be created by poorly integrated grain boundaries when metal–organic frameworks (MOFs) or other fillers are introduced into a polymeric matrix. These voids also increase permeability and reduce selectivity [23, 31]. Configurational order and percolating channeling are key for both permeability and selectivity [24, 32]. To overcome these adverse impacts, then membrane can be surface functionalized or passivated with grain boundary defects, which will enhance the efficiency of the membrane system.

3.8.3 Dislocations: Influence on Molecular Pathways

Dislocations, which are linear defects in the crystal lattice, lead to the regions of atomically distorted arrangement that can influence molecular transport [17, 22]. These defects usually introduce additional permeation pathways, increasing the overall permeability of the material [26, 33]. However, dislocations increase selectivity by providing nonspecific molecular transport sites, in this case, particularly at the interfaces between polymers and filler materials in MMMs.

In applications involving gas separation, dislocations at the polymer matrix and inorganic filler interface could lead to the formation of micro-voids additionally reducing the selectivity [10, 23]. These gaps immensely increase permeability because they provide direct paths for molecular transit [22]. The disadvantage of the voids, however, lies in their uncontrolled behavior that generally lowers the selectivity, enabling the non-discriminating transport of different molecular species [10, 26].

These challenges are overcome through advanced deposition techniques, which minimize void formation, allow for better integration of components, and enhance selective permeability [35]. Also, the dislocation density is controllable, by processing techniques, such as thermal annealing allows the formation of these defects to be reduced, thus optimizing the selective permeability of the membrane [24, 32].

3.8.4 Phase Mismatches: Challenges in Selective Transport

Phase mismatches arise due to significant differences in lattice parameters or crystal structure between adjacent phases [17, 22]. The resulting mismatch generates interfacial regions of elevated energy that impede molecular transport, undermining permeability and selectivity [10, 26].

In gas separation membranes, such polymeric-inorganic phase mismatches can potentially break selective gas transport, and thus deteriorate molecular sieving [23, 31]. This problem can be addressed by introducing surface treatments and interfacial layers within polymeric membranes that compensate for differences in structure between the different phases that form the membrane. These approaches enhance compatibility between components, enabling the retention of high selectivity with minimal decline in permeability [35].

3.9 Role of Free Surface Defects on the Enhanced Permeability

Free surface defects occur on the open surfaces of materials and can create dangling bonds as unsaturated atomic sites, exposing highly energetic sites for molecular interaction. The incorporation of defects could enhance permeability, enabling more sites for the adsorption of gas molecules. They can, however, possibly reduce selectivity by promoting the nonspecific movement of various other species.

Surface passivation methods, such as chemical functionalization, are used to control the density of free surface defects and decrease these adverse effects. This dramatically improves the permeability and selectivity of the material, making it a very effective material for industry applications such as water filtration and gas separation.

3.9.1 Catalysis and Surface Reactivity: Free Surface Defects

Defects at the interface significantly influence the overall catalytic performance, including activity, selectivity, and stability of the materials [10, 23]. Such defects alter the electronic and structural characteristics of catalytic surfaces [17, 30], thus regulating their performance against reactions such as hydrogenation, oxidation, and photocatalysis. Understanding these defects is fundamental to improving catalysts used in environmental remediation, chemical synthesis, and energy conversion [33].

3.9.1.1 Grain Boundaries: Enhanced Catalytic Activity

Grain boundaries often exhibit particular chemical characteristics as a result of their modified atomic arrangement and higher defect density. The boundaries act as active sites for catalytic reactions, offering enhanced reactivity in comparison to the surfaces of bulk materials.

Grain boundaries have been suggested as key sites for the activation of CO_2 molecules in metal oxide catalysts utilized for CO_2 hydrogenation, ultimately leading to improved catalysts for CO_2 conversion. The disordered grain boundary structure promotes active molecule adsorption, thus accelerating the surface reaction. The controlled crystallization and passivation of grain boundaries have been implemented to amplify these effects, resulting in improved catalytic activity in reactions such as water splitting and CO_2 reduction.

3.9.2 Voids: Balancing Surface Area and Stability

Voids form volumetric defects that induce free space in the catalyst structure, which might increase the surface area available for chemical reactions to occur [22]. Indeed, the development of some excess pores could compromise the structural integrity of the catalyst, resulting in poor stability and/or performance under reaction conditions [10, 32].

In a solid-state catalyst [33, 34], the gaps could block ion transport and result in mechanical instability at the metal-support interface under high-temperature conditions. Although small, dealt voids enhance catalytic activity by providing additional reaction sites, uncontrolled voids most often lead to structural failure. Advanced synthesis methods, such as HIP and ALD, have been addressed to reduce the void formation and enhance the interfacial characteristics of these materials without sacrificing the stability of such electrodes [17].

3.9.3 Phase Mismatches: Effects on Charge Movement and Efficiency

Phase mismatches, particularly at the boundaries of semicoherent and incoherent phases, create local disturbances that modify the electronic behavior of catalytic materials [17, 22]. These defects often act as active sites for catalytic reactions, in which case they can also facilitate side reactions that reduce the overall selectivity of the process [10].

Researchers have proved that the electronic structure of titanium dioxide TiO_2-based catalysts can be altered by forming faults at the phase boundaries of the interface in the band gap [32]. Having these defect states around can make it easier for hydrogen species to attach, which speeds up the hydrogenation reaction. However, mismatches may also facilitate a nonselective mechanism, which compromises the purity of the product. Epitaxial growth and thermal annealing methods have been used to reduce the number of flaws. The result is an improved balance between catalytic activity and selectivity [31, 33].

In addition, the phase mismatches can introduce the un-equalization of energy bands at the interface [17, 22]. Such misalignments result in the formation of barriers that hinder charge transfer, which may deleteriously affect catalysis reactions, particularly those that involve electron transfer processes [23, 32]. In heterojunction catalysts, phase misalignments can make it harder for photogenerated electron–hole pairs to separate, which lowers the efficiency of photocatalysis [31, 33].

To deal with these effects, it is important to carefully choose materials whose lattice parameters are compatible and to use interfacial buffer layers [31]. These layers not only enhance charge separation but also improve the catalytic activity. It resolves phase mismatch by depositing a thin conductive layer over TiO_2-based photocatalyst's interface. This improves charge transfer and accelerates the rate of photocatalytic reactions.

3.9.4 Free Surface Defects: Increasing Surface Reactivity

Exposed surfaces of catalysts reveal free surface defects [22], leading to the generation of high-energy sites. Such sites can serve as active centers for reagent adsorption and therefore enhance the catalytic activity of the material [10]. However, the presence of numerous surface defects can facilitate charge carrier recombination, thus reducing the overall efficiency of the catalyst [33, 35].

Interfacial defects play an important role in controlling mechanical, thermal, electrical, and catalytic properties of the concerned materials and thus offer to open avenues for the development of new solutions for various technology applications. Understanding the mutual instruct ensures more adept materials, ubiquitous for energy, electronics, and medicine. Recently unraveling previous advances in characterization and modeling methods is expected that precise control of interfacial defects will be one of the crucial tool for the synthesis of intelligent and sustainable materials. These improvements will enhance the performance of current materials and create new opportunities for innovative applications in upcoming fields.

3.10 Conclusions

This chapter highlights the importance of interfacial defects. The existence of these discontinuities and atomic misfits is responsible for the generation of grain boundaries, phase boundaries, twin boundaries and heterointerfaces. These defects have been found to be strongly affected by the processing conditions, thermodynamics and kinetics. To achieve a holistic understanding of their contribution to material response, their phases must be systematically classified according to both crystallographic and energetical attributes. Such knowledge would provide the platform for designing materials with optimal properties for high performance in semiconductors, metallurgy and clean energy conversion and storage.

References

1 Callister, W.D. and Rethwisch, D.G. (2022). *Fundamentals of Materials Science and Engineering: An Integrated Approach*. Wiley.

2 Chen, Z., Zheng, Y., Löfler, L. et al. (2021). Real-time atomic-resolution observation of coherent twin boundary migration in CrN. *Acta Mater.* 208: 116732.

3 Ali, A., Rahimian Koloor, S.S., Alshehri, A.H., and Arockiarajan, A. (2023). Carbon nanotube characteristics and enhancement effects on the mechanical features of polymer-based materials and structures – a review. *J. Mater. Res. Technol.* [Internet] 24: 6495–6521. Available from: https://www.sciencedirect.com/science/article/pii/S2238785423007652.

4 Han, H., Zhao, L., Wu, X. et al. (2024). Advancements in thermoelectric materials: optimization strategies for enhancing energy conversion. *J. Mater. Chem. A* [Internet] 12 (36): 24041–24083. Available from: https://doi.org/10.1039/D4TA03666B.

5 Shao, S., Wang, J., Beyerlein, I.J., and Misra, A. (2015). Glide dislocation nucleation from dislocation nodes at semi-coherent {111} Cu–Ni interfaces. *Acta Mater.* [Internet] 98: 206–220. Available from: https://www.sciencedirect.com/science/article/pii/S1359645415005157.

6 Sharma, H., Kumar, A., Rana, S. et al. (2023). Critical review on advancements on the fiber-reinforced composites: role of fiber/matrix modification on the performance of the fibrous composites. *J. Mater. Res. Technol.* [Internet] 26:

2975–3002. Available from: https://www.sciencedirect.com/science/article/pii/S2238785423018483.

7 Chen, Y., Prasad, V., Yasar, M. et al. (2024). Enhancing interfacial performance and fracture toughness of carbon fibre reinforced thermoplastic composites. *Compos. Part A Appl. Sci. Manuf.* [Internet] 187: 108434. Available from: https://www.sciencedirect.com/science/article/pii/S1359835X24004317.

8 Fu, S., Sun, Z., Huang, P. et al. (2019). Some basic aspects of polymer nanocomposites: a critical review. *Nano Mater. Sci.* 1 (1): 2–30.

9 Zhang, J., Tang, X., Hong, Y. et al. (2023). Carbon-based single-atom catalysts in advanced oxidation reactions for water remediation: from materials to reaction pathways. *Eco Environ. Health* 2 (2): 47–60.

10 Ghaderi, M., Bi, H., and Dam-Johansen, K. (2024). Advanced materials for smart protective coatings: unleashing the potential of metal/covalent organic frameworks, 2D nanomaterials and carbonaceous structures. *Adv. Colloid Interf. Sci.* 323: 103055.

11 Daniel, R., Meindlhumer, M., Zalesak, J. et al. (2020). Multi-scale interface design of strong and damage resistant hierarchical nanostructured materials. *Mater. Des.* 196: 109169.

12 Symington, A.R., Molinari, M., Dawson, J.A. et al. (2021). Elucidating the nature of grain boundary resistance in lithium lanthanum titanate. *J. Mater. Chem. A* [Internet] 9 (10): 6487–6498. [cited 2024 Nov 25]. Available from: https://pubs.rsc.org/en/content/articlehtml/2021/ta/d0ta11539h.

13 Schneider, M., George, E.P., Manescau, T.J. et al. (2020). Analysis of strengthening due to grain boundaries and annealing twin boundaries in the CrCoNi medium-entropy alloy. *Int. J. Plast.* 124: 155–169.

14 Paul, M.J., Kruzic, J.J., Ramamurty, U., and Gludovatz, B. (2024). The importance of fracture toughness evaluation for additively manufactured metals. *Acta Mater.* 276: 120061.

15 Ma, Y., Yuan, F., Yang, M. et al. (2018). Dynamic shear deformation of a CrCoNi medium-entropy alloy with heterogeneous grain structures. *Acta Mater.* 148: 407–418.

16 Dahmen, U. (2003). Phase transformations, crystallographic aspects. In: *Encyclopedia of Physical Science and Technology*, 3e, 821–853. Academy Press.

17 Chen, E.Y., Dingreville, R., and Deo, C. (2018). Misfit dislocation networks in semi-coherent miscible phase boundaries: an example for U–Zr interfaces. *Comput. Mater. Sci.* 154: 194–203.

18 Nie, K., Hong, Y., Qiu, J. et al. (2018). Interfaces between cathode and electrolyte in solid state lithium batteries: challenges and perspectives. *Front. Chem.* 6 (DEC): 422093.

19 Shan, Y., Zhang, G., Shi, Y., and Pang, H. (2023). Synthesis and catalytic application of defective MOF materials. *Cell Rep. Phys. Sci.* 4 (3): 101301.

20 Mahajan, S. (2013). Critique of mechanisms of formation of deformation, annealing and growth twins: face-centered cubic metals and alloys. *Scr. Mater.* 68 (2): 95–99.

21 Jiang, L., Radmilović, V.R., Sabisch, J.E.C. et al. (2021). Twin nucleation from a single <c+a> dislocation in hexagonal close-packed crystals. *Acta Mater.* 202: 35–41.

22 Smallman, R.E. and Ngan, A.H.W. (2014). Chapter 10: Surfaces, grain boundaries and interfaces. In: *Modern Physical Metallurgy*, 8e, 415–442. Butterworth-Heinemann.

23 Wieszczycka, K., Staszak, K., Woźniak-Budych, M.J. et al. (2021). Surface functionalization – the way for advanced applications of smart materials. *Coord. Chem. Rev.* 436: 213846.

24 Wang, L., Lai, K.C., Huang, L. et al. (2020). Low-index surface energies, cleavage energies, and surface relaxations for crystalline NiAl from first-principles calculations. *Surf. Sci.* 695: 121532.

25 Marcus, P., Protopopoff, É., and Maurice, V. (2019). Chapter 5: Surface chemistry and passivation of metals and alloys. In: *Mechanics - Microstructure - Corrosion Coupling: Concepts, Experiments, Modeling and Cases*, 91–120. Elsevier.

26 Hazarika, G. and Ingole, P.G. (2024). Nano-enabled gas separation membranes: advancing sustainability in the energy-environment Nexus. *Sci. Total Environ.* 944: 173264.

27 Liu, Y., Li, S., Meng, S. et al. (2023). Defects-enriched two-dimensional ultrathin g-C_3N_4/In_2O_3 nanoparticles for effective NO_2 detection at room temperature. *Sensors Actuators B Chem.* 396: 134558.

28 Pande, C.S., Rath, B.B., and Imam, M.A. (2004). Effect of annealing twins on Hall–Petch relation in polycrystalline materials. *Mater. Sci. Eng. A* 367 (1–2): 171–175.

29 Taji, I., Hajilou, T., Karimi, S. et al. (2022). Role of grain boundaries in hydrogen embrittlement of alloy 725: single and bi-crystal microcantilever bending study. *Int. J. Hydrog. Energy* 47 (25): 12771–12781.

30 Scott, E.A., Hattar, K., Rost, C.M. et al. (2018). Phonon scattering effects from point and extended defects on thermal conductivity studied via ion irradiation of crystals with self-impurities. *Phys. Rev. Mater.* [Internet] 2 (9): 095001. [cited 2024 Nov 25]. Available from: https://journals.aps.org/prmaterials/abstract/10.1103/PhysRevMaterials.2.095001.

31 Sun, Y., Liu, Y., Li, R. et al. (2022 May). Strategies to improve the thermoelectric figure of merit in thermoelectric functional materials. *Front. Chem.* 19 (10): 865281. Available from: https://www.frontiersin.org/journals/chemistry/articles/10.3389/fchem.2022.865281/full.

32 Zhou, X., Yue, X., Dong, Y. et al. (2021). Enhancing electrochemical performance of electrode material via combining defect and heterojunction engineering for supercapacitors. *J. Colloid Interface Sci.* 599: 68–78.

33 Wang, Y., Yu, W., Wang, C. et al. (2024). Defects in photoreduction reactions: fundamentals, classification, and catalytic energy conversion. *eScience.* 4 (3): 100228.

34 Rogalski, A., Wang, F., Wang, J. et al. (2024). The perovskite optoelectronic devices - a look at the future. *Small Methods* [Internet] [cited 2024 Nov 25]; Available from: https://pubmed.ncbi.nlm.nih.gov/39235586/.

35 Sherkar, T.S., Momblona, C., Gil-Escrig, L. et al. (2017). Recombination in perovskite solar cells: significance of grain boundaries, interface traps, and defect ions. *ACS Energy Lett.* [Internet] 2 (5): 1214–1222. [cited 2024 Nov 25]. Available from: https://pubs.acs.org/doi/full/10.1021/acsenergylett.7b00236.

4

Thermodynamics and Kinetics of Formation of Surface and Interfacial Defects

Juan Hernández-Tecorralco[1] *and Carlos M. Ramos-Castillo*[2]

[1] *Universidad Nacional Autónoma de México, Instituto de Física, Departamento de Física Química, Apartado Postal 20-364, 01000 México, México*
[2] *Centro de Investigación y Desarrollo Tecnológico en Electroquímica S.C., Sanfandila Pedro Escobedo, 76703, Querétaro, México*

4.1 Defects in Thermodynamic Equilibrium

Defects are intrinsic to all materials and have a profound impact on their physical, chemical, and mechanical properties. Among these, point defects such as vacancies, interstitials, and substitutional atoms are particularly critical due to their influence on properties like conductivity, ionic transport, and chemical reactivity. In oxides, for example, charged defects play a pivotal role. Schottky defects, which involve paired vacancies of cations and anions, and Frenkel defects, which consist of a vacancy and an interstitial ion, are the most significant. These defects govern essential processes such as ionic conductivity in solid-state electrolytes for batteries and fuel cells and also affect the electrical and catalytic properties of semiconducting oxides. Point defects often arise from stoichiometric imbalances, such as excess or deficit of cations or anions, or from the introduction of dopants and impurities. The nature and concentration of these defects can either enhance or degrade the performance of the material. A thermodynamic framework offers a powerful tool for predicting defect concentrations in equilibrium by treating defects as dilute solutions. This method enables us to quantify how defect behavior changes with temperature, pressure, and chemical environment, providing valuable insights for tailoring material properties in applications ranging from energy storage to electronics and catalysis.

4.1.1 Configurational Entropy

In a simple solid, the atomic arrangement in the crystal lattice determines its macroscopic properties. The formation of defects, such as vacancies, introduces disorder and increases the configurational entropy of the system. This entropy change is crucial in understanding defect formation and its influence on the thermodynamic stability of the material under various conditions. This chapter will first discuss the simplest case of defect formation using statistical thermodynamics.

Surface and Interfacial Defects in Nanomaterials for Sustainable Energy Production and Storage, First Edition. Edited by Noé Arjona, Lorena Álvarez Contreras, and Minerva Guerra Balcázar.

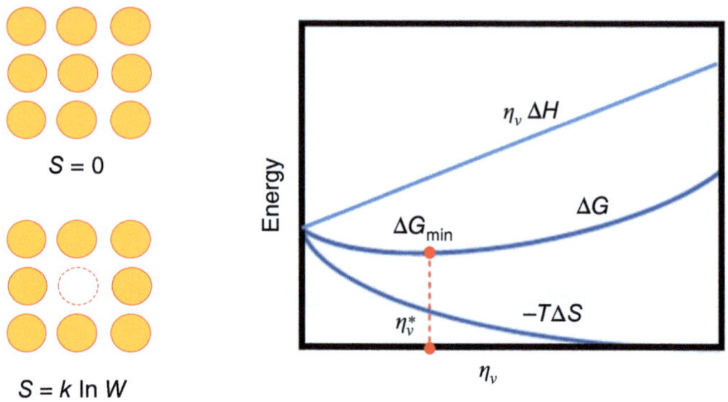

Figure 4.1 Schematic representation of point defects and the origin of configurational entropy in a perfect crystal.

Consider a perfect crystal of an elemental material with N atoms as the initial material. Let n_v represent the number of vacancies formed. Vacancies are created when an atom moves from a site within the crystal to the surface, increasing the total number of sites to $N + n_v$.

In Figure 4.1, the left side shows a crystal lattice with atoms represented as orange circles. The top arrangement depicts a perfect, defect-free crystal, where the entropy (S) is zero, because only one microstate is available, corresponding to the perfect configuration. The bottom arrangement introduces a vacancy, represented by a missing atom, which increases the entropy to $S = k \ln W$, where W is the number of possible microstates. This reflects the increase in the number of possible configurations due to the vacancy, leading to a higher entropy in the system [1]. The right side of the figure displays a plot of energy vs the number of vacancies (n_v). The curve represents the total Gibbs-free energy (ΔG) as a function of the vacancy concentration. The plot features crucial terms: $n_v \Delta H$ which shows the energy contribution of vacancies, and $-T\Delta S$, the entropy term. The red dashed line at n_v^* indicates the equilibrium number of vacancies, where the Gibbs-free energy reaches its minimum value (ΔG_{min}). This point represents the optimal concentration of vacancies that balances the enthalpy and entropy contributions, minimizing the total free energy of the system. The enthalpy change associated with the formation of n_v vacancies is $n_v \Delta H$, representing the energy cost of creating vacant sites. The entropy change contains two components:

1) The vibrational entropy change (ΔS_{vib}) due to vibrations around each vacancy. The total vibrational entropy change is proportional to n_v, i.e. $n_v \Delta S_{vib}$.
2) The configurational entropy change (ΔS_{conf}), reflecting the distribution of vacancies among $N + n_v$ sites. The configurational entropy is defined through the thermodynamic probability W:

$$W = \frac{(N + n_v)!}{N!\, n_v!} \tag{4.1}$$

Using Boltzmann's relation, the configurational entropy is:

$$S_{\text{conf}} = k \ln W \tag{4.2}$$

where k is Boltzmann's constant.

For large N and n_v, typical in real crystals, we apply the Stirling's approximation ($\ln x! \approx x \ln x - x$) to simplify S_{conf}:

$$\Delta S_{\text{conf}} = k \left(N \ln \frac{N + n_v}{N} + n_v \ln \frac{N + n_v}{n_v} \right) \tag{4.3}$$

where the change in Gibbs-free energy associated with the formation of n_v vacancies is:

$$\Delta G = n_v \Delta H - T \Delta S \tag{4.4}$$

Expanding the entropy term:

$$\Delta G = n_v \left(\Delta H - T \Delta S_{\text{vib}} \right) - kT \left(N \ln \frac{N + n_v}{N} + n_v \ln \frac{N + n_v}{n_v} \right) \tag{4.5}$$

At equilibrium, G is minimized with respect to n_v, i.e.

$$\frac{d\Delta G}{dn_v} = 0 \tag{4.6}$$

This yields:

$$\Delta H - T \Delta S_{\text{vib}} + kT \ln \frac{n_v}{N + n_v} = 0 \tag{4.7}$$

The fraction of total sites that are vacant is:

$$\frac{n_v}{N + n_v} = \exp\left(\frac{\Delta S_{\text{vib}}}{k} \right) \exp\left(-\frac{\Delta H}{kT} \right) \tag{4.8}$$

The equilibrium constant for the formation of vacancies in a crystal can be written in terms of the activities of the vacancies, denoted as a_{V_E}. This activity is related to the fraction of lattice sites that are vacant, and it is expressed as:

$$K = a_{V_E} = X_{V_E} = \frac{n_v}{N + n_v} \tag{4.9}$$

In this equation, n_v represents the number of vacancies, and N is the total number of lattice sites. The term X_{V_E}, also called the vacancy concentration, gives the fraction of vacant sites relative to the total number of sites in the system. It describes the equilibrium distribution of vacancies within the crystal lattice [2].

Further, to relate this to the thermodynamic properties, we use the expression for the equilibrium constant, which is a function of the entropy and enthalpy changes:

$$K = \exp\left(\frac{\Delta S_{\text{vib}}}{k} \right) \exp\left(-\frac{\Delta H}{kT} \right) \tag{4.10}$$

Here, ΔS_{vib} is the change in vibrational entropy due to the formation of vacancies, ΔH is the enthalpy change, k is the Boltzmann constant, and T is the temperature. For molar quantities, replace k with $R = N_A k$, where N_A is Avogadro's number.

The equation shows how the equilibrium vacancy concentration depends on both the entropy and enthalpy contributions to the system's free energy.

Finally, X_{V_E} is a measure of the vacancy fraction, representing the ratio of vacant sites to the total number of available lattice sites, essentially quantifying the presence of vacancies in the system at equilibrium.

The standard change in Gibbs free energy is:

$$K = \exp\left(\frac{\Delta S^0_{\text{vib}}}{R}\right)\exp\left(-\frac{\Delta H^0}{RT}\right) = \exp\left(-\frac{\Delta G^0}{RT}\right) \tag{4.11}$$

This equation is key to determining the equilibrium vacancy concentration in materials. It relates to the standard changes in entropy (ΔS^0_{vib}) and enthalpy (ΔH^0) to the Gibbs-free energy change (ΔG^0). Using this equation, the equilibrium constant K can be calculated, which describes the vacancy distribution at a given temperature. It provides insights into the thermodynamics of defects, aiding in material design by predicting how vacancy concentrations influence properties such as diffusion and ionic conductivity [3].

4.1.2 Thermodynamics of Defect Reactions

Defect reactions resemble chemical reactions, allowing thermodynamic variables such as pO_2 and temperature to relate to free energy changes. This enables the creation of equilibrium diagrams linking defect concentration and pO_2.

The free energy change for a chemical system is:

$$\Delta G = \Delta G^\circ + RT \ln Q \tag{4.12}$$

At equilibrium ($\Delta G = 0$):

$$\Delta G^\circ = -RT \ln K \tag{4.13}$$

Additionally:

$$\Delta G = \Delta H - T\Delta S \tag{4.14}$$

Combining:

$$K = K_0 \exp\left(-\frac{\Delta H^\circ}{RT}\right), \quad K_0 = \exp\left(\frac{\Delta S^\circ}{R}\right) \tag{4.15}$$

This approach treats defects as solutes in a solvent.

4.1.3 Charged Point Defects and Their Ionization

Charged point defects play a crucial role in the thermodynamic and electronic properties of materials, particularly in ionic compounds. These defects can ionize, leading to the formation of complementary electronic defects that help to maintain charge neutrality in the system. The nature of these defects and their interactions significantly affect the material's behavior under various conditions, such as high temperatures or changing atmospheric conditions [4].

The primary types of charged point defects are:

1) **Cation and Anion Vacancies:** These defects occur when a metal cation (V_M) or an oxygen anion (V_O) is missing from its lattice site, creating a vacancy. These vacancies are crucial for ionic conduction and can significantly influence the material's mechanical properties.
2) **Vacancies and Interstitials:** These involve a vacancy in the lattice (e.g. V_O or V_M) and an interstitial ion of the same species, such as oxygen interstitials (O_i) or metal interstitials (M_i). Interstitial defects occur when ions occupy positions between the regular lattice sites, leading to localized stress and potential changes in conductivity.

The ionization of these defects occurs when they acquire or lose electrons, forming complementary electronic defects. For example, a cation vacancy (V_M) can ionize by attracting an electron to form a positively charged vacancy, while the creation of an anion vacancy (V_O) often results in an excess of negative charge that must be compensated by the movement of electrons within the material.

The key types of charged point defects that emerge from such ionization processes include:

- **Schottky Disorder:** This type of disorder arises when both cation and anion vacancies (V_M and V_O) are created simultaneously. Schottky defects maintain charge neutrality by ensuring that, for each cation vacancy, there is a corresponding anion vacancy. This type of disorder is common in stoichiometric ionic compounds, where the loss of ions from the lattice is balanced by an equal number of vacancies in both sublattices. Schottky disorder leads to a reduction in the material's density as both cation and anion vacancies are created.
- **Frenkel Disorder:** In Frenkel disorder, an ion (typically a cation or anion) leaves its normal lattice site and occupies an interstitial position, creating both a vacancy and an interstitial defect. In the case of oxygen (V_O and O_i) or metal (V_M and M_i) Frenkel defects, the vacancies and interstitials are of the same species, which means that the material remains electrically neutral as the defect formation involves the movement of ions within the lattice without introducing any net charge. Frenkel defects are typically found in materials with large ions or at high temperatures, where the atoms have enough thermal energy to migrate to interstitial sites [5].

These charged point defects and their ionization behavior significantly influence various properties of the material including electrical conductivity, diffusion, and mechanical stability. For example, Schottky defects, by creating vacancies in both sublattices, can enhance ionic conductivity, as vacancies serve as pathways for ion movement. On the other hand, Frenkel defects lead to interstitials that frequently alter the electronic structure and mechanical characteristics of the material, owing to localized lattice distortions.

Understanding the formation and behavior of these defects is essential for designing materials with specific properties such as solid electrolytes, semiconductors, and materials that are used in high-temperature environments, where defect formation is prominent.

4.1.4 Schottky Defects

Schottky defects involve the creation of vacancies in both the cation and anion sublattices of an ionic compound, maintaining charge neutrality. The defect reaction for Schottky formation in an oxide can be expressed as:

$$\mathrm{MO} \rightarrow V_{\mathrm{M}}'' + V_{\mathrm{O}}^{\cdot\cdot} + \mathrm{M}_{\mathrm{metal}} + \frac{1}{2}\mathrm{O}_2 \tag{4.16}$$

Here, both a metal cation vacancy (V_{M}'') and an oxygen vacancy ($V_{\mathrm{O}}^{\cdot\cdot}$) are created and the oxygen gas ($\frac{1}{2}\mathrm{O}_2$) is released to maintain charge balance.

The equilibrium constant for the Schottky defect formation, denoted as K_{S}, is given by:

$$K_{\mathrm{S}} = [V_{\mathrm{M}}''][V_{\mathrm{O}}^{\cdot\cdot}] \cdot (p\mathrm{O}_2)^{1/2} \tag{4.17}$$

where $p\mathrm{O}_2$ is the oxygen partial pressure. The term $(p\mathrm{O}_2)^{1/2}$ indicates the dependence of defect formation on oxygen availability. This can also be expressed in terms of the Gibbs-free energy change for the reaction:

$$K_{\mathrm{S}} = \exp\left(-\frac{\Delta G_{\mathrm{S}}}{RT}\right) \tag{4.18}$$

Here, ΔG_{S} represents the standard Gibbs-free energy change for Schottky defect formation, and the equation reflects the temperature dependence of the defect concentration.

4.1.5 Oxygen Partial Pressure Dependence of Schottky Defects

The concentration of Schottky defects is directly influenced by the oxygen partial pressure $p\mathrm{O}_2$. As $p\mathrm{O}_2$ increases, the concentration of oxygen vacancies ($V_{\mathrm{O}}^{\cdot\cdot}$) also increases, leading to a higher number of Schottky defects. This relationship is crucial for understanding how the material's defect structure varies with oxygen availability, which in turn affects properties such as ionic conductivity.

For applications where controlling vacancy concentrations is essential, such as in solid-state electrolytes, manipulating the oxygen partial pressure can effectively tune the material's properties by influencing the density of Schottky defects.

4.1.6 Frenkel Defects

Frenkel defects involve the creation of a vacancy and an interstitial ion of the same kind, ensuring charge neutrality. For an oxide MO, the defect reaction can be expressed as:

$$\mathrm{O}_{\mathrm{lattice}} \rightarrow V_{\mathrm{O}}^{\cdot\cdot} + O_{\mathrm{i}}'' \tag{4.19}$$

The equilibrium constant for the Frenkel reaction is:

$$K_{\mathrm{F}} = [V_{\mathrm{O}}^{\cdot\cdot}][O_{\mathrm{i}}''] \tag{4.20}$$

Similarly, for cations, the reaction is:

$$M_{\mathrm{lattice}} \rightarrow V_{\mathrm{M}}'' + M_{\mathrm{i}}^{\cdot\cdot} \tag{4.21}$$

And the equilibrium constant is given as:

$$K_F = [V''_M][M_i^{\cdot\cdot}] \tag{4.22}$$

In both cases, the equilibrium constant can be expressed in terms of the Gibbs-free energy change:

$$K_F = \exp\left(-\frac{\Delta G_F}{RT}\right) \tag{4.23}$$

Here, ΔG_F represents the standard Gibbs-free energy change associated with the formation of Frenkel defects.

Understanding the thermodynamics of point defects in oxides has significant implications for various applications [6], including the following:

- **Ionic Conductivity:** Point defects such as vacancies and interstitial ions facilitate ionic transport in solid-state electrolytes, which are crucial for fuel cells and batteries.
- **Chemical Reactivity:** Defects affect the surface activity and catalytic properties of oxides, enabling applications in heterogeneous catalysis.
- **Mechanical Properties:** The presence of defects can influence the mechanical strength and fracture behavior of ceramic materials.
- **Electronic Properties:** Charged defects impact the electrical conductivity and semiconducting behavior of oxides, essential for electronic and optoelectronic devices.

4.2 The Kinetics of Defect Formation

Surface and interfacial defects [7] are lattice imperfections within materials, which intrinsically influence their properties. The first case corresponds to defects present on the surface of the material, including vacancies, adatoms, discontinuities, steps, and islands, among others. The second case refers to defects at the boundaries within the material in which most common are grain boundaries, phase boundaries, twin boundaries, and stacking faults. The study of these imperfections allows us to tailor and enhance the properties of materials for applications in areas such as energy production and storage. For instance, these defects can serve as adsorption or catalytic sites to improve the chemical reactions on the surfaces. In addition, they could contribute to tuning the ion conductivities, a key variable in energy storage applications.

In materials science, kinetics refers to the study of the pathways and rates at which initially nonequilibrium systems move towards equilibrium [8–10]. It covers the mechanisms and time scales of processes such as diffusion, defect migration, and phase transitions. Also, describing the kinetics and mechanism of defect formation involves several key quantities such as formation energies, migration energies, reaction pathways, and rate constants. This section examines these phenomena on the surface and interfacial defect formation as well as their mathematical description, accompanied by relevant examples.

4.2.1 Arrhenius Equation and Thermally Activated Processes

In material science, several processes are temperature dependent and their rates increase exponentially with temperature. In the field of physical chemistry, the Arrhenius equation is a key formula that describes how the rate constant k of a chemical reaction depends on the temperature,

$$k = A \cdot \exp\left(-\frac{E_a}{k_B T}\right) \tag{4.24}$$

where k denotes the rate constant, A is a pre-exponential factor, E_a represents the activation energy corresponding to the energy barrier required to initiate the reaction through thermal excitation, k_B is Boltzmann's constant, and T is the temperature. At lower temperatures, the exponential component dominates, leading to a significant reduction in k, while higher temperatures increase k. Taking the natural logarithm on both sides of the equation 4.24 results in a linear equation as follows:

$$\ln(k) = \ln(A) - \frac{E_a}{k_B} \cdot \frac{1}{T} \tag{4.25}$$

Figure 4.2a shows the plot of Eq. (4.25), which exhibits a linear relationship between the logarithm of the rate constant $\ln(k)$ and the inverse of temperature $1/T$, where the slope provides the activation energy E_a. This linearization procedure is widely used in experiments to determine kinetic parameters. Figure 4.2b shows the activation energy necessary to initiate the process. In a typical reaction pathway vs energy plot, it is also called the barrier energy. In the context of defect migration, the path in Figure 4.2b shows the trajectory to which a chemical species must move from one stable position to an adjacent equivalent location and the energy required to achieve this process.

The Arrhenius equation is commonly applied to model processes such as diffusion phenomena or phase transitions. Also, the defect formation follows an Arrhenius relationship as we shall see later. In essence, both the concentration of defects and the rate of diffusion are exponentially sensitive to temperature changes. This mathematical behavior arises because the concentration of defects grows exponentially

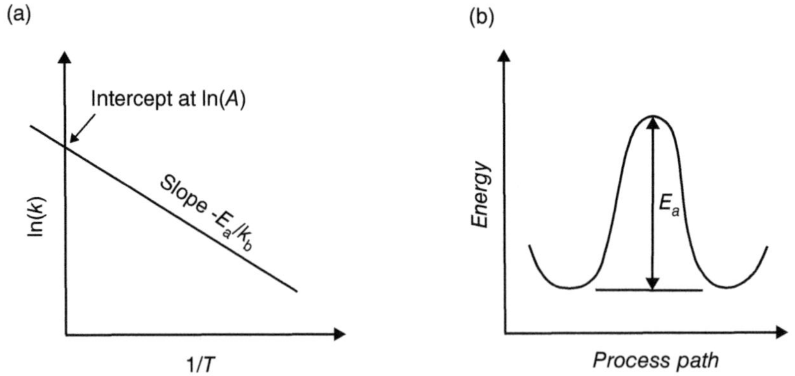

Figure 4.2 (a) Linear relationship of the logarithmic form of the Arrhenius equation and (b) activation energy to transition from one stable state to a neighboring equivalent location.

with increasing temperature. In addition, these defects facilitate mass transport, leading to a solid–state diffusion rate that also rises exponentially with temperature.

4.2.2 Mechanism of Defect Formation

The formation of surface and interfacial defects is related to kinetic processes under the influence of external conditions such as temperature and pressure. Although thermodynamics drives these mechanisms, they are regulated by kinetic restrictions. The temperature could affect the long-range lattice order in the crystals and spontaneously form defects in materials. As an example, we discuss the mechanism of vacancy formation on surfaces and apply the Arrhenius relation to study the kinetics involved in their creation. The same approach can be applied to the other point defects in the solid state. Figure 4.3a shows a square arrangement of atoms, at a given temperature, a certain fraction of the atoms on a surface have sufficient thermal energy to escape from their site and produce a vacancy.

Also, a vacancy can disappear when it is occupied by free atoms or by the migration of chemical species in their surroundings, where a vacancy disappears but another arises as shown in Figure 4.3b. In general, defect formation and migrations are closely related phenomena since they occur simultaneously. To characterize their kinetics of creation and annihilation, we need to define the rate of formation of vacancies k_c and the rate of annihilation of vacancies k_a using Arrhenius forms,

$$k_c = A^c \cdot \exp\left(-\frac{E_a^c}{k_B T}\right) \tag{4.26}$$

and

$$k_a = A^a \cdot \exp\left(-\frac{E_a^a}{k_B T}\right) \tag{4.27}$$

where A^f and A^a are the pre-exponential constants for each rate process, and E_a^c and E_a^a are the barrier energies to create or annihilate the vacancies on the surface. Once we have defined both rate functions, we can write a general rate equation that is time dependent as follows:

$$\frac{dN_v}{dt} = k_c \cdot N - k_a \cdot N_v \tag{4.28}$$

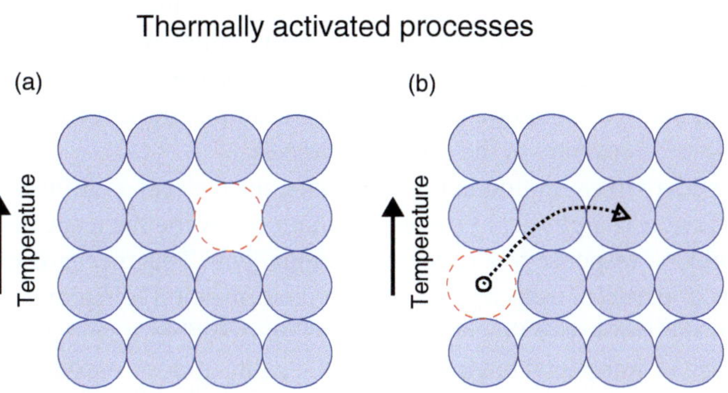

Figure 4.3 Thermally activated processes: (a) vacancy formation and (b) defect migration.

where N_v is the total number of defects on the surface and N is the total number of available sites where defects can be created. In equilibrium, the flux of defect formation equals the flux of annihilation:

$$k_c \cdot N = k_a \cdot N_v \tag{4.29}$$

and therefore,

$$\frac{N_v}{N} = \frac{A^c \cdot \exp\left(-\frac{E_a^c}{k_B T}\right)}{A^a \cdot \exp\left(-\frac{E_a^a}{k_B T}\right)} = \frac{A^c}{A^a} \exp\left(-\frac{(E_a^c - E_a^a)}{k_B T}\right) \tag{4.30}$$

If we define $N_v/N = C$, $A^c/A^a = C_0$, and $(E_a^c - E_a^a) = E_f$, we obtain the following equation:

$$C = C_0 \cdot \exp\left(-\frac{E_f}{k_B T}\right) \tag{4.31}$$

where C is the concentration of defects, C_0 is a pre-exponential constant, E_f is the defect formation energy, k_B is the Boltzmann constant, and T is the temperature.

Analyzing the kinetics of defect formation for vacancies, we derive Eq. (4.31), which provides a straightforward expression for the equilibrium defect concentration. Using experimental data on concentration and temperature, it is then possible to estimate E_f. Knowing E_f, it is possible to interpolate and determine the concentration of any other temperature using the logarithmic form of the Arrhenius equation. Experimentally, it is possible to characterize the concentration of defects by measuring ionic and electronic conductivity using methods such as electrochemical impedance spectroscopy [11] or the Hebb–Wagner polarization method [12, 13].

4.2.3 Diffusion and Fick's Laws

The periodic arrangement of atoms in solids is basically immobile, except for fluctuations due to local vibration at each atomic position. However, the presence of defects gives some level of internal mobility as mentioned earlier. To understand this mobility effect, we consider the case of surface defect vacancies. At enough temperature, atoms adjacent to a vacancy migrate into the empty site, creating a new vacancy that, of course, can be occupied by another neighboring atom (see Figure 4.3b). The overall process has no change in the free enthalpy since the initial and final configurations are identical. In order to initialize the process, we need to overcome an energy barrier between the two contiguous states. In this example, the direction in which the material moves is contrary to the flow of vacancies.

To describe the kinetics of mass motion in the solid state, the concept of diffusion is employed [14]. Diffusion is a physical process in which some chemical species flow within a material is in response to a concentration gradient. This mechanism involves the transport of particles from regions of high concentration to regions of low concentration. In the context of this chapter, diffusion describes how defects move through a surface or interfacial environment. This study aims at measuring the concentration C of these defects as a function of distance (and time), and the degree of diffusion can be quantified by a diffusion coefficient D. This coefficient describes the rate at which defects propagate through the material and can be obtained by applying one of the two diffusion equations, namely Fick's first law for steady-state diffusion or Fick's second law for nonsteady-state diffusion.

Fick's first law for steady-state diffusion states that the diffusive flux, or flow rate, J, is proportional to the concentration gradient ∇C:

$$J = -D\nabla C \tag{4.32}$$

In Eq. (4.27), D represents the diffusion coefficient. The negative sign in Eq. (4.27) indicates that diffusion takes place toward lower concentration zones. In steady-state diffusion, the concentration profile remains constant over time. Fick's second law of nonsteady-state diffusion describes the temporal evolution of concentration:

$$\frac{\partial C}{\partial t} = D\nabla^2 C \tag{4.33}$$

Equation (4.33) is widely used to model dynamic diffusion processes. In a standard diffusion problem, such as defects on a surface, the initial condition is typically assumed to be that the concentration is a step function. At the time $t = 0$ in a one-dimensional system, the concentration is C_0 within the range $0 \leq x \leq L$, while for $x > L$, the concentration is zero.

The general solution to this one-dimensional diffusion equation with a step initial condition is:

$$C(x,t) = C_0 \operatorname{erfc}\left(\frac{x}{2\sqrt{Dt}}\right) \tag{4.34}$$

In Eq. (4.35), erfc(z) is the complementary error function, defined as $\operatorname{erfc}(z) = 1 - \operatorname{erf}(z)$.

Analysis of the solution (4.35) reveals that concentration $C(x, t)$ decreases as distance x increases. The term $\frac{x}{2\sqrt{Dt}}$ in the argument for the error function represents the normalized position relative to the diffusion length $\sqrt{Dt}$, which increases over time. At $t = 0$, the concentration profile is a step function, and as time increases, the concentration spreads out in both directions from the origin, following a Gaussian shape. As $t \to \infty$, the concentration becomes more evenly distributed on the surface, and the profile becomes flatter.

The diffusion coefficient D in Eqs. (4.32) and (4.33) is a material-specific property, which depends on the temperature, the type of diffusing chemical species, and the crystalline lattice. The relation of the diffusion coefficient with the temperature follows an Arrhenius equation:

$$D = D_0 e^{-\frac{E_a}{k_B T}} \tag{4.35}$$

where D_0 is a pre-exponential constant, E_a is the activation energy, k_B is Boltzmann's constant, and T is the temperature. At higher temperatures, diffusion occurs faster. In general, the magnitude of the diffusion coefficients depends on the route of transport. Diffusion at surfaces or in interfacial defects is much faster than in bulk because there are paths with less resistance to mass movement. In addition, the activation energy is lower in the former cases.

4.3 Summary

All materials inherently contain defects, which significantly influence their physical and chemical properties. A thermodynamic and kinetic analysis of these defects allows us to describe their properties both at equilibrium and out-of-equilibrium. As we show in this chapter, the formation and migration of defects are governed by a combination of thermodynamic and kinetic principles. Understanding these mechanisms is vital for tailoring the properties of materials for advanced applications. This chapter also provides an overview of the thermodynamic and kinetic aspects of surface and interfacial defect formation, which are influenced by external effects such as thermal processes. The concepts that are examined in this chapter include enthalpy, entropy, Gibbs-free energy, and chemical equilibrium equations concerning defect formation. The kinetics of defect formation is related to an Arrhenius expression, showing an exponential increase with rising temperature. Also, these defects play a critical role in the diffusivity in the solid state, which can be described by Fick's laws, where the diffusivity increases exponentially following another Arrhenius-like equation reflecting the temperature-dependent behavior. Both defect formation and diffusion processes directly affect the ionic or electronic conductivities in materials. These conductivities are sensitive to defect concentration and act as a fundamental parameter for understanding diffusion processes at surfaces and interfaces.

References

1 Ubic, R. (2024). *Point Defects*, 347–372. Cham: Springer International Publishing. https://doi.org/10.1007/978-3-031-49752-0_16.

2 Kröger, F.A. and Vink, H.J. (1956). *Relations Between the Concentrations of Imperfections in Crystalline Solids*, vol. 3, 307–435. New York: Academic Press.

3 Kröger, F.A. (1973). *The Chemistry of Imperfect Crystals*, Preparation, Purification, Crystal Growth and Phase Theory, 2e, vol. 1. Amsterdam; London: North-Holland Publishing Company New York: American Elsevier Publishing Company, Inc.

4 Kröger, F.A. (1974b). *Defect Thermodynamics-Historical*, 3–24. Boston, MA: Springer US. ISBN: 978-1-4615-8723-1, https://doi.org/10.1007/978-1-4615-8723-1_1.

5 Corish, J. (1994). *Thermodynamic Properties of Defective Solids*, 413–434. Netherlands, Dordrecht: Springer. https://doi.org/10.1007/978-94-011-1942-9_19.

6 Martin, M. (2005). *Diffusion in Oxides*, 209–247. Berlin, Heidelberg: Springer-Verlag. ISBN: 978-3-540-30970-3, https://doi.org/10.1007/3-540-30970-5_5.

7 Pond, R.C. and Hirth, J.P. (1994). Defects at surfaces and interfaces. *Solid State Phys.* 47: 287–365.

8 Maier, J. (1993). Defect chemistry: composition, transport, and reactions in the solid state; part II: kinetics. *Angew. Chem. Int. Ed. Engl.* 32 (4): 528–542.

9 Shackelford, J. (2022). *Introduction to Materials Science for Engineers Global Edition*. Pearson International. ISBN: 9781292440996.

10 Shewmon, P. (1989). *Diffusion in Solids*. Springer.

11 Lazanas, A.Ch. and Prodromidis, M.I. (2023). Electrochemical impedance spectroscopy–a tutorial. *ACS Meas. Sci. Au* 3 (3): 162–193.

12 Hebb, M.H. (1952). Electrical conductivity of silver sulfide. *J. Chem. Phys.* 20 (1): 185–190.

13 Wagner, C. (1956). Galvanische Zellen mit festen Elektrolyten mit gemischter Stromleitung. *Zeitschrift für Elektrochemie, Berichte der Bunsengesellschaft für physikalische Chemie* 60: 4–7. https://doi.org/10.1002/bbpc.19560600104.

14 Crank, J. (1979). *The Mathematics of Diffusion*, Oxford Science Publications. Clarendon Press. ISBN: 9780198534112.

5

Defects as Catalytic Sites in Energy Chemistry

Beatriz Ruiz Camacho, Adriana Medina Ramírez, and José de Jesús Ramírez Minguela

Department of Chemical Engineering, Division of Natural and Exact Sciences, University of Guanajuato, Noria Alta S/n Guanajuato 36050, Mexico

5.1 Defects as Active Sites

In this chapter, we address defect issues from their classification, synthesis methodologies, characterization techniques, and application as active sites for energy chemistry. The reaction kinetic rate and selectivity in a chemical reaction depend on the physical and chemical properties of catalytic material that affect the surface adsorption and desorption of reactive species. Defect engineering on catalytic activity has become an important direction in the catalyst's application as defects can affect the geometric and electronic characteristics of catalysts, improving the activity, adsorption properties, and selectivity. Ma et al. [1] reported that the optimization of introduction defects in catalytic materials can influence (i) the adsorption energy of intermediates, (ii) the reaction energy barriers, (iii) paths and mechanism, (iv) selectivity, and (v) catalytic efficiency. Defect engineering offers a powerful means to optimize catalytic properties. Defects increase the number of active sites in the system and thereby increase the total electrocatalytic activity. During the creation of defects, porous structures or channels are typically created, which also improve the mass transfer and kinetic of the reaction. The formation of lattice defects has shown to be an effective approach to increase the reaction rate [2].

Different types of defects can alter electronic/transport properties and lattice parameters. Depending on where the defects are present in the crystal, they can be classified as surface defects or bulk-phase defects. Figure 5.1 shows the classification (left) and schematic representation of defects (right) according to the different positions. Defects can be divided into intrinsic defects (anion vacancies, cation vacancies, anti-site atoms, and interstitial defects) and extrinsic defects (doping or substitution atoms) [3, 4]. A Frenkel defect (dislocation defect) is a type of point defect that occurs when an atom or ion is displaced from its lattice point and moved to a nearby space between atoms. Oxygen vacancies (OV) are a type of point defect that is produced by losing oxygen atoms in the metal oxide lattice. Vacancies or

Surface and Interfacial Defects in Nanomaterials for Sustainable Energy Production and Storage, First Edition. Edited by Noé Arjona, Lorena Álvarez Contreras, and Minerva Guerra Balcázar.

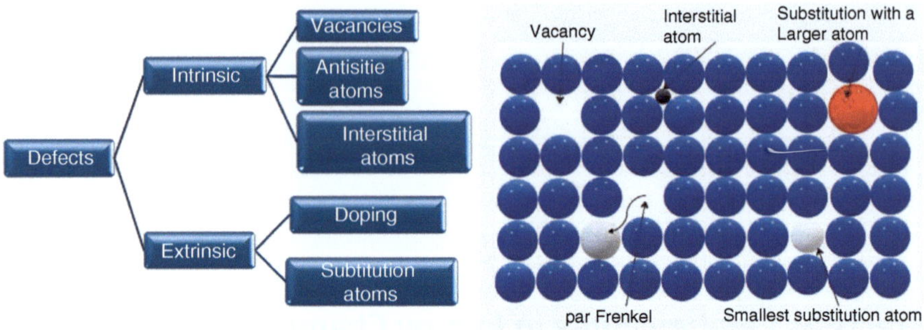

Figure 5.1 Classification of defects according to different positions (left) and schematic representation of defects (right).

doping generates more active sites to promote the reaction by removing, replacing, or introducing certain atoms in the structure, improving the catalytic performance and selectivity of the active site by modulating the local electronic environment [5]. For example, substitutional defects modify the electronic structure, while dislocations and grain boundaries create active sites for catalytic applications. Vacancy on a metal oxide surface can provide a favorable adsorption site for reactants in catalysis [3]. According to the literature, the anion and cation vacancies are presented as oxygen vacancy and metal vacancy. Anion vacancies in metal compounds are usually used as electron donors in the forbidden band. Cationic defects can be used as electron acceptors to effectively regulate the electron and band structure of the catalysts and to improve electrical conductivity [6]. However, recently, studies on defect engineering have found that not all the structural defects can improve catalytic activity. To elucidate the relationship between the structure of defects and catalytic performance, it is necessary to combine experiments and theoretical calculations like density functional theory (DFT).

5.2 Defects as Active Sites for Electrochemical Reactions

According to the literature, defects present in catalytic materials can be used as active sites as the defects (i) act as an active center to promote the reaction; (ii) can also modify the electronic structure; (iii) act as an anchoring site to improve catalytic performance; (iv) increase electron density; (v) modify the hydrophobicity of electrocatalysts; and (vi) change the reaction pathways to improve the adsorption energy, efficiency, conversion, and selectivity. The defect engineering modulates the structure and physical/chemical properties of electrocatalysts, improving charge transfer, adsorption/desorption process, selectivity and reducing the energy barriers improving electrocatalytic efficiency [1]. Defect engineering can provide plentiful catalytic sites for electrochemical redox reactions and is one of the most promising strategies for carbon material modification [7–9]. Table 5.1 shows a summary of the types of defects used as catalytic sites for different chemical and electrochemical reactions.

Table 5.1 Doping as active site for different catalytic reactions ORR, EOR, HER, NRR, and CO_2RR.

Reaction	Catalysts	Types	Synthesis method	Refs.
ORR	N-B-carbon	Doping	Annealing	[10]
ORR/OER	$Na_2Ta_8O_{21}/C$	OV	Annealing	[11]
ORR/OER	NSP/C	Doping	Bottom-up	[12]
HER/OER	$NiFeCoO_4$-Na	OV	Co-precipitation	[13]
Nitrogen photoreduction	TiO_2	OV	Chemical reduction	[14]
Hydrogen production	B-doped Pd	Doping	Chemical reduction	[15]
CO oxidation	Au/CeO_2	Oxygen vacancy	Deposition–precipitation	[16]
Water splitting	P-$NiSe_2$	Doping	Annealing process	[17]
CO_2 photoreduction	$ZnIn_2S_4$	OV Doping	Hydrothermal	[18]
OER	S-CeO_2	OV and doping	Hydrothermal	[19]
OER	Ru-CoCu	OV and doping	Hydrothermal	[20]
OER/RRO	N-P-Co-C	Doping	Annealing process	[21]
OER	Co/Se-MoS_2	Co-doping	Solvothermal method	[22]

5.2.1 Oxygen Reduction Reaction (ORR)

The oxygen reduction reaction (ORR) is a complex and multielectron reaction that can occur in the cathode of a fuel cell (FC). ORR mechanism is determined by the catalytic material used as the cathode. The ORR pathways depend on the pH of the electrolyte (Eqs. 5.1 and 5.2).

$$\text{Acid electrolyte: } 4H^+ + O_2 + 4e^- \rightarrow 2H_2O \tag{5.1}$$

$$\text{Alkaline electrolyte: } O_2 + 2H_2O + 4e^- = 4OH^- \tag{5.2}$$

Traditionally, platinum and its alloys are the most active catalysts for ORR in FC. However, the high cost of platinum limits its commercial scale. In recent years, extensive research has been carried out on noble metal-free electrocatalysts for ORR based on carbon doped with heteroatoms (N, S, P, B, etc.) [23]. Metal-free carbon materials base heteroatom-doped carbon can be prepared using different types of carbon (graphene, CNT, mesoporous carbon, aerogel carbon, etc.), which exhibited interesting and favorable characteristics for the ORR as carbon doping results in different types of defects such as point defects in graphene including vacancies, edge defects, and pentagon defects showing an improvement in the catalytic activity for ORR by altering their electronic structures. Specifically, the incorporation of nitrogen (N) in the carbonaceous skeleton modifies the structural and electronic properties of carbon, increasing significantly the active sites. However, the influence of the heteroatoms such as nitrogen in the ORR is controversial, and the actual ORR active sites (pyridinic-N, pyrrolic-N, or graphitic-N) are not clear. Figure 5.2 displays

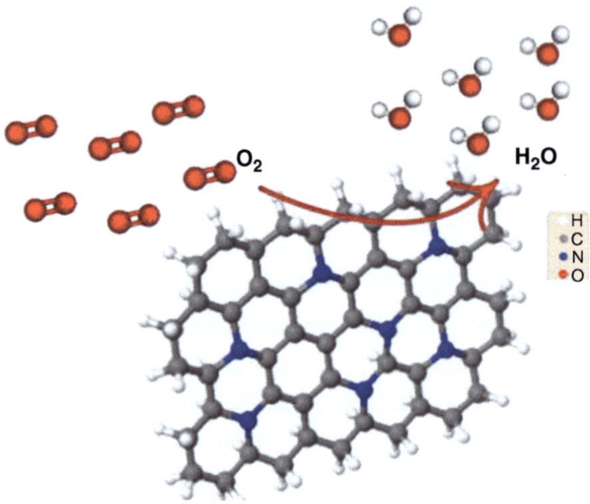

Figure 5.2 Schematic diagram of the oxygen reduction reaction on N-doped carbon based electrocatalysts in acid medium.

a schematic diagram of the oxygen reduction reaction on N-doping carbon based electrocatalysts in acid medium.

Furthermore, boron-doping metal can occupy an interstitial or substitutional site of the host material, for example, Pd-B is used for electrochemical reactions like ORR showing a higher catalytic activity and stability than commercial Pt/C or Pd/C [24].

5.2.2 Hydrogen and Oxygen Evolution Reactions (HER/OER)

Hydrogen is a cleaner fuel of renewable energy for solving the global environmental problem. Hydrogen production from electrochemical water splitting ($2H_2O = 2H_2 + O_2$) is a promising alternative different from natural gas catalytic reforming and photocatalysis. This reaction is a complex reaction, so that energy (i.e. electrical, thermal, or light energy) is required to cleave H – O – H bonds. Figure 5.3 shows a schematic representation of the electrochemical water splitting to produce oxygen and hydrogen through (a) the HER and (b) the OER. Eq. 5.3 represents a general mechanism for the HER taking into account that intermediates formed in HER depend on the electrolyte in acidic (Eq. 5.4) or alkaline solution (Eq. 5.5) [25].

$$2H^+ \rightarrow {}^*H \rightarrow H_2 \tag{5.3}$$

$$\text{Acid electrolyte: } 4H^+ + 4e^- \rightarrow 2H_2 \tag{5.4}$$

$$\text{Alkaline electrolyte: } 2H_2O + 2e^- \rightarrow 2OH^- + H_2 \tag{5.5}$$

The introduction of defects reduces the energy barrier during the water splitting, optimizes the adsorption/coupling process of *H, and accelerates HER efficiency. Jesudass et al. [13] reported the use of spinel-type Ni-Fe-CoO_4-Na prepared by a simple co-precipitation method as bifunctionally efficient electrocatalyst for water splitting. The presence of defects created by Na^+ ion intrusion abundantly increased the electrocatalytic active sites. On the other hand, Yang et al. [16] reported a

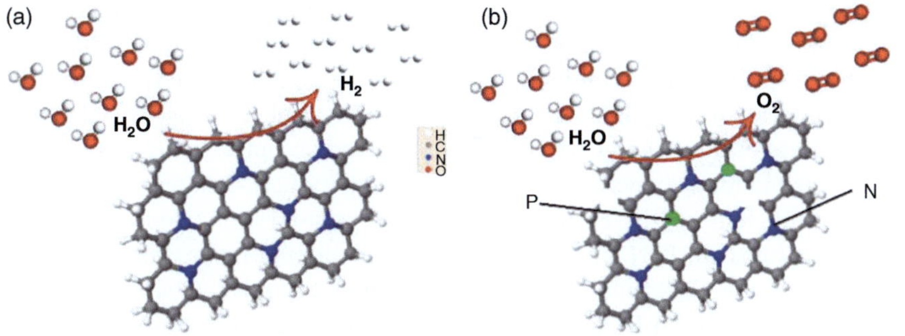

Figure 5.3 Schematic representation of the electrochemical water splitting: (a) HER and (b) OER on N- and P-doped carbon.

phosphorus-doped nickel selenide (P-$NiSe_2$) as efficient and active electrocatalysts for water splitting. The enhancement in the electrochemical activity can be ascribed to the P atom doping which leads to the high valence state of Ni as an active site toward OER. Moreover, the P doping makes the d-band center of Ni atoms in P-$NiSe_2$ move close to Fermi level, facilitating the HER kinetics with respect to proton adsorption and hydrogen desorption. DFT calculations about Se and Co co-doped MoS_2 materials exhibit an improvement in the HER activity in acid medium compared to both Se and Co single-doped MoS_2. Co/Se-MoS_2 exhibited a synergistic effect between the Se dopant on the surface and Co dopant in the inner MoS_2 layer. The Se doping facilitates the formation of active edge sites, while the Co activates the basal plane [22]. Engineering vacancy defects in the carbon matrix can greatly adjust the coordination degree to accelerate the adsorption, transformation, and desorption of reactants and intermediates, thus decreasing the energy consumption in HER and OER [26].

OER is an important reaction in electrochemistry fields like water splitting and batteries. Several OER catalysts are investigated to reduce overpotential and to improve catalytic activity to increase the reaction rate. Transition metal oxides (TMOs) are attractive for their applications due to their low cost and good catalytic activity. IrO_2 and RuO_2 are state-of-the-art catalysts used for water splitting in acid medium, specifically for OER. However, the introduction of defects in TMO materials can improve its catalytic activity for OER due to the synergistic effects found with the development of surface vacancies or heteroatom doping, which enhance the number of active sites by broadening catalytic active area. It also optimizes the surface-adsorption ability for reaction intermediates and improves the surface hydrophilic property [2, 27].

Equations (5.6) and (5.7) represent the general reaction mechanisms for the OER in different electrolytes.

$$\text{Acid electrolyte: } 2H_2O \rightarrow 4H^+ + O_2 + 4e^- \quad (5.6)$$

$$\text{Basic electrolyte: } 4OH^- \rightarrow O_2 + 2H_2O + 4e^- \quad (5.7)$$

OER can be catalyzed with electrocatalysts like Ru-doped Co–Cu with oxygen-rich vacancies that show efficient activity and stability for OER performance. The oxygen vacancy defects contribute to enhance the active sites due to the introduction of Ru atom shift in the valence states of Co displacement of the location of the d-band center with respect to the Fermi level, resulting in the enhancement of adsorption energy for intermediates, which is confirmed with DFT calculations. The electron transfer process occurs when Co^{2+} acts as an electron acceptor, which is believed to be responsible for enhancing electrocatalytic activity [20]. Rajapriya et al. [19] reported an efficient anodic electrocatalyst for OER-based CeO_2 doped with different heteroatoms (B, P, N, and S) finding that the doping heteroatoms into the CeO_2 improved the electronic conductivity, increases the reactive sites and electrocatalytic activity for OER due to the enhancement of the Ce^{3+} concentration at the surface, oxygen vacancies, and defective sites. S-CeO_2 exhibited the best electrochemical activity compared to others (B-CeO_2, P-CeO_2, and N-CeO_2). The superior performance of S-doped CeO_2 electrocatalyst was attributed to the petal-like sheets possessing high surface area that provides more active sites for OER.

According to the literature, the types of dopants and the distance between dopants also strongly affect the catalytic activity of the doped materials. Zhao and Xia [28] reported the results of N and P co-doping a carbon material as a highly active metal-free carbon-based bifunctional catalyst for ORR and OER in fuel cells, metal-air batteries, and water-splitting systems. Nitrogen doping has been used to improve OER activity of TMOS, improving the electronic conductivity and OH adsorption strength, therefore accelerating reaction [29].

Transition metal (Mo, Mn Fe, Co, and Ni) oxides exhibit oxygen vacancies defects, which play a crucial role in electrochemical applications for OER. The oxygen vacancies facilitate the charge redistribution of oxide materials, modify their adsorption energy, and improve their conductivity [8]. The modification of the electronic structure of a catalyst (e.g. d-band position for metals, valence orbital level for nonmetals) can increase the electron transfer and alter the adsorption energy of intermediates as well as alter the Fermi level of semiconductor oxides in order to increase their activity. In TMO perovskites, the band gaps of d^0 are affected by the divergence from the linearity of the framework structure.

5.2.3 CO_2 Reduction Reaction (CO_2RR)

CO_2 can be reduced to different carbon-containing chemicals (CO, HCOOH, CH_3OH, CH_4, C_2H_4, and others) under different mechanisms resulting in a promising strategy for transforming CO_2 into useful fuel. The CO_2RR is an alternative to achieving the carbon neutrality goals [1]. To reduce the concentration of CO_2 in the atmosphere, there are two main strategies: (i) the CO_2 collection and storage and (ii) the CO_2 conversion and utilization. The CO_2RR is a complex process that involves multi-protons and electron transfer process according to Eqs. (5.8)–(5.12) [30].

$$CO_2 + 2H^+ + 2e^- \rightarrow CO + H_2O \tag{5.8}$$

$$CO_2 + 2H^+ + 2e^- \rightarrow HCOOH \tag{5.9}$$

$$CO_2 + 6H^+ + 6e^- \rightarrow CH_3OH + H_2O \tag{5.10}$$

$$CO_2 + 8H^+ + 8e^- \rightarrow CH_4 + 2H_2O \tag{5.11}$$

$$2CO_2 + 12H^+ + 12e^- \rightarrow C_2H_4 + 4H_2O \tag{5.12}$$

Doped carbon materials have been widely used as catalysts for CO_2RR due to their unique characteristics of surface chemical state and regulation of the adsorption species. For example, the introduction of intrinsic vacancy type in metal-organic framework (MOF) supported on carbon can result in the breaking of the structural symmetry of the carbon, modifying the electrostatic potential and showing an unsaturated coordination structure to promote the adsorption of CO_2 and accelerate the production of the intermediate in the rate-determining step [31]. On the other hand, the doping carbon with nonmetallic elements with different electronegative properties may destroy the sp^2 hybrid carbon-based network, producing a redistribution of charge density. Duan et al. [32] reported that graphene doping with heteroatoms, such as B, N, and/or Ni, can modify the structure of graphene and facilitate the electrocatalytic CO_2RR due to a decrease in the CO_2 adsorption energy barrier. They described that doping boron stabilizes the negatively polarized O atoms of CO_2, enhancing the chemisorption of CO_2 onto the carbon surface [32]. Thus, the incorporation of heteroatoms with electron-donating or electron-withdrawing characteristics into carbon can significantly modulate the electronic properties of the different carbons, increase the number of active sites and the activity of each active site, and thus improve their catalytic activity toward the CO_2RR. However, nowadays, the roles and activity sequences of various defective structures in different reaction paths are still unclear and controversial, which needs to be further explored. Figure 5.4 shows the schematic representation of CO_2 reduction catalyzed by vacancy defects.

5.2.4 Nitrogen Photoreduction to Ammonia (NH_3)

N_2 photoreduction in ammonia (NH_3) synthesis is a clean and sustainable alternative process compared to the traditional Haber–Bosch process used for the

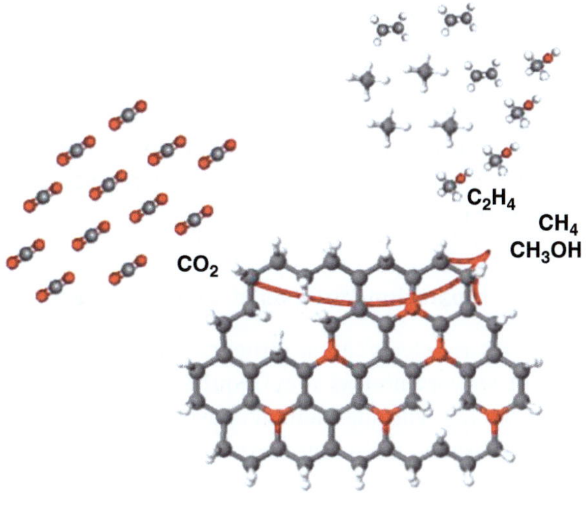

Figure 5.4 Schematic representation of CO_2 reduction catalyzed by oxygen vacancy defects.

synthesis of NH_3. Defect engineering of a semiconductor for N_2 photoreduction to NH_3 application enhances the range of light response, promotes the efficient separation and migration of photogenerated electron–hole pairs, and improves the adsorption sites and activation. Also, defects can be introduced into electronic states within the bandgap, which helps to narrow the bandgap and serve as mid-gap states for the photoexcited electron decreasing the band gap of semiconductors to enhance their performance for the nitrogen reduction to ammonia. According to the literature, surface defects on semiconductors may offer more active sites or transfer of photogenerated carriers to the adsorbate, accelerating carrier separation and adsorption of small target molecules on the photocatalyst. Defects can also act as reactive sites and capture photoinduced electrons, which can effectively restrain the photoinduced electron/hole recombination. For example, TiO_2 with abundant oxygen vacancies (anion defect) provide Ti^{3+} species as active sites that serve as adsorption sites for N_2 and promote the photocatalytic N_2 reduction to NH_3 (Eq. 5.13).

$$\text{Overall reaction: } N_{2\,(g)} + 3H_2O_{(l)} \rightarrow 2NH_{3(g)} + 3/2O_{2(g)} \tag{5.13}$$

Similarly, N, S, C, and F vacancies act as anion defects that improve the photocatalytic N_2 reduction to NH_3. On the other hand, heteroatom doping by the introduction of foreign atoms (metal or nonmetal) with different physical and chemical properties is another effective and feasible strategy for enhancing N_2 reduction. This is mainly due to the modification of electronic structure and electronic properties of semiconductor materials [14].

5.3 Synthesis Methods for Defects

Creating specific active sites on the catalyst allows precise modulation of its electronic structure and physicochemical characteristics. However, nowadays, the control of the type, concentration, and location of defects is a challenge since the existing synthesis methods are incapable of achieving defect control at such a high level [33]. According to the literature, the different synthesis methods to create defects in catalytic materials can be classified in post-treatment methods and in situ strategies (Figure 5.5). The post-treatment method to introduce defects involves two steps: first the synthesis and then a partial reduction. As opposed in the in situ methods, the synthesis of intrinsic defects are directly regulated during the synthesis process in one step [6].

5.3.1 Preparation Methods of Vacancy Defects

Vacancies as defects can be either cationic or anionic vacancies. Cationic vacancies are not so common as they require high formation energy compared to anion vacancies [2]. Various methods have been reported to generate oxygen vacancies in metal oxides, including thermal treatment, reduction processing, cation/anion

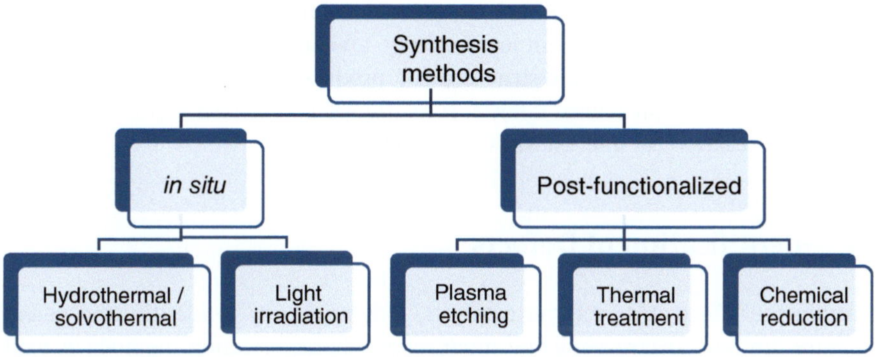

Figure 5.5 Classification of synthesis methods for defects as active sites.

doping, plasma treatment, and other technologies (laser, flame exfoliation, template strategy) [8]. The oxygen vacancies in oxides reduce the transitional metal cation (Mn^+) to lower valence state ($M^{(n-1)+}$). For example, the synthesis of oxygen vacancies on Zn-V_2O_5 catalysts with high purity and good crystallinity for cathode material in aqueous zinc batteries (AZBs) was reported [34]. The introduction of oxygen vacancy defects significantly influences the crystal structure of semiconductors by expanding the interlayer spacing, altering the internal electric field distribution, and reducing the diffusion barrier. However, the oxygen vacancies have the disadvantage of being unstable, and the degree of defect is difficult to regulate [35]. Furthermore, oxygen vacancies on CuO-CdO bimetallic catalyst for electrochemical CO_2 reduction (ECR) are synthesized by co-precipitation and post-thermal treatment showing that a low degree of oxygen defects in CuO is associated with an asymmetrical stretching of Cu-O and the interface between CdO showing an enhancing in the formation of *COOH intermediate and promoted the generation of ethylene products. According to the literature, atomic vacancies are commonly introduced electrochemically by annealing in reducing atmospheres and by plasma treatment. Further, the cationic vacancies exhibit the following advantages of (i) lowering the energy barriers for ion diffusion, (ii) increasing the electrical conductivity, (iii) promoting the charge transfer, and (iv) providing extra sites for charge storage [2].

5.3.2 Preparation Methods of Doping Defects

Doping defects occur when dopant atoms replace the occupant atoms in a crystal. Different p-block elements like nitrogen, boron, carbon, fluorine, iodine, phosphorus, sulfur, and others are typically used as doping material for defect engineering as nonmetallic atom-induced defects. Pyrolysis and annealing process are two main thermal treatment methods for the construction of heteroatom-doped carbon-based catalysts. N-doping is the most commonly used dopant in heteroatom doping as the pyridinic-N configuration has lone electron pairs, which can significantly enhance the binding of active sites on the catalyst. N-doped Co_3O_4 showed enhanced activity for OER due to the higher content of oxygen vacancies [2, 36]. On the other hand, B-doped carbon introduces asymmetric change and spin density. P doping allows to modulate the local electronic structure and produces a distortion of the carbon lattice creating more structural defects, thus improving the catalytic performance for different reactions of interest like ORR and OER [37].

5.4 Identification of Defects

To study the effect of defects on catalysts, it is necessary to identify the defect structure through spectroscopy and microscopic characterization methods as well as computational modeling [38]. The oxygen vacancies can be detected indirectly based on the crystal structure, transition metal, and oxygen electronic structure, as well as magnetic properties.

5.4.1 X-Ray Diffraction (XRD)

According to the literature, when other metal elements are doped into the material, the XRD characteristic peak will be broadened, which indicates that the crystal size of the material decreases, that is, there are more structural defects in the materials. Furthermore, the shift of XRD characteristic peak indicates the change of lattice parameters, which are usually corresponding to the lattice expansion or distortion. Furthermore, in perovskites ($Sr_2Fe_{1.3}Ni_{0.2}Mo_{0.5}O_{6-\delta}$), the shift of peaks to lower 2θ values can be related to the existence of oxygen vacancies due to the extra electrons generated together with the oxygen vacancies [8]. XRD technique can provide information about the existence of defects, but it is not possible to obtain information about the type, exact localization, and concentration of defects, thus XRD should be coupled with other techniques.

5.4.2 Electron Microscopy (TEM, HRTEM/STEM)

Transmission electron microscopy (TEM) allows us to observe information with respect to the type and density of defects. With high-resolution TEM (HRTEM), it is possible to see some lattice defects. However, nowadays distinguishing the difference between surface defects and bulk defects by TEM is not possible and it is also necessary to confirm the TEM results with other spectroscopic or resonance methods. HRTEM and scanning transmission electron microcopy (STEM) can be used to obtain the atomic arrangement of materials at atomic level which allow observed structural distortion related to the oxygen vacancies [8]. According to STEM technique, the structural defects (edge pentagon, heptagons, octagons, and combinations) in defective graphene are visualized in the edges of the graphene, while in the bulk regions, a perfect hexagon structure is observed. Additionally, STEM can detect heteroatom-modified carbon defects in graphene, and HRTEM could be utilized to verify the existence of oxygen defects in metal oxides [7].

5.4.3 X-Ray Photoelectrons Spectroscopy (XPS)

XPS is a surface-analysis method applied to the surface defect type, oxygen vacancies, or metal vacancies. The existing defects in materials can change the electronic structure and chemical environment of the materials, giving rise to differences in the binding energy or new peaks. The presence of oxygen vacancies on $NiCo_2O_4$ surface was probed by XPS based on the chemical environments of oxygen and the changes in cation valence states [39]. According to the literature, the O peak at 531.2 eV is assigned to the oxygen vacancies, while other peaks around could be associated with oxygen atoms bound to metals, hydroxyl groups, surface-adsorbed oxygen, or water molecules [8].

5.4.4 Raman Spectroscopy

Raman spectroscopy helps to obtain information of intrinsic defects of carbon materials. The level of defects can be obtained by the relative intensity ratio of D- to G-band (I_D/I_G). Also, the defects induce change in the surface chemical bond energy showing a shift in the Raman peak position of the material. The laser sources in Raman spectroscopy may affect the evolution and detection of defects. In the case of semiconductors like TiO_2, the peak position changes with the oxygen vacancies concentration.

5.4.5 Electron Paramagnetic Resonance (EPR) Spectroscopy

On the other hand, EPR technique can also be used to characterize defect structures, especially for oxygen defects in semiconductor oxides (Ti^{3+} or Zn^+ in TiO_2 and ZnO, respectively). This technique is based on detecting the magnetic moment of unpaired electrons, which can be used to qualitatively and quantitatively detect the unpaired electrons in materials with vacant defects. According to the literature, the EPR intensity increases a lot after introducing oxygen vacancies in the materials [8].

5.4.6 X-Ray Absorption Spectroscopy (XAS)

Vacancies in semiconductors at atomic scale can be also investigated by XAS analyzing the atomic coordination environment, bond length, and element valence. XAS can identify the energy depending on the structure for the X-ray absorption coefficient near the absorption edge of a particular element. A shift to lower energy values is associated with the existence of oxygen vacancies by probing the lower coordination number of material [40]. XAS can be subdivided into X-ray absorption near-edge structure (XANES) and extended X-ray absorption fine structure (EXAFS).

5.5 Conclusion and Perspectives

Despite the advantages of structural defects as active sites in catalysts applied to different chemical reactions for energy production and storage, there are still many challenges to control the creation of defects and establish a correlation between defects and catalytic activity. Some of the challenges are (i) new and greener methods for producing these large-scale defect catalysts, (ii) control of atomic doping at specific positions, (iii) control of defect types and concentrations, (iv) cheaper and more accessible characterization methods that operate in situ, and (v) control of the number of formation defects to avoid phase changes or structure breakage. Finally, the use of theoretical DFT simulations is required to investigate the impact of structural defects on the performance of the material.

References

1 Ma, W., Yao, J., Xie, F. et al. (2024). Optimizing electronic structure through point defect engineering for enhanced electrocatalytic energy conversion. *Green Energy Environ.* 10: 109–131.

2 Krishnan, A., Ajay, R., Anakha, J. et al. (2022). Understanding defect chemistry in TMOS involved electrocatalytic OER; an analysis for advancement. *Surf. Interfaces* 30: 101942–101964.

3 Mishra, R.K., Verma, K., and Sigh, D.S. (2024). Defect engineering in nanomaterials: impact, challenges, and applications. *Smart Mater. Manuf.* 2: 100052–100078.

4 Xie, C., Yan, D., Li, H. et al. (2020). Defect chemistry in heterogeneous catalysis: recognition, understanding, and utilization. *ACS Catal.* 10: 11082–11098.

5 Wag, Y., Yu, W., Wang, C. et al. (2024). Defects in photoreduction reactions: fundamentals, classification, and catalytic energy conversion. *eScience* 4: 100228–100249.

6 Gao, P., Chen, Z., Gong, Y. et al. (2020). The role of cation vacancies in electrode materials for enhanced electrochemical energy storage: synthesis, advanced characterization, and fundamentals. *Adv. Energy Mater.* 10 (14): 1903780–1903805.

7 Yan, X., Zhuan, L., Zhu, Z. et al. (2021). Defect engineering and characterization of active sites for efficient electrocatalysis. *Nanoscale* 13: 3327–3345.

8 Zhu, K., Shi, F., Zhu, X. et al. (2020). The roles of oxygen vacancies in electrocatalytic oxygen evolution reaction. *Nano Energy* 73: 104761–104779.

9 Zhao, Z., Chen, H., Zhang, W. et al. (2023). Defect engineering in carbon materials for electrochemical energy storage and catalytic conversion. *Mater. Adv.* 4: 835–867.

10 Kodithuwakku, U.S., Wamminayake, N., Thomas, M.P. et al. (2023). Unlocking efficiency in oxygen reduction reaction: synergistic edge dopants of nitrogen and boron in carbon nano onions. *Electrochim. Acta* 471: 143365–143374.

11 Ruiz-Cornejo, J.C., Sebastián, D., Martínez-Huerta, M.V. et al. (2019). Tantalum-based electrocatalysts prepared by a microemulsion method for the oxygen reduction and evolution reactions. *Electrochim. Acta* 317: 261–271.

12 Yao, S., Lyu, D., Wei, M. et al. (2021). N, S, P co-doped graphene-like carbon nanosheets developed via in situ engineering strategy of carbon pz-orbitals for highly efficient oxygen redox reaction. *FlatChem* 27: 100250–100262.

13 Jesudass, S.C., Surendran, S., Moon, D.J. et al. (2024). Defect engineered ternary metal spinel-type Ni-Fe-Co oxide as bifunctional electrocatalyst for overall electrochemical water splitting. *J. Colloid Interface Sci.* 663: 566–576.

14 Hirakawa, H., Hashimoto, M., Shiraishi, Y. et al. (2017). Photocatalytic conversion of nitrogen to ammonia with water on surface oxygen vacancies of titanium dioxide. *J. Am. Chem. Soc.* 139 (31): 10929–10936.

15 Jian, K., Zu, K., Zou, S. et al. (2014). B-doped Pd catalyst: boosting room-temperature hydrogen production from formic acid–formate solutions. *J. Am. Chem. Soc.* 136 (13): 4861–4864.

16 Guzman, J., Carrettin, S., and Corma, A. (2005). Spectroscopic evidence for the supply of reactive oxygen during CO oxidation catalyzed by gold supported on nanocrystalline CeO_2. *J. Am. Chem. Soc.* 27 (10): 3286–3287.

17 Yang, W., Wang, S., Zhao, K. et al. (2021). Phosphorus doped nickel selenide for full device water splitting. *J. Colloid Interface Sci.* 602 (15): 115–122.

18 Jiao, X., Chen, Z., Li, X. et al. (2017). Defect-mediated electron-hole separation in one-unit-cell $ZnIn_2S_4$ layers for boosted solar-driven CO_2 reduction. *J. Am. Chem. Soc.* 139 (22): 7586–7594.

19 Rajapriya, A., Keerthan, S., Rebekah, A. et al. (2021). Enriched oxygen vacancy promoted heteroatoms (B, P, N, and S) doped CeO_2: challenging electrocatalysts for oxygen evolution reaction (OER) in alkaline medium. *Int. J. Hydrog. Energy* 46 (75): 37281–37293.

20 Wang, Y., Jing, L., Jiang, W. et al. (2024). Rich oxygen vacancy and amorphous/crystalline ruthenium-doped CoCu-layered double hydroxide electrocatalysts for enhanced oxygen evolution reactions. *J. Colloid Interface Sci.* 671 (2024): 283–293.

21 Jo, H.-G. and Ahn, H.-J. (2020). Accelerating the oxygen reduction reaction and oxygen evolution reaction activities of N and P co-doped porous activated carbon for Li-O_2 batteries. *Catalysts* 10 (11): 1316–1330.

22 Zhen, Z., Yu, L. et al. (2020). Boosting hydrogen evolution on MoS_2 via co-confining selenium in surface and cobalt in inner layer. *Nat. Commun.* 11 (1): 3315–3325.

23 Tian, K., Wang, J., Cao, L. et al. (2020). Single-site pyrrolic-nitrogen-doped sp^2-hybridized carbon materials and their pseudocapacitance. *Nat. Commun.* 11 (1): 3884–3894.

24 Chen, T., Foo, C., and Edman Tsang, S.C. (2021). Interstitial and substitutional light elements in transition metals for heterogeneous catalysis. *Chem. Sci.* 12: 517–532.

25 Hu, E., Yao, Y., and Qian, C.G. (2021). Strategies for the enhanced water splitting activity over metal-organic frameworks-based electrocatalysts and photocatalysts. *Mater. Today Nano* 15: 100124–100162.

26 Chen, Y.-x., Zhao, X.-h., Dong, P. et al. (2024). Carbon-based electrocatalysts for water splitting at high-current-densities: a review. *New Carbon Mater.* 39 (1): 1–16.

27 Huang, J., Jiang, Y., An, T. et al. (2020). Increasing the active sites and intrinsic activity of transition metal chalcogenide electrocatalysts for enhanced water splitting. *J. Mater. Chem. A* 8: 25465–25498.

28 Zhao, Z. and Xia, Z. (2016). Design principles for dual-element-doped carbon nanomaterials as efficient bifunctional catalysts for oxygen reduction and evolution reactions. *ACS Catal.* 6 (3): 1553–1558.

29 Shi, Z., Wang, X., Ge, J. et al. (2020). Fundamental understanding of the acidic oxygen evolution reaction: mechanism study and state-of-the-art catalysts. *Nanoscale* 12: 13249–13275.

30 Tekalgne, M.A., Do, H.H., Hasani, A. et al. (2020). Two-dimensional materials and metal-organic frameworks for the CO_2 reduction reaction. *Mater. Today Adv.* 5: 100038–100055.

31 Lu, Y.-k., Chen, B.-x., Zhan, H.-y. et al. (2024). Defect engineering of carbon-base electrocatalysts for the CO_2 reduction reaction: a review. *New Carbon Mater.* 39 (1): 17–41.

32 Duan, X., Xu, J., Wei, Z. et al. (2017). Metal-free carbon materials for CO_2 electrochemical reduction. *Adv. Mater.* 29 (41): 1701784–1701804.

33 Li, C.C., Wang, T., Zhao, Z.J. et al. (2018). Promoted fixation of molecular nitrogen with surface oxygen vacancies on plasmon-enhanced TiO_2 photoelectrodes. *Angew. Chem. Int. Ed.* 57 (19): 5278–5282.

34 Han, B., Zheng, X., Ali, M. et al. (2024). Oxygen vacancy defects enable high performance V_2O_5 cathode for aqueous zinc batteries: a mini review. *J. Electroanal. Chem.* 971: 118571–118579.

35 Jia, T., Wang, L., Zhu, Z. et al. (2024). Modulating the degree of O vacancy defects to achieve selective control of electrochemical CO_2 reduction products. *Chin. Chem. Lett.* 35: 108692–108697.

36 Yin, C., Liu, Y., Xia, Q. et al. (2019). Oxygen vacancy-rich nitrogen-doped Co_3O_4 nanosheets as an efficient water-resistant catalyst for low temperature CO oxidation. *J. Colloid Interface Sci.* 553: 427–435.

37 Xue, D., Xia, H., Yan, W. et al. (2021). Defect engineering on carbon-based catalysts for electrocatalytic CO_2 reduction. *Nano-Mirco Lett.* 13 (1): 5–28.

38 Shen, H., Yang, M., Hao, L. et al. (2022). Photocatalytic nitrogen reduction to ammonia: insights into the role of defect engineering in photocatalysts. *Nano Res.* 15 (4): 2773–2809.

39 Bao, J., Zhang, X., Fan, B. et al. (2015). Ultrathin spinel-structured nanosheets rich in oxygen deficiencies for enhanced electrocatalytic water oxidation. *Angew. Chem.* 54 (25): 7399–7404.

40 Cai, Z., Bi, Y., Hu, E. et al. (2017). Single-crystalline ultrathin Co_3O_4 nanosheets with massive vacancy defects for enhanced electrocatalysis. *Adv. Energy Mater.* 1701694–1701702.

6

Advanced Characterization Techniques for Defect and Interface Engineering

José Béjar and Alfredo Aguilar-Elguezabal

Centro de Investigación en Materiales Avanzados, S.C., Av. Miguel de Cervantes Saavedra 120, Complejo Industrial Chihuahua, Chihuahua, C.P. 31136, México

6.1 Introduction

In the design of advanced materials for energy storage systems, the introduction of surface and interfacial defects has emerged as a key strategy. Chapters 2 and 3 describe the different superficial and interfacial defects in materials that confer different physical, chemical, and structural properties. Therefore, it is important to gain a deep understanding of the relationships between the atomic structures of defects and the performance displayed by the defective materials. However, the nanoscale dimensions and low concentrations of interfacial and surface defects make their characterization and determination difficult.

In this sense, different advanced characterization techniques have been developed to observe the presence of various defects, proving the conditions to understand the different processes occurring on the materials surface and their effect on energy production and storage systems. Important characterization techniques comprise various electronic microscopies including scanning electron microscopies (SEM), transmission electron microscopy (TEM), and atomic force microscopy (AFM). On the other hand, spectroscopic techniques provide extraordinary tools to study the defects. Spectroscopy techniques include X-ray diffraction (XRD), X-ray photoelectron spectroscopy (XPS), electron paramagnetic resonance (EPR), Raman spectroscopy (Raman), synchrotron X-ray absorption fine structure (XAFS), and electrochemical impedance spectroscopy (EIS). In this chapter, different characterization tools are summarized to understand the effect of different superficial and interfacial defects in physicochemical properties displayed by the materials employed in production and storage energy systems.

6.2 Electron Microscopy Techniques

Superficial and interfacial defects promote the missing of atoms in the lattice, dislocations, lattice modifications, etc. One of the most versatile ways to characterize

Surface and Interfacial Defects in Nanomaterials for Sustainable Energy Production and Storage, First Edition. Edited by Noé Arjona, Lorena Álvarez Contreras, and Minerva Guerra Balcázar.

and study these defects is through direct observation of the atomic numbers and the arrangement of each atom within the crystal lattice. For this purpose, transmission electron microscopy (TEM) and specifically high-resolution transmission electron microscopy (HRTEM) are forceful techniques used in qualitative and quantitative characterization, since they provide the possibility of generating images of these defects. In comparison with conventional microscopy, the images are created from interference in the image plane. The electrons interact with the sample, transfer energy, and cause changes in their wave properties. As the electron wave passes through the sample, they are deflected or scattered, which are then detected and used to form the image [1]. HRTEM allows the possibility to employ different modes such as scanning transmission electron microscopy (STEM), which can correct the spherical aberration, besides, it can be used with different detectors to obtain details and information about the material, for example, high-angle annular dark-field (HAADF), annular bright-field (ABF), annular dark-field (ADF), and aberration-corrected low-angle annular dark-field scanning transmission electron microscopy (ACLADDF-STEM) [2]. Additionally, the images obtained by STEM mode possess better resolution and sensitivity in addition to improved signal-to-noise ratios compared to conventional TEM. For example, Yuan et al. [3] reported a facile strategy to induce defects using a decarboxylation method, which allows to optimize the electronic structure of the transition metal by modifying the adsorption free energy of OOH on Co sites and boosting the oxygen reaction reduction. Figure 6.1a,b respectively shows the HRTEM and HAADF micrography for Co@DMOF-T material, which exhibits a pumice-like morphology. Furthermore, to demonstrate the atomic distribution of Co-N_4, Figure 6.1c shows the HAADF-STEM image of Co@DMOF-900, in which numerous isolated bright dots can be clearly distinguished within the onion-like carbon structure. On the other hand, the electron energy loss spectroscopy mapping (EELS, Figure 6.1d) confirms the uniform distribution for all elements. Finally, the ACLADDF-STEM image (Figure 6.1e) was utilized to examine defects within the material, revealing defects not only in carbon structures but also in the Co-N_4 sites. Liu et al. [4] reported a NiCo-bimetallic oxide with superior electrocatalytic activity toward the oxygen reactions due to their abundant oxygen vacancies and active sites. NiCo-oxides were supported on carbon nanofibers using electrospinning (Figure 6.1f). The PVP is decomposed into N, O-doped carbon fibers, since nitrogen sources are used (acetamide and thiourea), besides, a large number of defects are introduced. The nanofiber is covered with oxide nanoparticles (Figure 6.1g), which increases the contact between the electrocatalyst and the electrolyte. Figure 6.1h distinguishes the numerous interfacial domains between the two different crystals such as $NiCo_2O$@C, which possess different crystal structures and promote the formation of defects near the interface domain (yellow circle in Figure 6.1h). The great number of interfacial defects can act as atomic trap that could favor the adsorption of oxygen intermediates, improving the electrocatalytic activity toward oxygen reactions.

One of the most important limitations of transmission electron microscopy is the intricate sample preparation, which often complicates the efforts to accurately measure the thickness of the samples under analysis. Electron channeling contrast imaging (ECCI) technology using scanning electron microscopy (SEM) can solve this problem. Samples for ECCI are straightforward to prepare and require only

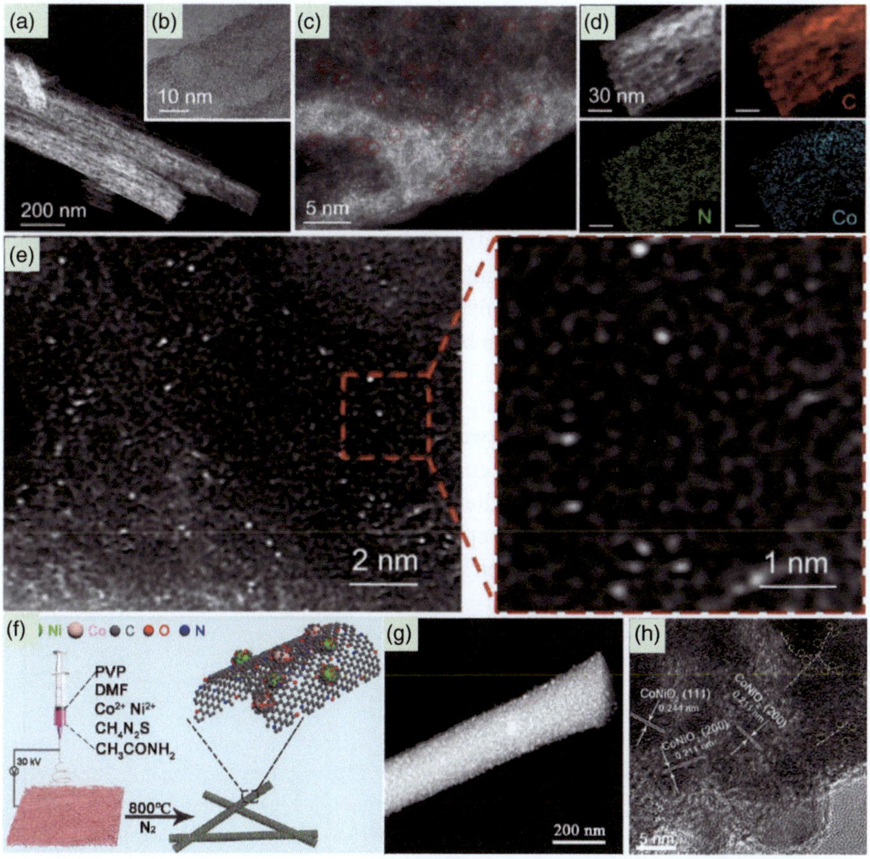

Figure 6.1 (a and b) HAADF-STEM and HRTEM images of Co@DMOF-900 catalyst, (c) AC HAADF-STEM, (d) AC HAADF-STEM image and corresponding EELS mapping images, (e) AC LAADF-STEM images of Co@DMOF-900. Source: Shuai Yuan et al. [3] / with permission of John Wiley & Sons., (f) scheme of synthesis of $NiCo_2O$@C catalyst, (g) HAADF-STEM, and (h) HRTEM images of $NiCo_2O$@C. Source: Chaojun Liu et al. [4] / with permission of Elsevier.

surface polishing to expose a large area for analysis, which enables the statistical evaluation of crystal defects. Besides, this process does not require the destruction of the sample, hence, it can be used in subsequent experiments [5].

Selecting the most suitable electron microscopy technique requires careful consideration of the specific objectives that is required for the study. It is essential to balance factors such as the needed resolution, the complexity of sample preparation, and the potential destructiveness of the analysis.

6.3 X-Ray Diffraction (XRD)

X-ray diffraction (XRD) is a widely utilized technique for characterizing materials, detecting changes in crystal structure, identifying new phases, and measuring expansions or contractions in cell parameters as well as shifts in characteristic XRD peaks. These alterations, which are critical for studying interfacial and surface defects, may arise from atomic vacancies, heteroatom doping, or similar phenomena. XRD studies the structure of matter by directing X-rays at a sample, where they interact with the atomic lattice and scatter in specific patterns. These diffraction patterns are determined by the arrangement of atoms, revealing information about crystal structure, phase composition, and lattice parameters. By analyzing peak positions, intensities, and shifts, XRD detects changes such as expansions, contractions, or defects. This nondestructive technique provides detailed insights into the atomic and molecular organization of materials. The interaction between X-rays and the sample produces constructive interference and generates a peak when it satisfies Bragg's law [6], given by

$$n\lambda = 2d\sin(\theta)$$

where n is an integer, θ is the angle between the surface of the crystal and the incident ray, and d is the lattice spacing between the layers of atoms.

As an example of the potential of the XRD technique, a case of atomic vacancies is reported by Han et al. [7], they prepared γ-Fe_2O_3 doped with Mo^{6+} to promote the Li-ion storage in defect spinel iron oxides. Mo ions substitute the Fe^{3+} into the

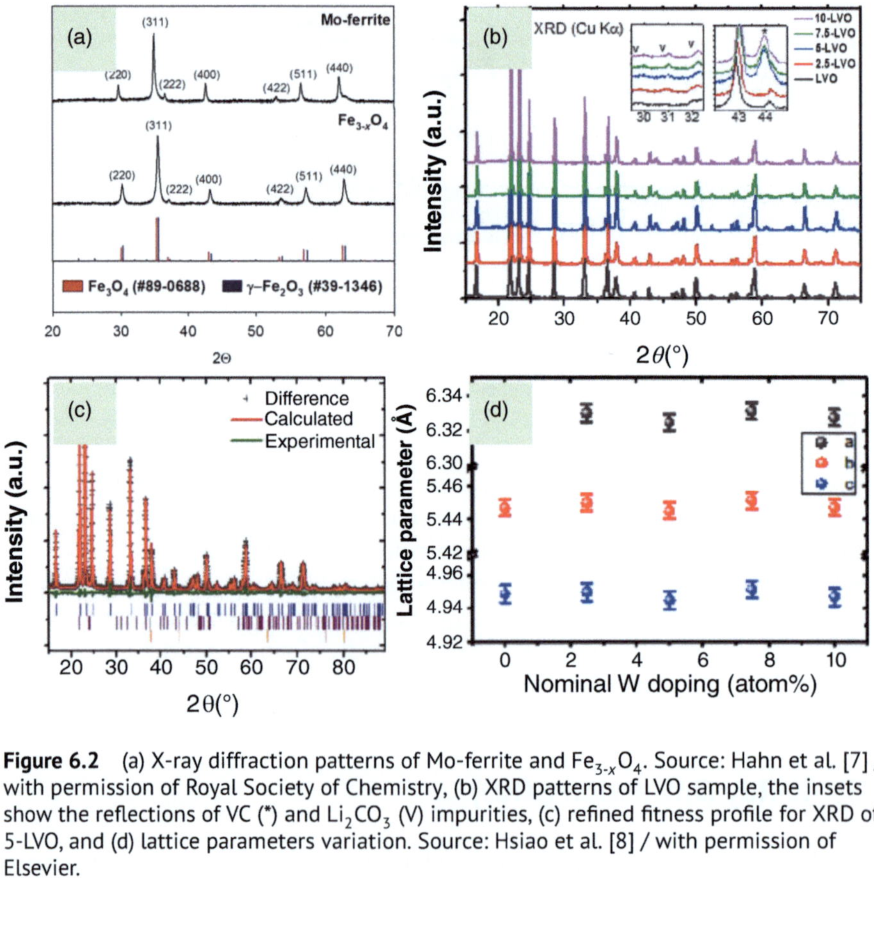

Figure 6.2 (a) X-ray diffraction patterns of Mo-ferrite and $Fe_{3-x}O_4$. Source: Hahn et al. [7] / with permission of Royal Society of Chemistry, (b) XRD patterns of LVO sample, the insets show the reflections of VC (*) and Li_2CO_3 (V) impurities, (c) refined fitness profile for XRD of 5-LVO, and (d) lattice parameters variation. Source: Hsiao et al. [8] / with permission of Elsevier.

crystal structure, expanding the size of the unit cell (Figure 6.2a), which increases the concentration of solid-state defects, resulting in superior electrochemical charge-storage properties in terms of both Li-ion capacity and insertion potential. For heteroatom doping, Hsiao et al. [8] reported a spray-drying method to synthesize Li_3VO_4 (LVO) doped with W^{6+}. Figure 6.2b displays the XRD patterns for pristine and doped LVO, and as can be seen, samples showed orthorhombic structure without important differences, which was corroborated by Rietveld refinement (Figure 6.2c). From the determination of lattice parameters (Figure 6.2d), authors found similar values across all samples, despite the differences in atomic radius between V^{5+} and W^{6+} (0.355 and 0.42 Å, respectively). This suggests that the doped structures maintain their crystalline framework while exhibiting enhanced electrical conductivity and improved ion migration into vacancies.

XRD is a useful technique for investigating the crystal structure of materials and detecting defects, offering valuable information to understand the relationships between atomic structure and material properties.

6.4 X-Ray Photoelectron Spectroscopy (XPS)

XPS is one of the most appropriate characterization techniques used for the study of superficial and interfacial defects. It enables quantitative surface analysis by accurately determining the elemental composition of a sample. Moreover, XPS provides valuable insights about the chemical valence states and electronic environments because variations in the electronic structure and bond energies lead to distinct shifts in binding energies observed in the spectra [9, 10]. The XPS principle determines the kinetic energy of photoelectrons ejected from the surface of a sample when employing an X-ray source with a constant energy, $h\nu$ (for Al Kα X-rays, $h\nu = 1486.7$ eV), in vacuum. The equilibrium between $h\nu$ and a kinetic energy of photoelectron (E_K) is expressed as:

$$h\nu = E_K + E_B + \phi$$

where E_B is the binding energy of electron to nucleus relative to the Fermi level and ϕ is a work function of specimen, in the case of solid. E_B and the chemical shift (difference from elemental state) are utilized for identification of an element and estimation of its chemical state in the specimen [2, 11]. Du et al. [12] reported a defected perovskite ($La_{0.8}Sr_{0.2}Co_{0.8}Fe_{0.2}O_{3-\sigma}$) LSCFO with oxygen vacancies, which were promoted by the co-doping with Sr and Fe atoms (Figure 6.3a, Co 2p), each oxygen vacancy generates two free electrons that can interact with the adjacent cations, lowering the binding energy [9], which was demonstrated by the shift in the signals for the Co 2p spectrum. Figure 6.3b displays the O 1s high-resolution XPS spectrum, where it is possible to remark the increasing in the O_2^{2-}/O^- signal for the defective perovskite, which corroborates the oxygen vacancies promoted by the heteroatom doping. Another relevant example is the study reported by Kang et al. [13], in which TiO_2 nanotubes were synthesized with oxygen vacancies induced by $NaBH_4$ treatment. Figure 6.3c,d represents the high-resolution XPS

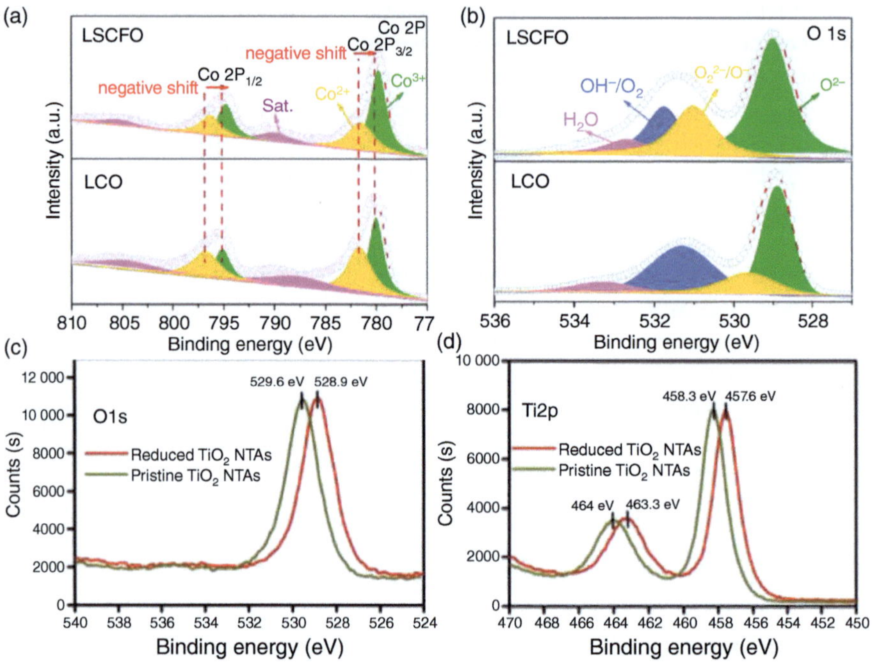

Figure 6.3 High-resolution XPS spectra for (a) Co 2p and (b) O 1s for LSCFO. Source: Du et al. [12] / with permission of American Chemical Society. High-resolution XPS spectra for (c) O 1s and (d) Ti 2p for TiO_2 nanotubes. Source: Kang et al. [13] / with permission of Royal Society of Chemistry.

spectra for O 1s and Ti 2p, respectively, where distinct shifts in binding energies for both species can be observed. These shifts indicate the migration of electrons from O and Ti to oxygen vacancies, which act as electron traps, reducing surface recombination centers and enhancing charge transfer efficiency [13].

By analyzing the shifts in binding energies and the decomposition of O 1s peaks and the valence states of transition metals, XPS allows the evaluation of the presence of oxygen vacancies, Schottky or Frenkel defects, and interactions with adsorbed species stabilized on the surface. These defects can act as active centers for surface reactions, promote charge separation in semiconductors, and improve ionic mobility in materials for batteries and fuel cells.

6.5 Raman Spectroscopy

Raman spectroscopy is a widely utilized technique for material characterization; however, its most significant applications are possibly in the study of carbon-based materials, since it is essential for characterizing various carbon allotropes [14], monitoring structural changes in graphitic materials [15], determining the chirality of single-wall carbon nanotubes [16], analyzing the structural properties of graphene [17], and quantifying the relative fractions of sp^2 and sp^3 hybridized carbon in diamond-like carbon materials [18]. Given its ability to probe the structural characteristics of a wide range of carbon-based materials, Raman spectroscopy is a suited tool for studying defects in these systems. The Raman spectra of graphite and other carbon materials are highly sensitive to structural imperfections that disrupt crystal symmetry in graphitic structures. These defects are commonly analyzed through characteristic Raman features, with the G and D bands being the most significant for assessing structural disorder. The G band is a Raman-active vibrational mode arising from the in-plane displacement of neighboring carbon atoms in opposite directions within the graphitic structure. It is typically observed around $1582\,cm^{-1}$ and serves as a key indicator of the degree of graphitization in carbon-based materials. In contrast, the D band, located at approximately $1360\,cm^{-1}$, is associated with structural disorder in graphitic materials. It originates from intervalley scattering processes enabled by the presence of defects, which break the long-range order of sp^2 carbon networks. The activation of the D band is attributed to a high density of phonon states near the Brillouin zone corner, resulting from finite-crystal size effects and structural imperfections. These defects can be introduced by oxygen-related functional groups, heteroatom doping, or other structural disruptions. Béjar et al. [19], reported various strategies for achieving N-doping in multi-walled carbon nanotubes (MWCNTs), including modifications in the synthesis method, the introduction of morphological defects, and the selection of nitrogen precursors. Raman spectroscopy analysis of all samples revealed the characteristic D band ($\sim 1350\,cm^{-1}$) and G band ($\sim 1580\,cm^{-1}$), with the I_D/I_G intensity ratio used to assess the degree of structural disorder (Figure 6.4a,b). The highest I_D/I_G values were observed in materials synthesized via thermal treatment, where morphological defects were introduced through the oxidative properties of H_2O_2. The combination of these morphological defects and nitrogen doping significantly enhanced the electrocatalytic activity toward the oxygen reduction reaction (ORR), demonstrating excellent performance in a Zn-air battery prototype. Ci et al. [20] reported a method to design an electrocatalyst (VS_2-VG@CC) through chemical vapor deposition of defective vanadium diselenide (VSe2) and vertical graphene (VG) on carbon cloth (CC). Raman spectroscopy was used to characterize the structure, composition, and chemical transformations of the VSe2-VG heterostructure, as well as monitor its behavior during the electrochemical processes of the electrode. Figure 6.5c–e displays the obtained spectra, where the peak at $469\,cm^{-1}$ is attributed to the LiPS (Li_2S_n $4 \leq n \leq 8$) confirming thus the VSe_2-VG anchoring. Concerning the peaks at $212–214\,cm^{-1}$, they are attributed to Se vacancies and can be used as active sites to capture sulfur materials. Results show that these peaks disappear during the discharging process, on the other hand, during the charge stage a certain amount of Se is released, regenerating signals at $212–214\,cm^{-1}$, they are attributed to Se vacancies and can be used as active sites to capture sulfur materials.

As can be observed, Raman spectroscopy serves as a useful tool in defect engineering within electrochemical science by providing real-time insights into structural modifications, defect formation, and chemical transformations of materials used in energy storage and conversion applications.

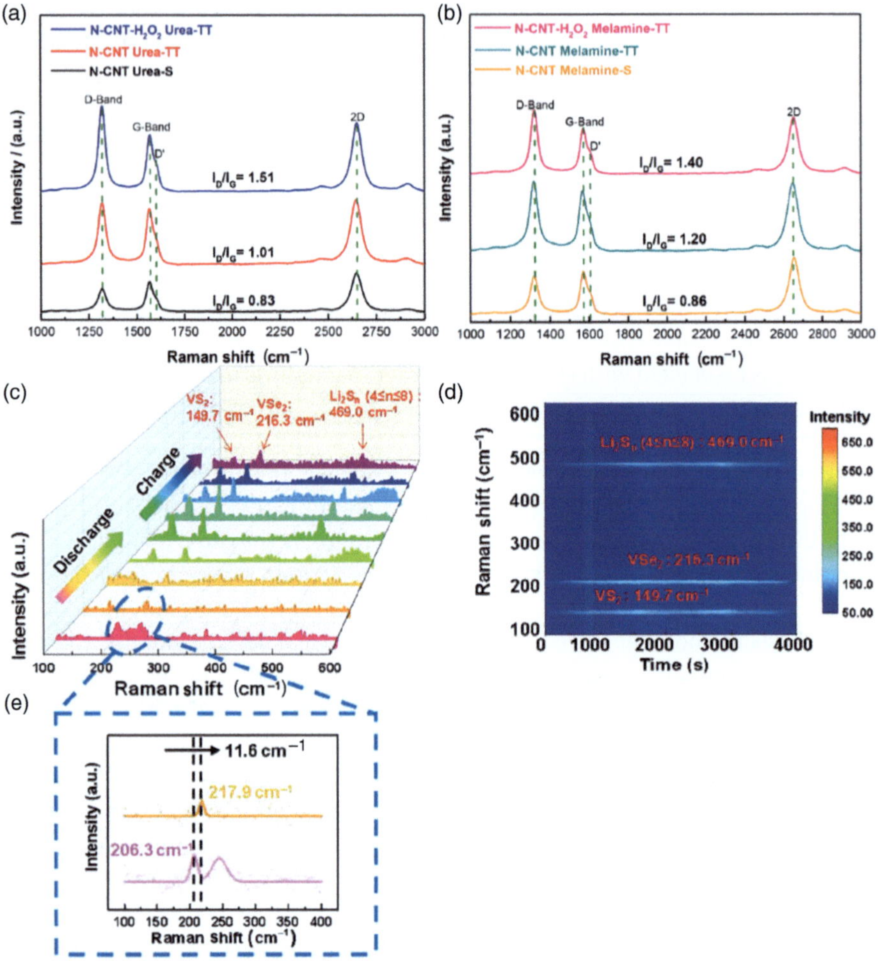

Figure 6.4 Raman spectra of N-doped carbon nanotubes using (a) urea as nitrogen source and (b) melamine as nitrogen source. Source: Béjar et al. [19] / with permission of Elsevier. (c) Raman spectra at different discharge and charge states, (d) Raman mapping, and (e) magnification of the A_{1g} signal of VSe_2–VG@CC/S electrode. Source: Ci et al. [20] / with permission of American Chemical Society.

6.6 Electron Paramagnetic Resonance (EPR)

Electron paramagnetic resonance (EPR) is a relevant spectroscopic technique used to investigate paramagnetic centers in materials. It operates using electromagnetic radiation within the microwave frequency range, positioned between radio and infrared frequencies in the electromagnetic spectrum. EPR shares fundamental principles with nuclear magnetic resonance (NMR), as both techniques probe the magnetic moments of subatomic particles. However, while NMR examines nuclear spins, EPR specifically detects unpaired electron spins, making it particularly useful for studying radicals, transition metal complexes, and defect centers. Performing

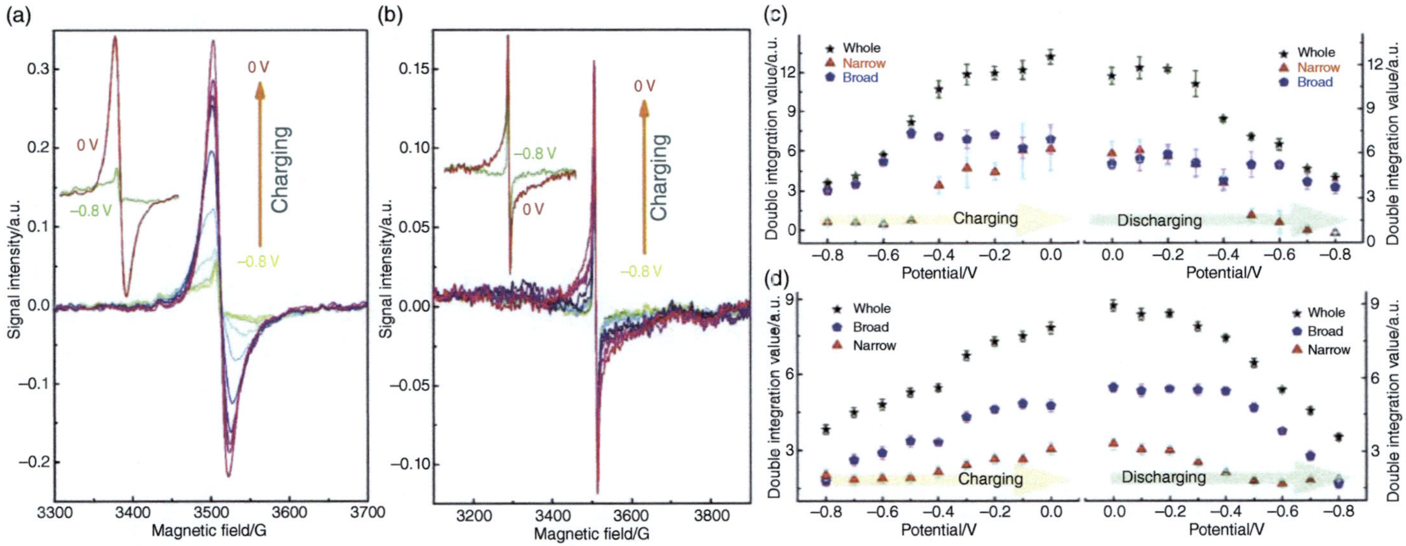

Figure 6.5 (a) EPR spectra change of rGo, (b) N-rGO over the potential from of −0.8 to 0 V in 6 M KOH for both materials, (c) summary of the electrochemically double integrated EPR signal of rGO, and (d) N-rGO. Source: Wang et al. [21] / Elsevier / CC BY 4.0.

EPR measurements requires three essential conditions: (i) the system under study must contain at least one unpaired electron, (ii) the application of an external magnetic field to lift the degeneracy of electron spin states, and (iii) the use of microwave irradiation to induce spin transitions between these states. These principles enable EPR to provide valuable insights into the electronic structure, local environment, and dynamic behavior of paramagnetic species, making it a crucial tool in electrochemistry and defects and interface engineering [22]. When an external magnetic field (B_0) is applied, the spin of an unpaired electron can occupy one of the two distinct states: parallel (α) or antiparallel (β) to the magnetic field axis. The parallel state (α) is associated with a magnetic quantum number $m_s = ½$, while the antiparallel state (β) corresponds to $m_s = -½$. Under the influence of the magnetic field, these two states become energetically nondegenerate, creating an energy separation ΔE. The application of microwave radiation (υ) can induce a transition between these states when the energy of the radiation matches the energy difference between them. This energy gap is governed by the fundamental resonance condition using applied magnetic field B_0, g-value g_e, and the Bohr magneton (μ_B) as fundamental constants, and is given by

$$\Delta E = h\nu = g_e \mu_B B_0$$

where ν is Planck's constant, g_e is the electron g-factor, μ_B is the Bohr magneton, and B_0 is the applied magnetic field strength. This equation, known as the resonance condition, is the cornerstone of electron paramagnetic resonance (EPR) spectroscopy, enabling the precise detection and characterization of paramagnetic species in various systems, including electrochemical environments [22, 23]. In the context of surface and interfacial defect characterization, EPR is a well-suited technique due to its sensitivity to unpaired electrons, which are often associated with oxygen vacancies. As previously discussed, EPR requires the presence of paramagnetic species, making it an ideal method for probing these types of structural defects. Oxygen vacancies, which introduce localized unpaired electrons, typically produce a characteristic EPR signal at $g \approx 2.004$. This signal is widely recognized as an indicator of defect states in metal oxides and other materials, providing crucial insights into their electronic structure, catalytic properties, and reactivity in electrochemical applications [24, 25]. Additionally, EPR can be used to study heteroatom doping such as N-doping in carbonaceous materials [26]. For example, Wang et al. [21] studied the performance of reduced graphene (rGO) and N-doped reduced graphene (N-rGO) in supercapacitors, where the N-rGO has approximately twice the double-layer capacitance as rGO. The EPR spectra for rGO and N-rGO are displayed in Figure 6.5a,b, respectively, showing a narrow and broad linewidth feature (Figure 6.5c,d). Broad components in the EPR spectrum for N-rGO showed greater variation over the potential range than rGO, suggesting an enhanced pseudocapacitive behavior arising from N-doping [21, 27].

EPR proves to be a crucial technique for characterizing oxygen vacancies and heteroatom doping, as demonstrated in metal oxides and N-doped graphene. Its ability to reveal defect-related electronic structures and electrochemical behavior highlights its importance in material optimization for energy storage and catalysis.

6.7 Fourier Transform Infrared (FTIR) Spectroscopy

Fourier-transform infrared (FTIR) spectroscopy is a characterization technique based on the interaction of infrared radiation with chemical bonds in matter, inducing characteristic molecular vibrations. When a molecule absorbs infrared radiation at specific wavelengths, the energy promotes transitions between vibrational states. The frequency of the absorption peaks corresponds to the energy gaps between vibrational ground and excited states, offering valuable insights into functional groups, molecular interactions, and structural modifications in materials [2]. Hence, using FTIR for investigating vibrational modes in defective materials enables the characterization of heteroatom doping, oxygen vacancies, and structural distortions. In TiO_2-based materials, for instance, FTIR can effectively detect oxygen vacancies, as absorption peaks in the 400–800 cm^{-1} range are attributed to the stretching vibrations of Ti—O—Ti and Ti—O bonds [2, 28]. Choudhury et al. [29] reported a redshift for doped TiO_2 nanoparticles, since the Ti—O peak shifts from 454 to 425 cm^{-1} for undoped and doped material, respectively (Figure 6.6a), and due to the charge neutrality, it is possible to infer that oxygen vacancies were introduced into the Co^{2+}-doped samples. Another example is reported by Chen et al. [30], where manganese oxide (K-OMS-S) doped with hexavalent molybdenum presents a structural distortion (change from tetragonal to monoclinic cell). The FTIR spectra are displayed in Figure 6.6b, where the K-OMS-S exhibits the typical signals for this type of manganese oxide. However, when the Mo is incorporated, the peaks suffer a redshift or blueshift, according to the amount of Mo in the K-OMS-2 structure, which corroborates the expected incorporation of Mo.

As can be seen, the sensitivity of FTIR spectroscopy to bond vibrations and lattice modifications makes it effective for understanding defect formation and doping effects during the design of functional materials.

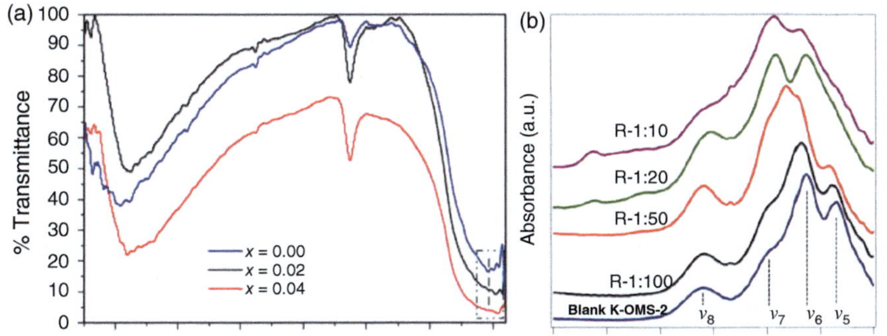

Figure 6.6 (a) FTIR spectra for undoped and doped TiO_2 nanoparticles. Source: Choudhury and Choudhury [29] / with permission of Elsevier. (b) FT-IR spectra of blank K-OMS-2 (blank) and [Mo] KOMS-2 materials. Source: Chen et al. [30] / with permission of American Chemical Society.

6.8 Conclusions

In this chapter, the most relevant advanced characterization methods were analyzed, emphasizing their critical role in monitoring structural changes induced by defect incorporation, since it is necessary to have a comprehensive understanding of the function of each kind of defect. Additionally, the same types of defects were examined using different techniques to provide alternative approaches for their identification, recognizing that access to a broad range of methods is often limited. The fundamental principles and recent studies related to each technique were discussed to aid in the result interpretation. Finally, further research is needed to develop more precise and quantitative methods for defect characterization, enabling a deeper understanding of their mechanisms and optimizing their role in enhancing material performance for advanced technologies.

References

1 Titus, D., James Jebaseelan Samuel, E., and Roopan, S.M. (2019). Nanoparticle characterization techniques. In: *Green Synthesis, Characterization and Applications of Nanoparticles* (ed. A.K. Shukla and S. Iravani), 303–319. Elsevier.

2 Su, Z., Liu, J., Li, M. et al. (2020). *Defect Engineering in Titanium-Based Oxides for Electrochemical Energy Storage Devices*, Electrochemical Energy Reviews, vol. 3, 286–343.

3 Yuan, S., Zhang, J., Hu, L. et al. (2021). Decarboxylation-induced defects in MOF-derived single cobalt Atom@Carbon electrocatalysts for efficient oxygen reduction. *Angew. Chem. Int. Ed.* 60 (40): 21685–21690.

4 Liu, C., Zuo, P., Jin, Y. et al. (2020). Defect-enriched carbon nanofibers encapsulating NiCo oxide for efficient oxygen electrocatalysis and rechargeable Zn-air batteries. *J. Power Sources* 473, 228604 (1–9).

5 Sun, L., Zhao, S., Gao, S. et al. (2025). Recent advances of metal vacancies in energy and environmental catalysis: synthesis, characterization, and roles. *Green Energy Environ.* 10 (1): 84–108.

6 Sharma, S. and Sharma, S.K. (2021). Diverse morphology zinc oxide films formulations and characterizations. In: *K. Awasthi (Ed.). Nanostructured Zinc Oxide: Synthesis, Properties and Applications*, 57–92. Elsevier.

7 Hahn, B.P., Long, J.W., Mansour, A.N. et al. (2011). Electrochemical Li-ion storage in defect spinel iron oxides: the critical role of cation vacancies. *Energy Environ. Sci.* 4 (4): 1495–1502.

8 Hsiao, Y.S., Huang, J.H., Weng, L.Y. et al. (2024). Advancing Li_3VO_4 as a high-performance anode material for use in lithium-ion batteries and lithium-ion capacitors. *Chem. Eng. J.* 489, 150973 (1–11).

9 Liu X, Liu X, Li C, Yang B, Wang L. Defect engineering of electrocatalysts for metal-based battery. *Chin. J. Catal.* 2023; 45, p. 27–87.

10 Krishna, D.N.G. and Philip, J. (2022). Review on surface-characterization applications of X-ray photoelectron spectroscopy (XPS): recent developments and challenges. *Appl. Surf. Sci. Adv.* 12: 100332. (1–30).

11 X-ray, K.H. (2016). Photoelectron spectroscopy. In: *Materials Science and Engineering of Carbon: Characterization* (ed. M. Inagaki and F. Kang), 153–171. Butterworth-Heinemann.

12 Du, D., Zheng, R., Chen, X. et al. (2021). Adjusting the covalency of metal-oxygen bonds in $LaCoO_3$ by Sr and Fe cation codoping to achieve highly efficient electrocatalysts for aprotic lithium-oxygen batteries. *ACS Appl. Mater. Interfaces* 13 (28): 33133–33146.

13 Kang, Q., Cao, J., Zhang, Y. et al. (2013). Reduced TiO_2 nanotube arrays for photoelectrochemical water splitting. *J. Mater. Chem. A* 1 (18): 5766–5774.

14 Thapliyal V, Alabdulkarim ME, Whelan DR, Mainali B, Maxwell JL. *A concise review of the Raman spectra of carbon allotropes. Diam. Relat. Mater.* 2022 127, p. 109180 (1–16).

15 Rebelo, S.L.H., Guedes, A., Szefczyk, M.E. et al. (2016). Progress in the Raman spectra analysis of covalently functionalized multiwalled carbon nanotubes: unraveling disorder in graphitic materials. *Phys. Chem. Chem. Phys.* 18 (18): 12784–12796.

16 Han, S., Hung, N.T., Xie, Y. et al. (2023). Observing axial chirality of chiral single-wall carbon nanotubes by helicity-dependent Raman spectra. *Nano Lett.* 23 (18): 8454–8459.

17 Paillet M, Parret R, Sauvajol JL, Colomban P. Graphene and related 2D materials: an overview of the Raman studies. *J. Raman Spectrosc.* 2018; 49, p. 8–12.

18 Rabadzhiyska, S., Kotlarski, G., Valkov, S. et al. (2022). Characterization of Diamond-like carbon films produced by electron- beam physical vapor deposition. In: *Materials Today: Proceedings*, vol. 67, 995–1000.

19 Béjar, J., Espinosa-Magaña, F., Avelar, J. et al. (2023). Rational design of nitrogen–doped carbon nanotubes by defect engineering for Zn–air batteries with high performance. *Carbon* 204: 411–426.

20 Ci, H., Cai, J., Ma, H. et al. (2020). Defective VSe_2-graphene heterostructures enabling in situ electrocatalyst evolution for lithium-sulfur batteries. *ACS Nano* 14 (9): 11929–11938.

21 Wang, B., Likodimos, V., Fielding, A.J., and Dryfe, R.A.W. (2020). In situ Electron paramagnetic resonance spectroelectrochemical study of graphene-based supercapacitors: comparison between chemically reduced graphene oxide and nitrogen-doped reduced graphene oxide. *Carbon* 160: 236–246.

22 Attar F, Yin H, Schumann SL, Langley J, Cox N, Zeng Z, et al. *Advanced electron paramagnetic resonance in chemical energy conversion: current status and future potential. Energy Environ. Sci.* 2024; 17, p. 3307–3328.

23 Sang, X., Xu, X., Bu, Z. et al. Application of electron paramagnetic resonance in an electrochemical energy storage system. *Magnetochemistry* 2023; 9(3), p. 63 (1–15).

24 Gao, S., Zhao, S., Tang, X. et al. (2024). Research on the application of defect engineering in the field of environmental catalysis. *Green Energy Environ.* 10: 1187–1209.

25 Zheng J, Lyu Y, Wang R, Xie C, Zhou H, Jiang SP, et al. Crystalline TiO_2 protective layer with graded oxygen defects for efficient and stable silicon-based photocathode. *Nat. Commun.* 2018; 9(1) p. 3572 (1–10).

26 Webster, R.D. (2023). Electrochemistry combined with electron paramagnetic resonance (EPR) spectroscopy for studying catalytic and energy storage processes. *Curr. Opin. Electrochem.* 40: 101308. (1–10).

27 Wang, B., Wang, W., Sun, K. et al. (2023). Developing in situ electron paramagnetic resonance characterization for understanding electron transfer of rechargeable batteries. *Nano Research,* 16 (10): 11992–12012.

28 Wang, P., Qi, C., Wen, P. et al. (2018). N co-doped nano-sized Tio_2 with high thermal stability and photocatalytic activity by mechanochemical method. *Nanomaterials* 8 (5): 294. (1–12).

29 Choudhury, B. and Choudhury, A. (2012). Luminescence characteristics of cobalt doped TiO_2 nanoparticles. *J. Lumin.* 132 (1): 178–184.

30 Chen, C.H., Njagi, E.C., Chen, S.Y. et al. (2015). Structural distortion of molybdenum-doped manganese oxide octahedral molecular sieves for enhanced catalytic performance. *Inorg. Chem.* 54 (21): 10163–10171.

7

Computational Modeling of Defects in Nanomaterials

Carlos M. Ramos-Castillo[1] *and Juan Hernández-Tecorralco*[2]

[1]*Centro de Investigación y Desarrollo Tecnológico en Electroquımica S.C., Sanfandila, Pedro Escobedo, Querétaro C. P. 76703, México*
[2]*Universidad Nacional Autónoma de México, Instituto de Física, Apartado Postal 20-364, México, 01000, México*

7.1 Defects Stability by Density Functional Theory

Defect engineering is a common strategy to enhance, modify, and tune the fundamental properties of nanomaterials. In experiments, different kinds of impurities are usually found and depending on the application, these defects are highly desired. Today, computational modeling with electronic structure methods plays a crucial role in understanding and predicting fundamental properties of both pristine and defective materials at the atomic level. These methods allow us to model their electronic behavior, providing a complementary and deeper insight into experimental measurements. In addition, these approaches guide and inspire experimental efforts in designing new platforms for energy production and storage applications. This section will address basic aspects of density functional theory (DFT) and other concepts relevant to the study of defective materials and their applications such as their formation energies and electronic structure. DFT is a ground-state theory and first-principle method that has been widely used for electronic structure calculations in condensed matter physics.

7.1.1 Fundamentals on Density Functional Theory

To understand the fundamental properties of materials, we can use quantum mechanics to describe their nature. Solving the time-independent Schrödinger equation allows us to determine the fundamental properties and physical observables of molecules or solids:

$$H\Psi(\mathbf{r},\mathbf{R}) = E\Psi(\mathbf{r},\mathbf{R}) \tag{7.1}$$

here, $\{\mathbf{r}\}$ and $\{\mathbf{R}\}$ are the coordinates of electrons and nuclei, respectively. The wave function $\Psi(\mathbf{r},\mathbf{R})$ depends on $3N$ and $3M$ variables, namely, N and M that represent

Surface and Interfacial Defects in Nanomaterials for Sustainable Energy Production and Storage, First Edition. Edited by Noé Arjona, Lorena Álvarez Contreras, and Minerva Guerra Balcázar.

the number of electrons and nuclei of the problem, respectively. The Hamiltonian[1] in Eq. (7.2) describes the dynamics of electrons and nuclei, including interactions between electrons, nuclei, and electron–nucleus:

$$H = -\frac{1}{2}\sum_{i}\nabla_i^2 - \sum_{i,I}\frac{Z_i}{|\mathbf{r}_i - \mathbf{R}_I|} + \frac{1}{2}\sum_{i\neq j}\frac{1}{|\mathbf{r}_i - \mathbf{r}_j|} - \frac{1}{2M_I}\sum_{I}\nabla_I^2 + \frac{1}{2}\sum_{I\neq J}\frac{Z_I Z_J}{|\mathbf{R}_I - \mathbf{R}_J|} \quad (7.2)$$

Decoupling the full Hamiltonian and the total wave function into electronic and ionic components simplifies the problem [1]. The idea involves approximating the total wave function as a product of electronic $\psi(\mathbf{r};\mathbf{R})$ and nuclear wave functions, $\Phi(\mathbf{R})$:

$$\Psi(\mathbf{r};\mathbf{R}) = \Phi(\mathbf{R})\psi(\mathbf{r};\mathbf{R}) \quad (7.3)$$

The new electronic Schrödinger equation is as follows:

$$H_{\text{el}}\psi(\mathbf{r};\mathbf{R}) = E(\mathbf{R})\psi(\mathbf{r};\mathbf{R}) \quad (7.4)$$

whose solutions now depend parametrically on the positions of the nuclei, $\{\mathbf{R}\}$. However, the electronic wave function, $\psi(\mathbf{r})$, still depends on $3N$ variables, and solving Eq. (7.4) becomes intractable for systems with many particles, even with the current computational resources. As an alternative to address this problem, DFT has become a reliable method for solving the electronic Schrödinger equation approximately. The idea is to replace the many-body Schrodinger equation, and the complete electron wave function, with a set of noninteracting single-particle equations, where the electron density, $n(\mathbf{r})$, is the fundamental variable. The electron density represents the probability of finding an electron in a point $\mathbf{r}$ in space and it is defined as follows:

$$n(\mathbf{r}) = N\int |\psi(\mathbf{r},\mathbf{r}_2,\ldots,\mathbf{r}_N)|^2 d\mathbf{r}_2 d\mathbf{r}_3 \ldots d\mathbf{r}_N \quad (7.5)$$

where $\psi(\mathbf{r},\mathbf{r}_2,\ldots,\mathbf{r}_N)$ is the electron wave function. DFT has its rigorous theoretical foundations in the work of Hohenberg and Kohn [2] and Kohn and Sham [3]. The first establishes two mathematical theorems: (i) the ground-state electron density, $n_0(\mathbf{r})$, uniquely determines the ground-state properties of the matter, for example, the energy, and (ii) the electron density that minimizes the energy is the ground-state density of the problem. The first theorem shows that any ground-state property like energy, is a functional[2] of the electron density, $E[n(\mathbf{r})]$. The second theorem involves the variational method, stating that any trial electron density, $n_t(\mathbf{r})$, must yield energy higher than or equal to the ground-state energy, with equality only when $n_t(\mathbf{r})$ is identical to the exact ground-state electron density, $n_0(\mathbf{r})$:

$$E[n_t(\mathbf{r})] \geq E_0[n_0(\mathbf{r})] \quad (7.6)$$

Both theorems show that we can employ the electron density instead of the many-body electron wavefunction as a variable to describe the system. Therefore,

1 In atomic units.
2 A function of functions.

for determining the ground-state electron density, we can predict and describe its fundamental properties. However, the exact form of the energy functional is still unknown. The second seminal work of Walter Kohn [3] addresses this problem by mapping the original many-body interacting electron system onto a reference system of noninteracting electrons subject to an effective potential. Under this approach, the energy functional can be written as:

$$E[n(\mathbf{r})] = T_{\mathrm{KS}}[n(\mathbf{r})] + \int V_{\mathrm{ext}}(\mathbf{r})n(\mathbf{r})d\mathbf{r} + E_{\mathrm{H}}[n(\mathbf{r})] + E_{\mathrm{xc}}[n(\mathbf{r})] \tag{7.7}$$

where the first term, T_{KS}, is the kinetic energy of the noninteracting electrons, V_{ext} is the external potential due to the nuclei, and E_{H} is the Hartree energy, which accounts the interaction of the electron density with itself. The last term of Eq. (7.7) is called exchange-correlation energy. Although the mapping between the interacting problem and the noninteracting one is in principle exact, E_{xc} is not fully known and holds the quantum mechanical and many-body effects of the electron exchange and correlation.

The Kohn–Sham (KS) equations are derived from minimizing the energy functional (Eq. 7.7) with respect to the electron density. The kinetic energy of the noninteracting system can be written in terms of single particle Kohn–Sham orbitals to obtain a Schrödinger-like equation,

$$\left(-\frac{1}{2}\nabla^2 + V_{\mathrm{ext}} + V_{\mathrm{H}} + V_{\mathrm{xc}}\right)\psi_i(\mathbf{r}) = \epsilon_i\psi_i(\mathbf{r}) \tag{7.8}$$

where $\psi_i(\mathbf{r})$ are now the Kohn–Sham orbitals and ϵ_i is their corresponding eigenvalues. Now, instead of solving one Schrödinger equation (Eq. 7.1), we solve N Kohn–Sham equations, one for each electron involved in the problem. The electron density for the noninteracting reference system is given as:

$$n(\mathbf{r}) = \sum_i^N |\psi_i(\mathbf{r})|^2 \tag{7.9}$$

and the effective Kohn–Sham potential is given as:

$$V_{\mathrm{KS}}[n(\mathbf{r})] = V_{\mathrm{ext}}(\mathbf{r}) + V_{\mathrm{H}}(\mathbf{r}) + V_{\mathrm{xc}}[n(\mathbf{r})] \tag{7.10}$$

As aforementioned the only unknown term is the exchange-correlation potential, which is defined as a functional derivative of the exchange-correlation energy, given as:

$$V_{\mathrm{xc}} = \frac{\delta E_{\mathrm{xc}}}{\delta n} \tag{7.11}$$

and this term is a crucial element of the theory. If we found the exact form of this functional theory, we could reproduce the true ground-state electron density and consequently the ground-state properties of any atomic system.

Currently, there are different approximations for V_{xc}, categorized by their complexity. The simplest form is the local density approximation (LDA), which assumes that the electron density is constant at any point in space. The next approach is the generalized gradient approximation (GGA), which incorporates the local gradient to the electron density. These are the most common approximations and although there are more complex forms of V_{xc}, there is not (until now) a unique exchange-correlation functional that accurately represents the true electron behavior in nature. Besides, these functionals also have shortcomings in describing localized electron states such as those found in transition metal oxides, rare-earth compounds, and magnetic materials. These systems contain d or f electrons that exhibit strong electron correlations, which LDA and GGA fail to account for due to the so-called self-interaction error (SIE). The SIE is an inherent problem in exchange-correlation functionals because they do not completely cancel the Coulomb interaction of an electron with itself. As a result, we encounter artificial electron delocalization and a convex curvature of total energy as a function of electron number, where the minimum could occur at fractional occupations [4]. The same effect occurs in the potential, V_{KS}, which should be discontinuous at each integer number of electrons. The absence of this discontinuity in the potential obtained by LDA or GGA functionals is considered as one of the reasons for the poor description of band gaps in insulators.

The usual strategy to address this problem is adding a Hubbard-like correction to the DFT Hamiltonian, DFT+U. The idea is to add an onsite Coulomb interaction only to the localized d or f orbitals. This approach restores the piece-wise linearity of the unreal convex curvature of total energy as a function of the electron number and could improve the band gap description of the systems. The formulation of DFT+U, as originally introduced by Anisimov et al. [5, 6], is written as follows:

$$E_{\mathrm{DFT}+U}[n(\mathbf{r})] = E_{\mathrm{DFT}}[n(\mathbf{r})] + E_{\mathrm{Hub}}[\{n_m^{I\sigma}\}] - E_{\mathrm{DC}}[\{n^{I\sigma}\}] \tag{7.12}$$

where E_{DFT} is the total energy calculated by solving the Kohn–Sham equations using approximations such as LDA or GGA, and E_{Hub} is the electron–electron interaction modeled through a Hubbard Hamiltonian. The term E_{DC} is subtracted from the total energy to prevent the double counting of the electron–electron interaction already contained in E_{DFT}. E_{DC} models the contribution of electronic interactions to the DFT energy as a mean-field approximation of E_{Hub}. In equation, $n_m^{I\sigma}$ represents the occupations of localized orbitals of the atom at the site I, state m, and spin σ, which experiences the Hubbard term. A simplified formulation of the rotationally invariant DFT + U proposed by Dudarev et al. [7] allows us to express the last two terms of the equation as follows:

$$E_{\mathrm{U}} = E_{\mathrm{Hub}} - E_{\mathrm{DC}} = \sum_{I,\sigma} \frac{U^I}{2} \,\mathrm{Tr}[\mathbf{n}^{I\sigma}(\mathbf{1} - \mathbf{n}^{I\sigma})] \tag{7.13}$$

where $\mathbf{n}^{I\sigma}$ represents the occupation matrices of the localized atomic manifold at site I and spin σ. The main challenge with DFT+U lies in determining the appropriate U value for a specific system. Initially, the scientific community treated U as an empirical parameter is often adjusted to match the experimental measurements. However, this approach can be impractical, as U values are not transferable between different materials. Nowadays, the state-of-the-art DFT+U implementations allow us to calculate the Hubbard parameters through a linear response method since the strength of the U correction is an intrinsic response property [8]. This method is fully *ab initio*, where the U values are calculated by applying small perturbations to the localized orbitals and measuring the system's response. This approach ensures that the U parameter is specific to the material and chemical environment under investigation, making it more reliable than empirically derived values. An extension of DFT+U includes an inter-site parameter V between different atomic sites.

$$E_{\text{DFT+U+V}} = E_{\text{DFT}} + E_{\text{U}} + E_{\text{V}} \tag{7.14}$$

These two Hubbard parameters favor the electron localization and the interaction between two different atomic sites and species, respectively. More details are discussed in refs. [9] and [10], which describe the current implementation for calculating self-consistently the Hubbard parameters via density functional perturbation theory. The use of these corrections is crucial for accurately describing the ground-state properties of materials with localized electrons, where standard DFT approaches fail.

Once we choose wisely the exchange-correlation potential (LDA, GGA, or anything else), the practical way to solve the KS equations is through an iterative method, the flowchart in Figure 7.1 shows the steps to follow. First, we start with a trial electron density to calculate the Kohn–Sham potential, $V_{\text{KS}}[n(\mathbf{r})]$, then we solve the Kohn–Sham equations and from the result, we calculate a new electron density using (7.9). We evaluate if the convergence criteria are fulfilled by comparing the initial and final electron densities in each iteration. If the criteria

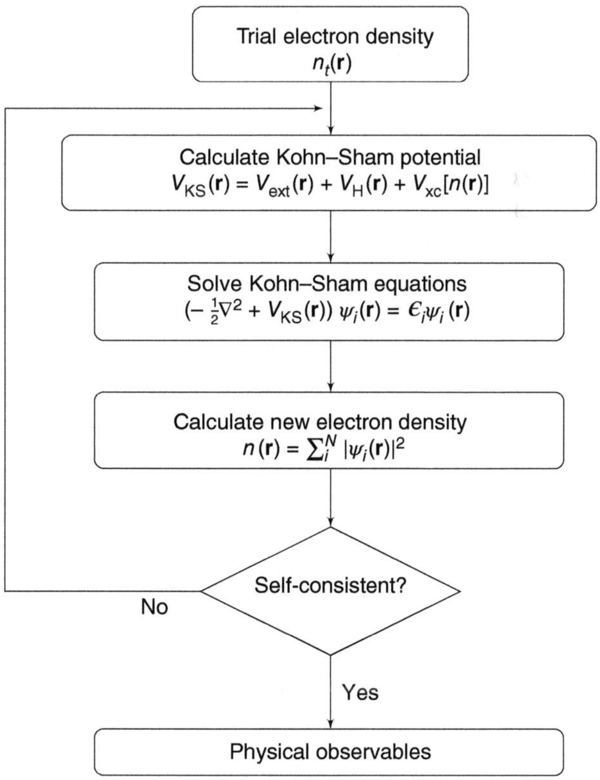

Figure 7.1 Flow chart of the iterative process to solve the Kohn–Sham equations.

are not satisfied, the process continues until convergence is achieved. The electron density in subsequent iterations is obtained by mixing the initial and final densities from the previous iteration step. Once the convergence is achieved, we can obtain a wide range of physical observables such as total energies, atomic forces, pressure, magnetic moments, vibrational frequencies, and dipole moments.

The great success of DFT lies in the fact that simple approximations perform remarkably well in determining and predicting the physical and chemical properties of most materials with extraordinary accuracy, especially when compared to experimental observations. Over the past 60 years, its popularity has grown due to its balance of accuracy and computational efficiency, making it widely used in the scientific community. The recent advances in computational infrastructure, along with the development of new approximations and implementations, have enabled the possibility of studying new phenomena across a wide range of materials. In Section 7.1.3, we will review how DFT can be used to investigate the fundamental properties of defective materials.

7.1.2 Defect Formation Energy

The incorporation of defects in pristine materials is one of the main strategies for modifying, controlling, and enhancing their properties. Understanding how these defects act in a specific material is highly desired in their design for different technological applications. There are different types of defects in crystalline lattices such as (i) vacancies, where an atom is missing from the periodic lattice, (ii) interstitials, where an extra atom is located in the free spaces within the atomic arrangement, (iii) substitutional impurities, where a foreign atom replaces an atom of the crystal host, and (iv) antisite defects, where atoms are located in incorrect sites within the material. These imperfections are known as point defects, a schematic representation is shown in Figure 7.2a.

Depending on the type, concentration, and distribution of defects within a crystalline lattice, these impurities can drastically modify the material properties such as electrical conductivity, thermal behavior, and catalytic activity, among others. DFT allows us to computationally investigate and model point defects at the atomic

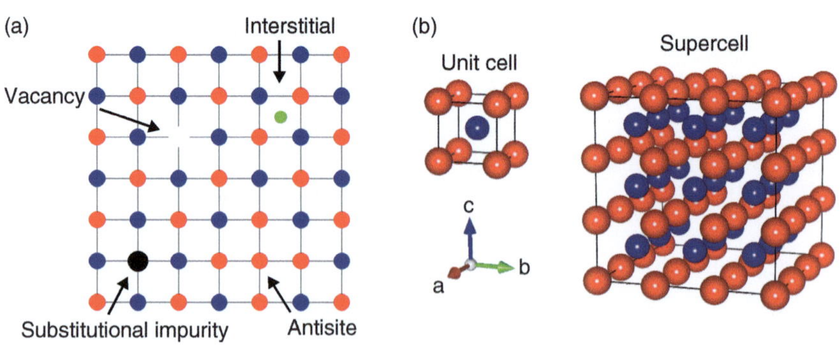

Figure 7.2 (a) Schematic representation of point defects. (b) Supercell approach.

scale, providing insights into their formation energy and electronic properties. Defect formation energy is a key concept as it provides valuable information about the stability of defects in materials. The defect formation energy, E_f, is a quantitative measure of how much energy is required to introduce a defect in a pristine material. The simplest expression of defect formation energy is given as:

$$E_f = E_{\text{defective}} - E_{\text{pristine}} + \sum_i n_i \mu_i \tag{7.15}$$

where $E_{\text{defective}}$ is the total energy of the material with the defect, E_{pristine} is the total energy of the perfect crystal, n_i is the number of atoms of type i added ($n_i > 0$) or removed ($n_i < 0$) in the pristine material, and μ_i is the chemical potential of atom i. To study point defects, we employ the use of the supercell method (see Figure 7.2b), which is the repetition of the crystalline unit cell along the lattice vectors multiple times. The size of the supercell should be determined by the type and concentration of defects being modeled. For low impurity concentrations, larger supercells are required. Also, we require larger supercells to avoid possible impurity–impurity interaction between periodic images, a consequence of the periodic boundary condition in solids. Once the supercell is constructed, we introduce the point defect and then we perform a geometrical optimization of the atomic positions to minimize the total energy of the defective material. The incorporation of defects generally induces local deformation in their vicinity. The solutions to the KS equations depend parametrically on the positions of the nuclei, so it is essential to ensure that the forces acting on each atom are zero. To achieve this, we perform geometric optimization using well-established algorithms such as the conjugate gradient method [11] or the Broyden–Fletcher–Goldfarb–Shanno (BFGS) method [12]. To evaluate Eq. (7.15), we calculate, the total energy of the defective supercell, $E_{\text{defective}}$, and the total energy of the defect-free supercell, E_{pristine}. Finally, the chemical potential for the atom i, μ_i is the total energy per atom in their elementary phase. A more detailed expression for calculating the defect formation energy in charged systems can be found in Refs. [13] and [14].

Estimating the defect formation energy has important implications for material design and also helps to guide the control of the type and concentration. Particularly in energy storage applications, the defect formation energies determine how easily defects are created, thus offering valuable insights into ionic conductivity in batteries. Lower defect formation energies correspond to lower activation energies for ion migration, thereby enhancing the ionic conductivity needed for efficient performance. On the other hand, higher defect formation energies can improve the stability of electrode materials by making them more resistant to structural changes such as mechanical failure or phase transitions, that may be induced by continuous charging and discharging cycles.

7.1.3 Electronic Structure of Defects

Electrons in crystalline materials are arranged in energy bands, and their behavior across a range of energies as a function of the momentum (wave vector k in reciprocal space) represents the band structure. The two most important energy bands in

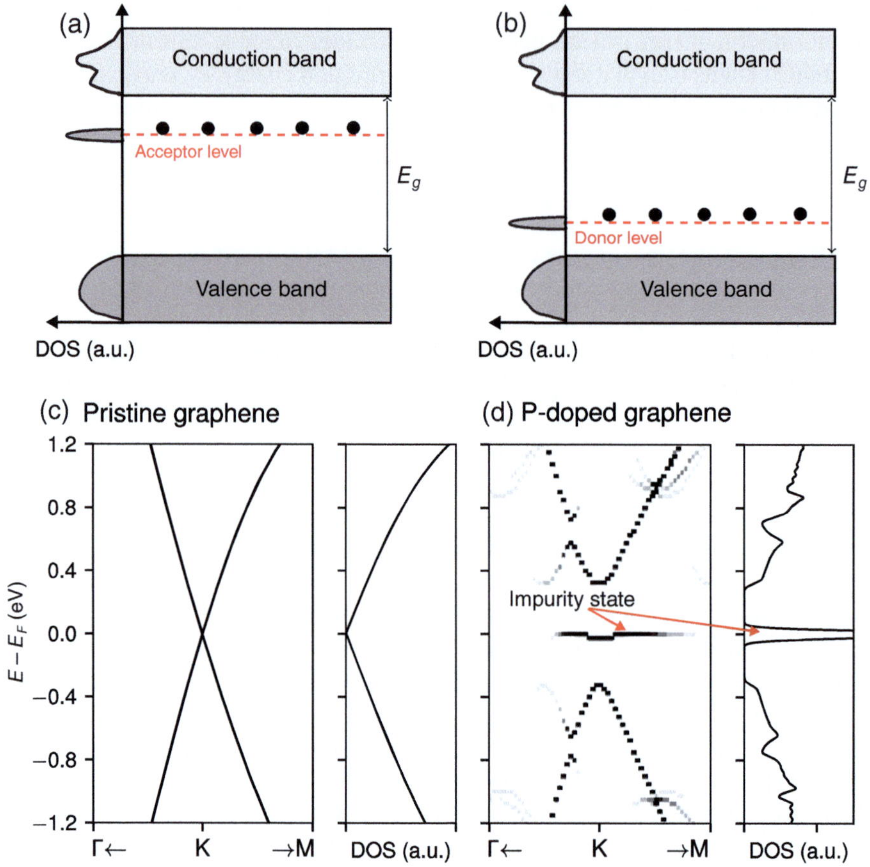

Figure 7.3 Schematic representation of the electronic structure of defects: (a) acceptor level and (b) donor level. The band structure and density of states for (c) pristine graphene and (d) phosphorus-doped graphene are also shown. Notice that the incorporation of phosphorus into graphene opens a gap and leads to the appearance of an impurity state at the Fermi level.

crystals are the valence band and the conduction band (see Figure 7.3a,b). The electron occupation of these bands helps identify the material's nature. A filled valence band with a forbidden region before an empty conduction band corresponds to a semiconductor or insulator. This forbidden region is called the band gap. Materials with partially filled bands are metals. The density of states (DOS) measures the number of available electronic states at each energy level. The DOS is closely related to the electronic band structure, and their analyses are complementary. An interesting consequence of the defect's incorporation is the change in the electronic structure of materials. Defects introduce electronic perturbations in the band structure and usually create new electronic states within the band gap in semiconductors or insulators, and even modify the bands in metals. In semiconductors, such defect-induced states are particularly significant because they create intermediate energy levels within the gap that can act as donor or acceptor states (red dashed line in Figure 7.3a,b), enhancing electrical conductivity, altering optical properties, or even influencing chemical reactivity.

Therefore, understanding the role of defects in the electronic structure of materials is essential to predict and tune their physical and chemical properties. Using DFT, we can accurately characterize these defect states through the calculation of the band structure and the DOS. The electronic properties must be studied using the supercell approach as described in order to simulate various impurity concentrations and avoid impurity–impurity interactions. As an example, we present a real case of one of the most groundbreaking materials in recent years, graphene. This material is the first two-dimensional crystal and has numerous technological applications today. We compare the pristine and defective electronic properties to visualize the impact of the incorporation of defects in the crystalline lattice. The calculated band structure and DOS are presented in Figure 7.3c. Graphene has a linear relationship of its electronic bands around the high-symmetry K point, giving it a semi-metal character, where the valence and conduction bands touch at a single point at the Fermi level (set to zero in Figure 7.3c). This linear dispersion forms conical surfaces in the Brillouin zone, known as Dirac cones. Both the bands and DOS show the linear dispersion close to the Fermi level. Chemical functionalization of graphene with impurities is one of the strategies used to tune its fundamental properties. In particular, the use of substitutional phosphorus, a common impurity for n-doping in the silicon electronic industry, induces drastic changes in its electronic structure. Figure 7.3d shows the unfolded band structure of phosphorus-doped graphene. Incorporation of this impurity breaks the Dirac cones, opening a gap at the K point, and inducing the formation of parabolic bands. Moreover, a new localized state appears just at the Fermi energy. This is an impurity state that is formed by the incorporation of phosphorus and leads to a localized state in DOS. This impurity state has very interesting consequences for graphene, as it can induce magnetism through an electronic instability caused by the sharp peak at the Fermi level [15, 16]. These impurity states also act as active sites for adsorption or catalysis.

As discussed previously in the defect formation energy calculations, the analysis of the band structure and DOS is important for designing the materials for energy production and storage applications, since their study provides insight into how defects impact positively on the fundamental properties of the materials for practical applications.

7.2 Electronic Descriptors in Catalysis

Recent years have seen a surge of interest in catalysis, particularly in sustainable energy conversion. A key factor in enhancing catalytic efficiency and selectivity is the ability to design catalysts at the atomic scale. Defect engineering in nanomaterials offers a powerful approach to manipulate electronic properties for optimized catalytic performance. By introducing and controlling defects such as vacancies, interstitials, or dopants, researchers can create highly reactive sites and also improve reactions like oxygen reduction, hydrogen evolution, and carbon dioxide reduction

[17–19]. This section explores predictive modeling of catalytic properties in defective materials, focusing on the principles and computational methods that inform the design of next-generation catalysts.

7.2.1 The d-Band Center in Defective Materials

Point-defects can significantly alter the electronic structure and influence the catalytic activity. These defects introduce localized states within the band structure, modify DOS near the Fermi level, and alter charge distribution, all of which are crucial in determining reactivity. For instance, defects in transition metal oxides, such as oxygen vacancies, also affect the relative positions of both the d- and p-band centers. Shifts in the p-band center, especially in materials with significant covalent bonding, influence the overlap between metal d-states and adsorbate states. The presence of defects such as vacancies can shift the p-band center, which consequently affects the bonding and antibonding states formed during adsorption [20].

Density of states (DOS) calculations are essential for understanding the influence of defects on electronic properties. The analysis of the DOS reveals shift in the electronic structure near the Fermi level that correlate with catalytic activity. The d-band center model serves as a valuable tool for predicting catalytic behavior in transition metal catalysts, as the position of the d-band center relative to the Fermi level directly influences the interaction strength of adsorbates. Defects can induce shifts in the d-band center, thereby tuning the binding strength of intermediates.

The d-band center (ϵ_{d}) is calculated from the projected DOS of the d-band as:

$$\epsilon_{\mathrm{d}} = \frac{\displaystyle\int_{-\infty}^{E_{\mathrm{F}}} \epsilon\, n_{\mathrm{d}}(\epsilon)\, \mathrm{d}\epsilon}{\displaystyle\int_{-\infty}^{E_{\mathrm{F}}} n_{\mathrm{d}}(\epsilon)\, \mathrm{d}\epsilon} \tag{7.16}$$

where ϵ is the energy of the electronic states, $n_{\mathrm{d}}(\epsilon)$ is the DOS of the d-band as a function of energy ϵ, and E_{F} is the Fermi energy level. The numerator represents the weighted average energy of the states within the d-band, while the denominator normalizes it by the total number of d-states. This descriptor is particularly effective for transition metal catalysts, where the position of the d-band center influences adsorption energies [21]. A shift in the d-band center, caused by defects, can tune the binding strength of reaction intermediates. The *d*-band center can be calculated using structure-based methods that are based on DFT.

7.2.2 The Newns–Anderson Model for Chemisorption

Chemisorption at solid surfaces is fundamental for describing reactions at solid–gas or solid–liquid interfaces, as it provides insights into the key electronic structure properties that determine a surface's chemical reactivity. A pivotal framework for understanding these adsorption processes is the Newns–Anderson model [22], which specifically addresses interactions on transition metal surfaces. In this model, adsorption occurs when an atom or molecule (adsorbate) binds near the surface, sharing electrons with the metal. This interaction perturbs the adsorbate's energy levels, depending on bond strength. The model posits that the adsorbate's one-electron states interact with the metal's nearly continuous valence band. Broad, delocalized metal states lead to a single broadened resonance, while narrow, localized states create distinct bonding and anti-bonding resonances.

A noteworthy application of the Newns–Anderson model occurs when a single d-state interacts with the a-state of the adsorbate. This interaction can be conceptualized as a two-state model for chemisorption, wherein the electronic states of the adsorbate interact with a single d-orbital of the metal surface. This process results in the formation of bonding and antibonding states, encapsulating the fundamental mechanisms of adsorption.

The Hamiltonian formalism provides a mathematical framework to describe these interactions, expressed as follows:

$$\hat{H} = \varepsilon_a |a\rangle\langle a| + \varepsilon_d |d\rangle\langle d| + V\left(|d\rangle\langle a| + |a\rangle\langle d|\right) \tag{7.17}$$

In this expression, ε_a and ε_d denote the energy levels of the adsorbate and the d-states of the metal, respectively, while V represents the coupling strength between these states. This Hamiltonian captures the essential interactions that influence how adsorbates bond to metal surfaces, laying the groundwork for further explorations of surface chemistry.

The energies of the hybridized states, denoted as $\varepsilon_\pm$, are derived from the Hamiltonian. They can be calculated using the following relation:

$$\varepsilon_\pm = \frac{\varepsilon_d + \varepsilon_a \pm \sqrt{(\varepsilon_d - \varepsilon_a)^2 + 4(V+S)^2}}{2} \tag{7.18}$$

Here, the terms $\varepsilon_\pm$ illustrate how the energies of the bonding and antibonding states are affected by both the energy levels of the d-orbitals and the adsorbate as well as the coupling term V, which accounts for the interaction between these states. The parameter S is the overlap integral between the wavefunctions of the involved states, which further influences the resulting energies. Typically, S is considered proportional to V.

The contribution of the d-orbitals to the adsorption energy, ΔE_d^A, can be approximated by:

$$\Delta E_d^A = \varepsilon_+ + f_- \varepsilon_- - \varepsilon_a - f_d^{site} \varepsilon_d \tag{7.19}$$

In this equation, f_- represents the filling of antibonding orbitals, reflecting how hybridization alters the occupation of these states following adsorption. The term f_d^{site} captures the occupation of the d-band at the surface site of the metal. Together, these factors elucidate the changes in energy contributions resulting from the interaction between the adsorbate and the metal.

The energy of the bonding and antibonding states, as described in Eq. (7.18), can be approximated using perturbation methods in the limit of weak coupling $\left(|V| \ll |\varepsilon_d - \varepsilon_a|\right)$. By expanding the square root to second order in $\frac{V}{|\varepsilon_d - \varepsilon_a|}$ and substituting into Eq. (7.19), we can derive an expression that facilitates the calculation of the adsorption energy as follows:

$$\Delta E_{\mathrm{d}}^{\mathrm{A}} \simeq -(1-f_{-})\frac{V^2}{|\varepsilon_{\mathrm{d}}-\varepsilon_{\mathrm{a}}|} + (1+f_{-})a_{\mathrm{s}}V^2 + (f_{-}-f_{\mathrm{d}}^{\mathrm{site}})\varepsilon_{\mathrm{d}} \tag{7.20}$$

The model expression in Eq. (7.20) encapsulates information regarding the adsorbate–surface interaction, making it challenging to isolate from the individual systems [23]. To eliminate any dependency on the composite system, earlier derivations have assumed that the electron density of the adsorbate after interaction approximates that of the clean surface, i.e. $f_{-} \approx f_{\mathrm{d}}^{\mathrm{site}} = f$. Under this assumption, Eq. (7.20) simplifies the expression derived in the original d-band model:

$$\Delta E_{\mathrm{d}}^{\mathrm{A}} \simeq -(1-f)\frac{V^2}{|\varepsilon_{\mathrm{d}}-\varepsilon_{\mathrm{a}}|} + (1+f)a_{\mathrm{s}}V^2 \tag{7.21}$$

The adsorption energy is influenced by various factors, including the energy levels of the *d*-states, the adsorbate, coupling strength V, orbital overlap $a_{\mathrm{s}}V$, and band filling f. The first term represents hybridization energy, contributing to attraction when antibonding states are not fully filled, while the second term accounts for repulsion from orthogonalization between the adsorbate and metal *d*-states. This relationship highlights how defects can shift band centers and alter fillings, significantly affecting adsorption energies. Defects or modifications in the metal's electronic structure can lead to partial filling of antibonding states, thus strengthening the bond between the adsorbate and the surface [24].

The Newns–Anderson model effectively describes the interactions between adsorbates and metal surfaces, enhancing our understanding on the mechanisms affecting adsorption energies and catalytic behavior, particularly in materials with defects or altered electronic structures. Despite its simplicity and limitations, the model establishes a direct relationship between descriptors like the d-band center and orbital filling with chemisorption energy. This concept has also been applied beyond metallic systems, guiding the design of new catalysts, including metal oxides.

Figure 7.4a demonstrates how oxygen vacancies in metal oxides enhance chemisorption and catalytic activity by shifting the d-band and 2*p*-band energies upwards toward the Fermi level, forming bonding and antibonding states crucial for adsorbate–metal surface interactions (Figure 7.4b). This shift influences the position and occupancy of the antibonding level upon adsorption, affecting the d-band's contribution to adsorption energy. Thus, generating vacancies effectively tunes the band center and adsorption energy of transition metal oxides. Empirical observations show a correlation between adsorption energy and catalytic activity across various materials, supported by the Brønsted–Evans–Polanyi principle, which indicates that the activation energy for catalytic processes is proportional to the Gibbs-free energy of adsorption.

Figure 7.4c shows the relationship between catalytic activity and adsorption energy in a volcano plot, consistent with the Sabatier principle. This principle explains catalyst performance, exemplified by platinum, which excels in hydrogen evolution due to its optimal binding energy. Platinum binds hydrogen atoms strongly

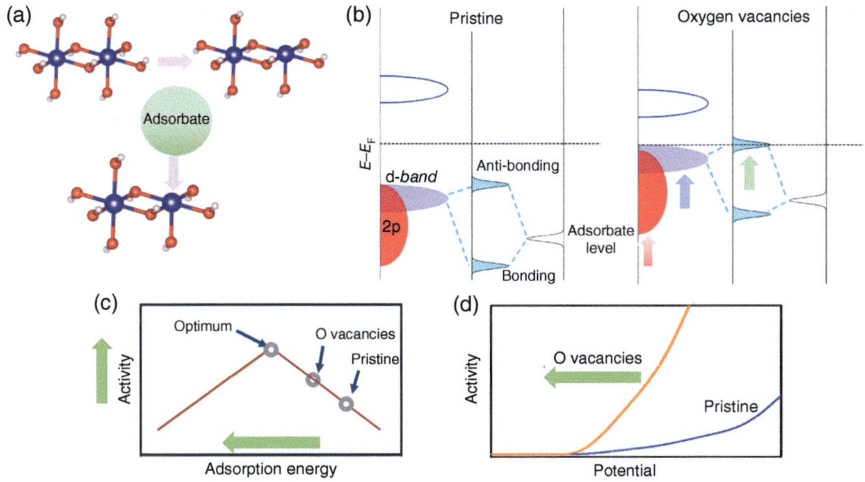

Figure 7.4 Schematics illustrating the effects of transition metal oxides on electronic structure, adsorption, and electrocatalytic activity: (a) oxygen vacancies and the adsorption process on metal oxides, (b) the influence of vacancies on the d-band, 2p-band, and the formation of bonding and antibonding energy levels during the adsorption process, (c) a volcano plot correlating catalytic activity with adsorption energy for pristine and defected materials, and (d) the impact of oxygen vacancies on electrocatalytic activity and potential in an electrochemical process.

enough to dissociate H_2 but weakly enough to release them efficiently. Defects in materials, such as nitrogen dopants in carbon-based catalysts or vacancies in transition metal oxides, enhance catalytic performance by fine-tuning the adsorption energy of intermediates, optimizing binding, and improving efficiency. However, the Sabatier principle highlights the need to balance active sites and defect density. While defects can enhance activity through improved adsorption, excessive defects may cause overly strong binding, reducing catalyst efficiency. Therefore, designing an optimal catalyst requires selecting the appropriate type and concentration of defects to align with the Sabatier principle. As illustrated in Figure 7.4d, this balance enhances electrocatalytic activity by shifting the potential and improving electrochemical performance, resulting in a catalyst with high activity and stability through a finely tuned interaction between adsorbate and active sites.

7.2.3 Computational Hydrogen Electrode

Generally, research in computational modeling of reactions in nanomaterials emphasizes the electronic properties of electrode surfaces and the nature of surface reactions, as these establish the structure-performance relationships of catalysts. To support this focus, numerical modeling techniques have been developed to enhance understanding of electrocatalytic processes. The *computational hydrogen electrode (CHE)* model [25] has revolutionized the study of electrocatalytic properties by providing a consistent framework for calculating reaction-free energy in electrochemical processes using DFT. By simplifying these complex reactions, the

CHE model enables direct comparisons of catalytic activity, facilitating the design of efficient electrocatalysts for energy conversion applications.

The CHE model streamlines reaction-free energy calculations by referencing the electrochemical potential of hydrogen at the standard hydrogen electrode. The chemical potential of protons and electrons relates to the free energy of molecular hydrogen H_2 at 0 V on the reversible hydrogen electrode (RHE) scale:

$$\mu_{H^+} + \mu_{e^-} = \frac{1}{2}\mu_{H_2} \tag{7.22}$$

This equation indicates that the chemical potential of protons and electrons in solution is half the free energy of a hydrogen molecule under standard conditions. Furthermore, the Brønsted–Evans–Polanyi principle shows that reaction barriers correlate with the Gibbs-free energy change [26], enabling the evaluation of electrocatalytic processes based on thermodynamic properties and simplifying computations. The CHE approach is widely utilized to identify active sites, clarify reaction mechanisms, and design electrocatalysts by linking these sites to surfaces, defects, and modifiers that influence the binding energies of intermediates. Consequently, the CHE model has advanced the computational design of electrocatalysts for critical reactions like the oxygen evolution reaction (OER), hydrogen evolution reaction (HER), and oxygen reduction reaction (ORR), emphasizing the determination of true active species and monitoring reactive intermediates [27].

The OER is a crucial half-reaction in electrochemical processes, especially in water splitting and fuel cells. It involves oxidizing water to produce oxygen gas, protons, and electrons:

$$2H_2O \rightarrow O_2 + 4H^+ + 4e^- \tag{7.23}$$

The OER is characterized by its high overpotential, limiting the efficiency of electrochemical systems. Advancements in materials science and computational modeling are vital for developing catalysts that reduce overpotential and enhance OER kinetics. A universal design principle for electrocatalysts aims to provide experimental guidance and elucidate mechanisms, with reaction-free energy widely adopted as a descriptor to evaluate the catalytic activity or selectivity of newly developed electrocatalysts.

The OER pathway is described by:

$$H_2O + * \rightarrow OH^* + H^+ + e^- \tag{7.24}$$

$$OH^* \rightarrow O^* + H^+ + e^- \tag{7.25}$$

$$O^* + H_2O \rightarrow OOH^* + H^+ + e^- \tag{7.26}$$

$$OOH^* \rightarrow * + O_2 + H^+ + e^- \tag{7.27}$$

Figure 7.5a shows the schematic diagram of the Gibbs energy profile for the idealized OER reaction, where each step of the reaction requires an energy of 1.23 eV. The overall process requires 4.92 eV, which implies a transfer of 4 e^-. The Gibbs-free energy changes for these steps are:

$$\Delta G_1 = \Delta G_{OH} \tag{7.28}$$

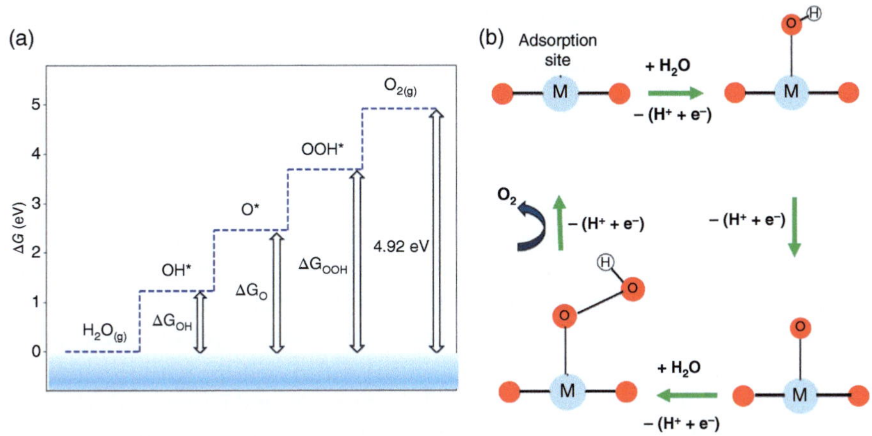

Figure 7.5 (a) Schematic diagram illustrating the ideal OER process. (b) Bonding modes for the adsorption of oxygenated species on surface metal sites.

$$\Delta G_2 = \Delta G_O - \Delta G_{OH} \tag{7.29}$$

$$\Delta G_3 = \Delta G_{OOH} - \Delta G_O \tag{7.30}$$

$$\Delta G_4 = 4.92\ \text{eV} - \Delta G_{OOH} \tag{7.31}$$

where,

$$\Delta G_x = \Delta E_x^{DFT} + \Delta ZPE - T\Delta S, \quad x : OH^*, O^*, OOH^* \tag{7.32}$$

For each oxygenated species (OH, OOH, O), the equation describes the free energy of adsorption (ΔG_x) onto the electrocatalyst material. Here, ΔE^{DFT} signifies the change in energy derived from DFT calculations associated with the adsorption of each species, which is crucial for understanding the energetic landscape of the OER. The term ΔZPE accounts for the change in zero-point energy. Meanwhile, $-T\Delta S$ represents the change in entropy, influenced by reductions in translational, rotational, and vibrational degrees of freedom upon adsorption. Collectively, this equation facilitates the assessment of the thermodynamic feasibility of adsorption for each species in OER, where a negative ΔG_{ads} implies that the adsorption process is spontaneous and thermodynamically favorable, thereby contributing to enhanced catalytic activity.

In total, four protons and four electrons are transferred during the OER process. Figure 7.5b illustrates the bonding modes for the oxygenated species on the metal sites' surface. DFT-based calculations show similar results for a wide range of systems, however, the calculated adsorption energies vary depending on the selectivity of each catalyst that has a specific oxygenated species. As a consequence, the calculated energies usually deviate from the ideal value of 1.23 eV for each step of the reaction. Typically, the difference from the ideal value of 1.23 eV is known as the overpotential. Particularly in the case of the OER, the rate-limiting step of the reaction is the one that possesses the highest overpotential relative to the equilibrium value.

7.2.4 Example: Layered Double Hydroxide with Oxygen Vacancies

Nickel–iron layered double hydroxide (NiFe LDH) is a promising candidate for energy storage and electrocatalysis due to its favorable redox properties [28]. However, its OER activity remains inferior to noble metal oxides like IrO_2 and RuO_2 [29]. Strategies such as nanostructuring, cation substitution, and surface defect engineering are employed to improve performance. Recent DFT studies indicate that oxygen vacancies (O_v) significantly affect the electronic structure and electrocatalytic activity of NiFe LDH. These vacancies, resulting from under-coordinated Ni atoms, introduce defect states that enhance OER performance. Furthermore, combining NiFe LDH with heteroatom-doped carbon can improve conductivity and catalytic efficiency [17].

A prior study on LDH identified four metal site types: Ni and Fe within the lattice and at the edges, with edge sites on the facet exhibiting the highest reactivity [30]. A more recent work [31] examines the edge Fe sites to evaluate how vacancies affect intermediates: Fe*, Fe–OH*, Fe–O*, and Fe–OOH*. The findings, including adsorption energies and the theoretical overpotential η^{OER}, are presented in Figure 7.6. Figure 7.6a displays the free energy diagram for NiFe LDH without vacancies, where OH* having adsorption at −0.6 eV is the most favorable. The formation of O* necessitates 2.57 eV, marking the rate-determining step (RDS), with ΔG_{max} indicated by

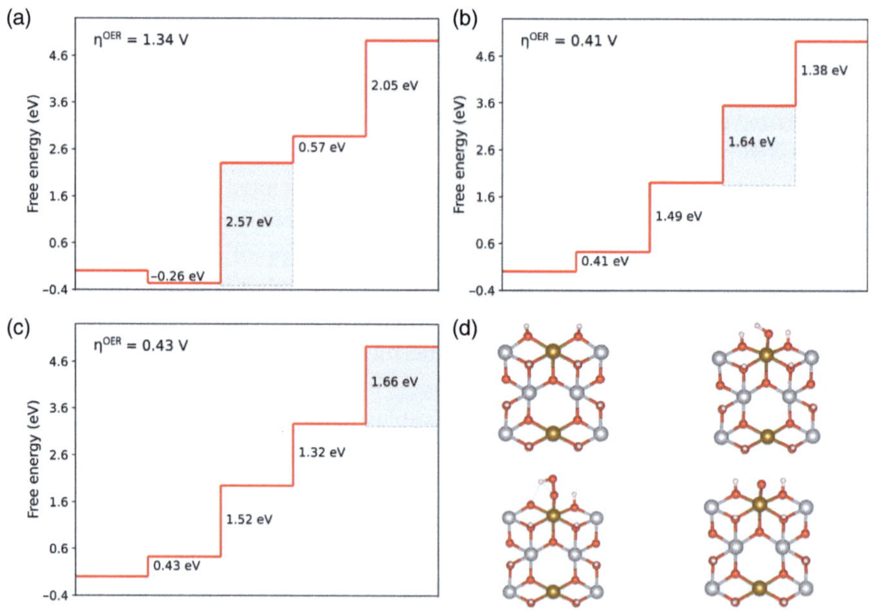

Figure 7.6 (a) Free energy diagram for OER in pristine NiFe LDH. (b) and (c) Free energy diagrams for OER in defected NiFe LDH (x = 1.3%) with vacancies positioned differently relative to the Fe active site. (d) Relaxed structures for the adsorption of oxygenated species (OH, O, and OOH) on the Fe active site in defected LDH. Source: Ramos-Castillo et al. [31] / American Chemical Society.

a gray square. The theoretical overpotential is $\eta^{\mathrm{OER}} = 1.34$ V, determined from the difference between ΔG_{max} and the equilibrium potential (1.23 V).

Figure 7.6b shows that a single vacancy destabilizes OH* adsorption to 0.41 eV, which is less favorable than in Figure 7.6a. The RDS shifts to OOH* formation, with $\Delta G_{\mathrm{max}} = 1.64$ eV and $\eta^{\mathrm{OER}} = 0.41$ V. Figure 7.6c reveals that a different vacancy position near the active Fe site results in $\eta^{\mathrm{OER}} = 0.43$ V, where OOH* desorption is the RDS, indicating the influence of vacancy location. Finally, Figure 7.6d presents the relaxed atomic structures for all reaction steps from Figure 7.6c. Although theoretical η^{OER} indicates catalytic activity, it cannot be directly compared to experimental overpotentials, which depend on current density. These findings highlight the significance of defect engineering in enhancing electrocatalytic activity in NiFe LDH materials.

7.2.5 Solvent Effects

The continuum solvent model (CSM) provides an effective framework for understanding solvation effects in various chemical processes, particularly in catalysis. This model simplifies the complex interactions between solutes and solvent by representing the solvent as a continuous medium, allowing for efficient computational modeling of solvation energy and electrostatic interactions. Central to CSM is the Poisson equation, which describes how the electrostatic potential is influenced by charge distributions within the solvent and solute. The CSM enables the calculation of solvation-free energies and provides insights into the stabilization of reactive intermediates [32, 33].

Recent advancements have led to more sophisticated models, such as solvated jellium [34], which incorporate in addition explicit solvent molecules while retaining computational efficiency. These models address limitations inherent in traditional CSM by allowing for better representation of the solvent's impact on reaction dynamics. Solvated jellium models utilize an effective medium approximation, treating the solvent as a uniform background with variable dielectric properties. This approach has shown great potential in catalysis, as it captures critical solvation phenomena, such as the influence of solvent polarity on catalytic activity and selectivity.

Solvation models, particularly CSM and advanced solvated jellium approaches, play a crucial role in understanding and optimizing catalytic processes by elucidating solvation effects on reaction mechanisms and energy profiles.

7.2.6 Outlook

The role of computational methods in defect engineering within nanomaterials is vital for advancing both materials science and catalytic technologies. Defects, such as vacancies, dopants, and dislocations, fundamentally alter the electronic, structural, and catalytic properties of nanomaterials. Techniques like DFT allow detailed atomistic insights into these defects, predicting their formation energies, charge states, and local atomic rearrangements. This precision in modeling facilitates the intentional design and manipulation of defects to create materials optimized for specific energy-related applications, particularly in catalysis.

In electrocatalysis, computational models such as CHE have been revolutionary in understanding key reaction mechanisms including the OER. These models enable accurate predictions of free energy profiles and adsorption energies, crucial for identifying active sites and optimizing defect-engineering strategies. For example, the introduction of oxygen vacancies in LDHs like NiFe LDH has been shown to significantly enhance catalytic activity, showcasing the power of defect engineering.

A synergistic approach between theory and experiment is essential for further progress. Experimental validation of theoretical predictions, particularly in identifying reaction pathways and catalytic products, provides critical feedback on catalyst effectiveness. Conversely, theoretical insights guide the selection and engineering of defects for the rational design of high-performance catalysts.

As computational power continues to grow, combined with the use of sophisticated solvent models like CSM, the capacity to predict and design defect-driven nanomaterials will only expand, driving innovation in catalysis and sustainable energy conversion.

References

1 Born, M. and Oppenheimer, R. (1927). Zur quantentheorie der molekeln. *Ann. Phys.* 389 (20): 457–484.
2 Hohenberg, P. and Kohn, W. (1964). Inhomogeneous electron gas. *Phys. Rev.* 136: B864–B871.
3 Kohn, W. and Sham, L.J. (1965). Self-consistent equations including exchange and correlation effects. *Phys. Rev.* 140: A1133–A1138.
4 Cohen, A.J., Mori-Sánchez, P., and Yang, W. (2008). Insights into current limitations of density functional theory. *Science* 321 (5890): 792–794.
5 Anisimov, V.I., Zaanen, J., and Andersen, O.K. (1991). Band theory and Mott insulators: Hubbard *U* instead of Stoner *I*. *Phys. Rev. B* 44: 943–954.
6 Anisimov, V.I., Aryasetiawan, F., and Lichtenstein, A.I. (1997). First-principles calculations of the electronic structure and spectra of strongly correlated systems: the LDA+ *U* method. *J. Phys. Condens. Matter* 9 (4): 767.
7 Dudarev, S.L., Botton, G.A., Savrasov, S.Y. et al. (1998). Electron-energy-loss spectra and the structural stability of nickel oxide: an LSDA + U study. *Phys. Rev. B* 57: 1505–1509.
8 Cococcioni, M. and de Gironcoli, S. (2005). Linear response approach to the calculation of the effective interaction parameters in the LDA + U method. *Phys. Rev. B* 71: 035105.
9 Campo, V.L. and Cococcioni, M. (2010). Extended DFT + *U* + *V* method with on-site and inter-site electronic interactions. *J. Phys. Condens. Matter* 22 (5): 055602.

10 Timrov, I., Marzari, N., and Cococcioni, M. (2021). Self-consistent Hubbard parameters from density-functional perturbation theory in the ultrasoft and projector-augmented wave formulations. *Phys. Rev. B* 103: 045141.

11 Payne, M.C., Teter, M.P., Allan, D.C. et al. (1992). Iterative minimization techniques for ab initio total-energy calculations: molecular dynamics and conjugate gradients. *Rev. Mod. Phys.* 64: 1045–1097.

12 Broyden, C.G. (1970). The convergence of a class of double-rank minimization algorithms 1. General considerations. *IMA J. Appl. Math.* 6 (1): 76–90.

13 Freysoldt, C., Grabowski, B., Hickel, T. et al. (2014). First-principles calculations for point defects in solids. *Rev. Mod. Phys.* 86: 253–305.

14 Zhang, S.B. and Northrup, J.E. (1991). Chemical potential dependence of defect formation energies in GaAs: application to Ga self-diffusion. *Phys. Rev. Lett.* 67: 2339–2342.

15 Hernández-Tecorralco, J., Meza-Montes, L., Cifuentes-Quintal, M.E., and de Coss, R. (2020). Effects of biaxial strain on the impurity-induced magnetism in P-doped graphene and N-doped silicene: a first principles study. *J. Phys. Condens. Matter* 32 (25): 255801.

16 Hernández-Tecorralco, J., Meza-Montes, L., Cifuentes-Quintal, M.E., and de Coss, R. (2022). Understanding the *sp* magnetism in substitutional doped graphene. *Phys. Rev. B* 105: 224425.

17 Ambríz-Peláez, O., Béjar, J., Ramos-Castillo, C.M. et al. (2022). Defected NiFe layered double hydroxides on N-doped carbon nanotubes as efficient bifunctional electrocatalyst for rechargeable zinc–air batteries. *Appl. Surf. Sci.* 601: 154253. https://doi.org/10.1016/j.apsusc.2022.154253.

18 Devi, P., Verma, R., and Singh, J.P. (2022). Advancement in electrochemical, photocatalytic, and photoelectrochemical CO_2 reduction: recent progress in the role of oxygen vacancies in catalyst design. *J. CO_2 Util.* 65: 102211. https://doi.org/10.1016/j.jcou.2022.102211.

19 Wu, Z., Shi, Y., Wei, Y. et al. (2024). Early transition metal oxides with oxygen vacancies as new thermodynamically stable materials for electrocatalysis. *Appl. Catal., B* 358: 124368. https://doi.org/10.1016/j.apcatb.2024.124368.

20 Lee, H., Gwon, O., Choi, K. et al. (2020). Enhancing bifunctional electrocatalytic activities via metal d-band center lift induced by oxygen vacancy on the subsurface of perovskites. *ACS Catal.* 10 (8): 4664–4670. https://doi.org/10.1021/acscatal.0c01104.

21 Ruban, A., Hammer, B., Stoltze, P. et al. (1997). Surface electronic structure and reactivity of transition and noble metals: communication presented at the First Francqui Colloquium, Brussels, 19–20 February 1996. *J. Mol. Catal. A: Chem.* 115 (3): 421–429. https://doi.org/10.1016/S1381-1169(96)00348-2.

22 Newns, D.M. (1969). Self-consistent model of hydrogen chemisorption. *Phys. Rev.* 178: 1123–1135. https://doi.org/10.1103/PhysRev.178.1123.

23 Saini, S., Stenlid, J.H., and Abild-Pedersen, F. (2022). Electronic structure factors and the importance of adsorbate effects in chemisorption on surface alloys. *npj Comput. Mater.* 8 (1): 163. https://doi.org/10.1038/s41524-022-00846-z.

24 Hammer, B. and Nørskov, J.K. (2000). Theoretical surface science and catalysis–calculations and concepts. In: *Advances in Catalysis*, Impact of Surface Science on Catalysis, vol. 45, 71–129. Academic Press. https://doi.org/10.1016/S0360-0564(02)45013-4. https://www.sciencedirect.com/science/article/pii/S0360056402450134.

25 Nørskov, J.K., Rossmeisl, J., Logadottir, A. et al. (2004). Origin of the overpotential for oxygen reduction at a fuel-cell cathode. *J. Phys. Chem. B* 108 (46): 17886–17892. https://doi.org/10.1021/jp047349j.

26 Wang, S., Temel, B., Shen, J. et al. (2011). Universal Brønsted–Evans–Polanyi relations for c–c, c–o, c–n, n–o, n–n, and o–o dissociation reactions. *Catal. Lett.* 141: 370–373. https://doi.org/10.1007/s10562-010-0477-y.

27 Zhao, Y., Saseendran, D.P.A., Huang, C. et al. (2023). Oxygen evolution/reduction reaction catalysts: from in situ monitoring and reaction mechanisms to rational design. *Chem. Rev.* 123 (9): 6257–6358. https://doi.org/10.1021/acs.chemrev.2c00515.

28 Hameed, A., Batool, M., Liu, Z. et al. (2022). Layered double hydroxide-derived nanomaterials for efficient electrocatalytic water splitting: recent progress and future perspective. *ACS Energy Lett.* 7 (10): 3311–3328. https://doi.org/10.1021/acsenergylett.2c01362.

29 Fang, Y.-H. and Liu, Z.-P. (2010). Mechanism and Tafel lines of electro-oxidation of water to oxygen on RuO2(110). *J. Am. Chem. Soc.* 132 (51): 18214–18222. https://doi.org/10.1021/ja1069272.

30 Peng, C., Ran, N., Wan, G. et al. (2020). Engineering active Fe sites on nickel–iron layered double hydroxide through component segregation for oxygen evolution reaction. *ChemSusChem* 13 (4): 811–818. https://doi.org/10.1002/cssc.201902841.

31 Ramos-Castillo, C.M., Contreras, L.Á., Arjona, N., and Guerra-Balcázar, M. (2024). Defect engineering of oxygen vacancies in ultrathin NiFe-layered double hydroxides: insights from density functional theory. *J. Phys. Chem. C* 128 (10): 4161–4170. https://doi.org/10.1021/acs.jpcc.3c07521.

32 Held, A. and Walter, M. (2014). Simplified continuum solvent model with a smooth cavity based on volumetric data. *J. Chem. Phys.* 141 (17): 174108. https://doi.org/10.1063/1.4900838.

33 Im, W., Beglov, D., and Roux, B. (1998). Continuum solvation model: computation of electrostatic forces from numerical solutions to the Poisson-Boltzmann equation. *Comput. Phys. Commun.* 111 (1): 59–75. https://doi.org/10.1016/S0010-4655(98)00016-2.

34 Kastlunger, G., Lindgren, P., and Peterson, A.A. (2018). Controlled-potential simulation of elementary electrochemical reactions: proton discharge on metal surfaces. *J. Phys. Chem. C* 122 (24): 12771–12781. https://doi.org/10.1021/acs.jpcc.8b02465.

8

Defect Healing and Control Strategies in Energy Systems

César Coello-Mauléón[1], *Carlos Guzmán-Martínez*[*], *and Noé Arjona*[2]

[1] *Centro de Investigación y Desarrollo Tecnológico en Electroquímica, Posgraduate Department, Parque Tecnológico Querétaro st., Pedro Escobedo, Querétaro, C.P. 76703, México*
[2] *Centro de Investigación en Materiales Avanzados, Department of Materials Physics, Miguel de Cervantes St., Complejo Industrial Chihuahua, Chihuahua, Chih., C.P. 31136, México*

8.1 Introduction to Self-Healing Systems

Defect engineering has emerged as a powerful tool for tuning the properties of materials used in energy storage and conversion systems. By introducing and controlling defects such as vacancies, dislocations, grain boundaries, or dopants, researchers can significantly enhance conductivity, catalytic activity, mechanical stability, and electrochemical performance by tunning this defect. However, despite these advancements, materials remain susceptible to degradation over time due to prolonged electrochemical cycling, mechanical stress, and environmental exposure. These degradation processes lead to capacity fading, loss of efficiency, and device failure, limiting the practical implementation of high-performance energy storage technologies.

In this context, self-healing mechanisms have gained considerable attention as an innovative strategy to mitigate performance losses and extend the lifespan of energy-related materials and devices. Self-healing refers to the intrinsic or extrinsic ability of a material to autonomously or externally restore its structure and functionality after suffering from mechanical, thermal, or electrochemical damage. Inspired by biological systems, where self-repair is fundamental for sustaining life, researchers have sought to integrate similar principles into synthetic materials to develop more resilient energy systems. The application of self-healing in energy storage and conversion appears on multiple areas including batteries, supercapacitors, fuel cells, and electrocatalysts. For example, in lithium- and sodium-ion batteries, dendrite formation on metal anodes remains a critical issue that compromises safety and performance. Recent studies have demonstrated that electrolyte additives and dynamic polymer-based coatings can induce self-healing behavior, suppressing dendrite growth and enhancing cycling stability. Similarly,

* Deceased author.

Surface and Interfacial Defects in Nanomaterials for Sustainable Energy Production and Storage, First Edition. Edited by Noé Arjona, Lorena Álvarez Contreras, and Minerva Guerra Balcázar.

in supercapacitors, electrode materials such as graphene-based composites or metal oxides have been engineered with reversible bonding interactions that allow for the restoration of structural integrity after mechanical damage.

Another critical area of self-healing in energy systems is the regeneration of active sites in electrocatalysts, particularly for fuel cells and metal-air batteries. Catalysts used in oxygen reduction (ORR) and oxygen evolution reactions (OER) often undergo deactivation due to surface poisoning, metal dissolution, or particle aggregation. Advanced strategies such as defect-mediated self-repair, reversible metal–ligand interactions, and dynamic surface restructuring, have been proposed to counteract these effects and maintain catalytic performance over extended operation. This chapter hunts through into the fundamental principles of self-healing in energy materials, providing an analysis of the underlying mechanisms that govern self-repair in different systems. We will explore various design strategies, including the role of defect engineering in enabling self-healing behavior and the use of smart nanomaterials with adaptive properties. By exploiting the ability to autonomously or externally repair damage, future energy storage systems will not only exhibit longer operational lifetimes but also contribute to reducing electronic waste, improving energy efficiency, and promoting the transition toward a more sustainable energy landscape.

8.2 Thermodynamics and Kinetics Implication on Self-Healing Systems

The thermodynamic basis principles of self-healing in energy storage systems are complex and multifaceted, requiring a detailed exploration between energy, entropy, and kinetics. Self-healing mechanisms in these systems depend critically on favorable thermodynamic driving forces that enable the reversible restoration of their functional properties. These processes often occur in response to external stimuli such as changes in temperature, light, or pH, and are driven by supramolecular interactions of reversible chemical bonds [1]. In this section, we dig into the thermodynamic basis of self-healing energy systems and the influence of the different external stimuli.

At the heart of any self-healing process is the minimization of Gibbs free energy (ΔG), which serves as the thermodynamic criterion for spontaneous processes. For a self-healing reaction to occur, the change in Gibbs free energy must be negative ($\Delta G < 0$), indicating a thermodynamically favorable process. Mathematically, this is expressed as:

$$\Delta G = \Delta H - T\Delta S$$

where ΔH is the enthalpy change, T is the temperature, and ΔS is the entropy change of the system. In many self-healing materials, the entropy plays a critical role, the disordered nature of supramolecular assemblies usually leads to an increase in entropy upon bond dissociation. For example, in polymeric systems with dynamic bonds like on all-solid-state batteries, the healing process requires the reorganization of polymer chains and reformation of bonds, which can be entropically favorable. Thus, when the entropy term exceeds the enthalpy term, the self-healing process becomes spontaneous, even at relatively low temperatures [2]. These entropy contributions in these systems are particularly significant due to the molecular dynamics involved in these supramolecular assemblies and their dynamics bonds. These assemblies depend on reversible interactions that can respond dynamically to environmental changes, i.e. external stimuli. For instance, in a self-healing battery electrolyte, the dissociation of ionic clusters or the reformation of dynamics hydrogen bonds contributes to an increase in configurational entropy, facilitating the repair of microcracks or remove unwanted defects [3] (Figure 8.1).

While thermodynamics supply the driving force for self-healing, kinetics decree the rate at which healing or reactions occur. Thus, the rate of self-healing depends on the kinetic constants of the reversible reactions as well as on the mobility and diffusion of the molecular components. For example, in a polymer-based self-healing system, the diffusion of polymer chains and the mobility of dynamic bonds are critical factors to take into account when designing a self-healing system. The kinetic constant of this self-healing reaction is influenced by the activation energy (Ea) and the temperature (T), as described by the Arrhenius equation and as a general rule, lower activation energies and higher temperatures enhance the reaction rate, as a result accelerate the self-healing process. Additionally, the mobility of the molecular assembly is influenced by parameters such as viscosity, molecular weight, and cross-link density, which determine the ability of molecular components to reorganize and repair damage. In systems where diffusion plays a significant role such as ionic liquids, all-solid-state membrane, or gel-based electrolytes, the viscosity of the medium becomes a key parameter. Lower viscosities facilitate faster diffusion

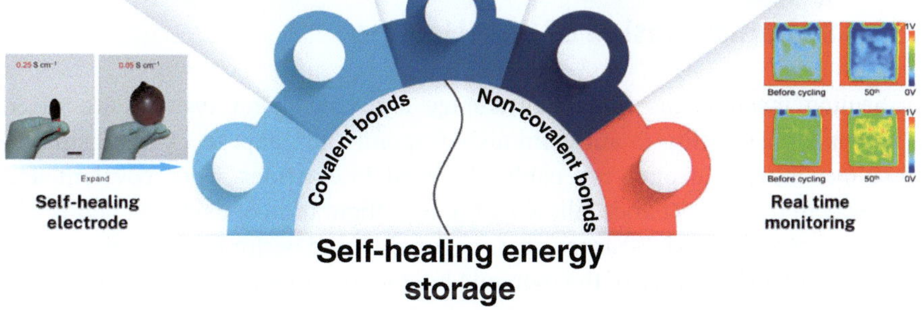

Figure 8.1 Schematic representation of self-healing in energy storage systems and its application to each of their components.

of reactive species, allowing interactions and more efficient self-healing. Contrarily, in highly viscous systems, the rate of self-healing may be limited or even committed, making necessary external stimuli to enhance mobility.

Many of these self-healing energy systems are often designed to respond to external stimuli, which act as an initiator for the healing process. These stimuli can modulate the thermodynamic and kinetic parameters of the complete system and allow fine control of the self-healing mechanism. The most common external stimuli involve pH, light, heat, and electric or magnetic fields. For example, changes in pH can cause the protonation of functional groups, as a result, it modulates the strength of non-covalent interactions or the stability of dynamic covalent bonds [4]. In other ways, high temperatures increase the kinetic energy of the reactive components, enhancing diffusion and reaction rates. Thermal energy can also break weaker interactions, such as van der Waals forces, at the same time, promoting the formation of stronger, more stable bonds [5]. Heat-activated self-healing systems are particularly luckily effective in applications where thermal cycling is inherent such as in electrodes exposed to repeated charge–discharge cycles. In some advanced self-healing materials, external electric or magnetic fields can be used to align molecular components or induce localized heating, driving the self-healing process [6].

Another way to perform self-healing is using a photochemical reaction that can provide the energy required to defeat activation barriers, allowing the reformation of broken bonds due to cracks. Light-responsive self-healing systems often incorporate chromophores or photo-switchable-sensible groups that allow reversible isomerization. For example, the use of azobenzene derivatives allows for reversible changes in molecular configuration, driving self-healing processes in response to ultraviolet or visible light [7].

The cooperation between external stimuli and the intrinsic property of the dynamic bonds of self-healing system determines the efficiency and reversibility of the healing process. By tailoring the sensitivity of the system to specific stimuli, researchers can design energy storage materials that exhibit self-healing under practical operating conditions. These principles discussed here have profound implications for the design and optimization of self-healing energy storage systems.

8.3 Mechanism Inside of Self-Healing

Self-healing mechanisms in energy storage and conversion systems are based on the material's ability to autonomously respond and adapt to external stimuli. These mechanisms exploit reversible chemical interactions, both covalent and non-covalent dynamics bonds, allowing the reconfiguration of damaged bonds and restoration of the material's structure, functionality, and performance. This chapter provides a general analysis of the dynamic interactions that form the heart of these systems, with a focus on dynamic covalent and non-covalent bonds.

The base of intrinsic self-healing reclines the ability of materials to undergo reversible changes in their chemical structure. Dynamic covalent bonds can break and reform under specific conditions or stimuli, this serves as a robust platform for autonomous repair. Disulfide bonds, for example, play a central role in self-healing systems due to their great reversible redox behavior. When exposed to oxidative or reductive environments, these bonds can reorganize, allowing the material to recover from damage due to cracks. This mechanism is particularly advantageous in redox flow batteries and polymeric membranes, where maintaining structural integrity is critical for performance [8, 9]. Another example of dynamic covalent chemistry is the well-known Diels–Alder reaction. This cycloaddition process between a diene and a dienophile yields a cyclohexene structure that can be reversed through the retro-Diels–Alder reaction, typically induced by temperature. Materials incorporating this reversible reaction exhibit exceptional self-healing capabilities, especially under controlled thermal conditions, making them highly appropriate for applications in thermal energy storage systems. Similarly, radical dimerization offers a pathway for dynamic covalent bond formation. Through ultraviolet light, radiation, or thermal energy, this radical species can recombine to restore damaged regions in coatings and electrodes, enhancing their durability and lifespan.

Dynamic covalent bonds are not limited to traditional chemical linkages. Systems based on dynamic exchange reactions, such as boronate esters and urea bonds, also exhibit great potential for the application in energy systems. These interactions depend totally on the chemical ability of their bonds to reorganize in response to environmental stimuli like changes in pH or temperature. For example, vitrimers-polymeric networks that utilize dynamics covalent exchanges can self-healing while enhancing their mechanical properties and extending their lifespan (Figure 8.2).

This is attributed to the presence of covalent adaptable networks, where reversible bond rearrangements occur through associative exchange reactions, such as transesterification, disulfide exchange, or imine metathesis. These reactions allow the material to undergo stress relaxation and defect healing without losing its cross-linked structure. Normally, with thermal or catalytic activation, bond exchange leads to topological rearrangement at the molecular level, maintaining network integrity while improving durability and mechanical performance.

Dynamics non-covalent interactions are weaker in energy compared to covalent bonds, even both are equally critical for self-healing mechanisms due to their reversibility and high sensitivity to different external stimuli. Hydrogen bonds, characterized by their electrostatic interaction between a hydrogen atom bonded to an electronegative atom and another electronegative atom, are the most versatile bond of this style. These bonds allow self-healing polymers to restore structural integrity after mechanical damage. In energy storage systems, such as supercapacitors and batteries, materials with multiple hydrogen bonds can recover from surface abrasions or internal fractures [10]. Metal–ligand interactions are another option of dynamic non-covalent bonding. These interactions are particularly effective in coordination complexes, where the metal center interacts reversibly with ligand molecules. Changes in pH, light, or electrochemical potential can modulate these bonds, facilitating repair at the molecular level.

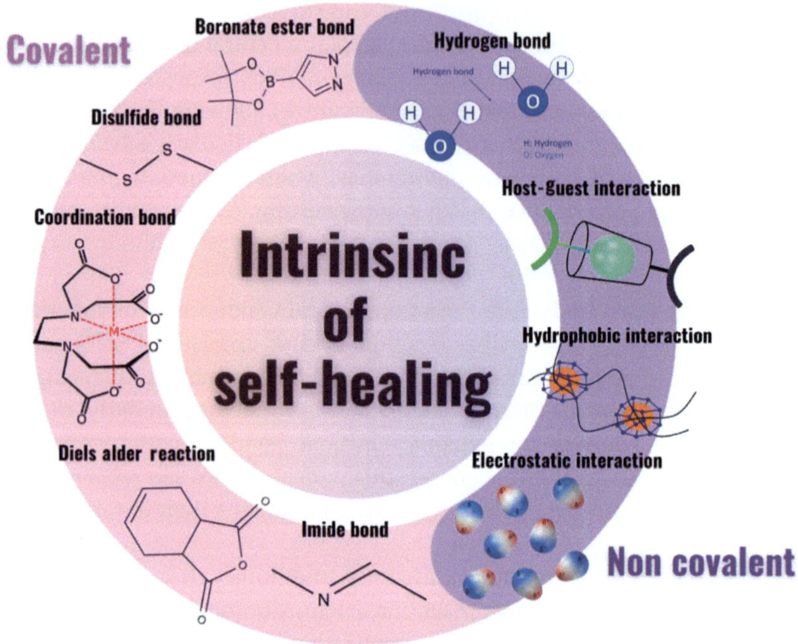

Figure 8.2 Schematic representation of the intrinsic chemistry of self-healing system mechanisms.

Host–guest interactions, based on the reversible binding of a host molecule and a guest molecule through forces like van der Waals interactions, hydrogen bonding, or hydrophobic effects, also play a crucial role. In solid-state electrolytes and polymeric matrices for energy storage, these interactions allow for the self-repair of structural components, maintaining performance over extended cycles. Magnetic materials further exemplify dynamic non-covalent interactions [11, 12]. In polymeric matrices embedded with magnetic particles, external magnetic fields can induce particle reorganization, filling cracks or defects caused by mechanical stress. Ionic bonds, formed between cations and anions, are particularly advantageous in polar solvent environments. Their ability to break and reform dynamically is essential for hydrogels and self-healing electrolytes, where ionic rearrangements restore both conductivity and mechanical integrity after structural disruptions. These properties make ionic bonds indispensable in next-generation energy storage devices, where robustness and reliability are critical.

The integration of dynamic covalent and non-covalent bonds within a single system creates synergistic effects that significantly amplify self-healing capacities. For example, redox flow batteries that combine polymers with disulfide bonds and dynamic metal–ligand complexes achieve both chemical stability and molecular-level defect repair. Similarly, supercapacitors employing polymeric matrices with hydrogen bonds and ionic bonds exhibit enhanced self-healing and durability, demonstrating their potential for sustainable energy storage.

The exploration of intrinsic self-healing mechanisms reveals how materials can leverage dynamic interactions to restore functionality autonomously. These advancements not only extend the lifespan of energy storage and conversion systems but also cover the way for designing sustainable and highly efficient devices. The integration of dynamic covalent and non-covalent bonds represents a promising strategy to address current challenges in the energy sector, highlighting the transformative potential of self-healing materials.

8.4 Coupled Self-Healing in Electrodes, Electrolytes, and Interfaces

Electrode degradation represents one of the most critical challenges for energy storage and conversion systems such as batteries, supercapacitors, and fuel cells. In these systems, electrodes act as the basis for electrochemical processes making their stability and performance a central issue for device longevity, efficiency, and performance. Self-healing electrode mechanisms have emerged as a breaking approach to address these challenges, allowing the restoration of structural and functional integrity during device operation without or with external stimuli. This section focuses on the application of self-healing concepts specifically to diverse electrodes, describing their development, implementation, benefits, and limitations as reported in the literature.

However, this system has primarily aimed to resolve issues such as mechanical stress, capacity fading, and safety concerns. For example, silicon is a high-capacity anode material for lithium-ion batteries but undergoes extreme volumetric changes (~300%) during lithiation and delithiation, leading to severe cracking and eventual pulverization. To mitigate this, researchers have incorporated self-healing polymers within silicon-based electrodes. For example, Weng et al. [13] demonstrated the use of a polymer binder containing a dynamic gradient of reversible hydrogen bond and disulfide crosslinks, which effectively repaired microcracks in silicon anodes, resulting in significantly enhanced cycle life and capacity retention. Such advancements underline the importance of molecular-level engineering in achieving self-healing functionality. Similar strategies have been explored for sulfur cathodes, where dynamic covalent chemistries allow continuous healing of internal structural defects. Zhou et al. [14] showcased the use of dynamic imine bonds in a conductive polymer binder for lithium–sulfur batteries. These imine bonds facilitated the autonomous repair of structural damages in sulfur cathodes, maintaining high performance over specific capacity retention over 98% after 80 cycles. Beyond polymeric solutions, hybrid systems combining nanoscale reinforcements with self-healing binders have demonstrated synergistic improvements in mechanical robustness and electrochemical properties. Such innovations underscore the multifaceted approaches being adopted to enhance the resilience of battery electrodes.

Supercapacitors rely heavily on electrode materials with high surface area and conductivity such as carbon-based structures. However, these materials are prone to mechanical degradation under long-term cycling or high-power operation. Self-healing strategies for supercapacitor electrodes have largely focused on incorporating self-healing hydrogels or composites to interface modification. For instance, Wu et al. [15] developed an oxide-reduced graphene-based supercapacitor electrode embedded within a self-healing hydrogel containing Zn^{2+} metal–ligand coordination bonds with the imidazole block of the ionogel copolymer. This design allowed the electrode to autonomously repair mechanical damage while maintaining superior electrochemical performance, achieving 98% capacitance retention after repeated mechanical damage. Furthermore, hierarchical structures such as carbon nanotube frameworks embedded in self-healing matrices have significantly enhanced the durability of supercapacitor electrodes. These designs leverage the inherent flexibility and conductivity of nanomaterials while incorporating a matrix capable of autonomously healing cracks and preserving mechanical integrity. Such hybrid systems are promising solutions to balance electrochemical performance and structural resilience in supercapacitors, offering substantial improvements in both reliability and operational lifespan.

Innovative designs have also utilized conductive microcapsules containing liquid metals. Upon electrode cracking or delamination, these microcapsules rupture, thus releasing liquid metal to fill voids and restoring electrical pathways. For example, Yang et al. [16] developed a self-healing GaLnSn-MS/Si hybrid electrode by incorporating microencapsulated gallium liquid with high electrical and mobility. By integrating this microcapsule into the hybrid electrode, the dispersed GaLnSn not just prevents oxidation but also enables uniform dispersion and enhanced fluidity. As a result, this anode does not require manual intervention when cracks form, allowing for self-healing. This approach has shown significant promise in extending the operational lifespan of both fuel cells and metal–air batteries. Another practical example in lithium–sulfur batteries is to construct poly(ether-urethane)-based electrolytes. Due to rearrangements of the dynamic covalent disulfide bonds and the hydrogen bonds between urethane groups, the interface of two integrated electrodes exhibited self-healing to overcome the problem of multiple interfacial contacts [17].

However, while self-healing electrodes hold immense promises, several challenges must be overcome to achieve their full potential like material compatibility, healing efficiency, scalability, long-term stability, and energy efficiency. Remembering and understanding that self-healing phenomena have synergistic effects between all its components, it is difficult to discern that the effect is due just to the electrode or the electrolyte. This is why some modifications are made to the electrode but at the same time, they affect the electrolyte i.e. interface (Figure 8.3).

In practical terms, we will now talk about the effects of electrolyte modification and how they affect the interface and in turn the entire device.

Electrolytes act as the medium for ion conduction between the anode and the cathode; ionic conduction is critical for the performance and lifespan of energy storage devices such as batteries, supercapacitors, and fuel cells. The introduction of self-healing electrolytes, which can repair damage or restore functionality after degradation, marks a significant advancement in electrochemical energy storage technologies. Self-healing electrolytes leverage advanced materials and chemistries

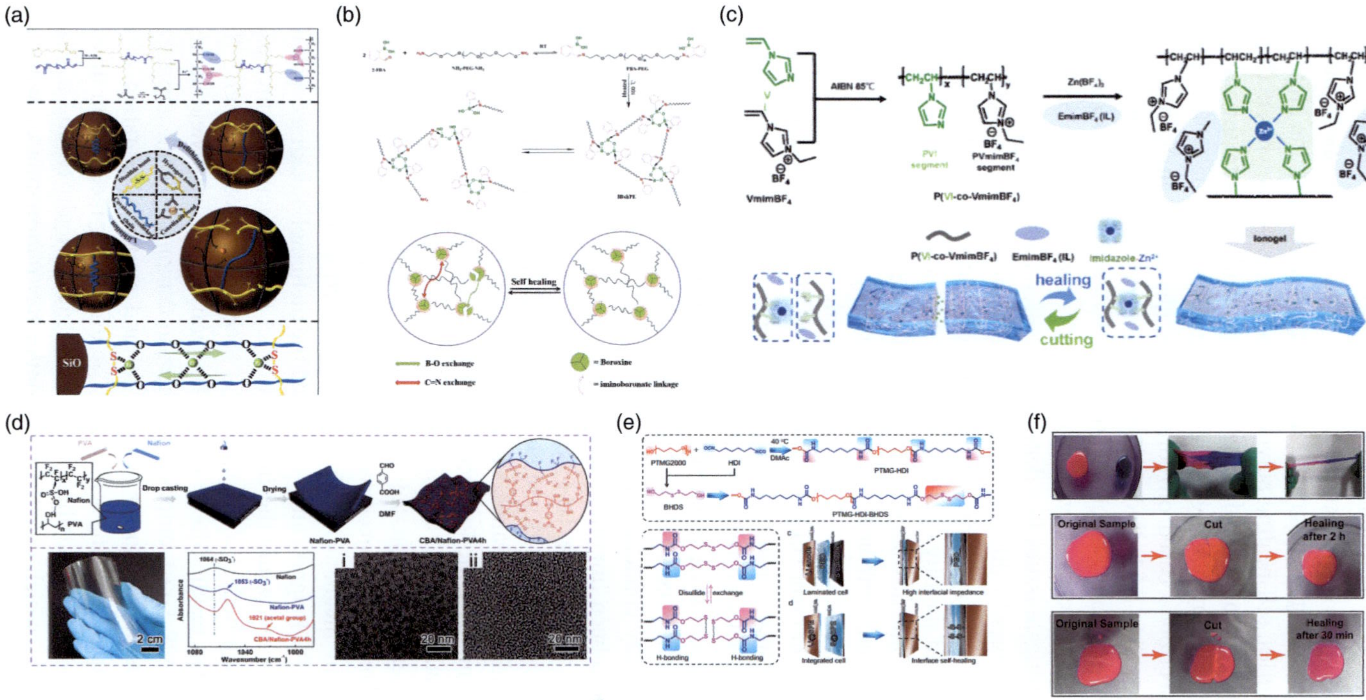

Figure 8.3 (a) Healing mechanism of dynamics reversible bonding. Source: Weng et al. [13] / with permission of Elsevier. (b) Self-healing mechanism of synergetic dynamic imine bond and boroxine bond. Source: Suping Zhou et al. [14] / with permission of Elsevier. (c) Schematic of polymerization copolymers and the design concept for spontaneous self-healing ionogel using Zn ions as a dynamic motif. Source: Jintian Wu et al. [15] / with permission of Elsevier. (d) Schematic illustration of the fabrication process for healable membranes. Source: Yixuan Li et al. [18] / with permission of John Wiley & Sons. (e) Optical images of the self-healing process. Source: Ke-hu Zhu et al. [10] / with permission of Elsevier. (f) Optical images showing the self-healing process of the shPE pieces under different ambient conditions. Source: Binghua Zhou et al. [19] / with permission of Royal Society of Chemistry.

to dynamically reassemble their molecular or supramolecular structures. However, polymer-based electrolytes functionalized with reversible covalent or non-covalent bonds, such as Diels–Alder reactions or hydrogen bonds, are the lead in this field. These dynamic bonds allow the electrolyte matrix to reform after mechanical or thermal damage, ensuring both structural integrity and efficient ion transport. In lithium-ion batteries, self-healing polymer electrolytes have shown remarkable promise, such that dynamic cross-linking agents incorporated into the polymer structure enhance ionic conductivity and mechanical resilience. A notable example is a self-healing electrolyte based on poly(ethylene glycol) dimethacrylate functionalized with reversible Diels–Alder linkages, as reported by Chen et al. [20]. This system achieved a healing efficiency exceeding 90% after mechanical damage and maintained an ionic conductivity of 10^{-3} S cm^{-1} under ambient conditions. Such innovations mitigate issues like dendrite formation and electrolyte degradation, common challenges in lithium-ion technology. Similarly, liquid electrolytes with additives promoting repair mechanisms or dynamic ionic interactions have demonstrated improved stability under mechanical stress [21].

Solid-state batteries have also benefited from self-healing advancements. Ceramic-based electrolytes, when combined with viscoelastic polymers or composite structures incorporating self-healing polymers and high ionic conductivity ceramics like lithium lanthanum zirconate (LLZO), address issues of interfacial delamination and cracking. For instance, Mishra et al. [22] described a composite solid electrolyte capable of repairing cracks, ensuring uninterrupted ion transport and significantly extending the operational lifespan of batteries. Such hybrid systems capitalize on the mechanical strength of ceramics and the self-repairing properties of polymers, thus addressing the key challenges in solid-state battery interfaces. Self-healing electrolytes have also been important and transformative in fuel cells, especially in proton exchange membrane fuel cells (PEMFCs). The repeated hydration and dehydration cycles in PEMFCs often lead to mechanical degradation of the membranes. Self-healing membranes, such as those derived from sulfonated block copolymers with dynamic ionic interactions, have shown significant potential in addressing these challenges. For example, Tai et al. [23] investigated the development of a self-healing sulfonated polyether ether ketone/polyvinyl alcohol (SPEEK/PVA) membrane reinforced with sulfonated silica (S–SiO_2), emphasizing its autonomous repair mechanisms and improved electrochemical performance. The incorporation of PVA significantly enhanced the methanol-blocking capability of the SPEEK membrane, while the introduction of S–SiO_2 facilitated increased water absorption and a higher concentration of accessible charged carriers. This structural modification resulted in an elevated proton conductivity, yielding high selectivity. The S/PVA/S–SiO_2 composite membrane demonstrated an impressive 83.6% recovery in methanol-blocking capability and an 89% restoration of its maximum power output following self-repair. The underlying mechanism driving this self-healing behavior was attributed to the presence of hydroxyl and sulfonic acid functional groups, which enabled the spontaneous reformation of hydrogen bonds, thereby restoring the membrane's integrity and functional properties. Another example is using Nafion and poly(vinyl alcohol) (PVA) complex modified with 4-carboxybenzaldehyde (CBA). Unlike extrinsic self-healing methods that use microcapsules or vascular networks, which can reduce proton conductivity and increase methanol permeability, this intrinsic self-healing approach relies on reversible hydrogen-bonding interactions between Nafion and CBA-modified PVA. The resulting CBA/Nafion–PVA membranes can autonomously heal mechanical damage, such as cuts and holes of 10 μm, restoring their original proton conductivity and methanol resistance under DMFC working conditions. The healing capability originates from the high chain mobility of Nafion and the dynamic hydrogen-bonding interactions within the composite [18].

As mentioned earlier, the self-healing strategies are not limited to bulk electrolytes or electrode surfaces, they also play a critical role in stabilizing electrolyte–electrode interfaces. This is especially relevant in lithium–sulfur and metal–air batteries, where interfacial instability caused by parasitic reactions and dendritic growth poses significant challenges. For example, Pei et al. [17] present a poly-(ether-urethane)-based solid polymer electrolyte (SPE) with interfacial self-healing capabilities, designed to enhance the SEI. This incorporates dynamic disulfide bonds and hydrogen bonds between urethane groups, the S–S bonds in the polymer have a relatively low dissociation energy, allowing them to reorganize without external stimuli and during battery operation. Additionally, the hydrogen bonds create a cohesive network that enhances the mechanical flexibility of the electrolyte. Resulting in long-term stability, operating for over 6000 hours, while full Li-S cell achieved 700 cycles with a capacity retention of 93% at 0.3 °C. In metal–air batteries, electrolyte instability often arises from oxygen reduction and evolution reactions. Self-healing liquid electrolytes containing reversible ionic complexes have shown potential in addressing these issues. These electrolytes dynamically reconfigure their structure to repair damage caused by oxygen-related by-products, maintaining ionic conductivity and chemical stability.

Additionally, gel-based self-healing electrolytes incorporating zwitterionic polymers or dynamic cross-linked networks have been developed for zinc-air batteries. These electrolytes not only facilitate ion transport but also suppress dendrite growth on the zinc anode, a common failure mechanism. Research by Wan et al. [24] demonstrated a zinc–air battery employing a self-healing polymer electrolyte with boronic ester bonds.

This system achieved over 1000 stable charge–discharge cycles by significantly extending the operational lifespan compared to conventional designs. Furthermore, self-healing electrolytes enhance the interface between the electrolyte and the air cathode by repairing microcracks induced during oxygen reduction and evolution reactions, ensuring consistent performance under cycling conditions. Despite these benefits, challenges remain. The integration of self-healing functionality can complicate electrolyte synthesis and processing, potentially increasing production costs. Additionally, ensuring that self-healing mechanisms do not hinder ionic conductivity or electrochemical stability is a critical design consideration. Balancing mechanical durability with ionic mobility remains a key hurdle. Future research should focus on optimizing this balance while employing advanced characterization techniques, such as in situ spectroscopy and imaging, to uncover the underlying mechanisms of self-healing processes. Such insights will be instrumental in designing next-generation materials.

8.5 Real-Time Monitoring

In situ characterization and real-time monitoring of self-healing mechanisms in energy storage systems, including electrolytes and cathodes, are critical for extending lifespan, enhancing efficiency, and ensuring safety and scientific knowledge. Observing these processes in real time provides invaluable insights into the chemical and physical transformations occurring within these components. However, this type of advanced characterization is usually complicated since it involves continuous monitoring during system operation. Hence, in this chapter, we will review some of the most commonly used techniques for this advanced characterization.

Advanced spectroscopic techniques have been at the vanguard of real-time monitoring. Electrochemical impedance spectroscopy (EIS) is a widely utilized tool for tracking ionic conductivity and charge transfer resistance in both electrolytes and cathodes. For self-healing electrolytes, EIS enables the observation of impedance changes as these materials undergo compositional or structural recovery. However, it is possible to go further, with more complex spectroscopic techniques. In situ Raman spectroscopy is another powerful technique, offering molecular-level insights into chemical interactions and phase changes (Figure 8.4).

For example, in cathodes with self-healing coatings, Raman spectroscopy has detected the reformation of protective layers during cycling, while for electrolytes, it has revealed dynamic bonding interactions that facilitate ionic transport recovery [25]. Nuclear magnetic resonance (NMR) spectroscopy has emerged as a great noninvasive method for studying the chemical environment and dynamics of both the cathodes and electrolytes. In self-healing electrolytes, real-time NMR studies have elucidated the diffusion and reorganization of ionic species. Similarly, in self-healing cathodes, NMR has been employed to track changes in the local chemical environment of active materials, offering insights into the mobility and stability of lithium or other charge carriers during the self-healing process [26].

Imaging techniques in other ways, scanning electron microscopy (SEM) and its advanced variants, such as environmental SEM (ESEM) and cryo-SEM, enable real-time observation of surface and structural changes under operational conditions. For cathodes, these methods provide direct visualization of crack healing or the recovery of structural integrity. In solid electrolytes, SEM has revealed the sealing of fractures, which is crucial for maintaining ionic pathways. Atomic force microscopy (AFM) has further contributed to the understanding of self-healing by monitoring topographical and mechanical changes in films and coatings. Conductive AFM, in particular, has been applied to correlate mechanical recovery with electrical conductivity in both cathodes and electrolytes [27].

Another powerful image technique is X-ray computed tomography (XCT) that has proven invaluable for three-dimensional monitoring of self-healing in both cathodes and electrolytes. In cathodes, XCT imaging has visualized the repair of cracks and voids, while in gel or hybrid electrolytes, it has offered quantitative assessments of volumetric recovery. Time-resolved XCT imaging provides a comprehensive view of the healing processes, offering spatial and temporal resolution [28].

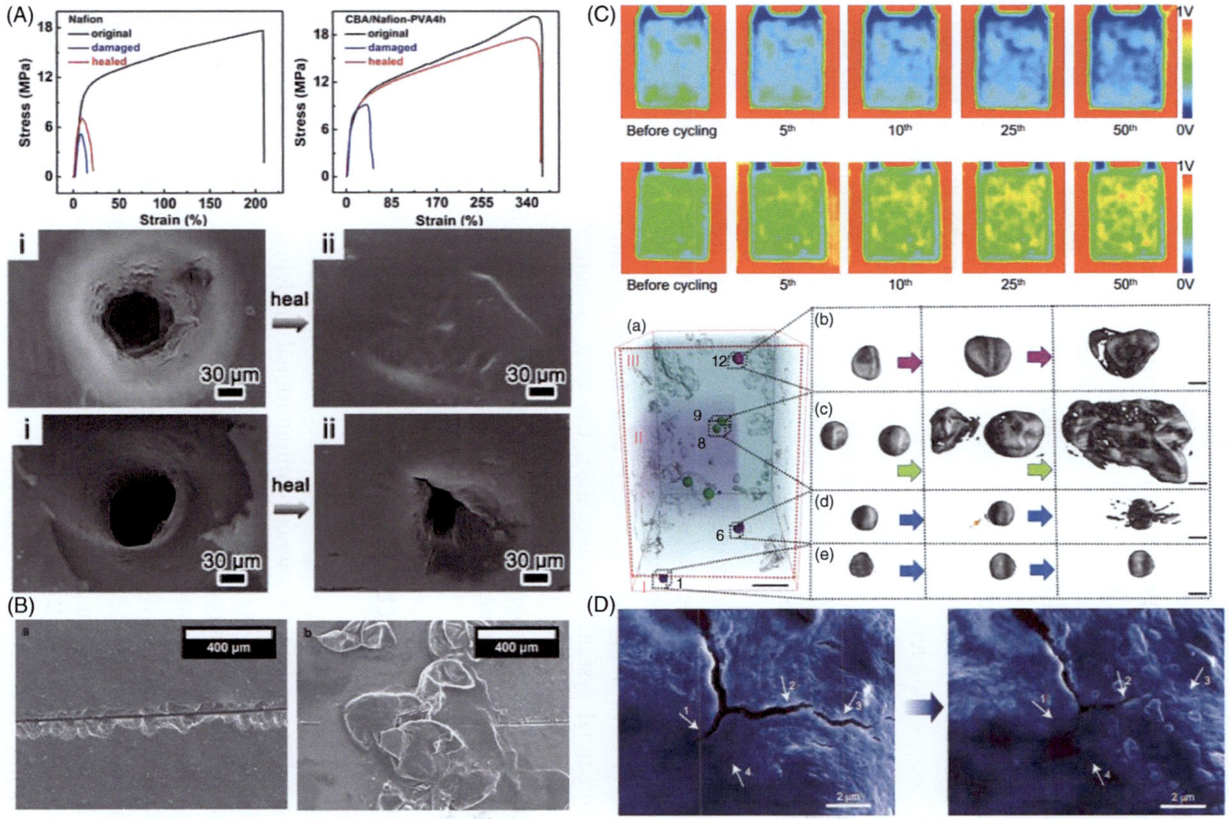

Figure 8.4 (A) Stress–strain resistance before and after healing processes. Source: De He Weng et al. [13] / with permission of Elsevier. (B) A line crack is introduced by fiber removal and release of carbon black onto the line crack. Source: Suping Zhou et al. [14] / with permission of Elsevier. (C) Ultrasonic transmission images of a pouch cell over different cycles and their voltage profiles. Source: Jintian Wu et al. [15] / with permission of Elsevier. (D) Cracks in the polymer layer in the lithiated state, after five hours of healing processes. Source: Yixuan Li et al. [18] / with permissions of John Wiley & Sons.

Synchrotron-based techniques like X-ray absorption spectroscopy (XAS) have been easily coupled to this type of system. XAS provides element-specific information on the oxidation state and local coordination environment of materials. In situ XAS has been employed to track dynamic changes in cathodes during self-healing, such as the recovery of oxidation states or coordination geometries in transition metal oxides. For electrolytes, XAS has revealed changes in coordination chemistry and electronic structure, providing a deeper understanding of the self-healing mechanisms. Small-angle X-ray scattering (SAXS) has been particularly effective for examining the mesoscale structures of self-healing polymer electrolytes and cathodes, shedding light on how morphological changes correlate with functionality [28, 29].

Emerging techniques like operando infrared spectroscopy and electrochemical noise analysis (ENA) are opening new avenues for real-time monitoring. Operando IR spectroscopy complements Raman by detecting bond formation and decomposition products in both cathodes and electrolytes during self-healing. ENA, on the other hand, captures subtle fluctuations in electrochemical systems, providing early indications of healing processes. These methods, along with advanced imaging and spectroscopic tools, provide a multifaceted approach to characterizing self-healing materials. The benefits of real-time characterization are manifold including the ability to optimize material design, extend device longevity, and gain a deeper understanding of the molecular and ionic interactions underpinning self-healing. However, challenges persist, such as the complexity of experimental setups, the influence of environmental conditions, and the integration of these techniques into commercial systems. Nevertheless, the combination of advanced spectroscopic, imaging, and electrochemical tools paves the way for more robust and efficient energy storage solutions, ensuring the viability of self-healing materials in next-generation technologies.

8.6 Future Perspectives

The future of self-healing energy systems depends on a multidisciplinary approach, bringing together materials science, computational modeling, and real-time diagnosis. These technologies offer an opportunity to enhance durability, efficiency and sustainability in energy storage, and conversion. However, translating these advances into large-scale applications remains a challenge, particularly in achieving the right balance between self-healing capability and essential properties like ionic conductivity, mechanical strength, and electrochemical stability. One of the key hurdles is integrating self-healing materials without compromising the devices performance. While polymer-based systems have shown great potential in autonomous repair, they often come at the cost of reduced mechanical robustness or conductivity. To address this, novel fabrication methods such as additive manufacturing, advanced coatings, and hybrid material design must be explored to embed self-repairing functionalities without disrupting existing device architectures.

Another critical factor in the advancement of self-healing energy systems is real-time monitoring and predictive maintenance. In situ characterization techniques, including operando spectroscopy and electrochemical noise analysis, are making it possible to detect early-stage degradation and anticipate failures before they impact performance since other in situ characterization techniques require synchrotron radiation and therefore it is more complicated to carry out. The incorporation of artificial intelligence and machine learning for predictive diagnostics will be instrumental in optimizing system reliability and extending the service life of energy technologies. Looking beyond energy storage, self-healing materials hold promise for a variety of fields including structural applications, biomedical technologies, and soft robotics. As research progresses, these materials could become a standard feature of next-generation energy systems, contributing to a more resilient and adaptive energy infrastructure. Despite the challenges, continued innovation in this field has the potential to redefine how energy devices are designed, manufactured, and maintained, paving the way for more sustainable and long-lasting energy solutions.

References

1 Wang, H., Wang, P., Feng, Y. et al. (2019). Recent advances on self-healing materials and batteries. *ChemElectroChem* 6 (6): 1605–1622.

2 Callen, H.B. (1985). *Thermodynamics and an Introduction to Thermostatics*. New York: John Wiley & Sons.

3 Wang, H., Wang, P., Feng, Y. et al. (2019). Recent advances on self-healing materials and batteries. *ChemElectroChem* 6 (6): 1605–1622.

4 Xue, C., Xu, X., Zhang, L. et al. (2022). Self-healing/pH-responsive/inherently antibacterial polysaccharide-based hydrogel for a photothermal strengthened wound dressing. *Colloids Surf. B Biointerfaces* 218: 112738.

5 Kong, D., Li, J., Guo, A. et al. (2019). Self-healing high temperature shape memory polymer. *Eur. Polym. J.* 120: 109279.

6 Song, H., Wang, Z., He, X., and Duan, J. (2017). Self-healing of damage inside metals triggered by electropulsing stimuli. *Sci. Rep.* 7 (1): 7097.

7 Salzano de Luna, M., Marturano, V., Manganelli, M. et al. (2020). Light-responsive and self-healing behavior of azobenzene-based supramolecular hydrogels. *J. Colloid Interface Sci.* 568: 16–24.

8 Huang, Y., Wang, J., Shi, Z. et al. (2022). Disulfide bond-embedded polyurethane solid polymer electrolytes with self-healing and shape-memory performance. *Polym. Chem.* 13 (42): 6002–6009.

9 Sáiz, L.M., Prolongo, M.G., Bonache, V. et al. (2023). Self-healing materials based on disulfide bond-containing acrylate networks. *Polym. Test.* 117: 107832.

10 Zhu, K., Han, X., Ye, S. et al. (2022). Flexible all-in-one supercapacitors enabled by self-healing and anti-freezing polymer hydrogel electrolyte. *J. Energy Storage* 53: 105096.

11 Sun, Q., Wu, X., He, Z. et al. (2025). A self-healing binder based on host-guest interaction for carbon-silicon anodes in lithium ion batteries. *J. Electrochem. Soc.* 172 (1): 010508.

12 Kwon, T., Jeong, Y.K., Deniz, E. et al. (2015). Dynamic cross-linking of polymeric binders based on host–guest interactions for silicon anodes in lithium ion batteries. *ACS Nano* 9 (11): 11317–11324.

13 Weng, D.H., Wu, L.Z., He, Z.F. et al. (2024). Robust self-healing ion-conductive interlocking dual-network binder based on gradient dynamic bonding for advanced SiO anodes in Li-Ion batteries. *Chem. Eng. J.* 499: 156360.

14 Zhou, S., Deng, K., Xu, Z. et al. (2022). Highly conductive self-healing polymer electrolytes based on synergetic dynamic bonds for highly safe lithium metal batteries. *Chem. Eng. J.* 442: 136083.

15 Wu, J., Huang, L., Wang, S. et al. (2023). Ionogel electrolyte with dynamic metal-ligand interactions enabled self-healable supercapacitor with high energy density. *Energy Storage Mater.* 57: 549–556.

16 Yang, J., Li, J., Yang, Z. et al. (2022). Self-healing silicon anode via the addition of GaInSn-encapsulated microcapsules. *ACS Appl. Energy Mater.* 5 (10): 12945–12952.

17 Pei, F., Wu, L., Zhang, Y. et al. (2024). Interfacial self-healing polymer electrolytes for long-cycle solid-state lithium-sulfur batteries. *Nat. Commun.* 15 (1): 351.

18 Li, Y., Liang, L., Liu, C. et al. (2018). Self-healing proton-exchange membranes composed of nafion–poly(vinyl alcohol) complexes for durable direct methanol fuel cells. *Adv. Mater.* 30 (25).

19 Zhou B., He. D, Hu, J. et al. (2018). A flexible, self-healing and highly stretchable polymer electrolyte via quadruple hydrogen bonding for lithium-ion batteries. *J. Mat. Chem.* A 6: 11725–11733.

20 Chen, X., Wudl, F., Mal, A.K. et al. (2003). New thermally remendable highly cross-linked polymeric materials. *Macromolecules* 36 (6): 1802–1807.

21 Ratwani, C.R., Kamali, A.R., and Abdelkader, A.M. (2023). Self-healing by diels-alder cycloaddition in advanced functional polymers: a review. *Prog. Mater. Sci.* 131: 101001.

22 Mishra, M., Hsu, C.-W., Chandra Rath, P. et al. (2020). Ga-doped lithium lanthanum zirconium oxide electrolyte for solid-state Li batteries. *Electrochim. Acta* 353: 136536.

23 Tai, M.H., Thiam, H.S., Tee, S.F. et al. (2024). Self-healing sulfonated poly(ether ether ketone)/polyvinyl alcohol membrane reinforced with sulfonated silica for direct methanol fuel cell. *Int. J. Hydrog. Energy* 104: 487–495.

24 Wan, L., Du, X., Guo, L. et al. (2023). Self-healing polymer electrolytes with nitrogen-boron coordinated boroxine for all-solid-state lithium metal batteries. *J. Energy Storage* 74: 109485.

25 Zedler, L., Hager, M.D., Schubert, U.S. et al. (2014). Monitoring the chemistry of self-healing by vibrational spectroscopy – current state and perspectives. *Mater. Today* 17 (2): 57–69.

26 Tsai, W.-T., Lu, Y.-H., and Liu, Y.-L. (2024). In situ self-healing of gel polymer electrolytes enhancing the cycling stability of lithium ion batteries. *ACS Sustain. Chem. Eng.* 12 (20): 7894–7902.

27 Wang, C., Wu, H., Chen, Z. et al. (2013). Self-healing chemistry enables the stable operation of silicon microparticle anodes for high-energy lithium-ion batteries. *Nat. Chem.* 5 (12): 1042–1048.

28 Pan, H., Li, Y., Zhang, H. et al. (2022). In situ investigation of the healing process in dual-microcapsule self-healing materials by the synchrotron radiation computed tomography. *Compos. Part A Appl. Sci. Manuf.* 158: 106955.

29 Kang, S., Jones, A.R., Moore, J.S. et al. (2014). Microencapsulated carbon black suspensions for restoration of electrical conductivity. *Adv. Funct. Mater.* 24 (20): 2947–2956.

9

Future Frontiers in Defect Science for Advanced Energy Technologies

Lorena Álvarez Contreras[1], Noé Arjona[2], and Minerva Guerra Balcázar[3]

[1] *Centro de Investigación en Materiales Avanzados S.C., Departamento de Química e Ingeniería de Materiales, Miguel de Cervantes 120, Chihuahua, C.P. 31136, México*
[2] *Centro de Investigación en Materiales Avanzados S.C., Departamento de Física, Miguel de Cervantes 120, Chihuahua, C.P. 31136, México*
[3] *Universidad Autónoma de Querétaro, División de Investigación y Posgrado, Facultad de Ingeniería, Cerro de Las campanas S/N, Querétaro, C.P. 76010, México*

9.1 Introduction

Recent decades have seen remarkable advancements in defect science, which now plays a central role in the design of high-performance energy materials. Defects, ranging from point defects and dislocations to vacancies and grain boundaries, offer a unique opportunity to control electronic, ionic, and catalytic properties in nanomaterials. Their strategic manipulation has led to significant improvements in energy-related devices, including enhanced electrocatalytic activity in batteries and fuel cells, increased ionic conductivity in solid-state electrolytes, and improved mechanical resistance in high-temperature materials [1].

Current research in defect science focuses on correlating the atomic-level structure of materials with their macroscopic properties. This is particularly important in the context of emerging energy storage and conversion applications such as metal–air batteries and liquid fuel cells, where defect-modulation interfaces lead reaction kinetics and long-term stability. Given the global need for cleaner and more reliable power sources [2], defect engineering has attracted attention as a cost-effective route for improving the performance of electrodes and electrocatalysts without the use of expensive noble metals [3].

At present, many research groups are converging on advanced in situ and operando techniques such as high-resolution microscopy and synchrotron-based spectroscopy to track defect behavior during real-time device operation. These experiments, combined with density functional theory (DFT) studies and machine learning algorithms, are assembling the way for rational defect design tailored to specific reactions and operating conditions [4]. Looking toward the future, there is a growing emphasis on scalable fabrication methods that can reliably incorporate

Surface and Interfacial Defects in Nanomaterials for Sustainable Energy Production and Storage, First Edition. Edited by Noé Arjona, Lorena Álvarez Contreras, and Minerva Guerra Balcázar.

desired defects into large-scale electrode materials. Another key goal is to ensure long-term stability by understanding how defects evolve during repeated cycling, high-temperature exposure, mechanical, electrical, or chemical degradation. Furthermore, multidisciplinary efforts spanning chemistry, physics, materials science, and engineering are vital for reducing the gap between fundamental discoveries and commercial applications. Academia, industry, and government agencies must work in synergy to accelerate the transition of defect-engineered materials from laboratory prototypes to market ready energy technologies. Within this framework, the integration of artificial intelligence, computational modeling, and sustainable manufacturing strategies will play a central role in pushing the boundaries of energy storage and conversion efficiency.

9.2 Evolving Paradigms: Trends and Prospects in Defect-Driven Nanomaterials

Defect science has emerged as a column in energy materials research, offering new strategies for tuning electronic band structures, charge-carrier dynamics, and catalytic efficiency. By precisely engineering defects, including vacancies, interstitial atoms, dopant sites, and grain boundaries, researchers have unlocked novel functionalities that enhance the performance and durability of the next generation of materials and energy devices [5, 6].

Frontier energy devices take advantage of these defects to induce phenomena to enhance several electrochemical processes. In metal–air batteries such as Zn–Air, controlled defects introduced into carbon-based or oxide-based catalysts can promote bifunctional activity, promoting both oxygen reduction and oxygen evolution, while mitigating issues related to intermediate poisoning effects [7–9]. Similarly, in solid oxide fuel cells (SOFCs), oxygen vacancies within doped ceria or zirconia electrolytes radically enhance ionic conductivity, lowering activation energy barriers and enabling high power densities at reduced operating temperatures [10].

Defects also play a key role in enhancing redox kinetics and ion transport in advanced battery chemistries. For instance, lithium–sulfur (Li–S) batteries can suffer from polysulfide shuttle effects; nevertheless, engineered defects in conductive hosts such as doped graphene and defect-rich carbon nanotubes provide anchoring sites that confine polysulfides, thereby inhibiting capacity fade and prolonging battery lifespan [11]. In redox flow batteries, tailored defects in electrode materials such as carbon cloth or metal foams, increase active surface areas and introduce additional catalytic sites, thus boosting reaction kinetics for vanadium or organic-based flow chemistries [12]. Moreover, defect engineering can promote thermal stability and mechanical strength in high-temperature fuel cells, in particular, perovskite-based cathodes and oxide electrolytes are highly susceptible to internal stresses and thermal expansion, both of which can be modulated by defect profiles to extend operating lifespans [13].

9.2.1 Advanced Characterization of Defects in Energy Materials

Uncovering the complex relationship between defects and material performance necessitates high-resolution and in situ/operando characterization. Transmission electron microscopy (TEM) and scanning transmission electron microscopy (STEM) offer direct visualization of dopant distributions and defect clusters [3, 14–16] by employing high-angle annular dark-field (HAADF-STEM, Figure 9.1) imaging in combination with energy-dispersive spectroscopy (EDS). Meanwhile, synchrotron X-ray absorption spectroscopy (XAS), X-ray absorption near-edge structure (XANES), and near-edge X-ray absorption fine structure (EXAFS) enable the identification of oxidation states and local coordination environments, which are critical for identifying active defect sites in complex oxides or sulfides. In addition, neutron scattering proves valuable for probing light elements such as hydrogen and lithium, within solid-state electrolytes, and clarifying diffusion pathways within solid electrolytes at varying defect concentrations.

On the other hand, synthesis methods play a fundamental role. Recent advances in atomic layer deposition (ALD), molecular beam epitaxy (MBE), electrospinning, and hydrothermal processing have allowed for unprecedented control over defect concentration, distribution, and stabilization. Ion implantation has further emerged as a precise tool for introducing controlled dislocations and vacancy-rich regions, which serve as catalytic active sites for electrochemical reactions. Figure 9.2 summarizes various synthesis strategies and defect configurations engineered for energy applications. These methodologies not only enable defect concentration tuning but also ensure structural stability under prolonged electrochemical cycling and high-temperature operation.

9.2.2 Expanding the Role of Defect Engineering Beyond Catalysis

While defect science has traditionally been associated with catalytic applications, new frontiers are emerging across multiple energy disciplines. In photoelectrochemical (PEC) water splitting, defect engineering of semiconductors such as TiO_2, Fe_2O_3, or ZnO can optimize bandgap positions and charge transport, thereby improving overall photon-to-fuel conversion [18]. Thermoelectric materials also benefit from the deliberate introduction of dislocations or point defects in compounds like skutterudites and half-Heusler compounds, to scatter phonons without significantly impeding electronic conduction [19]. Furthermore, piezoelectric and pyroelectric energy harvesting rely on perovskite materials such as $BaTiO_3$ and $Pb(Zr,Ti)O_3$, where engineered defect sites optimize polarization switching under mechanical or thermal stress, thereby enhancing energy harvesting efficiency [13]. Through these broader applications, defect engineering continues to demonstrate its transformative potential in defining the next generation of high-performance, versatile energy technologies.

Defect engineering has surpassed its traditional focus on catalysis and semiconductor doping, emerging as a fundamental approach across diverse energy platforms where advancements in ionic transport, mechanical strength, and

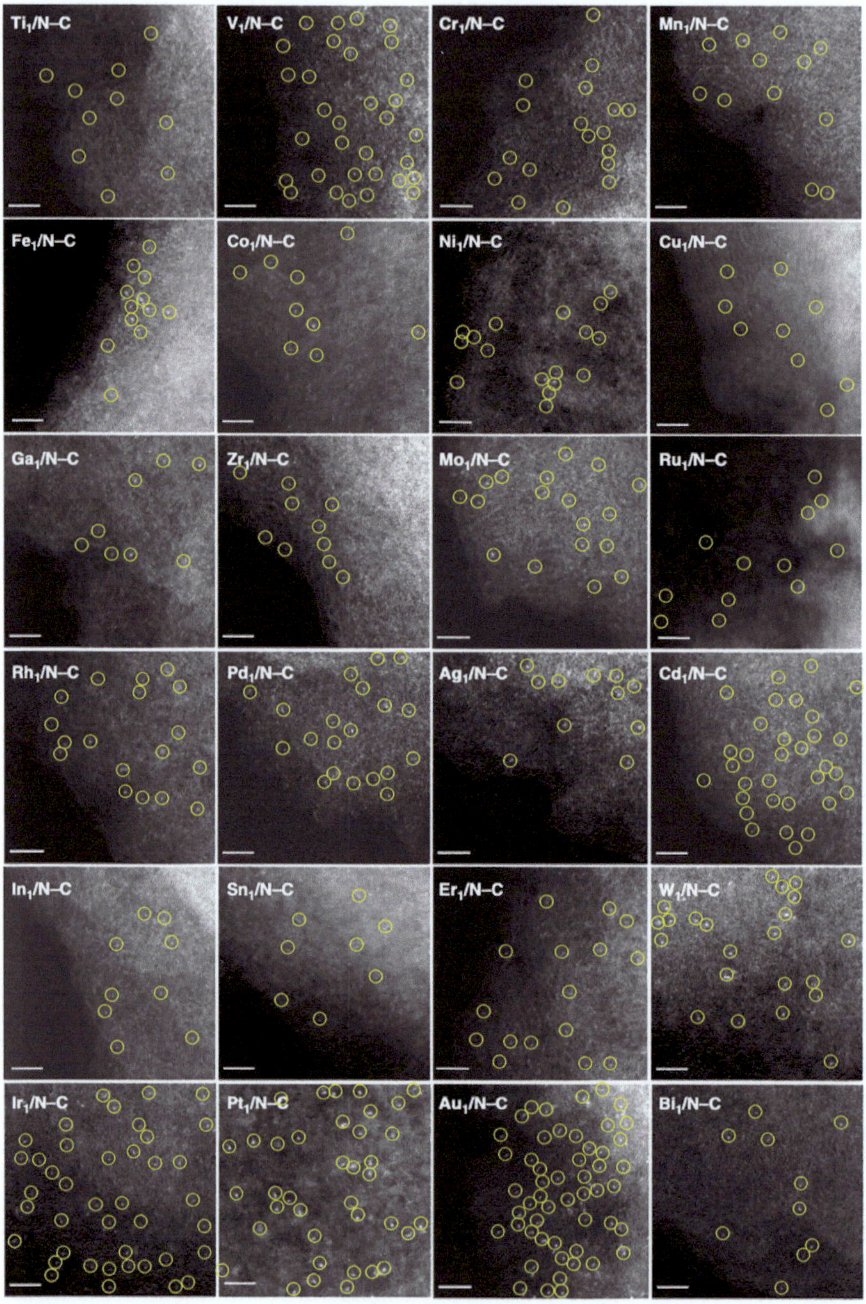

Figure 9.1 AC HAADF-STEM images of M_1/N–C (M = Ti, V, Cr, Mn, Fe, Co, Ni, Cu, Ga, Zr, Mo, Ru, Rh, Pd, Ag, Cd, In, Sn, W, Ir, Pt, Au, and Bi) reveal the atomic dispersion of metal sites within the nitrogen-doped carbon framework. The presence of single atoms (SAs) is distinctly marked by yellow circles, emphasizing their spatial distribution. The scale bar represents 2 nm. Source: Xiaohui He et al. [16] / Springer Nature / CC BY 4.0.

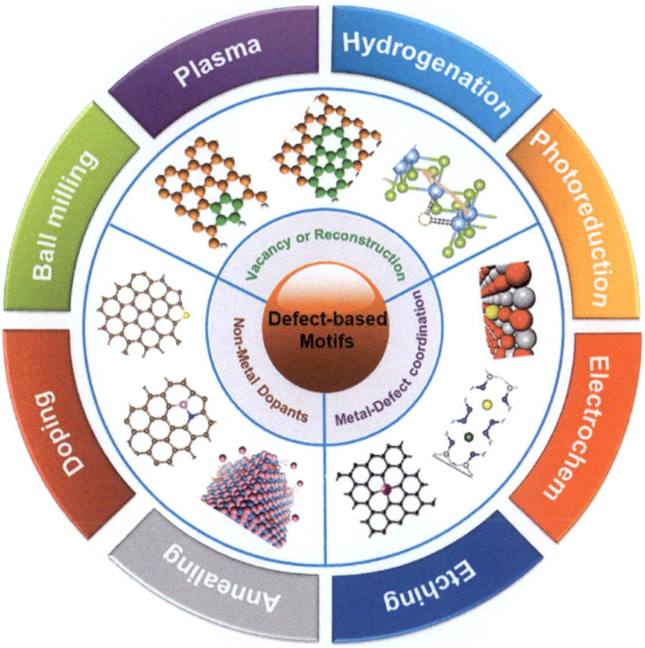

Figure 9.2 Schematic representation of defect-based patterns in nanomaterials and their corresponding preparation methods. Source: Yi Jia et al. [17] / with permission of Elsevier.

thermal management are crucial. For instance, in SOFCs, controlling oxygen vacancies in materials such as gadolinia-doped ceria or yttria-stabilized zirconia can enable reduced operating temperatures while maintaining high ionic conductivity. Similarly, lithium–sulfur batteries benefit from defect-rich carbonaceous hosts that chemically anchor soluble polysulfides, mitigating capacity diminish via the so-called shuttle effect. Even redox flow batteries, often used for large-scale stationary energy storage, can be optimized by incorporating designed defects in carbon cloth or metal foam electrodes to enhance reaction kinetics and electrolyte access. A central challenge in these evolving paradigms is ensuring that artificially introduced defects remain stable under the extremes of electrochemical cycling, temperature variations, and mechanical stress.

9.2.3 Real-Time Tracking of Defect Evolution

In situ and operando characterization techniques have redefined the study of defect evolution in batteries and fuel cells by enabling high-resolution, real-time analysis under both synthesis and operational conditions. These advanced methodologies offer unparalleled insights into defect dynamics, facilitating precise control over defect concentrations, spatial distributions, and interactive phenomena to optimize electrochemical performance. Over the past decade, their application has profoundly advanced the understanding of dynamic processes within battery

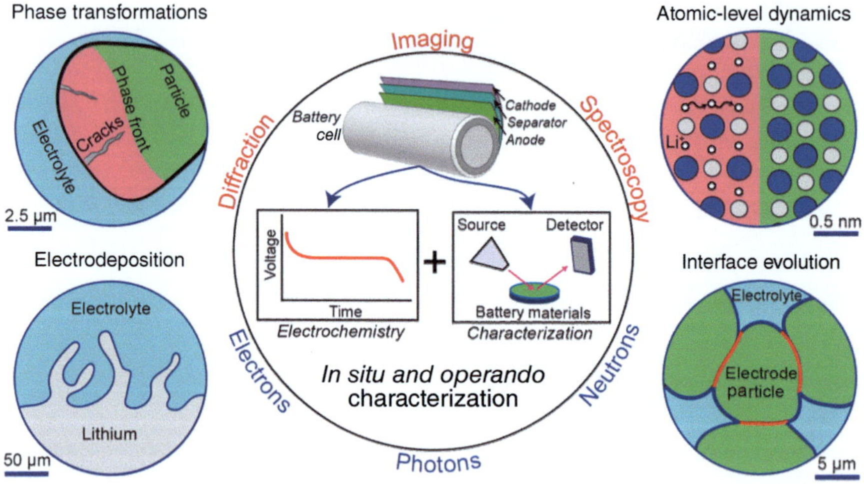

Figure 9.3 Comprehensive schematic view of the dynamic evolution of battery materials as observed through in situ and operando characterization techniques. Source: Boebinger et al. [20] / with permission of American Chemical Society.

materials, elucidating the complex interplay of phase transformations, interfacial reactions, and mechanical degradation during charge–discharge cycles. By capturing the structural, chemical, and morphological evolution of electrodes in real time, these techniques provide a fundamental framework for enhancing performance and extending the lifespan of next-generation energy storage systems (Figure 9.3). In situ TEM and STEM have permitted atomic-scale visualization of defect nucleation, propagation, and dissolution in alloy and conversion electrodes, while operando synchrotron-based X-ray diffraction and XAS have elucidated local coordination changes and oxidation state variations across a range of electrode materials. Complementary techniques such as operando Raman and infrared spectroscopy further capture transient reaction intermediates linked to defect sites, providing critical mechanistic understandings that update both material design and computational modeling.

These high-resolution techniques have been particularly impactful in revealing the evolution of solid–electrolyte interphases (SEI) and the degradation mechanisms in high-capacity alloying anodes and intercalation electrodes and are now extending their reach to emerging battery technologies such as solid-state batteries and alkali metal systems. The convergence of advanced characterization techniques with data analytics and machine learning has enabled a multiscale correlation of spatial and temporal phenomena, establishing a comprehensive paradigm that not only elucidates the fundamental mechanisms governing battery operation nonetheless but also informs the rational design of next-generation energy storage systems with enhanced performance and durability. This paradigm shift, underpinned by robust in situ/operando methodologies, is set to drive significant advancements in battery technology, establishing the way for improved energy density, enhanced safety, and lower production costs while contributing to the global effort in reducing greenhouse gas emissions.

In summary, defect engineering represents a transformative paradigm in energy materials research, offering novel pathways to enhance functionality, durability, and efficiency. By integrating state-of-the-art synthesis, in situ/operando characterization, and computational modeling, the next generation of defect-modulated materials is ready to drive development in energy storage and conversion, facilitating the transition toward sustainable and high-performance energy systems.

9.3 Intersection with Other Disciplines: Collaborations and Synergies

The integration of defect science with complementary disciplines such as computational materials science, bioinspired nanofabrication, and large-scale industrial engineering has catalyzed transformative progress in advanced energy technologies. By strategically theory-guided experimentation and data-driven methodologies, researchers are now capable of designing defect-rich nanomaterials with unprecedented precision and scalability. The multifaceted conjunction, emphasizing (i) computational materials science and machine learning approaches, (ii) bioinspired nanofabrication, and (iii) industrial manufacturing, all converge upon defect science as a central pillar of innovation.

9.3.1 Computational Materials Science and Data-Driven Approaches

Advances in computational materials science, applying first-principle simulations and high-throughput screening, now enable the identification of metastable defect configurations that enhance charge transport, catalytic activity, and mechanical strength. In particularly, the synergy between density functional theory (DFT), molecular dynamics (MD), and emerging machine learning (ML) techniques is revolutionizing the capacity to predict and optimize defect behavior across multiple length scales.

For instance, ML models can rapidly screen thousands of dopant-defect configurations, efficiently identifying those that minimize vacancy formation energies or optimize electronic alignments to enhance functionality [21]. When coupled with multiscale simulations approaches encompassing from quantum-level defect interactions (DFT, Monte Carlo) to macroscopic device performance modeling (phase-field, continuum mechanics) provide a comprehensive understanding of defect dynamics under realistic operational conditions. As summarized in Table 9.1, each modeling scale addresses distinct defect phenomena like DFT, Monte Carlo, and genetic algorithms facilitate atomic-level doping strategies, whereas phase-field simulations resolve defect cluster evolution during morphological transformations. Ultimately, the tight integration of computational and experimental methodologies accelerates the design of defect-engineered electrodes, electrolytes, and catalysts, ensuring they meet the stringent performance and durability requirements of frontier energy applications.

Table 9.1 Multiscale and ML-driven modeling strategies.

Scale	Technique	Defect relevance
Atomic	DFT, Monte Carlo, genetic algorithms	Defect formation energies, doping strategies
Nanometric	Molecular dynamics, reactive force fields	Ion diffusion, defects clustering, stress fields
Mesoscopic	Phase-field simulations	Nucleation and growth of phases, morphological evolution
Macroscopic	Continuum modeling, ML prediction	Device lifetime, mechanical integrity, optimum operating conditions

Through this integrated and data-driven paradigm, researchers can iteratively refine materials design and synthesis strategies, ensuring that defect concentrations are precisely engineered to optimize specific energy conversion and storage processes. The incorporation of big data analytics, machine learning algorithms, and high-throughput screening is increasingly transcending conventional trial-and-error methodologies, enabling a predictive and rational approach to defect modulation. This transition not only accelerates defect-driven material innovations but also establishes a systematic platform for the scalable design of advanced energy materials with tailored functionality and enhanced performance.

9.3.2 Bioinspired Design and Nanofabrication

Bioinspired nanofabrication strategies, well known by hierarchical self-assembly and defect-templated growth, provide innovative pathways for the development of structurally adaptive materials capable of dynamically modulating their physicochemical properties under operational conditions. Nature serves as an exceptional pattern for defect engineering, with hierarchical architectures observed in biological systems inspiring advanced strategies for optimizing energy materials.

By emulating multiscale vascular networks in plants or the cellular porosity of bone, researchers have designed electrodes with enhanced ion and electron transport capabilities, while simultaneously promoting the formation of defect-rich catalytic sites [10]. An alternative approach uses biochemical assembly, where peptides, viral capsids, and other biomolecules selectively bind to metals and oxides, forming self-organized arrays with precisely modulated defect distributions dictated by molecular geometry [22]. This level of spatial control enables fine-tuning of surface energy, electronic states, and interfacial reactivity, crucial for optimizing electrochemical processes such as the oxygen reduction reaction in metal–air batteries or sulfur redox chemistry in lithium–sulfur systems. The key bioinspired strategies widely used are:

- **Hierarchical Porosity**: Electrodes fabricated using biomimetic templates exhibit tunable porosity that optimizes ion and electron transport, particularly in metal–air batteries and supercapacitors.
- **Biochemical Assembly:** Peptides, viral capsids, and other biomolecules selectively bind to metals and oxides, forming self-assembled arrays with controlled defect distributions tailored for enhanced catalytic activity.
- **Synthetic Biology and Self-healing Materials:** Advances in bioengineering have enabled the development of defect-adaptive materials that dynamically respond to electrochemical and mechanical stress, improving device longevity and performance stability.

The synergistic integration of bioinspired methodologies with cutting-edge nanofabrication technique, including additive manufacturing, plasma-assisted synthesis, automated thin-film deposition, ALD, electrospinning, and three-dimensional patterning, is enabling the precise engineering of defect architectures with unprecedented functional tunability. These advancements are driving the development of defect-modulated surfaces, with profound implications for electrocatalysts, solid-state electrolytes, and hybrid organic–inorganic structures. By offering precise control over defect distributions, electronic states, and interfacial reactivity, this interdisciplinary approach is laying the foundation for next-generation energy materials with enhanced efficiency, stability, and scalability.

9.3.3 Industrial Partnerships and Translational Research

In recent years, an increasing number of industrial participants, ranging from major automotive manufacturers, aerospace, and renewable energy sectors, have actively integrated defect-engineered materials into next-generation power sources. This strategic approach not only diminishes reliance on expensive and geopolitically sensitive metals (such as platinum-group elements) but also extends the operational lifetimes of energy devices, thereby offering a compelling pathway to reduce both manufacturing and operational costs, also more friendly environmental materials. Collaborative interaction to synergize the expertise of academia, national laboratories, and industrial research and development have become indispensable in translating laboratory-scale innovations into pilot-scale demonstrations. Critical efforts in this translational framework include scaling synthesis methods such as ALD and solvothermal processing from gram level experiments to kilogram scale production, performing rigorous cost–benefit analyses to assess the trade-offs between enhanced performance and material complexity, and implementing comprehensive lifecycle assessments to ensure compliance with circular economy principles. Figure 9.4 summarizes the cross-disciplinary interactions among computational materials science, bioinspired nanofabrication, and industrial manufacturing, elements that collectively drive the maturation and commercialization of defect-engineered technologies. This schematic represents

Figure 9.4 Conceptual diagram illustrating cross-disciplinary interactions in defect engineering.

the integrated nature of modern defect engineering. At its core, defect science interfaces bidirectionally with three critical domains: computation and ML (encompassing theoretical predictions via DFT, molecular dynamics, and machine learning approaches), bioinspired fabrication (highlighting the use of biomimetic and advanced nanofabrication techniques to design defect architectures), and industrial deployment (focusing on scaling synthesis methods, economic assessments, and lifecycle analyses to bring defect-engineered materials to market). Arrows indicate the dynamic and reciprocal interactions among these fields. Through these coordinated efforts, defect engineering is poised to serve as a transformative force in the development of sustainable, high-efficiency energy systems on a global scale.

9.4 Roadmap for Future Research in Surface and Interfacial Defects in Nanomaterials

A comprehensive roadmap for defect engineering in next-generation energy technologies must balance fundamental studies on defect formation and stability with clear translational goals that address performance, cost efficiency, and sustainability at scale. By integrating advanced characterization techniques, multiscale simulations, and environmentally responsible design principles, researchers can systematically tailor defect profiles to challenge the specific limitations inherent to each technology. These efforts encompass a broad spectrum of systems, from metal–air and lithium–sulfur batteries to solid oxide and protonic ceramic fuel cells, redox flow devices, thermoelectric modules, and photoelectrochemical systems.

Metal–air batteries (e.g. Li, Na, Mg, Al) benefit from rigorous control of defect densities in carbon-based or metallic oxides catalysts, which enhance bifunctional oxygen catalysis while mitigating dendrite formation through defect-tolerant anode designs. Similarly, lithium–sulfur and analogous multivalent chemistries (Na–S, Mg–S) require defect-engineered host materials that block soluble intermediates such as polysulfides and maintain electronic conductivity under high rate cycling conditions. In advanced fuel cells, particularly SOFCs and PCFCs, refined doping profiles in electrolyte compositions (e.g. gadolinia-doped ceria) optimize ion transport, while developing stable electrode–electrolyte interfaces at elevated temperatures. Redox flow batteries likewise profit from controlled defects in porous carbon or metal foam electrodes to enhance reaction kinetics for vanadium or organic-based redox couples, whereas thermoelectric materials take advantage of engineered dislocations, stacking faults, or nanoparticle inclusions to fine-tune phonon scattering and boost energy conversion efficiencies. Finally, photoelectrochemical systems, defect states in semiconducting photoelectrodes to bolster light absorption, facilitate charge separation, and enhance overall solar-to-fuel conversion efficiency [23–25].

Successfully deploying defect-engineered materials on a commercial scale requires deeper insights into how these defects evolve during real-world operation [26]. In situ TEM and SEM provide direct visualization of morphological changes and defect migration under battery cycling or high-temperature fuel cell conditions. Synchrotron-based X-ray diffraction and X-ray absorption spectroscopy track phase transitions and shifts in local coordination changes that influence catalytic or ionic transport phenomena. Operando Raman and IR spectroscopy enable the identification of transient reaction intermediates bound to defect sites, elucidating mechanistic pathways that either accelerate or impede redox reactions. These advanced techniques overall establish a dynamic feedback loop, where high-resolution characterization guides iterative improvements in synthesis and computational modeling, enabling the rational design of defect architectures for enhanced stability and performance.

As defect-rich materials and device configurations become increasingly complex, systematic data analysis emerges as a fundamental tool for rapid optimization. Automated defect recognition using machine vision algorithms accelerates the classification of dislocation networks, vacancy clusters, and dopant distributions from extensive electron microscopy datasets. Beyond image analysis, predictive modeling leverages Bayesian optimization or active learning to propose synthesis parameters ranging from annealing temperatures and doping concentrations that may yield optimal electrochemical properties. Machine learning applications tailored to frontier energy technologies, including convolutional neural networks (CNNs) that perform microstructure analysis, random forest models for correlating composition with electrochemical performance, and Bayesian optimization techniques for guiding synthesis strategies.

Beyond performance enhancement, a robust roadmap for defect engineering must align with ecological, economic, and social responsibilities. Green synthesis procedures that minimize the use of hazardous solvents, high-temperature processing, and energy-intensive manufacturing steps can significantly reduce help the environmental footprint of producing defect-engineered materials. Additionally, ensuring that these materials remain recyclable is critical, particularly in metal–air and lithium–sulfur batteries, where reactive anodes and sulfur-based cathodes introduce unique challenges. Incorporating lifecycle assessments (LCA) and environmental product declarations (EPDs) into early-stage material design fosters alignment with global sustainability objectives while simultaneously shaping regulatory policies governing end-of-life strategies and resource utilization.

By integrating targeted defect engineering, real-time characterization, machine learning-driven optimization, and sustainable production, researchers can drive transformative advancements in energy technology, improving cost efficiency, reliability, and environmental impact. Through systematic control of defect structures suited to each device architecture, the scientific community stands poised to redefine the frontier of energy storage and conversion for decades to come.

9.4.1 Future Research Thrusts in Defect Engineering

Moving forward, four critical research directions will shape the future of defect science in energy materials:

(i) **Dynamic Defect Control:** Understanding defect evolution under realistic operating conditions remains a key priority. Further refinement of operando methodologies is needed to capture rapid kinetics, nonequilibrium states, and defect interactions that influence macroscopic performance metrics.

(ii) **Multiscale Modeling and Data-driven Design:** The next generation of defect-engineered materials will leverage advanced quantum-mechanical simulations coupled with machine learning models to predict defect configurations that optimize ion transport, catalytic efficiency, and electronic conductivity. High-throughput synthesis strategies will incorporate these predictive insights to accelerate material discovery.

(iii) **Sustainable Synthesis and Circular Economy Considerations:** Reducing the ecological footprint of defect-engineered materials will be paramount. Green chemistry approaches, solvent-free processing, responsible nanotechnology, and energy-efficient synthesis techniques will be prioritized to ensure environmental compliance. Additionally, recyclability assessments of defect-modified electrodes and catalysts will be systematically explored to support circular economy initiatives.

(iv) **Industrial Scale-up and Integration:** Bridging the gap between fundamental research and large-scale commercial deployment remains an enormous challenge. Emerging fabrication techniques must preserve defect tunability while ensuring scalability, cost-effectiveness, and long-term stability, ultimately facilitating the widespread adoption of defect-engineered materials in advanced energy technologies.

9.5 Conclusions

Defect engineering has evolved from a theoretical construct to a transformative approach in materials design for energy storage and conversion. By precisely tuning defect panoramas, researchers can enhance electrochemical activity, optimize charge transport, and improve structural resistance, leading to next-generation batteries, fuel cells, and catalytic systems with superior performance. The synergy between computational modeling, operando characterization, and industrial fabrication is driving unprecedented progress, yet challenges remain in bridging lab-scale discoveries with commercial feasibility. Moving forward, advancements in dynamic defect tracking, sustainable synthesis, and machine learning-driven optimization will unlock the full potential of defect-engineered materials, ultimately redefining the future of energy technology.

This paradigm shift underscores the crucial role of defects as not merely imperfections, but as strategic enablers of high-efficiency, durable, and cost-effective energy systems. Through continued interdisciplinary collaboration, defect science is poised to address the most pressing challenges in global energy sustainability, consolidating its place at the forefront of next-generation material innovation.

References

1 Fuhr, A.S., Sumpter, B.G., and Ganesh, P. (2023). Defects go green: using defects in nanomaterials for renewable energy and environmental sustainability. *Front. Nanotechnol.* 5: 1291338. https://doi.org/10.3389/fnano.2023.1291338.

2 Larcher, D. and Tarascon, J.-M. (2015). Towards greener and more sustainable batteries for electrical energy storage. *Nat. Chem.* 7 (1): 19–29. https://doi.org/10.1038/nchem.2085.

3 Mishraa, R.K., Vermab, K., and Singh, D.S. (2024). Defect engineering in nanomaterials: impact, challenges, and applications. *Smart Mater. Manuf.* 2: 100052. https://doi.org/10.1016/j.smmf.2024.100052.

4 Jing, H., Zhu, P., Zheng, X. et al. (2022). Theory-oriented screening and discovery of advanced energy transformation materials in electrocatalysis. *Adv. Powder Mater.* 1: 100013. https://doi.org/10.1016/j.apmate.2021.10.004.

5 Gurylev, V. (2021). General principles of defect engineering. In: *Nanostructured Photocatalyst via Defect Engineering*. https://doi.org/10.1007/978-3-030- 81911-8_2.

6 Jangizehi, A., Schmid, F., Besenius, P. et al. (2020). Defects and defect engineering in soft matter. *Soft Matter* 16: 10809–10859. https://doi.org/10.1039/d0sm01371d.

7 Liu, Y., Junjie, L., Shaofeng, X. et al. (2022). Carbon-based composites for rechargeable zinc-air batteries: a mini review. *Front. Chem.* 10: 1074984. https://doi.org/10.3389/fchem.2022.1074984.

8 Ying, J.P., Zheng, D., Meng, S.B. et al. (2022). Advanced design strategies for multi-dimensional structured carbon materials for high-performance Zn-air batteries. *New Carbon Mater.* 37 (4): 641–657. https://doi.org/10.1016/s1872-5805(22)60623-1.

9 Dong, F., Wu, M., Zhang, G. et al. (2020). Defect engineering of carbon-based electrocatalysts for rechargeable zinc-air batteries. *Chem. Asian J.* 15 (22): 3737–3751. https://doi.org/10.1002/asia.202001031.

10 Mahato, N., Banerjee, A., Gupta, A. et al. (2015). Progress in material selection for solid oxide fuel cell technology: a review. *Prog. Mater. Sci.* 72: 141–337. https://doi.org/10.1016/j.pmatsci.2015.01.001.

11 Zhou, L., Danilov, D.L., Eichel, R.A., and Notten, P.H. (2021). Host materials anchoring polysulfides in Li–S batteries reviewed. *Adv. Energy Mater.* 11: 2001304. https://doi.org/10.1002/aenm.202001304.

12 Kim, K.J., Kim, Y.-J., Kim, J.-H., and Park, M.-S. (2011). The effects of surface modification on carbon felt electrodes for use in vanadium redox flow batteries. *Mater. Chem. Phys.* 131: 547–553. https://doi.org/10.1016/j.matchemphys.2011.10.022.

13 Feng, Y., Jiagang, W., Chi, Q. et al. (2020). Defects and aliovalent doping engineering in electroceramics. *Chem. Rev.* 120: 1710–1787. https://doi.org/10.1021/acs.chemrev.9b00507.

14 Liu, D., Shadike, Z., Lin, R. et al. (2019). Review of recent development of in situ/operando characterization techniques for lithium battery research. *Adv. Mater.* 31: 1806620. https://doi.org/10.1002/adma.201806620.

15 Wu, X., Gu, Y., Ge, R. et al. (2022). Electron irradiation-induced defects for reliability improvement in monolayer MoS_2-based conductive-point memory devices. *npj 2D Mater. Appl.* 6 (1): 31. https://doi.org/10.1038/ s41699-022-00306-8.

16 He, X. et al. (2019). A versatile route to fabricate single atom catalysts with high chemoselectivity and regioselectivity in hydrogenation. *Nat. Commun.* 10 (1): 3663. https://doi.org/10.1038/s41467-019-11619-6.

17 Jia, Y., Jiang, K., Wang, H., and Yao, X. (2019). The role of defect sites in nanomaterials for electrocatalytic energy conversion. *Chem* 5: 1371–1397. https://doi.org/10.1016/j.chempr.2019.02.008.

18 Kaur, G., Divya, S.A., Khan, V.R. et al. (2021). Nano-hetero-structured thin films, ZnO/Ag-(α)Fe_2O_3, with n/n junction, as efficient photoanode for renewable hydrogen generation via photoelectrochemical water splitting. *Renew. Energy* 164: 156–170. https://doi.org/10.1016/j.renene.2020.09.060.

19 Zhou, X., Yanci Yan, X.L., Zhu, H. et al. (2018). Routes for high-performance thermoelectric materials. *Mater. Today* 21 (9): 974–988. https://doi.org/10.1016/j.mattod.2018.03.039.

20 Boebinger, M.G., Lewis, J.A., Sandoval, S.E., and McDowell, M.T. (2020). Understanding transformations in battery materials using in situ and operando experiments: progress and outlook. *ACS Energy Lett.* 5: 335–345. https://doi.org/10.1021/acsenergylett.9b02514.

21 Shen, Y. and Zhu, S. (2023). Machine learning mechanical properties of defect-engineered hexagonal boron nitride. *Comput. Mater. Sci.* 220 (5): 112030. https://doi.org/10.1016/j.commatsci.2023.112030.

22 Diab, M. and Mokari, T. (2018). Bioinspired hierarchical porous structures for engineering advanced functional inorganic materials. *Adv. Mater.* 1706349. https://doi.org/10.1002/adma.201706349.

23 Liu, X., Yan Li, X.X., Zhou, L., and Mai, L. (2021). Rechargeable metal (Li, Na, Mg, Al)-sulfur batteries: materials and advances. *J. Energy Chem.* 61: 104–134. https://doi.org/10.1016/j.jechem.2021.02.028.

24 PrernaVinchhi, A.R., Mallik, K., and Pati, R. (2024). Gd-doped ceria with extraordinary oxygen-ion conductivity for low temperature solid oxide fuel cells. *Sci. Rep.* 14: 19010. https://doi.org/10.1038/s41598-024-59030-6.

25 Young, C., Liao, Z.-Q., Li, D.-R. et al. (2024). Enhancing vanadium redox flow battery performance with ZIF-67-derived cobalt-based electrode materials. *Molecules* 29: 5061. https://doi.org/10.3390/molecules2921506.

26 Loy, A.C.M., Ng, W.L., and Bhattacharya, S. (2024). Advanced characterization techniques for the development of Subatomic scale catalysts: one step closer to industrial scale fabrication. *Mat. Today Catal.* 4: 100033. https://doi.org/10.1016/j.mtcata.2023.100033.

10

Defects and Interface Engineering of MXenes: Heterojunction Hybrid Catalysts for Hydrogen Production

Divyadharshini Satheesh[1], Gouranga Maharana[2], Rekha Pachaiappan[3], Kovendhan Manavalan[1], and D. Paul Joseph[2]

[1] *SRM Institute of Science and Technology, Department of Physics and Nanotechnology, Kattankulathur, Chengalpattu, 603203, Tamilnadu, India*
[2] *National Institute of Technology, Department of Physics, Hanamkonda, Warangal, Telangana 506004, India*
[3] *University of Chile, Energy Center, FCFM, Santiago 8370450, Chile*

10.1 Defects

"Defect engineering" is a division of material science that deals with the tuning of properties of various defects in several crystalline solids. The material properties are stabilized by the elemental components and the fashion in which the atoms/molecules are arranged and bonded it. Defects are developed by means of factors that influence the electrical, optical, thermal, and magnetic characteristics of materials which have a prominent impact on the physicochemical characteristics of materials including their catalytic behavior. The physicochemical aspects of catalytic materials are critical for catalytic reaction as they influence the surface adsorption, desorption, and diffusion aspects. The classifications based on the dimensionality, 0D to 3D defects, property tailoring involve in the removal/substitution of foreign atoms that provide control over the material properties. The presence of several types of defects in a lattice can be classified according to their dimension based on their irregularities as follows: point defects for 0D materials (vacancies and foreign atom doping), line defects for 1D materials (screw and edge dislocations), planar defects for 2D materials (twin and grain boundaries), and volume defects as 3D defects (voids and cracks) [1].

10.1.1 Point Defects

Point defect can be classified as limited occurrence with the absence of lattice extension with no dependence on direction or dimension. The point defect is further classified into two types, intrinsic and extrinsic, where the intrinsic refers to the deviation of lattice structure from its regular formation without the addition of any external atoms or ions categorized as vacancies and self-interstitials. The extrinsic

Surface and Interfacial Defects in Nanomaterials for Sustainable Energy Production and Storage, First Edition. Edited by Noé Arjona, Lorena Álvarez Contreras, and Minerva Guerra Balcázar.

defects are caused due to the existence of impurity or intentional insertion of dopant elements into the material's lattice which can be further classified into substitutional type and interstitial type [2].

10.1.2 Line Defects

This defect or 1D defects in materials with two cases, the first, when a lattice plane indicates an incomplete row of atoms and develops stress between two planes and the second if a vacancy is identical in a single direction, which can make easily within crystal lattice and it generally associated with the occurrence of a phenomena called as slip which implies sliding of atomic planes. Dislocations occurring in higher densities are considered to be the main donor to all mechanical properties of the materials. The Burger vector is set as the benchmark for quantifying distortion due to dislocations. There are two main types of dislocations that are witnessed in materials namely edge and screw dislocations which can be attributed by the Burger vector in parallel and perpendicular, respectively, on their respective dislocation line. Edge dislocation generally represents when adding extra atomic plane(s) of atom into a regular crystal creating a strain. Screw dislocations will not undergo climb, since climb involves diffusion-mediated movement which is perpendicular to the slip plane, as it is linked with edge dislocations. Since screw dislocations lack an extra half-plane of atoms, they do not experience this type of movement. However, the screw dislocation can glide (move laterally) in multiple slip planes due to its ability to undergo cross-slip. This shows that they are not strictly limited to the single horizontal plane but can move between the various slip planes with the same Burgers vector [2].

10.1.3 Planar Defects

The planar defects can be categorized as grain boundaries, stacking faults, and grain interfaces. The grain boundaries imply the interface between the two crystallites with varying crystallographic orientations. This implies that it forms a transition region between these crystallinities where the atoms are shifted from their usual lattice positions. As a result, the grain boundaries become a heterojunction structure with various atomic reactions at the solid-state level [2].

10.1.4 Volume Defects

Volume defects, also called as bulk defects, occur within the crystal lattice in three dimensions including voids or pores which are empty spaces or cavity occurring within the material. They can be classified into precipitates, dispersants, inclusions, and voids/cracks. Precipitates are assembles of impurity atoms aggregated via the method of preparation, as a result, it contributes to increasing its strength and decreasing the per location of dislocation which acts as a barrier for its motion. Dispersant takes the second phase role which is introduced into the material to tune and modify the primary phase. It can be seen as larger sized particles with sizes varying from 10 to 100 μm. Inclusion is generally to be considered as an undesired feature whose intrusion and presence decline the useful feature of materials. Voids occur due to the presence of gas bubbles or aggregation of vacancies which are trapped during the process of solidification. These defects lead to detrimental effects on the efficiency of the materials for example the mechanical strength becomes weaker in solids [3].

10.2 Interface Engineering: A Brief Introduction

Interface engineering is an exciting strategy of materials science, where the properties of the materials are not merely determined by their inherent composition but also by the interface that exists between their components. The understanding of interfacial engineering is still the cornerstone of materials science which offers an idea of tailoring materials by enhancing the functionalization and optimization of materials by engineering the structure, composition which has control over the materials behavior such as materials strength and electrical conductivity. Interfacial engineering is divided into interactions, structures, and behavior occurring at the interface of the materials [4].

10.2.1 Interfacial Engineering in Nanomaterials

At the nanoscale level, the surface and interface of the nanoparticles are detected by the behavior, interactions, and applications. The dimension of nanomaterials which typically ranges from 1 to 100 nm shows a unique property when compared to the bulk components, this high surface-to-volume ratio of nanomaterials makes the surface more reactive and sensitive to the environmental conditions. The interface engineering focuses on obtaining a unique structure, which is suitable for specific applications like functionalizing the surface of nanoparticles with ligands by enhancing their stability, and compatibility with various matrices [4].

10.3 Influence of Defects and Interfaces on the Characteristics of Materials

Defects and interfaces play a key role in catalysis since they can significantly alter the surface of a catalyst where it undergoes chemical transformation, which leads to the formation of desired products. In general, the defects and interfaces can enhance the reactivity, which provides several active sites for the reactants to undergo a better chemical reaction. The defects can create more reactive sites than the other coordinate sites in its priority. During the crystallite formation, two key processes occur: nucleation and growth. Crystallites often contain various defects and interfacial sites on their surfaces, which can greatly affect their catalytic performance. (Figure 10.1) [6].

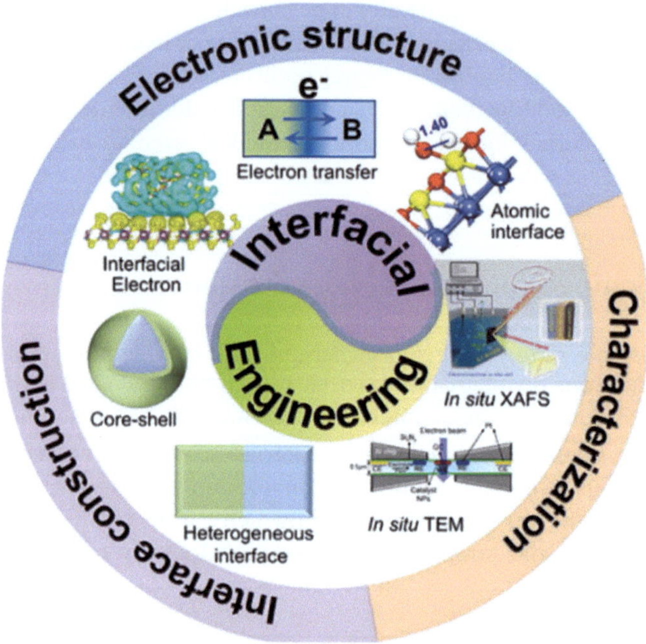

Figure 10.1 Schematics of types of interfacial engineering in materials. Source: Zhang et al. [5] / with permission of Elsevier.

10.3.1 Doping Strategy for Interface Technique

Doping is a simple and effective process used to tune the electronic structure by incorporating foreign atoms into the host materials. The heteroatoms doping effects include metallic- and nonmetallic atom doping. Metallic atoms can be doped with metal oxides, metal phosphides, metal chalcogenides, metal carbides, metal nitrates, and so forth. A distinctive two-dimensional (2D) metal dichalcogenides is MoS_2 that has been regarded as one of the most famous 2D materials studied for many energy and environmental applications. It is found that the size, crystal structure, and morphology, and the exposure of their edges may aid their activity and improve their application in hydrogen production [6].

10.3.2 Morphological Interface

Morphological interface is formed by combining components with varying morphologies in a single matrix within a single material. This strategy of designing the catalyst's morphology plays a vital role in enhancing its electrocatalytic performance. This design can help to improve the stability, conductivity, and selectivity of optimizing the properties of each component. The heterogeneous morphological structure encompasses various complexes of morphology combining different materials forming interfacial domain to achieve a synergistic effect for improving the performance. Morphological interface offers excellent stability, rapid electron mass transportation, and high wettability making them ideally suitable for long-term commercial applications [7].

10.4 Introduction to Hydrogen Production

10.4.1 What Is Hydrogen?

Hydrogen is a clean alternative to methane, which is also formed as natural gas, and one of the most abundant chemical elements in the periodic table but not available as such in pure form. This element is inexhaustible throughout the earth and the rest of the universe. According to the US Energy Information Administration report, 75% of the mass of the universe is made up of hydrogen. The chemical element symbolized by H in atomic form can be bound to exist in many forms like water, ammonia in plants, animals, and even humans. Without hydrogen, the life form as we know it would not be existing (Figure 10.2a) [11].

10.4.2 How Is Hydrogen Produced?

Hydrogen made with renewable energy like solar and wind energy is called renewable hydrogen and it plays a major role in our zero emission for the future which enables the industry and helps to minimize the replacement of fossil fuels in several energy-intension sectors. The generation of hydrogen can be categorized into three major types like green – renewable energy-based hydrogen production, purple – nuclear energy-based hydrogen generation, and blue – coal gasification and natural gas-based hydrogen generation which is combined with carbon captured and storage (CCS) hydrogen production. There are several technologies available for the production of hydrogen from fossil fuels which are hydrocarbon reforming and pyrolysis. A suitable form of H_2 generation must have high selectivity for hydrogen and high permeability to function with high flow and limited surfaces and high chemical and structural stability [12].

10.4.2.1 Steam-Reforming Method

The natural gas is an advanced and mature process for generating hydrogen that builds upon the existing natural gas pipeline delivery infrastructure. This is a type where a natural gas containing methane (CH_4) that can be utilized to generate hydrogen by thermal processes like steam-methane reformation and partial oxidation. Presently, most of the hydrogen is generated via steam-methane reformation in which high-temperature steam at 700–1000 °C is used to produce hydrogen from methane sources like natural gas. The steam-methane reforming is performed under 3–25 bar pressure in the presence of catalyst to generate hydrogen, carbon monoxide, and a relatively small amount of CO_2. The steam-methane reforming process occurs as follows:

$$CH_4 + H_2O\ (+\text{heat}) \rightarrow CO + 3H_2,\ \Delta H_{298} = 206\ \text{kJ/mol}, \tag{10.1}$$

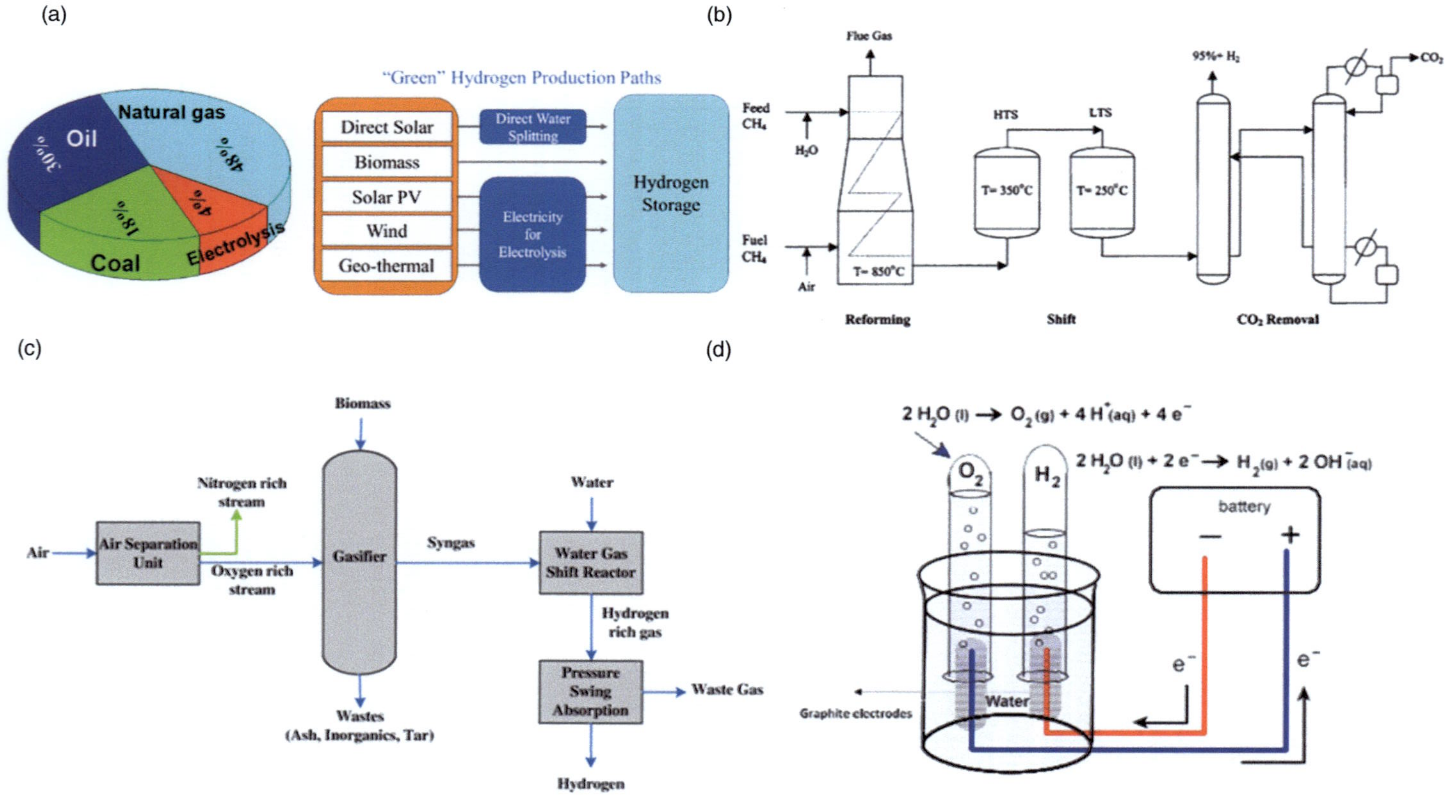

Figure 10.2 (a) Pie chart illustrating the percentile of hydrogen production from natural gas, oil, coal, and electrolysis. Production of hydrogen from (b) steam-reforming method, (c) bio-mass, and (d) electrolysis technique [8–10].

$$CH_4 + 2H_2O \rightarrow CO_2 + 4H_2, \Delta H_{298} = 164.9\ \text{kJ/mol}, \tag{10.2}$$

$$CO + H_2O \rightarrow CO_2 + H_2, \Delta H_{298} = 41\ \text{kJ/mol}. \tag{10.3}$$

Steam-reforming Eq. (10.1) and (10.2) are highly endothermic, so the forward process is benefited by high temperature. The water–gas shift reaction (Eq. 10.3) is mildly exothermic and it is benefited at lower temperatures. Additionally, this reaction is likely unaffected by the changing pressure (Figure 10.2b). In the presence of a selective CO_2 absorbent, the conversion of CH_4 to CO_2 through Eq. (10.2) is followed as it produces CO_2 through CO intermediate [13].

10.4.2.2 Hydrogen from Biomass

Hydrogen generation costs from biomass remain high. When gasification is chosen, the costs of H_2 generation from biomass range from 1.21 to 2.42 $/kg, for pyrolysis they are between 1.21 and 2.19 $/kg, which is thrice costlier than steam-methane reforming which is 0.75 $/kg. Presently 98% of global hydrogen used is generated from fossil fuels, another option based on renewable resources is biomass that is made from agri-food products, wastes, agricultural residues, energy crops, and forest-related products. Hydrogen generation from biomass permits an increase in quantity contributing to the economic capacity by increasing sources and greenhouse gas release. This process offers a promising route for sustainable hydrogen production, especially when combined with carbon capture and storage technologies (Figure 10.2c) [14].

10.4.2.3 Electrochemical Hydrogen Generation

The electrochemical energy conversion and storage technique is a long-lasting and promising energy system. The generation of less carbon dioxide by the electrocatalyst and its high energy densities, effectiveness, and affordability are critical for its widespread applications. Using electricity, water is electrolyzed and broken into hydrogen and oxygen. The overall mechanism is quite similar where the splitting of water into hydrogen and oxygen gases. Half-cell reaction provide an efficient and clean means for mass production of fuel generation.

In alkaline conditions, the reaction is

$$2H_2O + 2e^- \rightarrow H_2 + 2OH^- \tag{10.4}$$

In acidic conditions the relation is,

$$2H^+ + 2e^- \rightarrow H_2 \tag{10.5}$$

Noble materials like Pt/C and RuO_2 are the most effective hydrogen evolution reaction (HER) catalysts. Carbon-based materials such as g-C_3N_4 became the main focus of novel HER materials due to their electron–donor aspects as it is environmentally friendly and low in cost (Figure 10.2d) [15].

10.5 2D MXenes for Hydrogen Evolution Reactions

MXenes are a class of 2D materials made up of transition metal carbides, nitrides, or carbonitrides which was discovered in 2011 at Drexel University. MXenes provides

2D sheet-like structure with metallic-like electrical conductivity, which is largely missing from the series of 2D materials mostly they are semiconductors, semimetals, or dielectrics. These MXenes have higher strength and stiffness when compared to the other 2D materials. The MAX phase materials are considered to be hydrophilic or water dispersible with electrical conductivity whose rheology looks similar to that of the clay. MAX phases are unique class of layered ternary carbides and nitrides with general formula M_nAX_{n-1}, where M represents a transition metal such as titanium (Ti), vanadium (V), or niobium (Nb). A is Group 13 or 14 elements like aluminum (Al), silicon (Si), or gallium (Ga). Then X denotes the carbon or nitrogen. These types of materials exhibit a lamellar structure, which is similar to graphite, by altering metallic and ceramic-like layers, and provide highly anisotropic properties. MAX phases are characterized by a combination of metallic properties like electrical and thermal conductivity and ceramic properties such as high-temperature stability and oxidation resistance. These characteristics make the MAX phase typically involved in the solid state or chemical vapor deposition (CVD) at elevated temperatures. MAX phases are used as precursor for the synthesis of MXenes a class of 2D materials produced by selectively etching the 'A' layer from the MAX phase structure [16]. There are a variety of MXenes that are available and an infinite number of composites are possible to be made, the Fermi level can be tuned by external means like combining with semiconducting materials. MXenes with different transition metal ions exhibit different types of work functions; a variety of surface modifications like introducing functional groups by the etching technique can pointedly modify the surface and their properties. The study of heterostructure with MXene with another electrocatalyst can pave the way for new research opportunities. Being a relatively newer area, limited studies are available on this topic. The attempts in this section will bring a new insight into making MXene-based materials with some heterostructure doping/composites and light up the effective HER application. These MXenes are made by selective etching of the A-layer from the MAX phase. Generally, the MXenes are indicated as $M_{n+1}X_nT_x$, where the range of n varies from 1 to 3 and T represents the functional group, which results in the interaction with acids during the etching process. As mentioned, the n value ranges from 1 to 3 and the formed MXenes possibly exist in three different phases like M_2X, M_3X_2, and M_4X_3 with 3 to 7 atomic layer thickness. $Ti_3C_2T_x$ is considered to be the first synthesized MXene after which several other MXene compositions have been reported as of now. Since the discovery of $Ti_3C_2T_x$ was identified by Gogotsi's team in 2011, these MXenes have garnered a huge attention among researchers, which includes high stability and superior electrical conductivity. So far, several MXenes compositions have been synthesized and various kinds of applications have been identified. With extensive investigations, MXenes are identified as a potential alternative for H_2 storage, batteries, sensors, catalysis, and supercapacitor applications [17].

10.5.1 Synthesis of MXenes

10.5.1.1 Fluoride Etching Technique

The MAX phase material is etched at ambient temperature under constant agitation, and then the layered MXene structure can be exfoliated by ultrasonication using methanol and isopropyl alcohol. The second method is reaction formed HF in situ from HCl and fluorides like LiF, NaF, KF, CsF, CaF, and tetra-n-butyl ammonium. The SEM study indicates the layers to resemble the accordion. Restacked layers with gaps in between the sheet are found, particularly after etching the MAX. Recently, fluoride-free etching process has also been reported by Zhang et al. to etch the MAX phase in alkaline media, although the drawback is preventing corrosion [18]. The terminal group formed on the surface is mostly –O and –OH. Electrochemical etching can also be used to eliminate the A-layer from the MAX phase materials, but selecting an appropriate acidic medium remains a challenge [19]. Various types of preparation techniques of MXenes are shown in Figure 10.2a.

10.5.1.2 Alkali Etching Method

Theoretically, alkali technique is suitable for etching, especially for Ti_3AlC_2 (as prepared by following Figure 10.2b) material due to its high reactivity with amphoteric aluminum (Al). During the alkali treatment, there are some difficulties in obtaining MXenes because of its reaction kinetics, including some metal oxides and hydroxides that are accumulated on the surface of the $TiAlC_2$. Li et al. found that KOH can be used as an etchant for MXenes synthesis resulting in nanoscale single-layer $Ti_3C_2(OH)_2$. On the other hand, NaOH-assisted hydrothermal etching of Ti_3AlC_2 was formed to make $Ti_3C_2T_x$ (where T is OH or O). This hydrothermal process was performed in argon atmosphere to minimize oxidation of the sample. The multilayer of Ti_3AlC_2 can be formed by selectively removing the Al in Ti_3AlC_2 with the help of NaOH solution at 270 °C. The morphology of this material is analyzed using TEM and HAADF-STEM confirmed sheet structure with nonuniform interlayer spacing of 1.2 nm which is higher than the pristine Ti_3AlC_2. This alkali etching process succeeds in removing the amphoteric and acidic atoms from the pristine MAX phase; this also yields an HF-free means for forming innovative MXenes. Figure 10.3c–e provides the electrochemical data, which evaluates the HER performance of different MXene-based catalysts. In the polarization curve (Figure 10.3c), the current density increases with a decrease in potential, indicating the HER onset and catalytic efficiency. Among the tested materials, E-$Ti_3C_2O_x$ shows the best HER performance, which is closely approaching the Pt/C benchmark. The lowest Tafel slope (33.1 mV/dec) for E-$Ti_3C_2O_x$ suggests a near-optimal HER mechanism (Figure 10.3d). The stability test (Figure 10.3e) shows a stable current (~6.5–7.0 mA) over 24 hours, confirming the durability of the catalyst [21].

10.5.1.3 Water-Free Etching Technique

Currently, this nonaqueous process includes molten salt etching and iodine-supported etch. The MAX phase reaction in molten salt was similar to the HF such as reaction of Ti_3AlC_2 with $ZnCl_2$, following the identification of MXenes prepared using molten Lewis's acid in 2019. Huang et al. performed an in-depth analysis exploration on the preparation of MXenes with general molten salts. The redox potential of Lewis acid cations was very high to oxidize the A-layer of the MAX phase. Some varieties of molten salts like $CdCl_2$, $FeCl_2$, $CuCl_2$, AgCl, and $NiCl_2$ were applied to etch the different MAX phases. Recently, the nonaqueous

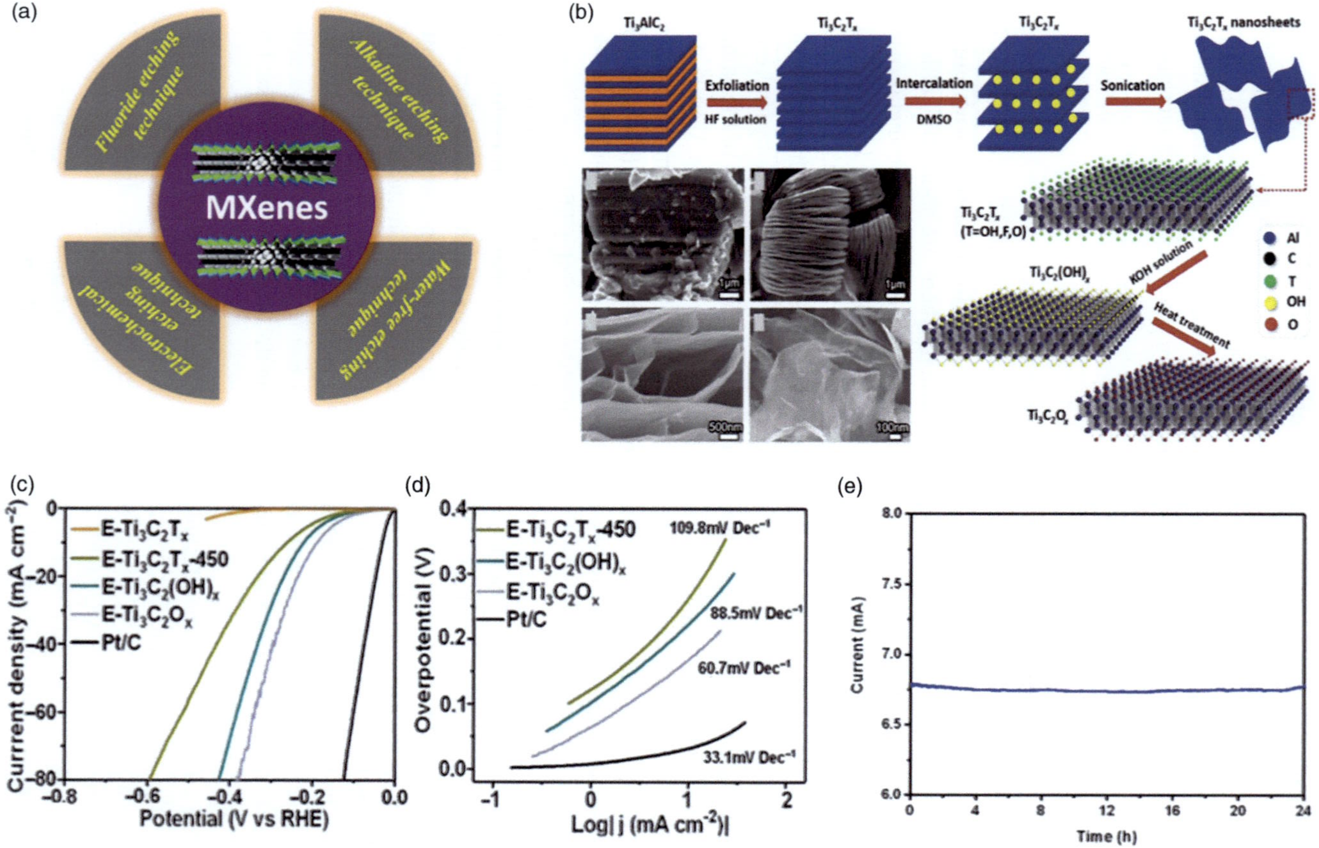

Figure 10.3 (a) Synthesis techniques of MXene with interfacial arrangements, (b) preparation of Ti_2AlC_2 exfoliated using HF technique with the in-set of SEM images showing layered arrangement of Ti_2AlC_2 after etching, (c) LSV polarization curve of few selected MXenes, (d) Tafel plot of the corresponding sample, and (e) chronoamperometric time-dependent curve at 150 mV vs RHE. Source: Shahzad et al. [20] / with permission of John Wiley & Sons.

iodine (I_2) supported etching route for the synthesis of 2D MXene with oxygen high terminal groups. This $Ti_3C_2I_x$ was initially made through iodine etching in anhydrous acetonitrile at 100 °C followed by stratification in HCl medium and, further, it is converted into flakes of medium size [21].

10.5.2 Single MXene for HER Studies

On the basis of adsorption characteristics, the MXenes can be categorized into three different ways: (i) the lowest adsorption site for O in the FCC site in both carbides and nitrides, (ii) the lowest adsorption site for O in the HCP site for both carbides and nitrides, and (iii) the lowest free energy adsorption site in the FCC site for carbides but the HCP site for the nitrides. The nitrides bind H faster than carbides. It is reported that based on the experimental and theoretical studies, MXenes are potential electrocatalysts for HER applications [15]. Also, this study shows the experimental synthesis of Mo_2CT_x and Ti_2CT_x, which is corroborated with the theoretical assumption; both theoretical and experimental approach results show better performance for Mo_2CT_x than Ti_2CT_x. To evaluate the HER activity of these two materials that are prepared by solvent drop-casted onto the glassy carbon electrode (GCE) and examined with three-electrode electrochemical set-ups, for comparison of bare, GCE is also measured along with the electrocatalyst. The bare GCE shows no assessable result within the potential window of performance. Ti_2CT_x shows considerable HER activity with high overpotential of 609 mV at 10 mA/cm^2 current density, this makes it a less efficient catalyst compared to Mo_2CTx, which shows an overpotential of 283 mV at 10 mA/cm^2. Calculations reveal the lower ΔG_H value for Mo_2CTx (0.048 eV) than Ti_2CTx (0.358 eV) correlating with higher activity of Mo_2CTx. According to Wenyu Yuan, MXene nanofibers are highly active catalysts for HER, which was optimized by preparing Ti_3C_2 nanofibers and Ti_3C_2 nanoflakes via hydrolyzation process. The HER studies were carried out in typical three-electrode system, with graphite rod as a counter electrode, Ag/AgCl (3.0 M KCl), and the working electrode as the catalyst. The results show that Ti_3C_2 NGs show a lower overpotential of 169 mV at a current density of 10 mA/cm^2, a depressed Tafel slope of 97 mV/dec, and low electrochemical impedance with less resistance. Also, this material shows high cyclic stability and long-term viability. Tran et al. [22] successfully prepared V_4AlC_3 by chemical exfoliation using aqueous hydrofluoric acid with potential applications. The $V_4C_3T_x$ MXene is discussed in the context of their HER performance in an acidic medium. There is a significant boost in the catalytic performance of $V_4C_3T_x$ over 100 cycles and the overpotential decreases almost by 200 mV and also decreases the Tafel slope from 236 to 168 mV/dec. The XPS studies before and after the catalytic cycling clearly showed the removal of oxides on the surface of the MXene. MXenes functionalized with oxygen on the basal plane are theoretically ideal for HER applications. This report shows the HER performance under acidic condition of $Ti_3C_2O_x$, which was synthesized by a simple chemical reaction. It exhibits superior HER activity at 190 mV and 10 mA/cm^2 with a long-term stability. Modifying the surface groups of MXene is demonstrated to be an effective strategy for enhancing its catalytic performance. The metallic Ti_2CT_x

(Figure 10.3) nanosheet with high F surface termination groups is highly efficient and stable electrocatalyst for HER mechanism. The electrocatalytic presence of HER of the rich F-terminated Ti_2CTx nanosheets was explored by electrochemical experiments in 0.5 M H_2SO_4 solutions with a low onset overpotential of 75 mV and large exchange current density of 0.41 mA/cm^2. This enhances the performance due to the rich F termination, which improves protons adsorption and reduces charge-transfer resistance with the ultrathin structure and metallic behavior. These features lead to more active sites and high conductivity. Compared with the layered Ti_2CT_x and Ti_2AlC, the nanosheets exhibit better HER activity, faster kinetics, and higher electrochemical stability [23].

10.5.3 Heterostructures of MXene for HER

The MXene surface provides homogeneous growth and synergetic bonding with other nanostructured materials leading to minimal agglomeration and builds the opportunity for chemical bonding with other nanosized materials, resulting in heterostructure formation.

MXene hybrid system is rarely investigated and the combination of MXenes with metal selenides has not been reported much. Hanmei Jian et al. reported hybridized interface of $NiSe_2/Ti_3C_2T_x$ that exhibits unique properties for energy conversion and storage (Figure 10.4a) [25]. This composite material was synthesized by one-pot hydrothermal method and its composition and morphology analyzed using various techniques confirm the strong interfacial bonding between $NiSe_2$ on the homogeneously distributed ultrathin $Ti_3C_2T_x$ MXene-layered nanosheets on the surface (Figure 10.4b). The HER performance of the synthesized $NiSe_2/Ti_3C_3T_x$ composite catalyst were evaluated in standard three-electrode system were performed in 2 M KOH aqueous solution. The working electrode, fabricated using 6 mg of active material ($NiSe_2/Ti_3C_2T_x$ composite, $NiSe_2$ and $Ti_3C_2T_x$) and 15 μL of Nafion solution, was added into 160 μL of mixed solvent of DI water and isopropanol and sonicated for about an hour. Then the homogeneous ink was loaded onto a 3-mm diameter glassy carbon electrode (GCE). The whole electrochemical measurement was carried out in a 0.5 M H_2SO_4 electrolyte by using Ag/AgCl and graphite as a reference and counter electrode, respectively. The LSV curve was used to assess the mechanism (Figure 10.4c) and was compared with the 20% Pt/C (44 mV at 10 mA/cm^2) current density. The prepared samples exhibits a low overpotential of 200 mV at 10 mA/cm^2 and low Tafel slope of 37.7 mV/dec are attained by the composite $NiSe_2/Ti_3C_2T_x$ electrocatalyst material (Figure 10.4d). This combinational material can also be a great candidate for supercapacitor application, which shows high specific capacitance of 531.2 F/g at 1 A/g (Figure 10.4e), and the EIS data supports the material's conductivity, where the best catalyst reported [24] with less charge transfer resistance (R_{ct}) is shown in Figure 10.4f. The prepared composite sample was studied for both the HER and supercapacitor application by comparing with the individual components of $NiSe_2$ and $Ti_3C_2T_x$, in which the modified material with strong interfacial interaction can provide an enhanced conductivity, abundant active sites, and faster charge

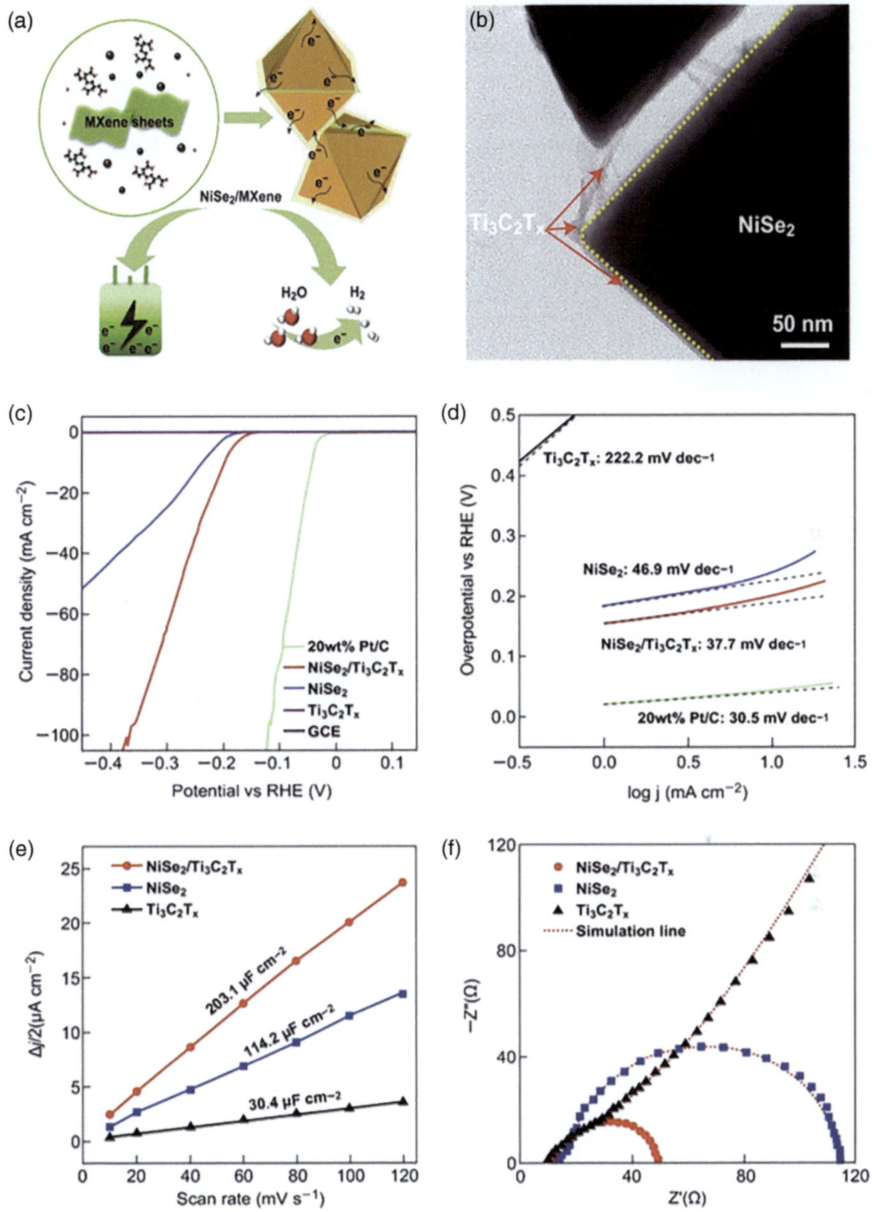

Figure 10.4 (a) Pictorial representation for the formation of octahedral $NiSe_2/Ti_3C_2T_x$ hybrid by one-pot hydrothermal method; (b) TEM images $Ti_3C_2T_x/NiSe_2$ hybrid interface in different magnifications; (c) HER catalytic performance in 0.5 M H_2SO_4 solution with LSV curve of pristine $NiSe_2$, hybrid $NiSe_2/Ti_3C_2T_x$, and pristine $Ti_3C_2T_x$, and 20 wt.% Pt/c, and the corresponding Tafel slope of the corresponding samples; (e) the double layer capacitance of different electrodes using CV technique; (f) EIS-Nyquist plot between 10 and 100 kHz range. Source: Jiang et al. [24] / Springer Nature / CC BY 4.0.

transfer highlights the possibilities of making MXene as a composite with metal selenides for multidisciplinary application like energy generation and storage.

A remarkable cyclic stability is again evidence for making TMDCs as a hybrid interface with MXene eventually that improves the HER activity than the individual counterparts. A self-supported non-precious MXene with Ni_3S_2 hybrid

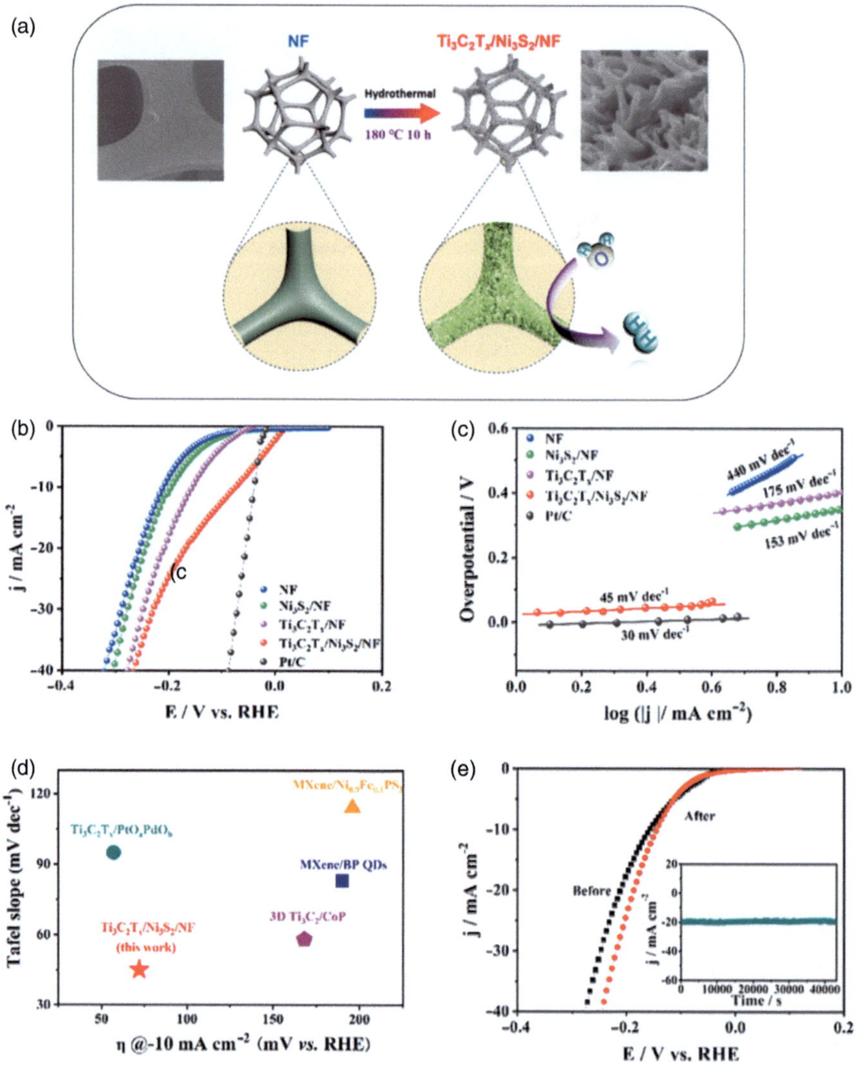

Figure 10.5 (a) Schematics of the formation of $Ti_3C_2T_x/Ni_3S_2$ on NF as a heterostructure electrocatalyst with SEM image of bare NF and $Ti_3C_2T_x$/NF; (b) polarization curve at a scan rate of 5 mV/s; (c) respective Tafel plot of the samples; (d) comparison of η_{10} and Tafel slope with $Ti_3C_2T_x/Ni_3S_2$/NF MXene-based catalysts; (e) polarization curves measured from the $Ti_3C_2T_x/Ni_3S_2$/NF at the scan rate of 50 mV/s before and after 2000 cycles, and the inset shows the chronoamperometric responses (J–t) measured from the $Ti_3C_2T_x/Ni_3S_2$/NF at a constant overpotential of −169 mV. Source: Tie et al. [26] / with permission of American Chemical Society.

electrocatalyst (Figure 10.5a) for efficient HER test has been conducted by Luna Tie and co-workers under alkaline condition. The HER performance was observed using three-electrode set-up with N_2-saturated 1 M KOH, where the sample was coated on nickel foam and used as a working electrode. The results show that the $Ti_3C_2T_x/Ni_3S_2$/NF interface with varying Ni_3S_2 concentrations has been reported (Figure 10.5b–e). The heterostructure displays the best HER performance with lower onset overpotential of 19 mV and a low overpotential of 72 mV achieved at 10 mA/cm^2 current density [26]. The Tafel plots of the respective polarization curve offer an insight into the fundamentals of HER kinetics governed by the Tafel step, in which recombination of adsorbed hydrogen takes place. The self-supported $Ti_3C_2T_x/Ni_3S_2$/NF gives a Tafel slope of 45 mV/dec, which suggests that this catalyst material works depending on the Heyrovsky mechanism. The researcher also reported that the hydrophile nature of the composite does not hold any H_2 on the surface of the electrode. The Tafel slope of the composite shows less value when compared to the other bare and individual counterparts. Table 10.1 shows a comparison of MXene with single and heterojunction MXene for overpotential and Tafel slope value with references.

10.5.4 Characteristics of MXenes in Electrocatalysis

The morphology of MXene plays a vital role in determining the electrocatalytic performance. The MXenes are generally prepared by selective etching technique

Table 10.1 Comparison of MXene-type materials and the electrochemical HER performance of various electrolytes.

Catalyst	Method of synthesis	Electrolyte	Over-potential (mV)	Tafel slope (mV/dec)	Refs.
Single-phase MXene					
Ti_2CTx	Etching	0.5 M H_2SO_4	169	97	[27]
Mo_2CT_x	Acid etching	0.5 M H_2SO_4	283	82	[28]
$V_4C_3T_x$	Chemically Exfoliated 2D Sheets	0.5 M H_2SO_4	200	236	[22]
$Ti_3C_2O_x$	Chemical reaction	0.5 M H_2SO_4	170	100	[29]
$Ti_3C_2O_x$		0.5 M H_2SO_4	190	60.7	[23]
Multi-phase MXene					
$NiS_2@Ti_3C_2T_x$	One-pot hydrothermal	0.2 M KOH	200	37.7	[25]
NiS_2@V-MXene	Hydrothermal synthesis	1.0 M KOH	179	85	[30]
$MoSe_2$@Ti-MXene		0.5 M H_2SO_4	180	91	[31]
$MoS_2@Nb_2CT_x$		0.5 M H_2SO_4	127	56.3	[32]
Co-$MoS_2@Mo_2CT_x$	Annealing process	1.0 M KOH	112	82	[33]
$MoS_2@Ti_3C_2T_x$ nano roll	Liquid nitrogen freezing and subsequent annealing	0.5 M H_2SO_4	152	70	[34]

from its layered MAX precursors, which is generally made of carbides or nitrides. There are several techniques available for etching MAX phase materials, including high-temperature reactions (up to 800 °C) using molten salts, gaseous halides, or etching the A-layer elements in a vacuum. The selective elimination of the A-layer in the MAX phase is a crucial step in the preparation of MXenes. The type of functional groups introduced on the MXene surface depends heavily on the etching process, including factors such as the etchant, etching temperature, and etching time. The hydrofluoric acid (HF) is generally used in the preparation the MXene for the etching of A layer, which contains active functional group may produce steric resistance. A mixture of NaF, KF, CsF, CaF_2, HCl, and H_2SO_4 can also be used as etchers to replace the HF. The direct etching of MXene sheet can be attained by sufficient concentration of HF solvent. Overall, better effect is associated with HF acid concentration, etching temperature, duration, and etching mode, which are quite essential parts of the research [35].

10.6 Conclusion

This chapter elaborated on the fundamentals of MXenes, the impact of defect and interface engineering of their structural characteristics, several preparation strategies, and their applicability in the field of electrochemical HER. The 2D MXenes have gained exclusive interest in the field of water-splitting and energy conversion applications due to their unique architecture that helps in enhanced electrical conductivity, intense redox surface-active sites, and hydrophilicity nature. The electronic structure of MXenes and their electrocatalytic activities can be stabilized using numerous techniques like surface functional group modification, heteroatomic doping, and so on. The defects and interface of the MXenes are carefully designed as a hybrid heterostructure because the MXene provides a platform of exposed numerous active sites, enhancing mass transfer that can ultimately enhance the activity resulting in beneficial HER performance. In spite of their widespread applicability of MXene-based materials in HER, still they have their own challenges, which is a concern and needs to be further explored.

10.7 Future Perspectives

(i) The exploration of new MXene composites, beyond the commonly synthesized materials like Ti_3C_2, Ti_2C, and Mo_2C, is essential for advancing energy conversion applications. The research may focus on expanding the compositional diversity of MXenes, incorporating novel elements or hybrid structures, and investigating their stability and performance in various electrochemical conditions. Thus, this approach will unlock new possibilities for enhancing the efficiency and scalability of MXene-based materials for sustainable energy solutions.

(ii) Strategies to minimize the charge carrier recombination are crucial for enhancing the HER performances of MXenes, which need to be synthetically explored.

(iii) Combining photocatalytic activity with the MXene electrocatalyst may further improve the applicability of MXene-based materials that yield the benefits of both processes and utilize light absorption for charge separation for effective water-splitting applications.
(iv) Electrocatalytic water-splitting reaction can be considered as key research in the current trend, which focuses more on tunning the MXene to boost the electrochemical performance and for enhancing efficient water splitting via stabilizing the surface sensitization by tuning the electronic characteristics that can rapidly promote the electrochemical reactions.
(v) Interdisciplinary approach of utilizing MXenes in various other fields could lead to newer applications and as a result, it can accelerate a huge impact on innovating solutions for various real-world problems.

References

1 Abrishamkar, A., Suárez-García, S., Sevim, S. et al. (2020). Pathway selection as a tool for crystal defect engineering: a case study with a functional coordination polymer. *Appl. Mater. Today* 20: 100632. https://doi.org/10.1016/j.apmt.2020.100632.

2 Jangizehi, A., Schmid, F., Besenius, P. et al. (2020). Defects and defect engineering in Soft Matter. *Soft Matter* 16 (48): 10809–10859. https://doi.org/10.1039/d0sm01371d.

3 Terrones, H., Lv, R., Terrones, M., and Dresselhaus, M.S. (2012). The role of defects and doping in 2D graphene sheets and 1D nanoribbons. *Rep. Prog. Phys.* 75 (6): https://doi.org/10.1088/0034-4885/75/6/062501.

4 Mishra, R.K., Verma, K., and sethi singh, D. (2024). Defect engineering in nanomaterials: impact, challenges, and applications. *Smart Mater. Manuf.* 2: 100052. https://doi.org/10.1016/j.smmf.2024.100052.

5 Zhang, Y., Lin, Y., Duan, T., and Song, L. (2021). Interfacial engineering of heterogeneous catalysts for electrocatalysis. *Mater. Today* 48: 115–134. https://doi.org/10.1016/j.mattod.2021.02.004.

6 Fu, S., Ma, Y., Yang, X. et al. (2023). Defect and interface engineering of hexagonal $Fe_2O_3/ZnCo_2O_4$ n-n heterojunction for efficient oxygen evolution reaction. *Appl. Catal. B Environ. Energy* 333: 122813. https://doi.org/10.1016/j.apcatb.2023.122813.

7 Zheng, G., Chen, M., Yang, C., and Liu, N. (2023). The interaction between interface morphology of particles with anisotropic effect of interface kinetics in an alloy melt. *Cryst. Res. Technol.* 58 (8): 2300030. https://doi.org/10.1002/crat.202300030.

8 Ahmad, A., Khan, S., Chhabra, T. et al. (2024). Synergic impact of renewable resources and advanced technologies for green hydrogen production: trends and perspectives. *Int. J. Hydrogen Energy* 67: 788–806. https://doi.org/10.1016/j.ijhydene.2023.06.337.

9 Barelli, L., Bidini, G., Gallorini, F., and Servili, S. (2008). Hydrogen production through sorption-enhanced steam methane reforming and membrane technology: a review. *Energy* 33 (4): 554–570. https://doi.org/10.1016/j.energy.2007.10.018.

10 Engge, Y., Maulana, F., Nurhuda, M., and Istiroyah. (2021). "Dissociation of water into hydrogen and oxygen through a combination of electrolysis and photocatalyst," *IOP Conf. Ser. Earth Environ. Sci.*, vol. 743, no. 1, https://doi.org/10.1088/1755-1315/743/1/012054.

11 Lima, T.M., Lima, C.G.S., da Costa, V.G. et al. (2024). An overview of hydrogen production from renewable sources and its main applications. *ChemistrySelect* 9 (35): e202401237. https://doi.org/10.1002/slct.202401237.

12 Worku, A.K., Ayele, D.W., Deepak, D.B. et al. (2024). Recent advances and challenges of hydrogen production technologies via renewable energy sources. *Adv. Energy Sustain. Res.* 5 (5): 2300273. https://doi.org/10.1002/aesr.202300273.

13 Dutta, A. (2024). Process considerations for the production of hydrogen via steam reforming of oxygenated gases from biomass pyrolysis and other conversion processes. *Adv. Sustain. Syst.* 8 (9): 2300241. https://doi.org/10.1002/adsu.202300241.

14 Garcia, I., Lopez, G., Santamaria, L. et al. (2024). Biomass source influence on hydrogen production through pyrolysis and in line oxidative steam reforming. *ChemSusChem* 17 (20): e202400325. https://doi.org/10.1002/cssc.202400325.

15 Inocêncio, C.V.M., Holade, Y., Morais, C. et al. (2023). Electrochemical hydrogen generation technology: challenges in electrodes materials for a sustainable energy. *Electrochem. Sci. Adv.* 3 (3): e2100206. https://doi.org/10.1002/elsa.202100206.

16 Tang, Y., Yang, C., Xingtao, X. et al. (2022). MXene nanoarchitectonics: defect-engineered 2D MXenes towards enhanced electrochemical water splitting. *Adv. Energy Mater.* 12 (12): 2103867. https://doi.org/10.1002/aenm.202103867.

17 Huang, B., Zhou, N., Chen, X. et al. (2018). Insights into the electrocatalytic hydrogen evolution reaction mechanism on two-dimensional transition-metal carbonitrides (MXene). *Chem. Eur. J.* 24 (69): 18479–18486. https://doi.org/10.1002/chem.201804686.

18 Zhang, B., Zhu, J., Shi, P. et al. (2019). Fluoride-free synthesis and microstructure evolution of novel two-dimensional $Ti_3C_2(OH)_2$ nanoribbons as high-performance anode materials for lithium-ion batteries. *Ceram. Int.* 45 (7, Part A): 8395–8405. https://doi.org/10.1016/j.ceramint.2019.01.148.

19 Xiao, Z., Xiao, X., Kong, L.B. et al. (2023). MXenes and MXene-based composites for energy conversion and storage applications. *J. Materiomics* 9 (6): 1067–1112. https://doi.org/10.1016/j.jmat.2023.04.013.

20 Shahzad, U., Marwani, H.M., Saeed, M. et al. (2023). Two-dimensional MXenes as emerging materials: a comprehensive review. *ChemistrySelect* 8 (25): e202300737. https://doi.org/10.1002/slct.202300737.

21 Jawaid, A., Hassan, A., Neher, G. et al. (2021). Halogen etch of Ti_3AlC_2 MAX phase for MXene fabrication. *ACS Nano* 15 (2): 2771–2777. https://doi.org/10.1021/acsnano.0c08630.

22 Tran, M.H., Schäfer, T., Shahraei, A. et al. (2018). Adding a new member to the MXene family: synthesis, structure, and electrocatalytic activity for the hydrogen evolution reaction of $V_4C_3T_x$. *ACS Appl. Energy Mater.* 1 (8): 3908–3914. https://doi.org/10.1021/acsaem.8b00652.

23 Jiang, Y., Sun, T., Xie, X. et al. (2019). Oxygen-functionalized ultrathin Ti_3C_2T MXene for enhanced electrocatalytic hydrogen evolution. *ChemSusChem* 12 (7): 1368–1373. https://doi.org/10.1002/cssc.201803032.

24 Jiang, H., Wang, Z., Yang, Q. et al. (2019). Ultrathin $Ti_3C_2T_x$ (MXene) nanosheet-wrapped $NiSe_2$ octahedral crystal for enhanced supercapacitor performance and synergetic electrocatalytic water splitting. *Nano-Micro Lett.* 11 (1): 31. https://doi.org/10.1007/s40820-019-0261-5.

25 Jiang, H., Wang, Z., Yang, Q. et al. (2019). Ultrathin $Ti_3C_2T_x$ (MXene) nanosheet-wrapped $NiSe_2$ octahedral crystal for enhanced supercapacitor performance and synergetic electrocatalytic water splitting. *Nano-Micro Lett.* 11 (1): 1–14. https://doi.org/10.1007/s40820-019-0261-5.

26 Tie, L., Li, N., Chongfei, Y. et al. (2019). Self-supported nonprecious MXene/Ni_3S_2 electrocatalysts for efficient hydrogen generation in alkaline media. *ACS Appl. Energy Mater.* 2 (9): 6931–6938. https://doi.org/10.1021/acsaem.9b01529.

27 Yuan, W., Cheng, L., An, Y. et al. (2018). MXene nanofibers as highly active catalysts for hydrogen evolution reaction. *ACS Sustainable Chem. Eng.* 6 (7): 8976–8982. https://doi.org/10.1021/acssuschemeng.8b01348.

28 Seh, Z.W., Fredrickson, K.D., Anasori, B. et al. (2016). Two-dimensional molybdenum carbide (MXene) as an efficient electrocatalyst for hydrogen evolution. *ACS Energy Lett.* 1 (3): 589–594. https://doi.org/10.1021/acsenergylett.6b00247.

29 Li, S., Tuo, P., Xie, J. et al. (2018). Ultrathin MXene nanosheets with rich fluorine termination groups realizing efficient electrocatalytic hydrogen evolution. *Nano Energy* 47: 512–518. https://doi.org/10.1016/j.nanoen.2018.03.022.

30 Kuang, P., He, M., Zhu, B. et al. (2019). 0D/2D NiS_2/V-MXene composite for electrocatalytic H_2 evolution. *J. Catal.* 375: 8–20. https://doi.org/10.1016/j.jcat.2019.05.019.

31 Huang, J.-J., Liu, X.-Q., Meng, F.-F. et al. (2020). A facile method to produce $MoSe_2$/MXene hybrid nanoflowers with enhanced electrocatalytic activity for hydrogen evolution. *J. Electroanal. Chem.* 856: 113727. https://doi.org/10.1016/j.jelechem.2019.113727.

32 Hu, L., Sun, Y., Gong, S.-J. et al. (2020). Experimental and theoretical investigation on MoS_2/MXene heterostructure as an efficient electrocatalyst for hydrogen evolution in both acidic and alkaline media. *New J. Chem.* 44 (19): 7902–7911. https://doi.org/10.1039/D0NJ00956C.

33 Liang, J., Ding, C., Liu, J. et al. (2019). Heterostructure engineering of Co-Doped MoS_2 coupled with Mo_2CT_x MXene for enhanced hydrogen evolution in alkaline media. *Nanoscale* 11 (22): 10992–11000. https://doi.org/10.1039/C9NR02085C.

34 Liu, J., Liu, Y., Danyun, X. et al. (2019). Hierarchical 'nanoroll' like $MoS_2/Ti_3C_2T_x$ hybrid with high electrocatalytic hydrogen evolution activity. *Appl. Catal., B* 241: 89–94. https://doi.org/10.1016/j.apcatb.2018.08.083.

35 Shen, B., Huang, H., Liu, H. et al. (2021). Bottom-up construction of three-dimensional porous MXene/nitrogen-doped graphene architectures as efficient hydrogen evolution electrocatalysts. *Int. J. Hydrogen Energy* 46 (58): 29984–29993. https://doi.org/10.1016/j.ijhydene.2021.06.157.

11

Defect and Interface Engineering in Electrocatalytic CO_2 Reduction

Narmadha Maharajan[1,2], Sampathkumar Prakasam[1,2], and Suresh Chinnathambi[1,2]

[1] *CSIR-Central Electrochemical Research, Institute, Electrodics and Electrocatalysis Division, Karaikudi, 630003, Tamil Nadu, India*
[2] *Academy of Scientific and Innovative Research (AcSIR), Ghaziabad, 201002, India*

11.1 Introduction

Recently, our Earth has faced two significant issues: global warming and climate change. The reason behind this is the rising population. As the population expands, various anthropogenic activities also increase, which leads to the excessive emission of CO_2 into the atmosphere. This excess CO_2 breaks the natural carbon cycle, bringing global climate change and environmental deterioration. Electrocatalytic CO_2 reduction reaction (CO_2RR) represents a significant methodology for transforming greenhouse gas CO_2 into commercially valuable products, thereby facilitating the achievement of carbon neutrality. The conversion process is quite difficult because the CO_2 molecule is thermodynamically more stable, it has a strong C=O bond energy (bond energy 750 KJ/mol) [1]. Hence, the thoughtful design of electrocatalysts aimed at CO_2 reduction holds considerable importance. The important issue in the electrocatalytic CO_2 conversion is the ability to identify and recognize the relationship between reaction intermediate adsorbed states on active sites and reaction kinetics. Defect chemistry plays a crucial role in increasing the inherent catalytic performance of electrocatalysts through tuned electronic and surface properties. The defects present widely in the nano electrocatalysts are commonly considered as active sites for the electrocatalytic process. Through defect engineering, the catalytic inertia associated with idealized structures is overcome, as it introduces distinctive electronic configurations and physicochemical characteristics to electrocatalysts, thus enhancing their catalytic performance. Consequently, defect engineering has become a widely embraced strategy for enhancing their electrocatalytic activities [2]. Recently, a diverse array of defective nanomaterials has been extensively investigated, revealing their considerable promise in the context of CO_2RR. There exists an imperative to deepen our understanding of the influence of defects on catalytic efficiency. Various defect types encompass metal compounds (such as vacancies, grain boundaries and lattice imperfections)

Surface and Interfacial Defects in Nanomaterials for Sustainable Energy Production and Storage, First Edition. Edited by Noé Arjona, Lorena Álvarez Contreras, and Minerva Guerra Balcázar.

and carbon materials (including intrinsic defects, doping of heteroatom, and individual-metal-atom sites) [3]. Poor selectivity and large overpotential requirements are the two major issues while facing reduction of CO_2. This can be shortened by adjusting heterogeneous electrocatalysts to control their size, morphology, grain boundaries, crystal facets, and surface defects/vacancies [4]. In this chapter, various methods available for creating defects and the impact of interface engineering in carbon dioxide reduction reaction (CO_2RR) are discussed in detail.

11.2 Types of Defects

Based on the dimension, defects on the solid material are categorized into different types such as zero-dimensional (0D) point defects that include (vacancy, reforming, and doping), one-dimensional (1D) line defects like dislocation, two-dimensional (2D) planar defects like grain boundaries, and three-dimensional (3D) volume defect, e.g. spatial lattice disorder and void (Figure 11.1).

Specific defects are described as follows:

- Zero-dimensional (0D) point defect: This includes vacancy and doping. Vacancies are created by the missing atoms in the lattice position. Insertion of foreign atoms into the lattice position of the host material is known as doping.
- One dimensional (1D) line defect: Dislocation.

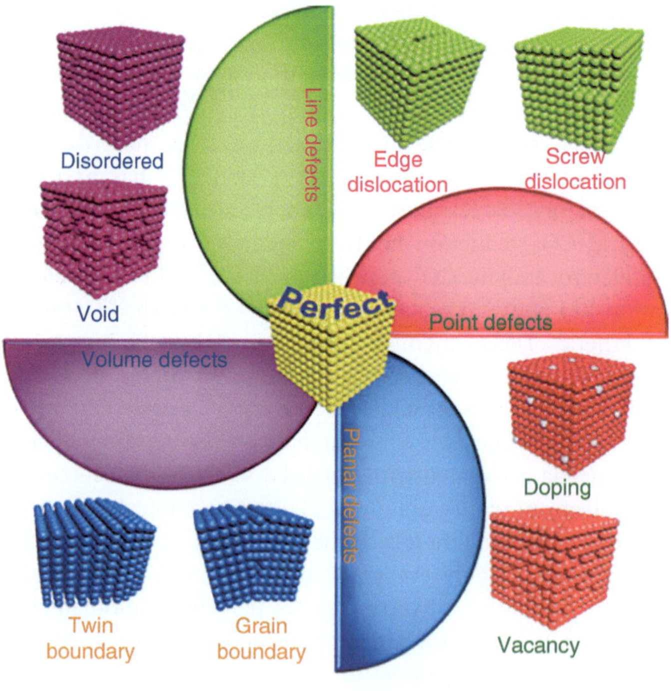

Figure 11.1 Schematic illustration of types of defects. Source: Zhang et al. [5] / with permission of Elsevier.

Dislocations refer to localized irregularities in atomic arrangements, which can be seen as the interface between the deformed and non-deformed sections of the crystal. The dislocation is further classified into two types such as edge and screw dislocation. Edge dislocations possess a Burgers vector that is oriented vertically to the displacement line, while screw disarticulations align equivalent to the direction of crystal displacement.

- Two-dimensional (2D) planar defect: Grain boundary.

Grain boundaries are regions where adjacent grains meet, resulting in atomic mismatches due to differing crystallographic orientations. Grain boundaries are classified into three types: (i) tilt boundary, (ii) twist boundary, and (iii) twin boundary. Tilt boundaries consist of associated edge dislocations, while screw borders comprise regions of screw dislocations. Twin boundaries can be formed by material deformation, displaying mirrored atomic arrangements across the boundary.

- Three-dimensional (3D) volume defect: Spatial lattice disorder and void defect.

Lattice disorder occurs when the regular organization of atoms is disrupted, leading to the formation of chaotic matrixes. Voids are minor areas characterized by a lack of numerous atoms, resembling the accumulation of several vacancies within bulk materials. The development of defects is largely influenced by the production process, excluding natural defects. In theory, the aforementioned defects can be precisely adjusted by modifying reaction parameters [6].

11.3 Methods to Create Defects

Various methods are available to create defects that include chemical reduction, thermal treatment, plasma etching, elemental doping, phase transformation, ion-exchange reaction, ball milling, lithium-induced conversion, and microwave-assisted hydrothermal reaction.

11.3.1 Thermal Treatment

Thermal treatment is an effective technique for creating defects by adjusting the reaction environment like introducing inert, reducing, or mixed gases and altering the reaction temperature. Various atmospheres and temperatures result in differing effects on defect formation. Treatments in inert atmospheres (e.g. Ar, N_2) generate surface defects and support the doping of foreign atoms. Additionally, these treatments can safeguard samples from decomposition at inappropriate temperatures. Experimental findings indicated that the presence of superficial imperfections and heteroatom doping was beneficial for generating abundant active sites, adjusting the electronic structure, and improving the electrochemical conductivity of the materials. Moreover, hydrothermal and solvothermal techniques are also categorized as thermal treatments.

11.3.2 Plasma Etching

For creating defects, the plasma etching method is considered to be a modest, rapid, and environmental technique. The bonds (e.g. ionic, hydrogen, and covalent bonds) within the host layer can be devastated by using this method which exhibits the deficiency of surface atoms from the lattice and the creation of diverse vacancies. However, diverse plasma atmospheres such as Ar, O_2, NH_3, and O_3 result in dissimilar effects on the materials.

11.3.3 Elemental Doping

Elemental doping is considered the best technique to alter the defect arrangement and improve the properties of metal surfaces. Doping can be attained by adopting two strategies: (1) diffusion doping and (2) ion exchange. Elemental doping plays a vital role in boosting the electrocatalytic CO_2 reduction reaction by regulating the band structure.

11.3.4 Defects Creation by Using Chemicals

Defects can also be created through chemical reduction. Reducing agents such as glycerol, $NaBH_4$, N_2H_2–H_2O, and ethylene glycol are used to create defects. $NaBH_4$ is used to generate imperfections, particularly in creating vacancies and increasing spacing interlayers.

11.3.5 Phase Transformation

The rearrangement of atoms is known as phase transformation, which facilitates crystal defects. For example, MoS_2, ZnS, and SnS_2 with plentiful S vacancies can be formed through phase alteration with their equivalent precursors.

11.3.6 Ion Exchange Membrane

An ion exchange membrane is used to fabricate defect-rich metal sulfides which result in porous structure and large defect sites [5].

11.4 Characterization of Defects

Identification of defects in the crystalline electrocatalysts is ideal for calculating the efficiency. These various physicochemical techniques are thus employed for identification. The complicated details of the crystalline structure of various materials can be analyzed using XRD. The deviation from the ideal crystal structure indicates changes in the diffraction pattern due to the presence of defects. Raman spectroscopy is a powerful technique for identifying defects and unwanted particles through changes in the position of the peak. Nuclear magnetic resonance (NMR) has been used to identify the defects in both organic and inorganic materials. This technique gives information based on the specific indigenous atmosphere of nuclei within a given material. Electron energy loss spectroscopy (EELS) is a highly sensitive technique that provides the best approach to investigating electronic structures and distinguishing spot defects by analyzing dissimilar spectrum signals. Further, X-ray Photoelectron Spectroscopy (XPS) was utilized for measuring the kinetic energy and intensity of emitted photoelectrons the surface impurities and defects were identified. Fourier transform infrared spectroscopy (FTIR) is another prevailing technique used to identify the molecular structure and functional group present in the material. Further, scanning tunneling microscopy (STM) and Auger electron spectroscopy (AES) act as precious tools for analyzing and characterizing atomic-scale defects and surface defects and impurities. In addition to all these precious instrumentation techniques, the utilization of density functional theory (DFT) is the most powerful approach to identifying by calculating the binding strength of the intermediates on the active sites using DFT [7].

11.5 Defect Engineering in Metal Electrocatalysts

Imperfect catalysts generally have distinctive electronic structures. The presence of defects can create binding sites for the reaction intermediate and facilitate the adsorption of reactants by enhancing their conversion efficiency. The rational creation of suitable defects is essential to improve the catalytic activity. The charge spreading and the adsorption of intermediate on a defective site depend on its microstructure, which holds considerable importance for the CO_2RR process. Based on the microstructures, the defect on the metal compound can be classified into lattice defects, grain boundaries, vacancies, and so on [3]. The defect in metal complexes can be introduced by creating oxygen vacancies through chemical reduction and thermal treatment. Generally, the defects on metal complexes include anion vacancies and cation vacancies. These vacancies can enhance catalytic performance through increased defect density. Also, another defect like grain boundaries and lattice distortion serves as active sites for catalysis [11]. Nowadays the facet dependent modification paves the way to achieve C_1 and C_2 products with higher efficiency. Copper is the only metal that effectively catalyzes CO_2RR to C_2+ products due to its specific binding energy with the *CO intermediate. Dazhong et al. coupled Cu (100) and Cu (110) facets that show 87% Faradaic efficiency (FE) for C_2 products [12].

11.5.1 Impact of Creating Vacancies in CO_2RR

Atomic vacancy represents the unoccupied electron site that arises from the loss of an atom from the surface of a perfect crystal. The presence of atomic vacancies on the crystal surface can trigger the reactant molecules by refining their adsorption strength. The most common types of atomic vacancy include oxygen vacancies, sulfide vacancies, selenium vacancies, and tellurium vacancies, which significantly affect the electronic structure of their neighboring atoms and create abundant active sites with short coordination numbers. Thereby increasing the availability of active sites for catalytic reactions and consequently improving the efficiency of electroreduction processes [11]. The presence of abundant zinc vacancy in ultrathin $ZnIn_2S_4$ nanosheets boosted the activity and selectivity of CO_2RR to formate with an FE of 94% [13].

11.5.1.1 Role of Anion and Cation Vacancy for Generation of Defects

Among various types of anion vacancies, oxygen vacancies are particularly prevalent due to their minimum formation energy, which can significantly alter the surface electronic configuration. Several superficial and governable approaches have been employed to engineer the formation of oxygen vacancies within the metal oxides. Wu, et al. synthesized defect-rich Bi/Bi_2O_3, which exhibits 90% FE for CO_2 to formate conversion. They specifically employ oxygen vacancy as a main type of defect in the Bi/Bi_2O_3 nanosheet. These vacancies and broken local spatial symmetry of the componential combination of Bi/Bi_2O_3 make structural disorders leading to the formation of under-coordinated sites. These sites further contribute to the catalytic activity by providing more active sites for the reaction [14]. Qi et al. synthesized an oxygen defect perovskite $(La_{0.5}Sr_{0.5})_{1-x}Fe_{0.9}Mn_{0.1}O_{3-\delta}$ oxide by hosting A-site cation deficiency. This A-site shortage leads to an increase in surface oxygen vacancies which significantly enhances the material's ability to adsorb CO_2 [15]. The creation of oxygen vacancy on the catalyst surface significantly improves the ability to adsorb CO_2. Not only the oxygen vacancy but the introduction of S vacancies also lowers the limiting potential [4]. Cation vacancy can significantly distress the characteristics of metallic compounds due to their exclusive electrical and orbital configuration. The limited focus has been devoted to cation vacancy due to their higher formation of energies compared with anion vacancy [6]. Hong et al. synthesized ZnS and proposed that the creation of cation vacancy (V_{Zn}) serves as new active sites and achieves formate with FE of >85%. The presence of Zn vacancy greatly lowers the energy barrier of CO_2RR and also suppresses the hydrogen evolution reaction (HER) (Figure 11.2a) [8].

11.5.2 Influence of Lattice Defects in CO_2RR

Lattice imperfection including lattice dislocations, expansion, and distortion has the potential to significantly modify the properties of metallic compounds. The unique electronic architecture and improved electrical conductivity offered by lattice imperfections facilitate the charge transfer throughout the CO_2RR, thereby enhancing catalytic activity [3]. For example, the research group led by Zhang developed in situ AgCo surface alloy electrocatalysts through a cold H_2-plasma activation technique. The resultant alloy demonstrated a remarkable FE of 72.3% during CO_2RR [16]. Both experimental and DFT analyses show that the distortion of the Ag lattice diminished the energy barrier for the formation of intermediates and improved C–C coupling. Chang et al. fabricated Fe nanoparticle-decorated (Ce, La, Sr) $(CrFe)O^{3-\delta}$ perovskite (Fe-CLSCF) was obtained by reducing the A-site-

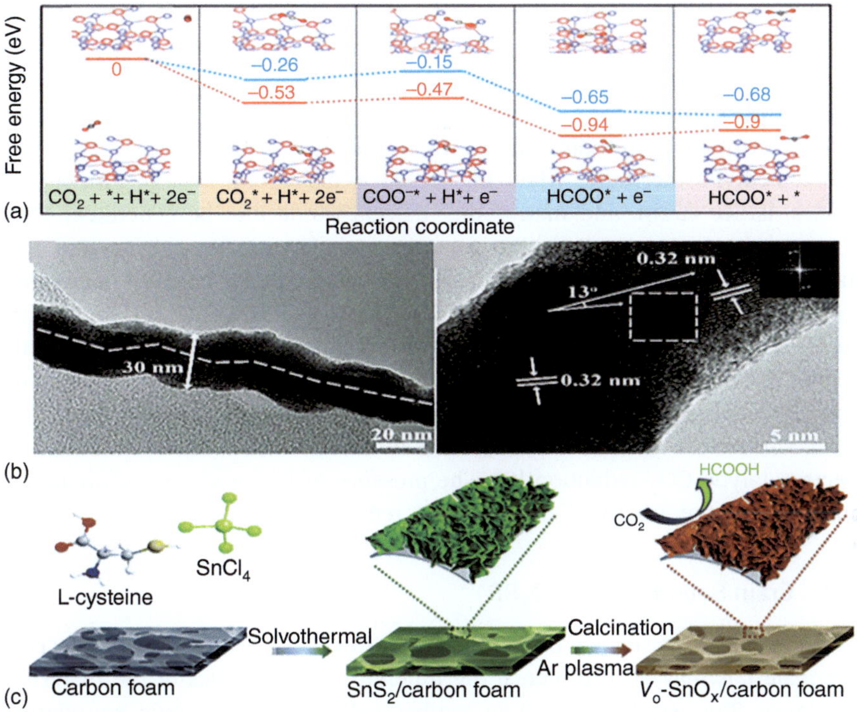

Figure 11.2 (a) Representing the free energy diagram pathways of CO_2 into formate on ZnS and VZn-ZnS, (b) TEM and HRTEM images of BiNW with grain boundary and crystal lattice dislocation, and (c) the fabrication process of oxygen-vacancy-rich SnO_x-carbon foam. Source: (a) Pang et al. [8] / with permission of American Chemical Society, (b) Zhang et al. [9] / with permission of Royal Society of Chemistry. (c) Li et al. [10] / with permission of Royal Society of Chemistry.

deficient $(Ce_{0.08}La_{0.52}Sr_{0.3})Cr_{0.5}Fe_{0.5}O_{3-\delta}$. Here the combination of lattice defects and Ce-doping on the A-site creates a synergetic effect. This results in positive oxygen vacancy and negative La vacancies, which together enhance the electrochemical performance and convert CO_2 to 97.2% FE of CO [17]. Lattice strain and lattice disorder are the two major lattice defects.

11.5.2.1 Lattice Strain

The phenomenon of lattice strain is predominantly associated with the ductile or compressive effect of heterogeneous atoms on the bulk atomic structure, which significantly impacts electrocatalytic reactions by modifying both the surface electronic structure of the catalysts and the adsorption capacity of the adsorbate [11]. Yang et al. constructed a hybrid electrocatalyst, Bi_2S_3–Bi_2O_3, with a large amount of defect (such as oxygen vacancies and lattice distortion) and abundant active sites. The resultant catalyst converts CO_2 into formate with 90% FE [18]. Shuang et al. also fabricated twisted bismuth nanowires with lattice dislocation on Cu foam. The resultant catalyst shows formate with an FE of 95% [19]. However, the introduction of lattice strain decreases the energy barrier for the intermediate formation.

11.5.2.2 Lattice Disorder

The lattice disorder is an indication, in which the periodic arrangement of atoms is interrupted. Furthermore, when the lattice is dislocated, a large number of vacancies, lattice strain, and many other defects would also be generated. This may create ample unsaturated sites for the catalytic process and increase the overall activity [11]. Xiaolong et al. developed lattice dislocated bismuth nanowires on copper foam (Cu foam@BiNW) through in situ electrochemical reduction (Figure 11.2b). The resultant catalyst shows Formate generation with an FE of 95%. This is because of the high inherent activity arising due to the existence of crystal lattice dislocations on the twisted Bi nanowires. Additionally, the large catalytic surface area with the porous structure helps to improve the efficiency of electrocatalytic CO_2 reduction [9]. The presence of lattice dislocation helps to stabilize the reaction intermediate in the CO_2 reduction process.

11.5.3 Grain Boundaries in CO_2RR

Beyond the role of vacancies, the grain boundaries (GBs) found within polycrystalline materials may also influence the inherent characteristics of the material. For instance, Tang et al. successfully synthesized many copper twin boundaries on polished copper electrodes. The atoms at grain boundaries substantially donate for the production of methane. Both experimental findings and DFT analyses show that the twin boundaries simplify the alteration of the absorbed CO* into CH_4. Twin boundaries in Cu FCC reverse stacking order forming a coherent twin boundary pattern. This structure provides low interfacial energy and enhanced properties [20]. Ying et al. proposed that GBs contribute to lowering the reaction energy of the potential limiting step in CO_2RR [4]. Defect control in metal–organic frameworks (MOFs) also enhances their catalytic performance in CO_2RR. Juan et al. synthesized CuTrZ (HTrz = 1H, 1, 2, 4) MOF with dissimilar sizes by directing the reaction kinetics. They claim that small CuTrz nanostructures are polycrystalline and rich in grain boundaries, which contribute to their superior performance in CO_2RR compared to larger, single-crystalline counterparts. It demonstrates the highest ethylene FE of 55.4 and 81.8% of C_2 products [21]. Hence, these findings imply that grain boundaries facilitate the charge density distribution, modify the resident electronic environment, and enhance the electrocatalytic performance.

11.6 Effect of Surface Defect Sites on CO_2RR

Defects play an important role in all solid-state materials. There are different strategies such as particle size, morphology, facet, lattice distortion, defect, copper vacancies (Cuv), or oxygen vacancies (Ov), and surface states of the elements have been adopted to enhance the electronic properties of the electrocatalysts. Surface defect sites are structural disorders that can be prudently introduced via numerous engineering strategies that influence site-specific area activities in the CO_2RR. Apart from the surface defect, most of the materials have native defects, which are diligently involved in catalytic activities. Moreover, additional defects such as composition, etching, pretreatment, and doping can be introduced to a catalyst to increase the performance. Generally, crystal defects and interfacial structure may exert a dynamic outcome on catalytic reactions by creating additional active sites for CO_2 adsorption.

11.6.1 Impact of Defects in Single/Bimetallic Materials

The single and bimetallic transition metals bring a greater significance for high product selectivity and stability. Among them, copper is considered to be the best catalyst for the reduction of CO_2 into the various C_2 hydrocarbons [4]. Selectivity is an important concept in CO_2RR, and the selectivity of the chosen hydrogenated product is inclined by proton concentration close to the electrode, the morphology, crystal structure, and the size of Cu-based materials. Jiwu et al. synthesized a bimetallic $A_2B_2O_7$ pyrochlore-type semiconductor for the conversion of CO_2 to HCOOH. Here the presence of more negative d-band levels of Bi and Sn bimetallic sites, which make weak coordination with O atoms in the sub-layer of $A_2B_2O_7$ oxides, was thermodynamically beneficial to the HCOOH production with 93% FE through asymmetric coordination activation [22]. Khedr et al. synthesized hydrogen interstitial (H-i) defects in iron pyrite (FeS_2). These hydrogen interstitials act as n-type dopants, which contribute extra electrons to the material, improving its electronic properties. The incorporation of H-i is shown to facilitate the methanol pathway in CO_2 reduction, thereby improving the efficiency of the reaction by inhibiting competing reactions like HER [23].

11.6.2 Defect-Induced Metal Alloys for CO_2RR

Nowadays alloying is one of the best options due to its excellent catalytic performance. Because of alloying, there has been a great change in the binding energies and stabilization degrees. The Ag–Co alloy with lattice distortion had the greatest FE for 72.3% for ethanol conversion. Introducing Co. into the Ag surface lowered the energy barrier for $*CO_2^{\delta-}$ creation, which was the crucial rate-determining step for the CO_2RR to ethanol [17]. Shahid et al. proposed that the Indium–Copper alloy selectively converts CO_2 to CO with 95% FE. Indium preferentially occupies edge sites on the Cu surface, which are critical for the electrochemical reactions. This positioning helps in stabilizing the COOH intermediate, a key species in the CO_2 reduction pathway, while simultaneously weakening the adsorption of H [24]. Hence, alloying can enhance the selectivity of the catalyst for CO_2RR.

11.6.3 Impact of Defects on Metal Oxides and its CO_2RR

Junliang et al. synthesized Au nanocrystals @ defective amorphous MnO_2 nanosheets core/shell nanostructure (Au NCs@a-MnO_2 NSs). Because of the tunable lateral size, the synthesized electrocatalysts exhibit excellent CO_2RR to CO with an FE of 90.5%. The presence of defective amorphous MnO_2 nanosheets used in the core/shell structure exhibits a high density of oxygen vacancies. This characteristic is attributed to the disordered structure of amorphous materials, which facilitates better interaction with CO_2 [25]. Jiazheng et al. reported that the creation of oxygen vacancy on Bi/Bi_2O_3 also enhances the catalytic activity and selectivity for CO_2 reduction to formate with more than 90% FE [26]. Level et al. developed a nickel-doped indium oxide nanocrystals confined in the carbon nanofiber (Ni-In_2O_3@C NFs) to produce formate with FE of 93.6%. The presence of nickel alters the adsorption energies of reaction intermediates, particularly OCHO*, which favors the formation of formate during the CO_2 reduction process [27]. Wang et al. introduced oxygen vacancy in In_2O_3 to improve the CO_2RR rate. The introduction of oxygen vacancy facilitates the significance of increasing the electrochemically active surface area and enhancing the intrinsic activity of the In_2O_3 catalyst [28].

11.6.4 Surface and Defect Aspects in Metal Sulfide Electrocatalysts

Metal sulfides (MS) have increasingly attracted scholarly attention in the domain of energy translation and storage due to their tunable electronic structures, semiconducting properties, and exceptional redox reversibility. Consequently, the incorporation of defects into MS is chosen to be an auspicious way to make extremely competent materials for energy conversion [5]. Concurrently, the engineering of defects in metal sulfides including vacancies and metal/heteroatom doping, enhances the electrocatalytic performance. This will cause increased current densities, FEs, and active reaction sites and also diminished work function in contrast to the original catalysts [4]. $CuInS_2$ nanofibrous structure was developed by Abdalaziz et al. through an electrospinning technique. The presence of sulfur increases the active surface area which is essential for improving FE and also the sulfur plays a role in stabilizing the nanofibrous structure. The resultant catalyst produces CO with 77% of FE [29].

11.6.5 Role of Defects in Metal Nitrides for CO_2RR

Transition metal nitrides show excellent catalytic performance in CO_2RR due to their excellent mechanical/electrochemical stability and metal-like properties. Yi et al. developed a metal oxide-supported N-CNT (MOx/N-CNT), which produces CO with 90.3% FE. The presence of N defect on N-CNT plays a significant role in providing local proton, which assists in the CO_2RR [30]. In general, N defects significantly enhance the performance of metal oxide catalysts in CO_2 reduction by providing active sites, stabilizing reactive species, facilitating electron transfer, and assisting with proton availability. The introduction of defects in MXenes also enhanced the catalytic performance in CO_2RR. The vacancy formed by removing X atoms (such as carbon or nitrogen) from the MXene also affects the electronic structure [31]. Wenzhe et al. prepared lattice strain calmed N-doped Cu (LSN-Cu) with rich defect sites and strong nitrogen incorporation via a facial N_2 plasma treatment. The resultant catalyst shows n-propanol formation with 54% FE. The compressive lattice strain introduced during the N_2 plasma treatment stabilizes nitrogen doping and enhances C–C coupling efficiency, which is important for achieving optimal C_3 production [32].

11.7 Impact of Defects in Carbon-Based Materials for CO_2RR

Carbon materials are extensively used in electrocatalysts due to their long stability, vast surface area, electrical conductivity, etc. Defect engineering in carbon material serves as an effective technique for modifying surface properties, thereby enhancing their activity in catalysis. Defects can be created through alkali activation, which can create defect edges in carbon aerogel but cost-effective methods are needed. High-temperature treatment is also used to create defects in carbon by introducing foreign atoms into graphene, which disturbs its structure. Generally, defects can be classified into intrinsic defects and those introduced externally [8]. Hu et al. synthesized Fe, N-Co-doped porous carbon sheets (rGO-PVP-Zeolitic imidazolate frameworks [ZIFc]) by pyrolyzing Fe-ZIF-8 and GO in the existence of PVP. The ZIF nanoparticles on the rGO surface shrink and transform into ZIF-derived carbon nanoparticles. This transformation leads to the creation of abundant macropores and mesopores, which are essential for enhancing the surface area and accessibility of active sites. The resultant porous carbon sheets exhibit a high FE of 98.6% for CO production at a specific potential, attributed to the effective defect mechanism that optimizes the active sites [33]. Hongqiang et al. prepared carbon foam-supported oxygen vacancy-rich SnO_X nanosheets for electrocatalytic reduction of CO_2 to formate with 86% selectivity (Figure 11.2c). Here the interconnected macroporous nature of carbon foam significantly improves the mass transfer of CO_2 and electrolyte to the active sites [10]. Compared with pristine CNTs, the incorporation of graphitic and pyridinic nitrogen moieties significantly reduces the overpotential and enhances the selectivity (approximately 80%) for the production of carbon monoxide (CO) [34]. Hence, the defect regulation in carbon material can significantly enhance the intrinsic conductivity of the material and promote electron transfer within the material.

11.8 Intrinsic Defect

The intrinsic defects are most common in carbon materials, which are made by the removal of atoms and lattice distortion. The symmetrical conformation of the inherent defects comprises of edges, vacancies, holes, topological defects, etc. These defects make the carbon materials full of unpaired π electrons, which facilitates the local electron density, quickens electron transfer, and drops the formation of energy barriers of key intermediates [3]. Heteroatom-doped carbon material especially N-doped carbon has been widely used as an efficient metal-free carbon-based catalyst for CO_2RR. The intrinsic carbon defects, especially sp^2 defects (octagonal and pentagonal), significantly enhance ECO_2RR performance, suggesting that these defects may serve as active sites. Wei et al. synthesized a sequence of imperfect N-doped carbon spheres (D-NC-X). The CO_2 reduction reaction of the catalyst is interrelated with the defect concentration but adversely correlated with the N content. The D-NC-X with the lowest N-content shows 94.5% FE for CO production. However, these results were controversial with the theory that N heteroatom sites are the active sites for CO_2RR. These findings insist on synthesizing a series of defective carbon catalysts without N heteroatom doping (D-C-X). This catalyst shows a positive correlation between ECO_2RR activity and the concentration of intrinsic carbon defects. The elimination of N heteroatom primes to the renovation of the remaining carbon atom, a process that leads to the creation of defects such as pentagonal and octagonal carbon atom arrangements and considerable carbon disorder [35]. Li-Li Ling et al. reported that potassium assisted fabrication of intrinsic defects in porous carbon. The K^+ defect –C achieves the ultrahigh CO FE of 99% [36]. Hence, the creation of intrinsic defect promotes the large number of active sites for CO_2RR.

11.9 Single-Metal Atom Sites

Recently catalysts comprising of single-metal atom have drawn much consideration because of their high atomic consumption rate and excellent catalytic activity. In contrast to active sites that are characterized by intrinsic defects or heteroatom doping, these distinctively isolated metallic entities are regarded as directly accessible active sites. The bulk material may be converted into nanoparticles, sub-nanoclusters, and ultimately single-metal atoms. This reduction of size results in an increase in the unsaturated coordination environment surrounding the metal species. Consequently, the surface area of the metallic component expands, rendering the metal sites increasingly reactive for chemical interactions with both support and adsorbates, thereby explaining the size-dependent phenomena observable in metal nanocatalysts. In the extreme scenario of the single-atom catalysts, the elevated reactivity of valance electrons, the quantum confinement of these electrons, and the sparse quantum energy levels associated with metal atoms lead up to the maximization of the surface free energy of the metallic species, thereby facilitating chemical interactions with the support and give rise to distinct chemical properties characteristics of SACs [37]. The performance of a single-atom catalyst can be influenced by various factors including the support material, loading amount, coordination numbers, and the presence of defects. The vacancy defect Ni-single-atom catalyst reduces CO_2 to CO with an FE of 90%. The incorporation of vacancy at the Ni–N_3 site plays a key role in boosting CO_2 reduction by diminishing the energy barriers associated with the formation of COOH* intermediate and the desorption of CO [38]. Metal–nitrogen–carbon (M-N-C) SACs have been widely examined in heterogeneous catalysis for CO_2RR due to their coordination and electronic state. The coordination number and structural conformations of N and C in M-N-C SACs can be systematically deployed through the introduction of defects.

The D-band center of metals inside a coordinately lacking M-N-C single-atom catalyst with improved electronic structure was earlier to the fermi level and possessed a better density of local electrons. That can quicken the electron transfer and optimizing the absorption and desorption dynamics between the electrocatalyst and the reactant [39].

11.10 Challenges and Perspectives in CO_2RR

The major challenges faced by CO_2RR include mass transport limitation, reaction efficiency, adsorption issues, electrode stability, selectivity, and overpotential. The transport of CO_2 to the catalyst surface is primarily through diffusion in a non-flowing electrolyte, which can limit the reaction efficiency. The efficiency of CO_2 reduction is influenced by factors such as binding tendencies, electron–proton coupling, and the solubility of CO_2. This factor can affect the formation of desired products. Strong adsorption of CO on metal surfaces (e.g. Cu) can lead to further reduction into C_1 and C_2 products. Weak adsorption (e.g. Zn, Ag, Au, Pd) may result in CO desorption without further reduction. Achieving high selectivity for desired products and minimizing overpotential requirements are significant challenges in the electrochemical CO_2RR process. The catalyst efficiency is mainly affected by sluggish reaction kinetics, high overpotentials, and poor selectivity due to the similar reduction potentials of the various reduction products. The electrode stability is also considered to be a key performance indicator for evaluating a catalyst's long-term applicability. Owing to the generation of byproducts, the electrolyte medium is termed to be a primary source of impurities in the working electrode. More often, the produced intermediates or byproducts are adsorbed on the electrode surface which restricts further reaction and affects the electrocatalyst performance [4]. These challenges are overcome by adapting suitable defect creation on the catalyst surface (Figure 11.3).

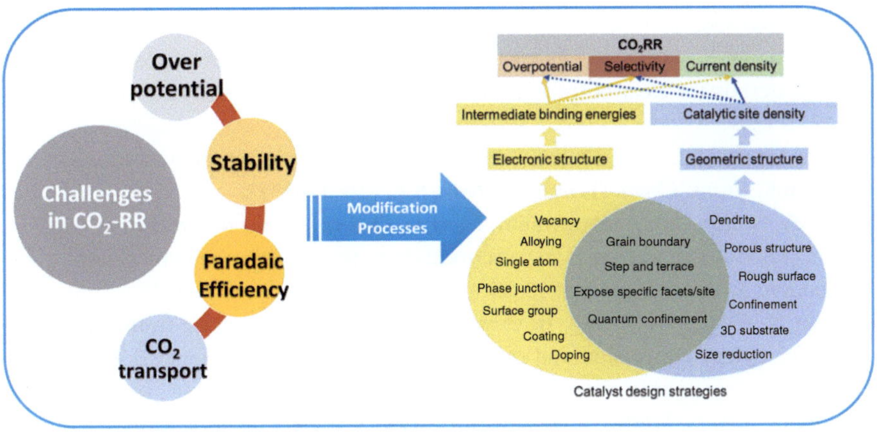

Figure 11.3 Schematic illustration of practical challenges and factors affecting CO_2 reduction.

11.11 Conclusion

The defect engineering in the catalyst that is employed for CO_2RR has gained more attention in recent years. These defect engineering can significantly alter the physicochemical properties of the catalyst by tuning the surface and making substantial improvements in catalytic performance. By creating an appropriate defect on the catalyst surface, we can overcome the challenges faced during CO_2RR. In this chapter, the various strategies for defect creation are explained in detail. The introduction of oxygen vacancy on the catalyst surface improves the adsorption ability of CO_2. The lattice strain decreases the energy barrier for the intermediate formation and also helps to stabilize the reaction intermediate on the catalyst surface. The grain boundaries favor charge density distribution or modify the local electronic environment. The alloying can also play a significant role in defect creation by enhancing the selectivity of CO_2RR. Moreover, defect engineering in metal surfaces alters the electronic structure of the catalyst and promotes CO_2 conversion with more efficiency. The defect creation in non-metallic catalysts also more significantly alters the electronic structure and facilitates the CO_2 conversion.

Acknowledgement

Narmadha Maharajan and Sampathkumar Prakasam gratefully acknowledge the Department of Science and Technology for Inspire Fellowship (IF-210717 & IF-190544). Suresh Chinnathambi acknowledges Dr. K. Ramesha, Director, CSIR-Central Electrochemical Research Institute, for the support.

References

1 Ji, Y., Du, J., and Chen, A. (2022). Review on heteroatom doping carbonaceous materials toward electrocatalytic carbon dioxide reduction. *Trans. Tianjin Univ.* 28 (4): 292–306.
2 Jia, Y., Jiang, K., Wang, H., and Yao, X. (2019). The role of defect sites in nanomaterials for electrocatalytic energy conversion. *Chem* (Elsevier Inc.) 5: 1371–1397.
3 Li, H., Li, R., Niu, J. et al. (2022). Defect chemistry of electrocatalysts for CO_2 reduction. *Front. Chem.* (Frontiers Media S.A.) 10: 1067327.
4 Balu, S., Hanan, A., Venkatesvaran, H. et al. (2023). Recent progress in surface-defect engineering strategies for electrocatalysts toward electrochemical CO_2 reduction: a review. *Catalysts* 13 (2): 393.
5 Zhang, Y., Zhang, Y., Zhang, H. et al. (2021s). Defect engineering in metal sulfides for energy conversion and storage. *Coord. Chem. Rev.* (Elsevier B.V.) 448: 214147.
6 Wang, Q., Lei, Y., Wang, D., and Li, Y. (2019). Defect engineering in earth-abundant electrocatalysts for CO_2 and N_2 reduction. *Energy Environ. Sci.* (Royal Society of Chemistry) 12: 1730–1750.
7 Muhammad, P., Zada, A., Rashid, J. et al. (2024). Defect engineering in nanocatalysts: from design and synthesis to applications. *Adv. Funct. Mater.* (John Wiley and Sons Inc) 34: 2314686.

8 Pang, H., Meng, X., Li, P. et al. (2019). Cation vacancy-initiated CO_2 photoreduction over ZnS for efficient formate production. *ACS Energy Lett.* 4 (6): 1387–1393.

9 Zhang, X., Sun, X., Guo, S.X. et al. (2019). Formation of lattice-dislocated bismuth nanowires on copper foam for enhanced electrocatalytic CO_2 reduction at low overpotential. *Energy Environ. Sci.* 12 (4): 1334–1340.

10 Li, H., Xiao, N., Wang, Y. et al. (2020). Promoting the electroreduction of CO_2 with oxygen vacancies on a plasma-activated SnO: X/carbon foam monolithic electrode. *J. Mater. Chem. A* 8 (4): 1779–1786.

11 Chen, Y., Chen, C., Cao, X. et al. (2023). Recent advances in defect and interface engineering for electroreduction of CO_2 and N_2. *Acta Phys. -Chim. Sin.* (Chinese Chemical Society) 39: 2212053.

12 Zhong, D., Zhao, Z.J., Zhao, Q. et al. (2021). Coupling of Cu(100) and (110) facets promotes carbon dioxide conversion to hydrocarbons and alcohols. *Angew. Chem. Int. Ed.* 60 (9): 4879–4885.

13 Wang, Z., Qi, R., Liu, D. et al. (2021). Exfoliated ultrathin $ZnIn_2S_4$ nanosheets with abundant zinc vacancies for enhanced CO_2 electroreduction to formate. *ChemSusChem* 14 (3): 852–859.

14 Wu, D., Huo, G., Chen, W.Y. et al. (2020). Boosting formate production at a high current density from CO_2 electroreduction on defect-rich hierarchical mesoporous Bi/Bi_2O_3 junction nanosheets. *Appl Catal B.* **271**:118957.

15 Qi, J., Bian, L., Ting, T. et al. (2023). Boosting electrochemical CO_2 directly electrolysis by tuning the surface oxygen defect of perovskite. *J. Power Sources.* 570:233032.

16 Zhang, Q., Tao, S., Du, J. et al. (2020). A cold plasma-activated situ AgCo surface alloy for enhancing the electroreduction of CO_2 to ethanol. *J. Mater. Chem. A.* 8 (17): 8410–8420.

17 Chang, H., Tian, W., Chen, H. et al. (2023). Improved CO_2 electrolysis by a Fe nanoparticle-decorated (Ce, La, Sr)(CrFe)$O_{3\text{-}\delta}$ perovskite using a combined strategy of lattice defect-building. *Electrochim. Acta.* 439:141699.

18 Yang, X., Deng, P., Liu, D. et al. (2020). Partial sulfuration-induced defect and interface tailoring on bismuth oxide for promoting electrocatalytic CO_2 reduction. *J. Mater. Chem. A.* 8 (5): 2472–2480.

19 Ma, S., Wu, K., Fan, S. et al. (2024). *Lattice-dislocated bismuth nanowires formed by in-situ chemical etching on copper foam for enhanced electrocatalytic CO_2 reduction.* Sep. Purif. Technol. **349**:127926.

20 Tang, C., Shi, J., Bai, X. et al. (2020). CO_2 reduction on copper's twin boundary. *ACS Catal.* 10 (3): 2026–2032.

21 Wang, J., Liu, J., Song, Y. et al. (2023). Simultaneous defect and size control of metal-organic framework nanostructures for highly efficient carbon dioxide electroreduction to multicarbon products. *ACS Mater. Lett.* 5 (8): 2121–2130.

22 Zhao, J., Wang, J., Xue, L. et al. (2024). Surface oxygen defect engineering of $A_2B_2O_7$ pyrochlore semiconductors boosts the electrocatalytic reduction of CO_2-to-HCOOH. *Small.* **20**: 2402459.

23 Khedr, G.E., Fawzy, S.M., and Allam, N.K. (2023). Two-dimensional n-type pyrite with tuned hydrogen interstitials as a highly selective CO_2 reduction catalyst. *J. CO_2 Util.* **78**(22):102619.

24 Xiang, H., Rasul, S., Hou, B. et al. (2020). Copper-indium binary catalyst on a gas diffusion electrode for high-performance CO_2 electrochemical reduction with record CO production efficiency. *ACS Appl. Mater. Interfaces* 12 (1): 601–608.

25 Zhang, J., Sun, W., Ding, L. et al. (2021). Au nanocrystals@defective amorphous MnO_2 nanosheets core/shell nanostructure with effective CO_2 adsorption and activation toward CO_2 electroreduction to CO. *ACS Sustain. Chem. Eng.* 9 (14): 5230–5239.

26 Ren, J., Long, X., Wang, X. et al. (2022). Defect-rich heterostructured bi-based catalysts for efficient CO_2 reduction reaction to formate in wide operable windows. *Energ. Technol.* 10 (10):2200561.

27 Chen, Z., Yu, G., Li, B. et al. (2021). In situ carbon encapsulation confined nickel-doped indium oxide nanocrystals for boosting CO_2 electroreduction to the industrial level. *ACS Catal.* 11 (23): 14596–14604.

28 Wang, S., Wu, Z., Xu, C. et al. (2022). Triple-phase interface engineering over an In_2O_3 electrode to boost carbon dioxide electroreduction. *ACS Appl. Mater. Interfaces* 14 (40): 45423–45432.

29 Aljabour, A., Apaydin, D.H., Coskun, H. et al. (2016). Improvement of catalytic activity by nanofibrous $CuInS_2$ for electrochemical CO_2 reduction. *ACS Appl. Mater. Interfaces* 8 (46): 31695–31701.

30 Cheng, Y., Chen, J., Yang, C. et al. (2023). Activation of transition metal (Fe, Co, and Ni)-oxide nanoclusters by nitrogen defects in carbon nanotube for selective CO_2 reduction reaction. *Energy Environ. Mater.* 6 (1):e12278.

31 Chen, H., Handoko, A.D., Wang, T. et al. (2020). Defect-enhanced CO_2 reduction catalytic performance in O-terminated MXenes. *ChemSusChem* 13 (21): 5690–5698.

32 Niu, W., Feng, J., Chen, J. et al. (2024). High-efficiency C_3 electrosynthesis on a lattice-strain-stabilized nitrogen-doped Cu surface. *Nat. Commun.* 15 (1):7070.

33 Hu, C., Mu, Y., Bai, S. et al. (2019). Polyvinyl pyrrolidone mediated fabrication of Fe, N-codoped porous carbon sheets for efficient electrocatalytic CO_2 reduction. *Carbon* 153: 609–616.

34 Sharma, P.P., Wu, J., Yadav, R.M. et al. (2015). Nitrogen-doped carbon nanotube arrays for high-efficiency electrochemical reduction of CO_2: on the understanding of defects, defect density, and selectivity. *Angew. Chem.* 127 (46): 13905–13909.

35 Wang, W., Shang, L., Chang, G. et al. (2019). Intrinsic carbon-defect-driven electrocatalytic reduction of carbon dioxide. *Adv. Mater.* 31 (19): 1808276.

36 Ling, L.L., Jiao, L., Liu, X. et al. (2022). Potassium-assisted fabrication of intrinsic defects in porous carbons for electrocatalytic CO_2 reduction. *Adv. Mater.* 34 (42): 2205933.

37 Yang, X.F., Wang, A., Qiao, B. et al. (2013). Single-atom catalysts: a new frontier in heterogeneous catalysis. *Acc. Chem. Res.* 46 (8): 1740–1748.

38 Rong, X., Wang, H., Lu, X. et al. (2020). Controlled synthesis of a vacancy-defect single-atom catalyst for boosting CO_2 electroreduction. *Angew. Chem.* 132 (5): 1977–1981.

39 Wei, S., Wu, H., Zhou, C. et al. (2024). Low-coordination M-N_3 active sites with high accessibility and efficiency for electrocatalytic O_2 and CO_2 reduction. *Cat. Sci. Technol.* (Royal Society of Chemistry) 14: 2684–2696.

12

Defect and Interface Engineering in Fuel Production

I. Velázquez-Hernández and M. Estévez

Centro de Física Aplicada y Tecnología Avanzada, Universidad Nacional Autónoma de México, Boulevard Juriquilla 3001, Querétaro 76230, México

12.1 Catalytic Defects in Alternative Fuel Synthesis

The escalating global energy demand, coupled with the urgent need to mitigate climate change, necessitates a fundamental shift toward sustainable and highly efficient fuel production technologies. This transition away from traditional fossil fuels has spurred extensive research into alternative energy sources and cleaner production methodologies. Catalysis, the acceleration of chemical reactions by specific substances, is pivotal in enabling these sustainable technologies. It facilitates faster and more selective chemical transformations, leading to enhanced product yields and simplified catalyst separation and recovery, which are crucial for industrial applications [1]. In this context, defect and interface engineering emerges as a critical strategy for precisely tailoring the physicochemical properties of catalytic materials at the atomic and nanoscale levels. This meticulous manipulation allows for significant improvements in catalyst performance across a broad spectrum of fuel synthesis pathways, including electrochemical, thermochemical, and photochemical processes. The objective of this chapter is to provide a comprehensive exploration of the fundamental principles underpinning defect and interface engineering, highlight recent innovative advances in the field, and outline potential future research directions. By synthesizing current knowledge, this work aims to emphasize the critical role of these advanced material design strategies in realizing a sustainable energy future. A preliminary review of existing research reveals a notable gap in understanding the direct influence of defects on electrochemical pathways specifically designed for fuel production, as well as a lack of detailed discussion on interface modulation strategies in fuel cells [2]. This information deficit underscores the novelty and critical importance of a comprehensive report that synthesizes knowledge from related but disparate areas to bridge this gap [3]. Furthermore, the recurring emphasis on achieving "sustainable and efficient catalysis" and the broader objective of "sustainable energy conversion and fuel production" within the research literature firmly establish the direct link between

Surface and Interfacial Defects in Nanomaterials for Sustainable Energy Production and Storage, First Edition. Edited by Noé Arjona, Lorena Álvarez Contreras, and Minerva Guerra Balcázar.

the technical aspects of defect and interface engineering and the overarching global imperative to address environmental and energy security challenges.

Defects within catalytic materials significantly influence their activity and selectivity. These defects, ranging from point imperfections such as vacancies, interstitial atoms, and substitutions to extended defects such as dislocations and grain boundaries, can serve as crucial active sites for reactant adsorption, bond breaking, and product formation. Defect engineering focuses on the precise manipulation of these imperfections to optimize catalytic performance [2]. For example, introducing oxygen vacancies in metal oxides can enhance CO_2 adsorption, promoting its conversion into fuels like alcohols, particularly methanol [4]. Additionally, surface defects can alter the electronic structure of the catalyst, thereby influencing reactant adsorption energy and reaction activation energy. It is crucial to recognize that defects can have both beneficial and detrimental effects on catalyst activity and stability, thus achieving optimal adsorption energy is fundamental for improved performance.

The synthesis method significantly influences the creation of materials with surface defects. Various strategies, including annealing, ion implantation, chemical etching, mechanical treatment, plasma treatment, and electrochemical treatment, are employed to create defective catalysts [2]. A recent example illustrating this concept involves the use of copper catalysts for the electrochemical reduction of CO_2 to valuable products such as ethylene and ethanol [4]. The relationship between surface defects and CO_2 reduction is based on integrating point defects (Sn doping) and planar defects (grain boundaries) in the Cu structure. This strategy optimizes electronic effects, suppressing H_2 evolution and favoring $*CO_2$ hydrogenation. Sn doping induces charge polarization near Sn sites, and grain boundaries enhance CO_2 adsorption. These combined defects enable high Faradaic efficiency (>95%) in CO_2 reduction over a wide potential window ($\Delta E = 1.3\,V$), primarily producing CO and HCOOH (formic acid) using the pc-SnCu catalyst. Moreover, in alkaline media, the catalyst also demonstrates high CO_2 conversion to ethylene (C_2H_4), with Faradaic efficiency exceeding 95% across a broad potential range [4].

12.2 Interfacial Considerations in Fuel Production

12.2.1 Heterogeneous Catalyst Interfaces

In the realm of heterogeneous catalysis, the intricate interface between the catalytic material and the reactant species is the locus where chemical transformations fundamentally occur [5]. This interface serves as the critical zone for molecular reaction, energy transfer, and the separation and migration of reaction products. The specific structure and physicochemical properties of this interface exert a profound influence on the fundamental processes of adsorption, activation, and the subsequent chemical reaction itself [5]. Interface engineering represents a deliberate and strategic approach to manipulate and optimize these interfacial characteristics with the overarching goal of enhancing the key performance metrics of the catalyst, including its activity, selectivity, and stability [5].

Particularly, surface defects located in the outermost regions of a catalyst frequently function as active sites where reactant molecules are preferentially adsorbed and undergo chemical transformations [2]. This enhanced reactivity originates due to the presence of unsaturated coordination sites (atoms with fewer neighboring atoms than in the bulk) and localized modifications in the electronic structure at these defects' locations. For instance, a metal oxide surface possessing a missing oxygen atom (an oxygen vacancy) can serve as a strong adsorption site for CO_2 molecules, thereby facilitating their subsequent reduction [6].

The correlation between oxygen vacancy (Ov) density and methanol synthesis is crucial for improving the performance of CO_2 hydrogenation. The Ov density in In_2O_3 directly affects CO_2 adsorption and the formation of essential intermediates such as HCOO*. A moderate concentration of Ov (In_2O_3-OvI model) favors methanol synthesis by reducing the energy barriers for HCOO* formation and the C-O bond cleavage of H_2COO* to H_2CO*. However, an excessive or insufficient Ov density can divert the reaction toward CO formation via the COOH*-mediated pathway, thereby diminishing methanol selectivity [6].

Defects also significantly influence the adsorption energy of both reactants and reaction intermediates [7]. This energy is critical for overall catalytic activity and selectivity, as described by the Sabatier principle, which postulates that optimal catalytic performance is achieved when the interaction between the catalyst and adsorbed species is neither too strong nor too weak [7]. Surface strain induced by defects can alter adsorption energies, thus affecting reaction rates in processes such as ammonia synthesis [8]. In electrocatalysis and photocatalysis, defects enhance charge transfer [9]. They act as trapping sites for charge carriers, facilitating their separation and prolonging their lifetime, which is essential for efficient photocatalytic reactions such as CO_2 reduction and water splitting [9]. Moreover, specific types of defects can strategically direct catalyst selectivity toward the production of desired fuels in complex synthesis reactions [2]. For example, sulfur vacancies in $AgInP_2S_6$ can selectively promote the CO_2 photoreduction pathway toward ethene formation [10].

Interfaces between phases (solid–liquid, solid–gas, and liquid–liquid) are determinants in fuel production efficiency [11]. Interfacial interactions impact reactant adsorption, charge transfer, and product diffusion, affecting the reaction rate and product selectivity. In heterogeneous catalysis, where the reaction occurs at the solid–gas or solid–liquid interface, interfacial properties (surface area, composition, and electronic structure) are crucial. Interface engineering, through tailoring catalyst morphology, controlling particle size, or introducing interfacial promoters, optimizes fuel production [12]. For example, increasing the interfacial surface area provides more active sites, while modifying the electronic structure enhances reactant adsorption or product desorption. A study demonstrated how NaZnFe and CuFe interfaces influence olefin selectivity and higher alcohol production, respectively [13]. The interface regulates mass and heat transfer, as well as reactant adsorption and product desorption. Interface engineering focuses on optimizing these interfaces using interfacial materials and controlling morphology and structure [12]. Strategies include the use of surfactants in liquid–liquid reactions [11],

Table 12.1 Interface modulation techniques and their impact on fuel production.

Interface type	Modulation strategy	Fuel synthesis reaction	Key enhancement	References
Metal–support	Potassium incorporation	Ethanol production from CO_2	Increased ethanol selectivity	[15]
Semiconductor–catalyst	Interface tailoring	Methanol production from CO_2	Enhanced methanol yield	[16]
Electrolyte	Cation effects	Electrochemical CO_2 reduction to C^{2+}	Favors the formation of multi-carbon products	[17]
Metal–support	Sulfur doping	Oxygen reduction reaction (fuel cells)	Outstanding sintering and corrosion resistance	[18]
MOF-based	Nanoparticle encapsulation	CO_2 conversion to alkynyl carboxylic acids	High yield under mild conditions	[19]
Metal–oxide	Interface engineering	CO_2 activation	Improved catalytic activity for downstream fuel synthesis	[20]
Gas–water	Interface area control	Volatile organic compounds (VOC) degradation	Enhanced mineralization of volatile organic compounds	[21]
Microalgae	Modulation strategies	Biofuel production (lipid accumulation)	Increased lipid accumulation and improved lipid quality	[22]

designing porous materials for solid–gas reactions [3], and controlling nanoparticle size at solid–liquid interfaces [14]. A comprehensive overview of interface modulation techniques is detailed in Table 12.1.

12.2.2 Metal–Support Interactions (MSIs)

For catalysts consisting of metal nanoparticles dispersed on a supporting material, the nature and strength of the interaction between the metal and the support are pivotal determinants of the catalyst's overall performance [23]. The support is not merely a passive platform for the active metal but actively participates in modulating its catalytic properties. MSI can exert a significant influence on the electronic structure of the metal nanoparticles, affecting their ability to adsorb and activate reactant molecules. Furthermore, MSI plays a crucial role in controlling dispersion and enhancing the resistance of the metal nanoparticles to sintering [23]. Strong metal–support interactions (SMSIs) represent a particularly important class of MSI, often leading to a substantial enhancement in both the catalytic activity and the long-term stability of the catalyst [18]. Sulfur doping in carbon support has been shown to induce a variety of SMSI phenomena [18]. The interfacial regions that form at the contact points between metal nanoparticles and the support material often exhibit synergistic effects [18, 24]. In the context of alternative fuel production, the interface between a metal catalyst and an oxide support is particularly significant for the crucial step of CO_2 activation [25].

12.3 Defect-Engineered Nanomaterials for Precision Fuel Synthesis

Engineered defect-containing nanomaterials offer precise control over fuel synthesis by influencing catalytic activity, selectivity, and stability [14]. The introduction of defects into nanomaterials can create active sites, modify electronic properties, and enhance reactant adsorption. Recently, nanomaterials with surface defects induced through defect engineering have been employed for precise fuel synthesis, such as defect-containing iron oxide nanoparticles to enhance solar energy conversion efficiency [14], defect-based copper nanomaterials for efficient CO_2 reduction across an ultra-wide potential window [4], and defect-engineered graphitic carbon nitride nanotextures for low-energy solar fuel production [10].

12.3.1 Defect-Mediated Pathways for Alternative Fuel Production

12.3.1.1 CO_2 Reduction

Oxygen vacancies in metal oxide catalysts are crucial for CO_2 reduction, facilitating adsorption, activation, and conversion to products like CO, formate, methanol, and ethanol [6]. For example, indium oxide with optimized oxygen vacancy density shows enhanced methanol selectivity [6]. Defects in g-C_3N_4, including vacancies and doping, enhance photocatalytic efficiency for CO_2 reduction and hydrogen production [10]. Defect engineering in Cu_{58} nanoclusters enables selective methanol synthesis [26]. Microwave heating creates defect-rich CuO electrodes for enhanced CO_2 reduction to C^{2+} products [17].

12.3.1.2 Methanol Synthesis

The functionality of catalytic systems is intrinsically linked to the presence and nature of defects. Oxygen vacancies within the lattice structure of ZnO have been identified as critical active sites in Au/ZnO catalytic systems [16]. These vacancies provide unsaturated coordination sites that facilitate the adsorption and activation of reactant molecules. Furthermore, the density of oxygen vacancies in In_2O_3 directly influences methanol selectivity during CO_2 hydrogenation [6]. Optimizing this density is essential for maximizing the yield of desired products. Precise engineering of defects in Cu_{58} nanoclusters has been shown to yield exceptionally high selectivity toward targeted catalytic reactions [26]. This demonstrates the power of defect control in tailoring catalyst performance. The strategic creation of defects within catalytic materials allows for the fine-tuning of electronic and structural properties, leading to enhanced reactant adsorption, improved charge transfer dynamics, and optimized product selectivity. These advancements are particularly crucial in the development of efficient and sustainable fuel production technologies.

12.3.1.3 Ammonia Synthesis

Recently, a Ba_3SiO_5-xN_yH_z-based catalyst that activates N_2 for ammonia synthesis through an anion vacancy-mediated mechanism, without the necessity of transition metals, was investigated. Surface defects, specifically anionic vacancies, play a fundamental role in ammonia synthesis [8]. These vacancies on the surface of the barium–silicon oxynitride–hydride orthosilicate catalyst facilitate the capture and activation of N_2 molecules. The flexible crystal structure of the material can accommodate a high electron density at these vacancies, enabling direct N_2 activation for hydrogenation and conversion into ammonia. Furthermore, the loading of ruthenium nanoparticles onto the catalyst enhances the formation of anionic vacancies at the Ru-support interface, even though ruthenium does not directly dissociate N_2. These surface defects are essential for the efficiency of the catalyst in ammonia synthesis without the requirement of transition metals [8].

12.3.1.4 Biodiesel Production

Defects in UiO-66(Zr) significantly enhance catalytic activity in biodiesel production, demonstrating the broad applicability of defect engineering across diverse fuel synthesis pathways [27]. Notably, oxygen vacancies emerge as a recurrent mechanistic factor, while the strategic selection of specific defect types is crucial for tailoring outcomes to targeted fuel production goals [27].

Similarly, interface engineering plays a pivotal role. For instance, potassium-loaded Cu/CoO_x interfaces enhance ethanol production from CO_2 hydrogenation [20], and Cu/Mo_2CTx interfaces improve CO_2 hydrogenation efficiency [20]. Beyond traditional catalytic processes, interface modulation can also increase lipid accumulation in microalgae for biofuel production [22]. The variety of interface modulation strategies underscores the versatility of this approach. Furthermore, the growing utilization of metal–organic frameworks (MOFs) highlights their unique advantages in this field. A significant focus on enhancing selectivity indicates that interface engineering is not only about boosting reaction rates but also about precisely directing reactions toward desired product formation [19].

These case studies illustrate the diverse applications of defect-engineered nanomaterials in enhancing the production of various biofuels (Table 12.2), highlighting the potential of this approach to contribute to a more sustainable energy future.

Table 12.2 Examples of defect-engineered nanomaterials in biofuel production.

Nanomaterial	Engineered defect	Biofuel	Improvement	References
TiO_2, ZnO	Crystal lattice defects	Biodiesel	Enhanced stability, faster catalysis, high yield	[28]
α-MoO_3	Oxygen vacancies, Mo^{5+} centers	Biodiesel	High conversion rate	[29]
Banana peel biochar	(Further research needed)	Biodiesel	High yield, reusability	[28]
SnO_2	Oxygen vacancies	Biohydrogen	Increased production rate and yield	[30]
Ni/Fe_2O_3	(Potentially interfacial defects)	Biohydrogen	Enhanced yield compared to individual components	[31]

12.4 Innovative Catalysts for Sustainable Fuel Synthesis

The strategic manipulation of defects and interfaces has catalyzed the development of innovative catalysts, significantly enhancing activity, selectivity, and stability for sustainable fuel synthesis. These advancements not only reduce the energy required for reactions but also improve product yields and minimize unwanted byproduct formation. A comprehensive summary of surface defects in various materials is presented in Table 12.3.

12.4.1 Single-Atom Catalysts (SACs)

An example of such innovative catalysts is SACs, where individual metal atoms are dispersed on a support. SACs offer high metal atom utilization and unique catalytic properties due to their atomically dispersed nature [15]. Defect engineering in SACs can further enhance their performance by introducing vacancies or dopants that modify the electronic coordination and environment of the individual metal atoms. However, SACs often face challenges in maintaining stable bonding with metal oxide supports, which is crucial for catalyst durability. Defect engineering can address this issue by introducing controlled defects in the support, improving the stability and sustainability of the catalyst [15].

Table 12.3 Examples of defect-engineered catalysts and their role in fuel synthesis.

Catalyst material	Type of defect engineered	Fuel synthesis reaction	Key performance enhancement	References
In_2O_3	Oxygen vacancy	CO_2 reduction to methanol	Increased selectivity for methanol	[6]
MoS_2	Sulfur vacancy	Hydrogen evolution reaction	Enhanced hydrogen production rate	[32]
g-C_3N_4	Nitrogen vacancy	Photocatalytic CO_2 reduction	Improved CO_2 conversion	[10]
Cu_{58} nanoclusters	Ligand vacancy	Electrochemical CO_2 reduction to methanol	Highly selective methanol synthesis	[26]
CuO	Grain boundary	Electrochemical CO^2 reduction to C^{2+}	Enhanced yield of C^{2+} products	[17]
TiO_2	Oxygen vacancy	Photocatalytic H_2 from woody biomass	Rapid conversion of biomass to H_2	[33]
Au/ZnO	Oxygen vacancy	Methanol synthesis	Increased number of active sites	[16]
Ba_3SiO_5-xN_yH_z	Anion vacancy	Ammonia synthesis	Activity without transition metals	[8]
UiO-66(Zr)	Structural defects	Biodiesel production	Enhanced catalytic activity	[27]

12.4.2 MXene-Based Catalysts

MXene-based catalysts, a novel class of two-dimensional materials with high electrical conductivity and surface area [34], represent a promising avenue for advanced catalysis. These materials can be engineered with defects through the precise control of vacancies, heteroatom doping, or surface functionalization, significantly enhancing their catalytic activity in crucial fuel synthesis reactions like hydrogen evolution reaction (HER) and carbon dioxide reduction [35]. A recent review highlighted diverse methods and catalysts for fuel production, with a particular emphasis on CO_2 and CH_4 conversion into valuable fuels [35]. CO_2 hydrogenation to methanol, for instance, is a clean process that reduces atmospheric CO_2 levels, where Cu/Mo_2CT_x MXene catalysts demonstrate high intrinsic methanol formation rates and selectivity [36]. The Fischer–Tropsch process, which produces various chemicals and fuels through CO hydrogenation, also shows promising results with Mo_2CT_x-derived catalysts in terms of activity and selectivity [37]. The importance of defects in MXene materials for enhancing catalytic performance in these reactions is clear: defect engineering significantly increases catalytic activity by creating additional active sites for reactant adsorption, improving reaction efficiency; modifying the functional groups on the MXene surface alters the catalyst's electronic structure, establishing strong interfacial interactions, with functional groups like −O, −OH, −F, and −H directly influencing catalytic activity [34, 35].

12.5 Integration of Defects in Electrochemical Fuel Production

In electrochemical fuel production systems, the interface between the electrode material and the electrolyte is of paramount importance as it governs the fundamental processes of charge transfer and the kinetics of the electrochemical reactions. The structure of the electrical double layer plays a critical role [38]. Interface engineering in electrochemical systems involves a range of strategies aimed at optimizing the performance of these interfaces. For instance, meticulous optimization of the electrocatalyst/electrolyte interface is particularly significant for achieving stable and high-performance operation [39].

The recurring concept of "synergistic effects" observed at metal–support interfaces and between different metallic components in bimetallic catalysts points to a fundamental principle in catalyst design. By strategically combining materials with complementary properties and functionalities at their interface, researchers can unlock performance enhancements that often surpass the additive effects of the individual components. The critical role of strong metal–support interactions in bolstering the thermal stability of supported metal catalysts and effectively preventing sintering is of paramount importance for ensuring the long-term operational viability of catalytic systems employed in fuel production. Furthermore, the significant influence of electrolyte interface modulation through various parameters in electrochemical fuel synthesis underscores the intricate and often subtle interplay between the electrode material and its surrounding electrochemical environment. Defects in electrocatalytic materials can significantly influence electrochemical pathways for fuel production [40]. For example, defects in ferroelectric materials, such as titanium vacancies, can affect photocatalysis efficiency by trapping electrons and increasing recombination. The introduction of nanolayers can improve water-splitting efficiency by eliminating these defects [40] .

Interface modulation in electrochemical fuel cells is also crucial for enhancing fuel synthesis. This can include designing electrodes with high surface area and porosity, as well as controlling the composition and structure of the electrode–electrolyte interface [40]. Interface engineering in Ni-based electrocatalysts enhances alkaline HER [39]. Controlling the gas–water interface enhances volatile organic compounds (VOC) degradation [21].

12.6 Conclusions

Defect and interface engineering emerges as a transformative field in sustainable fuel production, offering precise control over catalytic activity, selectivity, and stability through atomic-level manipulation. Advances in this field have demonstrated that both catalytic defects and interfacial considerations are crucial for optimizing reactant adsorption, charge transfer, and the durability of catalysts.

Despite inherent challenges, such as the precise control of defect concentration and location and the understanding of complex interactions at interfaces, the future is promising. Ongoing research, driven by advanced characterization techniques and computational models, is paving the way for the development of innovative catalysts and fuel cells.

Ultimately, the ability to design nanomaterials with specific defects and to modulate interfaces will enable the creation of more efficient and sustainable energy systems, significantly contributing to the transition toward a cleaner energy future.

References

1 Alam, M.I. and Saha, B. (2015). Catalysis for the production of sustainable chemicals and fuels from biomass. In: *Sustainable Catalytic Processes* (ed. B. Saha, M. Fan, and J. Wang), 99–123. Elsevier.

2 Luo, Y. and Wu, Y. (2023). Defect engineering of nanomaterials for catalysis. *Nanomaterials* (MDPI) 13.

3 Zhao, Z., Chen, H., Zhang, W. et al. (2023). Defect engineering in carbon materials for electrochemical energy storage and catalytic conversion. *Mater. Adv.* (Royal Society of Chemistry) 4: 835–867.

4 Wu, Q., Liu, C., Su, X. et al. (2023). Defect-engineered cu-based nanomaterials for efficient CO_2 reduction over ultrawide potential window. *ACS Nano* 17 (1): 402–410.

5 Xu, Z., Yang, Z., Lu, H. et al. (2024). Atomic defects engineering boosts urea synthesis toward carbon dioxide and nitrate coelectroreduction. *Nano Lett.* 24 (37): 11730–11737.
6 Wang, W., Huo, K., Wang, Y. et al. (2024). Rational control of oxygen vacancy density in In_2O_3 to boost methanol synthesis from CO_2 hydrogenation. *ACS Catal.* 14 (13): 9887–9900.
7 Zhao, X., Wang, J., and Niu, X. (2024). High-throughput screening approach for catalytic applications through regulation of adsorption energies via work function. *Langmuir* 40 (50): 26532–26539.
8 Zhang, Z., Miyashita, K., Wu, T. et al. (2025). Anion vacancies activate N_2 to ammonia on Ba–Si orthosilicate oxynitride-hydride. *Nat. Chem.*
9 Hoch, L.B., Szymanski, P., Ghuman, K.K. et al. (2016). Carrier dynamics and the role of surface defects: designing a photocatalyst for gas-phase CO_2 reduction. *Proc. Natl. Acad. Sci. USA* 113 (50): E8011–E8020.
10 Tahir, M., Sherryna, A., Khan, A.A. et al. (2022). Defect engineering in graphitic carbon nitride nanotextures for energy efficient solar fuels production: a review. *Energy Fuel.* (American Chemical Society) 36: 8948–8977.
11 Hong, J., Wang, Z., Li, J. et al. (2023). Effect of interface structure and behavior on the fluid flow characteristics and phase interaction in the petroleum industry: state of the art review and outlook. *Energy Fuel.* (American Chemical Society) 37: 9914–9937.
12 Li, F. and Jen, A.K.Y. (2022). Interface engineering in solution-processed thin-film solar cells. *Acc. Mater. Res.* 3 (3): 272–282.
13 Jiang, Q., Song, G., Zhai, Y. et al. (2023). Selective hydrogenation of CO_2 to Aromatics over composite catalyst comprising NaZnFe and polyethylene glycol-modified HZSM-5 with intra- and intercrystalline mesoporous structure. *Ind. Eng. Chem. Res.* 62 (23): 9188–9200.
14 Lak, A., Disch, S., and Bender, P. (2021). Embracing defects and disorder in magnetic nanoparticles. *Adv. Sci.* 8 (7).
15 Wang, D., Shan, H., Yin, W., and Li, H. (2024). Defect engineering of single-atom catalysts in biomass conversion. *Fuel.* (Elsevier Ltd) 355: 129439.
16 Chen, S. et al. (2021). Performance of Au/ZnO catalysts in CO_2 reduction to methanol: varying the Au loading / Au particle size. *Appl. Catal. A: Gen.* 624: https://doi.org/10.1016/j.apcata.2021.118318.
17 Chen, Q., Luo, X., Mu, G. et al. (2024). Microwave-assisted synthesis of a defect-rich CuO electrode for selective electrochemical CO_2 Reduction to C2+ Products. *ACS Sustain. Chem. Eng.* 12 (41): 15134–15146.
18 Yin, P., Yan, Q.Q., and Liang, H.W. (2023). Strong metal-support interactions through sulfur-anchoring of metal catalysts on carbon supports. *Angew. Chem. Int. Ed.* (John Wiley and Sons Inc) 62: e202302819.
19 Abdelkareem, M.A., Abbas, Q., Sayed, E.T. et al. (2024). Recent advances on metal-organic frameworks (MOFs) and their applications in energy conversion devices: comprehensive review. *Energy* 299: 131127.
20 Zhang, Y., Fan, G., Zheng, L., and Li, F. (2025). Synergistic surface-interface catalysis in potassium-loaded Cu/CoOx catalysts to boost ethanol production from CO_2 hydrogenation. *ACS Appl. Mater. Interfaces.*

21 Xu, Z., Chai, W., Cao, J. et al. (2021). Controlling the gas–water interface to enhance photocatalytic degradation of volatile organic compounds. *ACS ES&T Eng.* 1 (7): 1140–1148.

22 Sharma, P.K., Saharia, M., Srivstava, R. et al. (2018). Tailoring microalgae for efficient biofuel production. *Front. Mar. Sci.* 5 (NOV).

23 Song, T., Li, R., Wang, J. et al. (2024). Enhanced methanol synthesis over self-limited ZnOx overlayers on Cu nanoparticles formed via gas-phase migration route. *Angew. Chem. Int. Ed.* 63 (5).

24 Yang, B., Chen, X., Guo, L., and Zhang, L. (2024). Catalyst architecture for metal–support interactions and its effects on heterogeneous reactions. *J. Mater. Chem. A* [Internet] 12 (31): 19861–19884. Available from: https://doi.org/10.1039/D4TA03423F.

25 He, Q., Han, L., Lin, C., and Tao, K. (2024). A review on defect modulated electrocatalysts for the oxygen evolution reaction. *Nanoscale* [Internet] 16 (26): 12368–12379. Available from: https://doi.org/10.1039/D4NR01805B.

26 Biswas, S., Tanaka, T., Song, H. et al. (2025). Highly selective methanol synthesis using electrochemical CO_2 reduction with defect-engineered cu58 nanoclusters. *Small Sci.* 5: 2400465.

27 Abou-Elyazed, A.S., Ye, G., Sun, Y., and El-Nahas, A.M. (2019). A series of UiO-66(Zr)-structured materials with defects as heterogeneous catalysts for biodiesel production. *Ind. Eng. Chem. Res.* 58 (48): 21961–21971.

28 Velusamy, K., Devanand, J., Senthil Kumar, P. et al. (2021). A review on nano-catalysts and biochar-based catalysts for biofuel production. *Fuel* 306: 121632.

29 Lucena, S., de Medeiros, S.A., Menezes de Oliveira, A.L. et al. (2025). α-MoO_3 micro- and nanoparticles as catalysts for biofuel production. *ACS Appl. Nano Mater.* (American Chemical Society) 8 (9): 4339–4353.

30 Nadeem, F., Jiang, D., Tahir, N. et al. (2020). Defect engineering in SnO_2 nanomaterials: pathway to enhance the biohydrogen production from agricultural residue of corn stover. *Appl. Mater. Today* 21: 100850.

31 Mishra, P., Akaniro, I.R., Zhang, R. et al. (2024). Unlocking the potential of Ni/Fe_2O_3 bimetallic nanoparticles for fermentative biohydrogen production. *ACS ES&T Eng.* 4: 2424–2434.

32 Tan, Z.H., Kong, X.Y., Ng, B.J. et al. (2023). Recent advances in defect-engineered transition metal dichalcogenides for enhanced electrocatalytic hydrogen evolution: perfecting imperfections. *ACS Omega*. (American Chemical Society) 8: 1851–1863.

33 Huang, Z.X., Wang, Z.Y., Yuan, Y.J. et al. (2024). Defect engineering of ultrasmall TiO_2 nanoparticles enables highly efficient photocatalysts for solar H_2 production from woody biomass. *Nano Lett.* 24 (38): 11968–11975.

34 Zhu, Y., Wu, Z., Xie, X., and Zhang, N. (2020). A retrospective on MXene-based composites for solar fuel production. *Pure Appl. Chem.* 92 (12): 1953–1969.

35 Hamzehlouy, A. and Soroush, M. (2024). MXene-based catalysts: a review. *Mater. Today Catal.* 5: 100054.

36 Zhou, H., Chen, Z., López, A.V. et al. (2021). Engineering the Cu/Mo_2CT_x (MXene) interface to drive CO_2 hydrogenation to methanol. *Nat. Catal.* [Internet] 4 (10): 860–871. Available from: https://doi.org/10.1038/s41929-021-00684-0.

37 Kountoupi, E., Barrios, A.J., Chen, Z. et al. (2024). The impact of oxygen surface coverage and carbidic carbon on the activity and selectivity of two-dimensional molybdenum carbide (2D-Mo_2C) in fischer-tropsch synthesis. *ACS Catal.* 14 (3): 1834–1845.

38 Stamenkovic, V.R., Strmcnik, D., Lopes, P.P., and Markovic, N.M. (2017). Energy and fuels from electrochemical interfaces. *Nat. Mater.* [Internet] 16 (1): 57–69. Available from: https://doi.org/10.1038/nmat4738.

39 Zhang, X., Guo, Y., and Wang, C. (2024). Multi-interface engineering of nickel-based electrocatalysts for alkaline hydrogen evolution reaction. *Energy Mater.* 4: 400044.

40 Zhang, J., Liu, Y., Dittrich, T. et al. (2025). Unveiling charge utilization mechanisms in ferroelectric for water splitting. Nature. *Communications* 16 (1).

13

Defect and Interface Engineering in Electrochemical Valorization of Biomass to Value-Added Chemicals

Sampathkumar Prakasam[1,2], *Narmadha Maharajan*[1,2], *and Suresh Chinnathambi*[1,2]

[1] *Electrodics and Electrocatalysis Division, CSIR-Central Electrochemical Research, Institute, Karaikudi, 630003, Tamil Nadu, India*
[2] *Academy of Scientific and Innovative Research (AcSIR), Ghaziabad, 201002, India*

13.1 Introduction

The rapid growth of population and industrialization lead to excess consumption of fossil fuels (coal, oil, natural gas, etc.), which results in producing a huge amount of CO_2 which is responsible for climate change, air pollution, global warming, etc. According to a 2023 report from the World Meteorological Organization, the average global temperature in 2022 was increased by 1.15 °C, which is higher than that recorded in the 1900s. Furthermore, it will be increased by 1–3.7 °C by the end of this century [1]. In response to this issue, many efforts have been made to find alternative energy sources. Biomass is one of the best options because it is renewable, cheap, and distributed throughout the world. Producing chemicals and fuels from biomass can significantly reduce environmental pollution and achieve carbon neutrality, promoting a circular economy and enhancing waste management [2]. These biomass wastes originate from various sources, such as agricultural, forestry, industrial, municipal, and aquatic biomass (Figure 13.1a). The most abundant lignocellulosic biomass from crops and forestry crops consists of three primary components, including cellulose (25–55 wt.%), hemicellulose (10–50 wt.%), and lignin (6–30 wt.%) (Figure 13.1b). Biomass residues such as rice straw and husk contain ~0.84 gigatons per year on a global scale; however, 80% of these wastes are not efficiently utilized. Furthermore, the other biomass waste like sugarcane bagasse provides around 0.18 and 0.22 gigatons per year to the world [3]. Cellulose is a crystalline homopolysaccharide component consisting of β-D-glucopyranose followed by $\beta(1 \rightarrow 4)$ glycosidic bonds [4]. It leads to the formation of intermolecular and intramolecular hydrogen bonds, which make cellulose resistant during the conversion process and make it insoluble in common solvents [5]. During hydrolysis, the initial conversion product of cellulose is glucose and it is further converted into various value-added chemicals including

Surface and Interfacial Defects in Nanomaterials for Sustainable Energy Production and Storage, First Edition. Edited by Noé Arjona, Lorena Álvarez Contreras, and Minerva Guerra Balcázar.

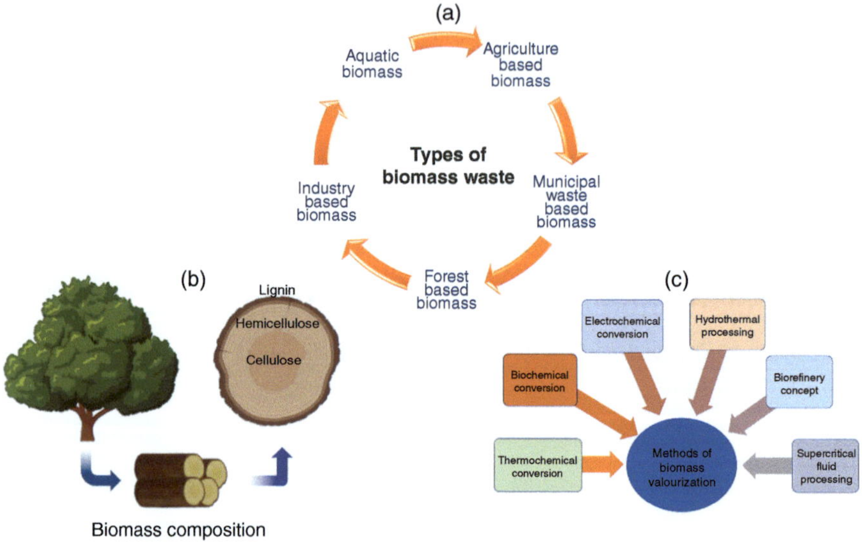

Figure 13.1 (a) Types of biomass wastes, (b) composition of biomass, and (c) methods of biomass valorization.

5-hydroxymethylfurfural (5-HMF) [6]. 5-HMF is the key platform chemical for the synthesis of several value-added chemicals, such as 2,5-dimethylfuran (DMF), 2,5-furandicarboxylic acid (FDCA), 2,5-bis(hydroxymethyl)furan (2,5-BHMF), and 2,5-dimethyltetrahydrofuran (2,5-DMTHF) [3]. Hemicellulose is the amorphous polysaccharide component composed of various heteropolymers, which include xylan, arabinoxylan, galactomannan, xyloglucan, and glucomannan. Recent research indicates that the depolymerization of hemicellulose present in raw biomass can be efficiently converted into C_5 and C_6 sugar monomers [7]. For example, xylan, a type of hemicellulose, can be hydrolyzed to yield xylose and arabinose. Additionally, these sugars can be further processed into various valuable chemicals, including xylitol, xylonic acid, and δ-valerolactone [3]. Lignin is an amorphous three-dimensional biopolymer with an aromatic structure. It is the second most abundant biopolymer after cellulose. Primarily, it consists of phenolic units and is mainly derived from three types of cinnamyl alcohols known as monolignols: p-coumaroyls (H), coniferyl (G), and sinapyl (S) [8]. Lignin by-products from the paper and pulp industry contribute over 50 gigatons annually around the world, with 98% of this substance mainly used as a low-value fuel [3]. The US Department of Energy (DOE) suggested highly important 12 sugar-based building blocks, such as succinic acid, maleic acid (MA), itaconic acid, FDCA, aspartic acid, fumaric acids, glucaric acid, 3-hydroxypropionic acid, glutamic acid, xylitol, sorbitol, glycerol, hydroxy butyrolactone, and levulinic acid [9]. These sugar based building blocks have significant ability to replace the usage of petroleum-based products. In addition to these, biomass waste contains a trace amount of lipids and proteins, lipids include both saturated and unsaturated fatty acids as well as glycerol. The primary product derived from these lipids is biodiesel, obtained through esterification, while glycerol is produced as a significant by-product [10]. In the case of proteins, it can be further converted into essential and nonessential amino acids, which can serve as raw materials for the production of a variety of polymers and value-added chemicals [11].

In the past decades, biomass valorization was frequently done by utilizing a thermocatalytic process, but it needs more operating temperature around 200–300 °C. Moreover, it has many key challenges, such as the need to minimize side reactions, consume more hydrogen, and ensure the hydrothermal stability of catalysts, which is crucial for managing the moisture content in biomass feedstocks [12]. Hence, it is essential to develop efficient and environmentally friendly methods for converting waste biomass into useful products. Among various valorization methods (Figure 13.1c), the electrochemical valorization method is a clean and eco-friendly process because we can preciously control the product selectivity by varying the electrocatalysts, electrode potential, reaction rate, and activation energy. Hence, this method can help to reduce carbon emissions, climate change, and environmental pollution [13]. In recent years, a variety of materials have been utilized as electrocatalysts for electrochemical biomass valorization. In addition, the stability, selectivity, and efficiency of the desired products were improved by changing various structural properties of the materials. However, many drawbacks in the electrocatalysts design and synthesis, that need to be addressed clearly. Therefore, it is needed to study the surface engineering of the materials in detail. Hence, this chapter aims to provide a comprehensive details of the electrochemical valorization of biomass based on the surface engineering and defect aspects of the electrocatalysts.

13.2 Defect Engineering and Its Types

The uneven arrangement of atoms and molecules in the crystal is known as defects, which include the presence of additional atoms or the absence of atoms in the unit cells. The introduction of defects into the surface of the material can significantly improve the physical and chemical properties, such as electrical, mechanical, and catalytic properties. The defects are classified into two types: point defects and extended defects [14].

13.2.1 Point Defects

Point defects refer to imperfections in the arrangement of atoms in a single lattice in the crystal structure. It is divided into three types: interstitial defects, vacancy defects, and substitutional defects. The interstitial defects are defined as when the atom occupies the sites between the lattice of a crystal. The absence of atoms for their lattice position, creating a vacancy in the crystal structure, is known as vacancy defect. When a foreign atom can replace a regular atom in the crystal structure, resulting in the formation of substitutional impurities, it is known as substitutional defects.

13.2.2 Extended Defects

The defects that occur in the large portion of the crystal structure are called as extended defects. In general, it is divided into two categories: dislocations and grain boundaries. When the stress is applied to the crystal lattice, one atom can shift its position to another atom resulting in a linear defect in the crystal structure known as dislocation. Grain boundaries are two-dimensional planar defects, the misalignment has emerged between adjacent gains in a polycrystalline material.

13.3 Biomass Valorization and Its Types

Biomass valorization involves the conversion of various biomass waste materials into value-added products, such as chemicals, biofuels, and biomaterials. Biomass valorization offers key advantages, such as reducing waste and pollution, promoting economic growth, maintaining carbon neutrality, minimizing land use, and decreasing the consumption of fossil fuels. In biomass waste conversion, there are many types of processes were adopted to convert into value-added chemicals (Figure 13.1c). Such as biorefinery concept, thermochemical conversion, biochemical conversion, hydrothermal processing, supercritical fluid processing and electrochemical valorization.

13.3.1 Biorefinery Concept

A biorefinery is an integrated concept – (thermochemical and biochemical process), which can convert biomass waste into a variety of value-added chemicals. For example, lignocellulosic biomass undergoes initial biomass treatment followed by enzymatic hydrolysis and fermentation to produce bioethanol. Further, the remaining biomass can be burned to produce electricity.

13.3.2 Thermochemical Conversion

The thermochemical conversion method utilizes heat to break down the molecules and convert them into highly value-added products, such as alcohols, furans, and glycerol. Further, these products can be valorized into various value-added chemicals. In general, this process typically consumes more energy but it offers efficient conversion. Thermochemical processes are of various types: pyrolysis, combustion, gasification, and hydrothermal liquefaction [12].

13.3.2.1 Pyrolysis, Combustion, and Gasification

Pyrolysis is a process that thermally decomposes biomass wastes in an oxygen-free environment at elevated temperatures to produce syngas, biochar, and bio-oils. The obtained products are further processed into valuable chemicals and fuels. Combustion is the process, in which biomass waste is directly burned to produce electricity. Gasification is a thermochemical process, where the biomass waste is partially burned in a controlled oxygen environment, producing a variety of syngases, such as carbon monoxide (CO), hydrogen (H_2), carbon dioxide (CO_2), and methane (CH_4). These gases can be further transformed into liquid fuels.

13.3.3 Biochemical Conversion

Biochemical conversion is a process that utilizes enzymes and microorganisms to break down biomass waste into valuable chemicals and fuels. Initially, it involves a hydrolysis process to convert complex polysaccharides into simpler sugars, which are further converted into useful chemicals. Different categories of biochemical conversion process include.

13.3.3.1 Fermentation, Anaerobic Digestion, and Enzymatic Hydrolysis

The yeast or bacterium is used to ferment the biomass sugars, resulting in the conversion of biofuels like bioethanol and butanol, which is known as fermentation. Anaerobic digestion involves the utilization of microorganisms to break down the biomass waste in an oxygen-free environment to produce CH_4, CO_2, and electricity. The enzymatic hydrolysis method utilizes enzymes to convert complex carbohydrates into simpler sugars, which can be converted into various biofuels.

13.3.4 Hydrothermal Processing

This process refers to thermochemical conversion, where the biomass waste is treated with hot water under high pressure and breaks down into valuable byproducts, such as bio-oil and biogas.

13.3.5 Supercritical Fluid Processing

This process utilizes supercritical fluids such as supercritical water, CO_2, and alcohol to convert the biomass waste into valuable chemicals. For example, in supercritical water gasification, water is heated to a supercritical state (above 374 °C and 22.06 MPa) to convert biomass waste into useful gases such as CH_4, H_2, and other syngases.

13.3.6 Electrochemical Valorization

Electrochemical valorization is a process, which utilizes electrons as the primary redox agent, to convert biomass waste into valuable chemicals. This process eliminates the need for external oxygen or hydrogen sources, making it an efficient process. This electrochemical method offers key benefits, such as the ability to operate the reaction under room temperature and atmospheric pressure, and the flexibility to scale up or down simply by modifying the cell design. The electrochemical reactions are highly sensitive to the electric double layer and can be fine-tuned by adjusting the parameters including applied potential, electrolytes, and electrocatalysts. The product selectivity and efficiency can be preciously controlled by changing these parameters [12].

13.4 Defects and Interface Engineering in Electrochemical Valorization of Biomass

In electrochemical valorization, the conversion efficiency and product selectivity are highly dependent on the presence of defects on the electrocatalyst surfaces. By changing the properties, it is possible to increase the active sites, enhance the conversion rates, reduce energy consumption, and improve both the stability and durability of the system. This section highlights the electrochemical valorization of cellulose, hemicellulose, and lignin through surface engineering and the impact of defects of the electrocatalysts.

13.4.1 Defect-Induced Chemical and Fuel Synthesis from Cellulose and Hemicellulose

Cellulose: Cellulose is a highly crystalline polysaccharide compound derived from plant waste. It is the most abundant biopolymer on earth. The −OH groups are ordered in equatorial position and connected with nearby anhydroglucose units via hydrogen bonds [6]. Electrochemical conversion processes typically involve the interruption in the crystalline structure of cellulose, breaking it into simple sugars, primarily glucose. This glucose can be further converted into a wide range of value-added chemicals. For example, the electro-oxidation of glucose can yield glucaric acid and gluconic acid. The dehydration of glucose can form levulinic acid, 5-5-hydroxymethylfurfural, etc. (Figure 13.2) [15].

Hemicellulose: Hemicellulose is a natural polysaccharide compound with a complex and heterogeneous structure. Hemicellulose is composed of pentoses, which include xylose, furoic acid, and xylitol. Hemicellulose is amorphous; therefore, it is more easily hydrolyzed than cellulose. During electrochemical processes, hemicellulose can be converted into a variety of valuable products, such as xylose, xylitol, furfural, 5-HMF, lactic acid, xylonic acid, and arabitol [16]. In this part, the electrochemical valorization of lignocellulosic biomass (cellulose and hemicellulose) into various chemicals, particularly polyols and furans will be discussed. In addition, the role of defect engineering on the material surface can improve the selectivity and conversion efficiency of products will be explored and the methods to optimize the production of desired products.

13.4.1.1 Synthesis of Polyols

Polyols are a type of alcoholic compound, which contain several −OH groups linked to the carbon atoms. Polyols are essential for the production of flexible polymers and also play the role of key building blocks for a variety of chemicals. They are also highly utilized compounds in industrial sectors, which include food, pharmaceuticals, cosmetics, and plastics. Some of the examples of polyols include glycerol, sorbitol, xylitol, propylene glycol, and polyethylene glycol. Generally, the electrochemical conversion of polyols usually involves oxidation or reduction to convert valuable chemicals. While this process has excellent benefits, it still possesses some

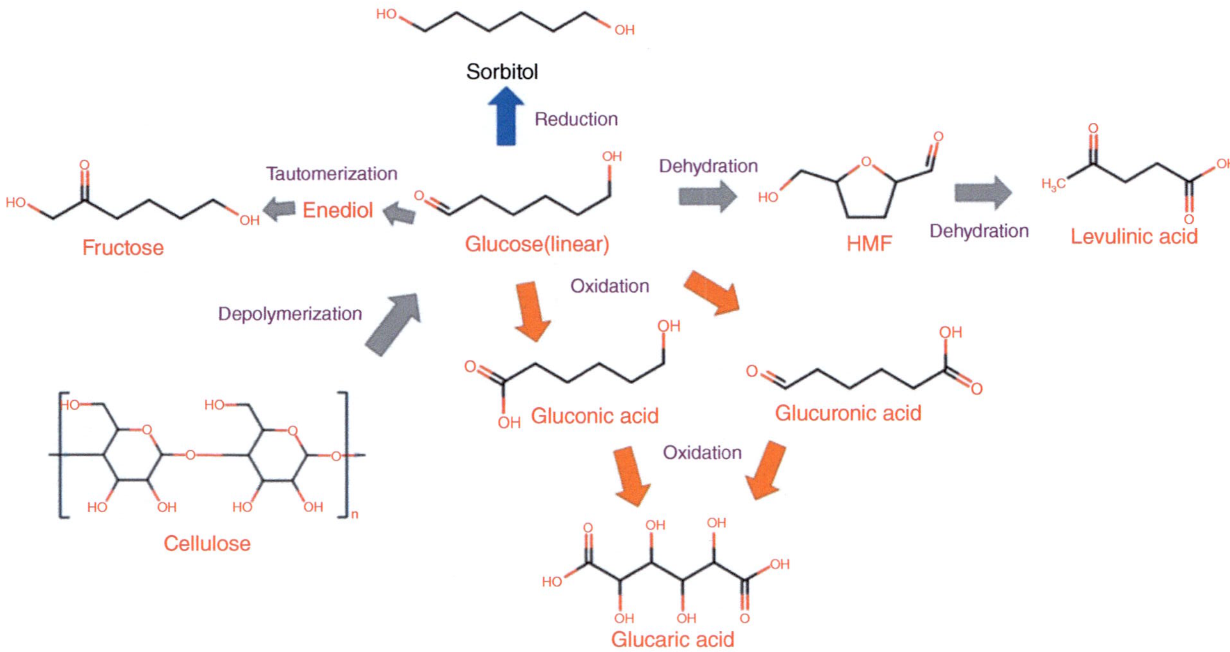

Figure 13.2 Electrochemical conversion of cellulose and its derivatives. Source: Lai et al. [15] / MDPI / CC BY 4.0.

challenges such as selectivity, catalyst durability, energy efficiency, and scalability. These are the significant hurdles that must be rectified [3]. This section demonstrated the recent research based on defect engineering for the electrochemical conversion of polyols into value-added products.

13.4.1.2 Electro-valorization of Glycerol

Glycerol, the simplest naturally occurring trihydroxy alcohol, was first identified by C.W. Scheele in the late eighteenth century and later named by M.E. Chevreul in the nineteenth century [17]. According to the forecasts by the Organization for Economic Cooperation and Development (OECD), the estimated global production of glycerol, a key by-product of biodiesel production, can reach 4 billion liters by 2026. Among biomass-derived polyols, glycerol oxidation reactions (GORs) have been extensively studied. The selectivity for specific products is influenced by various factors such as catalyst choice, reaction potential, and medium. During catalyst preparation, the defects and interfaces can impact the conversion efficiency and selectivity of the desired products. Youngkook et al. studied the bismuth-supported Pt/C (Bi–Pt/C) electrode for the electro-oxidation of glycerol to dihydroxyacetone with 100% selectivity (Figure 13.3a). The role of Bi enhanced the selectivity of secondary alcohols (Figure 13.3b) [18]. Zuyun et al. investigated bimetallic NiCo hydroxide as the catalyst for the electro-oxidation of glycerol to formate. The cobalt hydroxide alone shows poor selectivity because it has lower energy for oxygen anion removal, which leads to poor product desorption from the surface of the catalyst. But NiCo hydroxide, has higher energy for oxygen anion deintercalation, can promote the continued oxidation of glycerol to formate with a selectivity of 94.3% at an applied potential of 1.35 V (Figure 13.3c–e) [19]. Inderjeet et al. systematically studied the electrocatalytic oxidation of glycerol by utilizing different morphologies of palladium (Pd) nanocrystals such as Pd nanocubes (Pd–NCs), Pd truncated octahedra (Pd–TO), and Pd polycrystalline (Pd–PC) morphologies. During electro-oxidation, they applied a fixed potential of 1.2 V vs reversible hydrogen electrode (RHE), and the product conversion and selectivity were measured every 1 hour (Figure 13.3f). The results indicate that both the applied potential and the reaction time played a significant role in the product formation. After 1 hour reaction, under the same condition, the Pd–NC electrode exhibits a higher glycerol conversion of 47% than Pd–TO (30%) and Pd–PC (28%). These results suggest that the Pd–NC structure promotes faster electron transfer than other modified electrodes, which results in the higher selectivity of formic acid and C_2 products (Figure 13.3h) [20]. Muhammad et al. demonstrated the phosphorus (P)-doped P–Pd/CNT as an electrocatalyst for the electro-oxidation of glycerol to dihydroxyacetone with a selectivity of 90.8%, along with trace amounts of mesoxalic acid (<2%) and tartronic acid (<0.1%). The P doping helps to reduce the formation of carbonaceous intermediates and further enhances the glycerol oxidation process [21]. Xiaomei et al. utilized Pt NPs on sulfur-doped carbon nanotubes (SCNTs) as an electrocatalyst for the electro-oxidation of glycerol, formic acid, and carbon monoxide. From the results, they found that S doping enhances the spin density of nearby carbon atoms and can interact with platinum nanoparticles (Pt NPs)

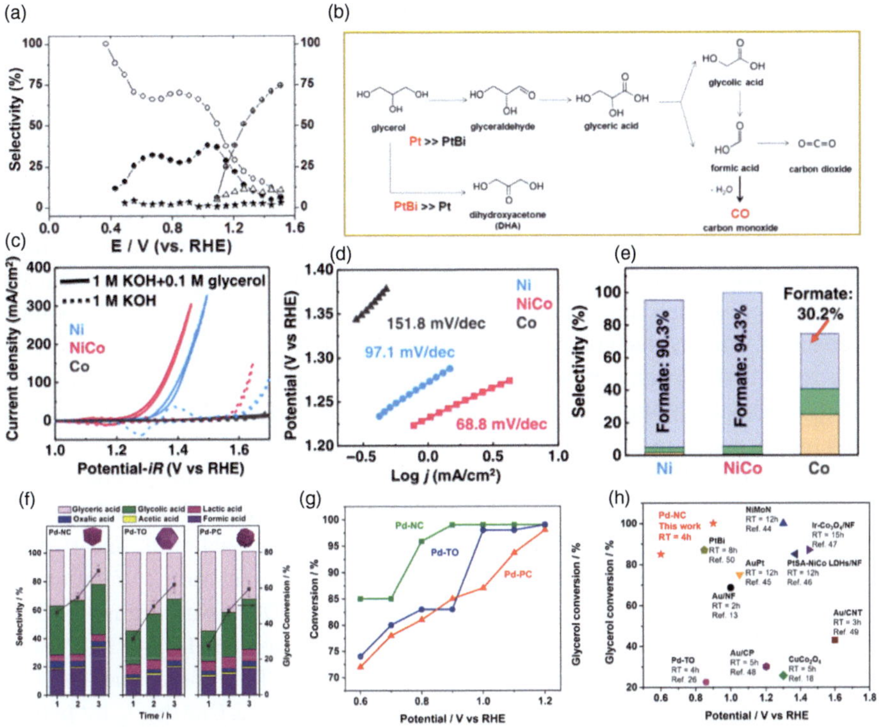

Figure 13.3 (a) Selectivity (%) products with the desired potential. ■ glycerol, ○ glyceraldehyde, ★ dihydroxyacetone, ● glyceric acid, Δ glycolic acid, ⊕ formic acid; (b) selective oxidation of glycerol; (c) glycerol oxidation on cyclic voltammetry; (d) Tafel curves; (e) selectivity, (f) glycerol conversion and selectivity at 1.2 V vs RHE over different morphologies of Pd catalysts; (g) comparison of glycerol oxidation performance over different catalyst surfaces in an alkaline solution at an applied potential. (h) A comparison of Pd–NC catalysts' results from the previous literature [18–20].

to improve the catalyst's performance and eliminate intermediate reactions [22]. Wang et al. demonstrated the electro-oxidation of glycerol to formate by utilizing a single-atom bismuth (Bi)-doped Co_3O_4. The Bi atom can replace the cobalt at the octahedral sites (Co^{3+}) of Co_3O_4, which enhances the reactivity of hydroxyl species (OH*) at the nearby tetrahedral Co sites (Co^{2+}). The presence of OH* promotes the oxidation of hydroxyl groups and the cleavage of C—C bonds, which can induce the efficient glycerol oxidation performance, with a current density of 400 mA cm^{-2} at 1.44 V and an excellent Faradaic efficiency of 97.05% for formate production [23]. Chencheng et al. studied the bimetallic Au–Pt catalyst for the electro-oxidation of glycerol to lactic acid. It achieves a higher Faradaic efficiency of 86% due to the synergistic effect of the metals [24].

13.4.1.3 Electro-valorization of Furans

Furans, a class of heterocyclic aromatic compounds, contain one oxygen atom and four carbon atoms. Furans are essential building blocks for synthesizing a wide range of value-added chemicals, pharmaceuticals, and biofuels [25].

Due to their flexible chemical structure and high reactivity, furans undergo many valuable chemicals during electrochemical valorization, which include furfural, 5-hydroxymethylfurfural, 2,5-dimethylfuran, and tetrahydrofuran.

13.4.1.4 5-Hydroxymethyl Furfural (5-HMF)

5-HMF is a versatile organic compound that is produced from the dehydration of biomass-derived carbohydrates, such as cellulose, glucose, fructose, sucrose, and starch [26]. The structure of 5-HMF includes both the hydroxymethyl and aldehyde functional groups connected with the furan ring. During electrochemical valorization, these functional groups of 5-HMF have significant potential for the conversion of value-added chemicals, such as 2,5- FDCA, 2,5 diformylfuran (DFF), 5, hydroxymethyl-2-furan carboxylic acid (HMFCA)), and reduction reactions, such as 2,5-dihydroxymethylfuran (DHMF), 2,5-hexanedione (HD), DMF, 2,5-dihydroxy methyl tetrahydrofuran (DHMTHF) [3], via oxidation.

Among various chemicals, FDCA is one of the top 12 value-added chemicals derived from biomass waste. The electrochemical oxidation of 5-HMF serves as an effective alternative method over traditional aerobic oxidation, and it has the potential to work under ambient conditions. Previous literature mentioned that the effective conversion of 5-HMF oxidation to FDCA was higher under high-pH conditions. Biao et al. studied the electro-oxidation of 5-HMF to FDCA by utilizing urchin-like $CuMn_2O_4$ nanostructured materials. To enhance the activity of $CuMn_2O_4$, they create oxygen vacancies on the surface of $CuMn_2O_4$ by using ammonia solution. Among various catalysts, $CuMn_2O_4$-40 exhibited excellent electrocatalytic performance, with the lowest Tafel slope value and a high current density of 20 mA/cm^2 at an applied potential of 1.31 V vs RHE in 1.0 M KOH (Figure 13.4). The enhanced performance is due to oxygen vacancies that lead to an increase in high-valence Mn^{4+}—O groups, which result in the higher Faradaic efficiency of 96% toward FDCA production [27]. Qin et al. utilized low-cost, nonmetal-doped carbon electrodes for the electro-oxidation of 5-HMF to FDCA. They used nitrogen and boron are the dopants, and the results show that nonmetal doping increases the Faradaic efficiency of 71% toward FDCA [28]. Yu-Feng et al. studied the cation-deficient (vacancies) NiFe layered double hydroxides (LDH) as an electrocatalyst for the electro-oxidation of 5-HMF to FDCA. From the electron paramagnetic resonance (EPR) results, they confirmed the presence of vacancies. The introduction of these vacancies on the LDH significantly alters the electronic structure and improves the electrochemical performance. At the fixed potential of 1.48 V, the catalyst achieved a high FDCA conversion of 96.8% with a Faradaic efficiency of 84.47% (Figure 13.5) [17]. Wu et al. studied the oxygen deficiency rich scandium oxide (V_O-Sc_2O_3) as an electrocatalyst for the electro-oxidation of 5-HMF to FDCA. The presence of oxygen vacancies improves the adsorption rate of the reactant and minimizes the energy barrier during the electrochemical reaction [29].

13.4.1.5 Electro-valorization of Furfural

Furfural is an organic compound like 5-HMF, and it serves as a starting material for furfuryl alcohol and furoic acid. Furfuryl alcohol and furoic acid are the

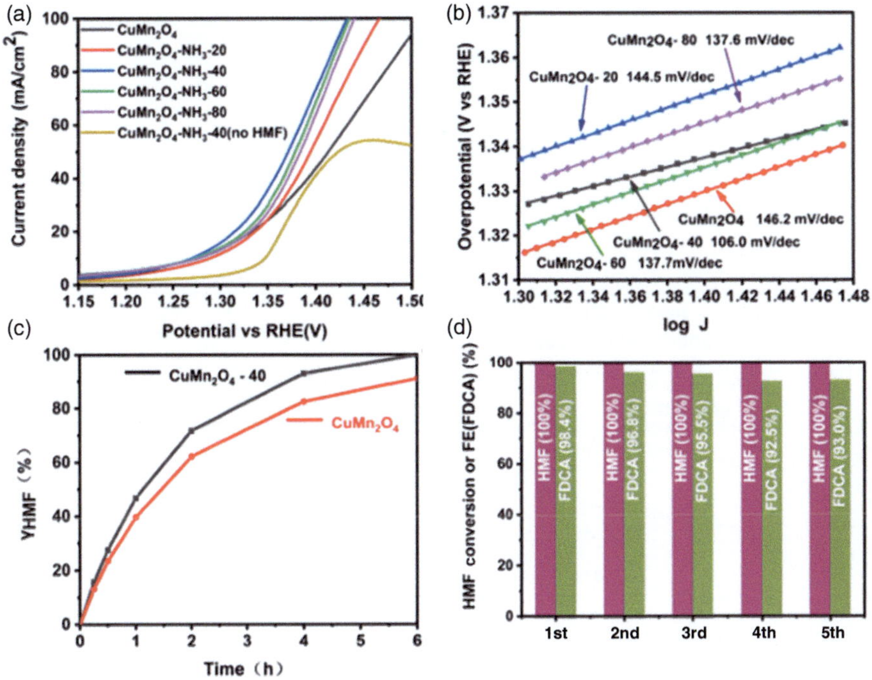

Figure 13.4 (a) The LSV curves of $CuMn_2O_4$-NH_3 and $CuMn_2O_4$ were recorded at a scan rate of 10 mV/s in 1 M KOH, with and without the addition of 10 mM 5-HMF. (b) Corresponding Tafel plots. (c) The variation in 5-HMF conversion (%) over time during chronoamperometric testing in a 1 M KOH solution with 10 mM 5-HMF at 1.31 V vs RHE. (d) The 5-HMF conversion (%) and Faradaic efficiency (FE %) of the product formed on the $CuMn_2O_4$-NH_3-40 electrode after 6 hours of reaction. Source: Zhuet al. [27] / with permission of American Chemical Society.

intermediates for pharmaceutical industries and polymer industries. During electrochemical valorization, furfural undergoes both oxidation and reduction. Nowadays, furfural achieves a higher conversion rate of FDCA compared to 5-HMF through biomass valorization, with a global production of 300 kt/yr [30]. Other than FDCA, the electro-oxidation of furfural is converted into a wide range of chemicals, such as furoic acid, 2,5-hydroxyfuranone, and MA, while, in case of reduction, it provides furfuryl alcohol, methyl furan, hydroxyfuran, etc. [3]. The design and optimization of material surfaces are essential for improving the electrochemical performance of desired products. For example, Zhenhao et al. demonstrated nanoporous copper (NP–Cu) for the electrocatalytic hydrogenation of furfural to furfuryl alcohol. The high specific surface area of NP–Cu provided abundant hydrogen adsorption sites. Therefore, it exhibits an efficient hydrogenation process and suppresses the hydrogen evolution reaction (HER). When compared to a commercial Cu sheet substrate, NP–Cu exhibits higher efficiency toward furfuryl alcohol (17–95%) [31]. Xian et al. developed Cu_3P nanosheets and Ni_2P nanocrystals onto commercial carbon fiber cloth (CFC) and studied the furfural hydrogenation reaction. The Cu_3P/CFC electrode provides an efficient furfural hydrogenation reaction with

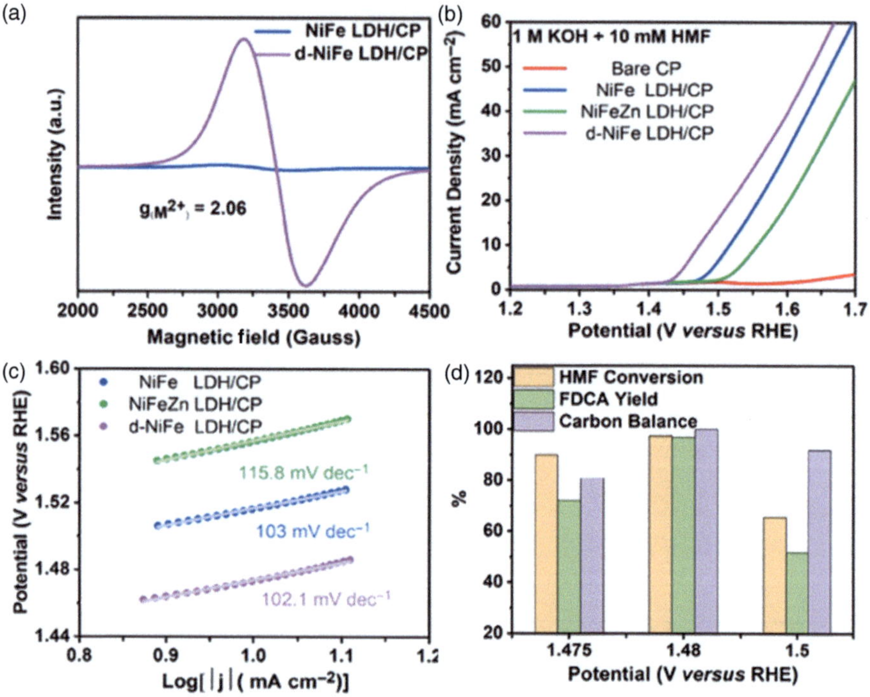

Figure 13.5 (a) EPR spectra of NiFe LDH/CP and d-NiFe LDH/CP, (b) LSV curves, and (c) corresponding Tafel plots in the presence of 10 mM 5-HMF. (d) Conversion of 5-HMF to yield of FDCA and carbon balance at various applied voltages were examined using d-NiFe LDH as the catalyst. Source: Adapted from Qi et al. [17].

the complete conversion of furfuryl alcohol with the Faradaic efficiency of 92–98% at −0.05 to −0.55 V potential vs RHE. Meantime, the Ni_2P/CFC exhibits effective oxidation furfural to furoic acid with the Faradaic efficiency of 90–98% at 1.2–1.7 V potential vs RHE in basic media. Density functional theory (DFT) studies evidenced that the excellent furfural oxidation is due to the formation of a high-valent Ni species Ni_2P catalyst during electrocatalysis. In the case of Cu_3P, efficient furfural hydrogenation is due to improved hydrogen coverage and controlled hydrogen desorption [32]. Peng et al. utilized single-atom copper with phosphorous-doped carbon (Cu_1/PC) as a catalyst for the electrochemical hydrogenation of furfural reaction. The modified material exhibits the excellent conversion of furfural to furfuryl alcohol and 2-methylfuran with high Faradaic efficiencies of 90% and 60%, respectively, at applied potentials of −0.75 V and −0.9 V vs RHE. The excellent selectivity and efficiency were attributed to the synergy between single-atom copper sites and the oxophilic phosphorus dopant in the catalyst. The single-atom copper sites enhanced efficiency and suppressed the HER; meanwhile, the P facilitated the two-step hydrogenation process to convert furfural into 2-methylfuran [33]. Stephen et al. studied the electrochemical oxidation of furfural to MA by using PbO_2 as an electrocatalyst under acidic conditions. The choice of this PbO_2 material is due to its higher oxidizing potential and good stability in an acidic medium. They compared the electrochemical performance of PbO_2 with MnO_2 and Pt anodes. The results confirmed that PbO_2 exhibited the highest MA yield of 65.1%, while MnO_2 and Pt alone exhibited poor yields because their oxidation potentials were closer toward water oxidation, which affected the MA conversion [34].

13.4.1.6 Chemical Intermediates and Fuels from Lignin

Lignin is the second largest biomass source after cellulose. Due to the presence of complex functional groups, high molecular weight, polymeric structure, and hydrophobic properties of lignin are very challenging through the electrochemical conversion method. Particularly, insolubility can restrict the effective interaction of reactants with the electrode surface, which results in poor reactivity and a greater number of possible side reactions. Therefore, the recent research focused on the rational design of electrocatalysts and interfaces to achieve a high conversion rate, Faradaic efficiency, and desired selectivity. In the selective conversion of lignin, the resulting chemicals typically include aromatic monomers such as phenol, 4-alkylphenols, syringol derivatives, guaiacol derivatives, vanillin, syringaldehyde, and various aromatic acids (Figure 13.6) [35].

For the significant conversion of lignin monomers to value-added chemicals, the introduction of a third element into the bimetallic alloys improves the reaction process. For example, Yulin et al. introduced boron as the third element in the Pt–Ni alloy and mesoporous carbon (PtNiB/CMK) composite for the electrochemical hydrogenation of guaiacol. The boron-introduced electrode exhibits a higher Faradaic efficiency of 86.2% than PtNi/CMK. The presence of boron significantly enhanced the adsorption energy of the reaction intermediate, which can improve the Faradaic efficiency [36]. Xukai et al. studied the platinum (Pt)-modified shrimp shell biochar (SSB) enriched with nitrogen as an electrocatalyst for phenol conversion. The as-synthesized electrocatalyst exhibited complete conversion of phenol into highly selective cyclohexanone and cyclohexanol with an efficiency of 98%. From DFT studies, they confirmed that the efficient conversion of phenols was attributed to C–O, Pt, and Pt–Nx active sites [37]. Peng et al. demonstrated the efficient electrochemical hydrogenation of phenol into cyclohexanol and cyclohexanone by using carbon-supported Ni_x-MoO_2 heterostructures. The Ni_{10}-MoO_2/C catalyst exhibits higher oxygen vacancies (V_O), which significantly enhances the

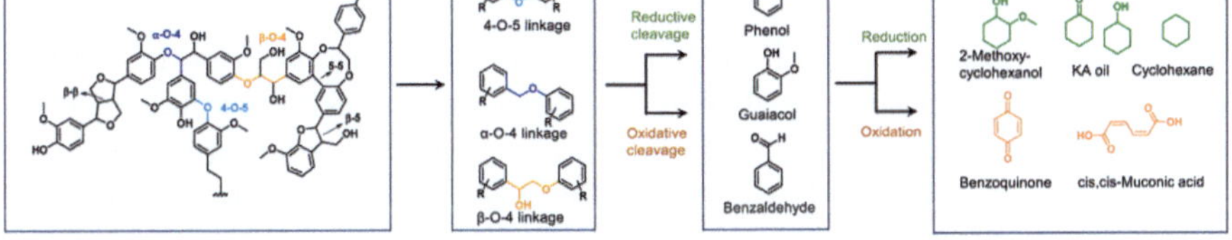

Figure 13.6 Schematic representation of lignin to chemical products by electrochemical biomass valorization. Source: Liu et al. [35] / American Chemical Society / CC BY-NC-ND 4.0.

adsorption of the cyclohexanone intermediate and promotes excessive hydrogenation. The Ni_{10}-MoO_2/C catalyst exhibits 95% selectivity toward cyclohexanol. With an increase in the nickel content (Ni_{20}-MoO_2/C), the density of oxygen vacancies was reduced, which facilitates better desorption of cyclohexanone with a selectivity of 86% [38]. Wang et al. studied the single-atom Pt-dispersed nitrogen-doped CNTs (Pt_1/N-CNTs) as an electrocatalyst for the selective C_α—C_β bond oxidative cleavage in lignin dimers. The Pt_1/N-CNT catalyst exhibits an excellent conversion rate with an 81% yield of benzaldehyde. Particularly, with the amount of 0.41 wt.% Pt, the Pt_1/N-CNT catalyst exhibits good results compared to both the bulk Pt electrode (which contains 100 wt.% Pt) and the commercial 20 wt.% Pt/C catalyst [39].

13.5 Challenges in Electrochemical Biomass Valorization

During electrochemical biomass valorization, there are significant challenges that have been addressed during the conversion of desirable products, such as (i) Selectivity: Biomass waste contains a wide range of organic compounds, such as alcohols, ketones, aldehydes, glycol, furanic compounds, and carbonyls. Based on the chemical properties, these compounds can be utilized in different ways. During electrochemical reactions, it is difficult to convert desired products. (ii) Side reaction: During electrochemical valorization, polymerization and cross-reactions are serious issues, which can lead to electrode fouling and create an impact on the selectivity of the final products. (iii) Faradaic efficiency: During electrochemical conversion, HER and OER are the important factors that affect the Faradaic efficiency. These reactions utilize the incoming electrons from the electrolyte. (iv) Impurity forbearance: There are several metal impurities that are present in the biomass and its intermediates, which can inhibit the electrode fouling as well as may lead to electrodeposition in some cases. Therefore, effective pretreatment is needed to remove the impurities in the biomass feedstocks. (v) Catalyst stability: Maintaining the electrocatalyst on the electrode surface over time is the main issue during electrochemical valorization because based on different pH, applied potential, and electrolyte composition, the oxidation or reduction of specific intermediates can result in catalyst dissolution, particle agglomeration, and corrosion.

13.6 Future Perspectives and Conclusions

In past decades, world has realized the environmental impacts of non-renewable energy sources, particularly fossil fuels. As it results in green greenhouse emissions and climate changes, we, therefore, need promising alternative methods to mitigate these impacts. Electrochemical biomass valorization is an excellent option to reduce the usage of non-renewable energy because it can convert waste biomass into valuable chemicals and electricity. The recent research explores a variety of materials such as metals, metal sulfides, and carbon-based materials for this conversion. The surface engineering of material is crucial for achieving the selective conversion of biomass intermediates into desired products. Among these, doping and interface engineering are highly effective because they provide additional active sites and increase the accommodation of biomass intermediates. Therefore, it can improve the long-term conversion efficiency, catalytic activity, and catalyst stability. This chapter addresses the recent advancements in electrocatalyst design, particularly, the role of doping and interface engineering of materials that can boost the electrocatalyst's performance for biomass valorization. It also addresses the key challenges in electrochemical biomass valorization. We hope that the present chapter can motivate further research on the electrochemical valorization of lignocellulosic biomass resources.

References

1 Negi, H., Suyal, D.C., Soni, R. et al. (2023). Indian scenario of biomass availability and its bioenergy-conversion potential. *Energies* 16 (15): 5805.

2 Erdiwansyah, G.A., Mamat, R., Bahagia, N.M. et al. (2024). Prospects for renewable energy sources from biomass waste in Indonesia. *Case Stud. Chem. Environ. Eng.* 10: 100880.

3 Lucas, F.W.S., Grim, R.G., Tacey, S.A. et al. (2021). Electrochemical routes for the valorization of biomass-derived feedstocks: from chemistry to application. *ACS Energy Lett.* 6 (4): 1205–1270.

4 Etale, A., Onyianta, A.J., Turner, S.R., and Eichhorn, S.J. (2023). Cellulose: a review of water interactions, applications in composites, and water treatment. *Chem. Rev.* 123 (5): 2016–2048.

5 Huang, Y.B. and Fu, Y. (2013). Hydrolysis of cellulose to glucose by solid acid catalysts. *Green Chem.* 15 (5): 1095–1111.

6 Shrotri, A., Kobayashi, H., and Fukuoka, A. (2018). Cellulose depolymerization over heterogeneous catalysts. *Acc. Chem. Res.* 51 (3): 761–768.

7 Isikgor, F.H. and Becer, C.R. (2015). Lignocellulosic biomass: a sustainable platform for the production of bio-based chemicals and polymers. *Polym. Chem.* 6 (25): 4497–4559.

8 Shah, S.W.A., Xu, Q., Ullah, M.W. et al. (2023). Lignin-based additive materials: a review of current status, challenges, and future perspectives. *Addit. Manuf.* 74: 103711.

9 Cho, E.J., Trinh, L.T.P., Song, Y. et al. (2020). Bioconversion of biomass waste into high value chemicals. *Bioresour. Technol.* 298: 122386.

10 Alishah Aratboni, H., Rafiei, N., Garcia-Granados, R. et al. (2019). Biomass and lipid induction strategies in microalgae for biofuel production and other applications. *Microb. Cell Factories* 18 (1): 178.

11 De Schouwer, F., Claes, L., Vandekerkhove, A. et al. (2019). Protein-rich biomass waste as a resource for future biorefineries: state of the art, challenges, and opportunities. *ChemSusChem* 12 (7): 1272–1303.

12 Du, L., Shao, Y.Y., Sun, J.M. et al. (2018). Electrocatalytic valorization of biomass-derived chemicals. *Cat. Sci. Technol.* 8 (13): 3216–3232.

13 Fan, Z.Y., Zhang, W.J., Li, L. et al. (2022). Recent developments in electrode materials for the selective upgrade of biomass-derived platform molecules into high-value-added chemicals and fuels. *Green Chem.* 24 (20): 7818–7868.

14 Luo, Y. and Wu, Y. (2023). Defect engineering of nanomaterials for catalysis. *Nanomaterials* 13 (6): 1116.

15 Lai, Z.I., Lee, L.Q., and Li, H. (2021). Electro reforming of biomass for value-added products. *Micromachines* 12 (11): 1405.

16 Huang, L.Z., Ma, M.G., Ji, X.X. et al. (2021). Recent developments and applications of hemicellulose from wheat straw: a review. *Front. Bioeng. Biotechnol.* 9: 690773.

17 Qi, Y.F., Wang, K.Y., Sun, Y. et al. (2022). Engineering the electronic structure of NiFe layered double hydroxide nanosheet array by implanting cationic vacancies for efficient electrochemical conversion of 5-hydroxymethylfurfural to 2,5-furandicarboxylic acid. *ACS Sustain. Chem. Eng.* 10 (1): 645–654.

18 Kwon, Y., Birdja, Y., Spanos, I. et al. (2012). Highly selective electro-oxidation of glycerol to dihydroxyacetone on platinum in the presence of bismuth. *ACS Catal.* 2 (5): 759–764.

19 He, Z., Hwang, J., Gong, Z. et al. (2022). Promoting biomass electrooxidation via modulating proton and oxygen anion deintercalation in hydroxide. *Nat. Commun.* 13 (1): 3777.

20 Chauhan, I., Bajpai, H., Ray, B. et al. (2024). Electrocatalytic glycerol conversion: a low-voltage pathway to efficient carbon-negative green hydrogen and value-added chemical production. *ACS Appl. Mater. Interfaces* 16 (20): 26130–26141.

21 Ahmad, M.S., Singh, S., Cheng, C.K. et al. (2020). Glycerol electro-oxidation to dihydroxyacetone on phosphorous-doped Pd/CNT nanoparticles in alkaline medium. *Catal. Commun.* 139: 105964.

22 Ning, X.M., Zhou, X.S., Luo, J. et al. (2019). Glycerol and formic acid electro-oxidation over Pt on S-doped carbon nanotubes: effect of carbon support and synthesis method on the metal-support interaction. *Electrochim. Acta* 319: 129–137.

23 Wang, Y., Zhu, Y.-Q., Xie, Z. et al. (2022). Efficient electrocatalytic oxidation of glycerol via promoted OH* generation over single-atom-bismuth-doped spinel Co_3O_4. *ACS Catal.* 12 (19): 12432–12443.

24 Dai, C.C., Sun, L.B., Liao, H.B. et al. (2017). Electrochemical production of lactic acid from glycerol oxidation catalyzed by AuPt nanoparticles. *J. Catal.* 356: 14–21.

25 Saeid, H., Al-sayed, H., and Bader, M. (2023). A review on biological and medicinal significance of furan. *Alq. J. Med. App. Sci.* 44–58.

26 Gao, Y.J., Ge, L., Xu, H.L. et al. (2023). Electrocatalytic refinery of biomass-based 5-hydroxymethylfurfural to fine chemicals. *ACS Catal.* 13 (17): 11204–11231.

27 Zhu, B., Qin, Y., Du, J.W. et al. (2021). Ammonia etching to generate oxygen vacancies on $CuMn_2O_4$ for highly efficient electrocatalytic oxidation of 5-hydroxymethylfurfural. *ACS Sustain. Chem. Eng.* 9 (35): 11790–11797.

28 Qin, Q., Heil, T., Schmidt, J. et al. (2019). Electrochemical fixation of nitrogen and its coupling with biomass valorization with a strongly adsorbing and defect optimized boron–carbon–nitrogen catalyst. *ACS Appl. Energy Mater.* 2 (11): 8359–8365.

29 Wu, Y., Ma, L., Wu, J. et al. (2024). High-surface area mesoporous Sc_2O_3 with abundant oxygen vacancies as new and advanced electrocatalyst for electrochemical biomass valorization. *Adv. Mater.* 36 (16): e2311698.

30 Mariscal, R., Maireles-Torres, P., Ojeda, M. et al. (2016). Furfural: a renewable and versatile platform molecule for the synthesis of chemicals and fuels. *Energy Environ. Sci.* 9 (4): 1144–1189.

31 Yang, Z.H., Chou, X.Y., Kan, H.Y. et al. (2022). Nanoporous copper catalysts for the fluidized electrocatalytic hydrogenation of furfural to furfuryl alcohol. *ACS Sustain. Chem. Eng.* 10 (22): 7418–7425.

32 Zhang, X., Han, M.M., Liu, G.Q. et al. (2019). Simultaneously high-rate furfural hydrogenation and oxidation upgrading on nanostructured transition metal phosphides through electrocatalytic conversion at ambient conditions. *Appl. Catal. B Environ.* 244: 899–908.

33 Zhou, P., Chen, Y., Luan, P. et al. (2021). Selective electrochemical hydrogenation of furfural to 2-methylfuran over a single atom Cu catalyst under mild pH conditions†. *Green Chem.* 23 (8): 3028–3038.

34 Kubota, S.R. and Choi, K.S. (2018). Electrochemical valorization of furfural to maleic acid. *ACS Sustain. Chem. Eng.* 6 (8): 9596–9600.

35 Liu, X., Wang, Y., and Duan, H. (2024). Recent progress in electrocatalytic conversion of lignin: from monomers, dimers, to raw lignin. *Precis Chem.* 2 (9): 428–446.

36 Zhou, Y., Gao, Y., Zhong, X. et al. (2019). Electrocatalytic upgrading of lignin-derived bio-oil based on surface-engineered PtNiB nanostructure. *Adv. Funct. Mater.* 29 (10): 1807651.

37 Lu, X., Wang, J., Peng, W. et al. (2023). Electrocatalytic hydrogenation of phenol by active sites on Pt-decorated shrimp shell biochar catalysts: performance and internal mechanism. *Fuel* 331: 125845.

38 Zhou, P., Guo, S.X., Li, L. et al. (2023). Selective electrochemical hydrogenation of phenol with earth-abundant Ni-MoO_2 heterostructured catalysts: effect of oxygen vacancy on product selectivity. *Angew. Chem. Int. Ed. Engl.* 62 (8): e202214881.

39 Cui, T., Ma, L., Wang, S. et al. (2021). Atomically dispersed Pt-N(3)C(1) sites enabling efficient and selective electrocatalytic C-C bond cleavage in lignin models under ambient conditions. *J. Am. Chem. Soc.* 143 (25): 9429–9439.

14

Defect and Interface Engineering in Fuel Cells

Minerva Guerra Balcázar, Carlos Guzmán Martínez, and Alejandra Álvarez López

Facultad de Ingeniería, División de Investigación y Posgrado, Universidad Autónoma de Querétaro, Querétaro, C. P. 76010, México

14.1 Impact of Defects on Electrocatalytic Activity

Defects in catalyst materials have become key factors in improving the performance of the fuel cell, especially in electrocatalytic processes such as oxygen reduction reaction (ORR) and hydrogen oxidation reaction (HOR). These structural defects, including vacancies, doping sites, and dislocations, change the electronic properties of the catalyst, allowing it to interact more effectively with active species. For example, nitrogen-doped carbon-based materials, which have been widely studied in fuel cells, rely on defects to create high-energy active sites that promote the adsorption and activation of oxygen or hydrogen intermediates. One of the most significant effects of defects is their ability to enhance the adsorption of the active species.

Defect engineering play a crucial role in the selectivity and stability of catalysts. It is possible to control the reaction pathways and reduce the activation energy of the important reactions of fuel cells. These effects are reflected in the improved catalytic efficiency and durability under operating conditions.

Theoretical models and experimental studies have confirmed their usefulness for understanding the relationship between defects and catalytic activity. Computational methods, such as density functional theory (DFT), provide insights into the electronic structure of defect-rich catalysts and predict how they interact with active species. These theoretical findings are corroborated by experimental techniques such as X-ray photoelectron spectroscopy (XPS) and transmission electron microscopy (TEM), which reveal the nature and distribution of defects in the catalysts. Together, these approaches enable the rational development of defect-engineered catalysts to ensure optimal fuel-cell performance. In summary, defect manipulation represents a transformative strategy for improving the efficiency, selectivity, and stability of fuel-cell electrocatalysts.

Surface and Interfacial Defects in Nanomaterials for Sustainable Energy Production and Storage, First Edition. Edited by Noé Arjona, Lorena Álvarez Contreras, and Minerva Guerra Balcázar.

14.2 Defects on Noble Metal-Based Catalysts

The use of noble metal catalysts in defect engineering represents a novel alternative for modulating surface properties and electronic structures. Arjona et al. utilized three-dimensional (3D) platinum and 3D palladium with various surface defects such as vacancies and terraces. Their findings indicated that the Pd/3D-C material demonstrated a lower overpotential owing to reduced energy activation. Furthermore, its increased stability was attributed to surface reorganization, specifically the formation of the (100) plane in Pd/3D-C, which resulted in high electrocatalytic activity and stability [1].

Other studies have highlighted the benefits of employing noble metal catalysts with engineered defects. For example, the introduction of step edges in noble metal-based catalysts has been shown to significantly impact the coordination environment and modify the d-band center. This modification enhances the performance of ORR by improving stability and electroactivity.

Moreover, research combining noble metals with transition metal oxides containing oxygen vacancies has yielded promising results for the alcohol oxidation reactions in fuel cells. Pd/C and Pd/O_v–MOs catalysts exhibited improved electrocatalytic activity for ethanol oxidation because of the presence of oxygen vacancies, which altered the adsorption energy of ethanol and improved the onset potential of the reaction.

In conclusion, defect engineering in noble metal catalysts, especially when combined with transition metal oxides, offers a transformative approach for enhancing the efficiency, stability, and selectivity of electrocatalytic processes in fuel cells. These advancements have significant potential for the development of more effective and durable fuel-cell systems (Figure 14.1).

In other studies, with the formation of Au/C hierarchical structures with numerous defects used for the electro-oxidation of glycerol, it was found that the contribution of the (100) plane of the gold electrocatalyst promotes the major OH adsorption and tolerance of poisoning, resulting in better electrocatalytic activity [2]. Another type of defect that is widely used in noble metal-based catalysts is step edges [3]. These types of defects have a significant effect on the coordination environment and modify the d-band center, resulting in better performance for ORR in terms of stability and electroactivity.

In alternative research, a combination of a noble metal and a transition metal oxide containing oxygen vacancies has been employed for the oxidation of short-chain alcohols intended for use as fuel in fuel cells. In this regard, Pd/C and Pd/O_ν–Mos were used for the oxidation of ethanol, where it was determined that $PdFe_2O_3$/C exhibits improved electrocatalytic activity due to the presence of oxygen vacancies. Additionally, it was demonstrated that changes in the onset potential of the reaction were related to modifications in the adsorption energy of ethanol [4]. In addition, a mixture of transition metal oxides and precious metals is also used for the oxidation of more complex alcohols, such as glycerol. In this regard, Zn, Mo, and Fe oxides, all of which contain oxygen vacancies, were combined with Pd. The results show that the use of different metal oxides with varying percentages

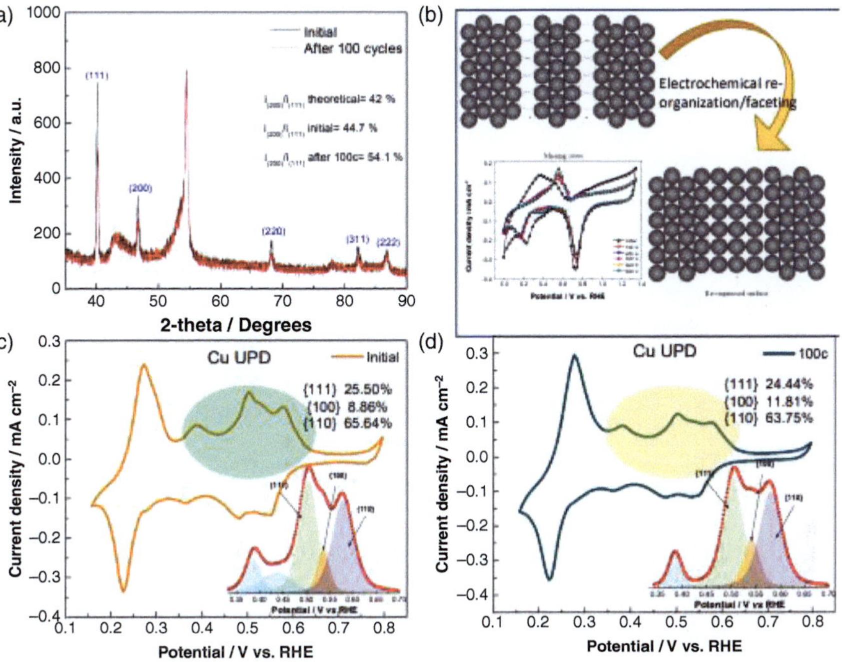

Figure 14.1 (a) X-ray patterns of Pd/3D-C, (b) electrochemical reorganization, and (c) and (d) Cu UPD (under potential deposition) for electrochemical reorganization evidence before and after the electrochemical cycling. Source: Arjona et al. [1] / with permission of Elsevier.

of vacancies leads to changes in the morphology and electronic interactions of the material, which in turn affect the electrocatalytic activity for crude glycerol oxidation [5]. Figure 14.2 shows a comparison between the oxidation of glycerol and crude glycerol using the tested materials. Notably, the presence of oxygen vacancies enhances the electrocatalytic activity and stability of the material. Additionally, the Pd/O_v-Fe_2O_3 material exhibited higher activity for the oxidation of crude glycerol, achieving a current density of 90 mA/g, compared with 60 mA/g for Pd/C alone. This can be explained by the additional adsorption and oxidation processes of the reaction intermediates, which may have limited the reaction rate. The mass activity data for this and other materials are presented in Figure 14.2-C, confirming the superior performance of the Pd/O_v-Fe_2O_3 material. Furthermore, this material is more stable, which is related to its ability to desorb and oxidize possible intermediates in the reaction, ensuring that the active sites remain available to carry out the desired reaction.

Most research on defective electrocatalysts is focused on the development of materials for ORR in fuel cells, which is considered the limiting reaction in these types of devices. Some works focus on the development of Pd@Pt4.5L nanocages with the partial removal of Pd/Pt core/shell icosahedra by changing the type or content of the etchant during the preparation process [6]. The selective internal etching of the electrocatalyst promotes the interdiffusion and alloying between Pt and Pd atoms,

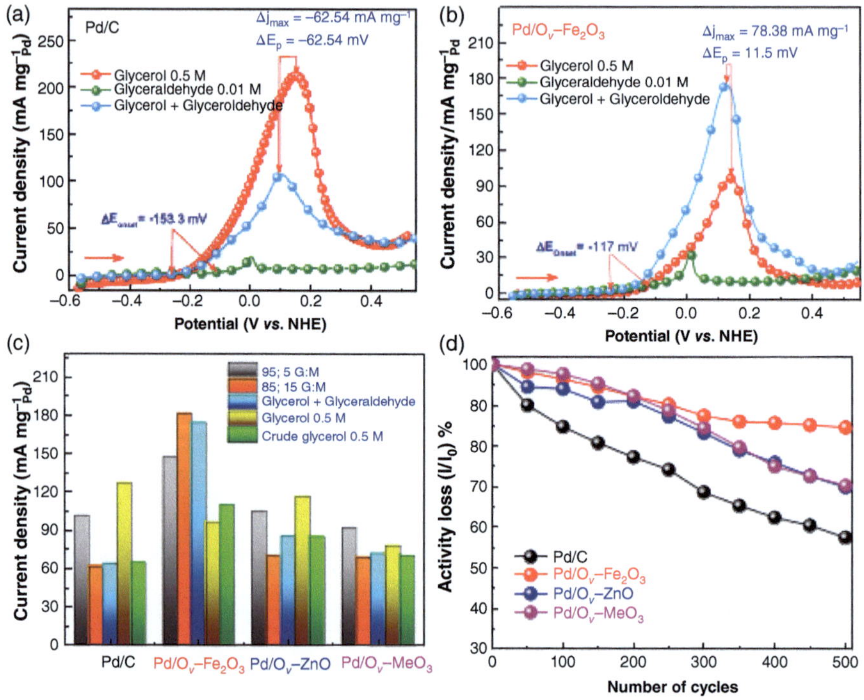

Figure 14.2 (a) Cyclic voltammetry ongoing scan 20 mV for electro-oxidation of crude glycerol, (b) Pd/O_v-Fe_2O_3 positive potential scan for the electro-oxidation of several molecules, (c) summary of electrochemical activity for several fuels, and (d) stability curves for the electrochemical oxidation of crude Glycerol 0.5 M at 50 °C in basic media. Source: Velazquez-Hernández et al. [5] / with permission of Elsevier.

forming twin defects and small holes on the catalyst wall. The twin defects on the electrocatalyst surface improve their ORR performance by regulating its d-band, while the holes increase the frequency of the three-phase interface for mass transport. As a result, electrochemical evaluations show that the mass activity, specific activity, and stability of Pd@$Pt_{4.5}$ nanocages are much higher than those of commercial Pt/C.

Other researchers have reported novel and effective stacking and carving strategies to synthesize the defect-rich surfaces of Pt-Cu and Pt-Fe nanocrystals, in which the formation of nuclei is fine-tuned by regulating the evolution of thermodynamically unstable structures [7]. A large number of edge steps, vacancies, kinks, and low-coordinate atomic steps were observed on the interlayer surfaces and connecting regions. These defects regulate the electronic surface structure of the catalyst, thereby improving its activity and stability for ORR applications.

Xia et al. recently prepared bundled $Pt_{1.5}Ni$ alloy nanocages containing Pt skin (PtNi-BNS), which are rich in step defects and grain boundaries and can be used in practical H_2–air fuel cells [8]. They obtained structures with steps and edges on the surface of PtNi BNS; the faceted structure on this material can contribute to changing the interaction of metal oxygen during ORR.

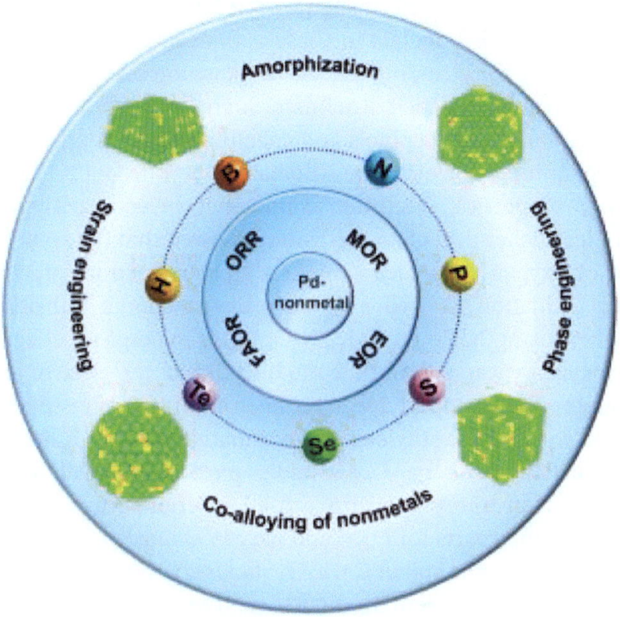

Figure 14.3 Fuel-cell applications of Pd–non-metal alloys. Formic acid oxidaion reaction, FAOR. Mengjun Wang et al. [10] / Springer Nature / CC BY 4.0.

Other types of materials are also used, such as Pd–non-metal alloys, particularly selenium. This type of electrocatalyst allows for the modulation of electrocatalytic properties by controlling the phases or through a process in which amorphous materials are obtained, which are rich in defects. These materials can undergo various synthesis processes or have diverse applications in the field of fuel cells like: FAOR (formic acid oxidation reaction), ORR and others. (Figure 14.3) because of the possibility of obtaining materials with different physicochemical properties and electronic structures [9].

Among the strategies to modify Pd–non-metal catalysts, the formation of amorphous materials presents a large number of surface atoms with low coordination numbers and is rich in defects [10]. Some researchers have prepared Pd–Ni–P nanoparticles and found that prolonged phosphorylation produces amorphous nanoparticles. This process can reduce the distance between active sites and positively affect the ethanol oxidation reaction (EOR). This improvement was also attributed to the amorphous nature of the nanoparticles, as their degree of disorder and defects allowed them to carry out the reaction of interest more efficiently [11].

14.3 Defects in Alternative Non-platinum Catalysts

The extensive use of noble metals in liquid fuel cells has limited their application because of the high cost and low stability of catalysts made from noble metals such as Pt, Pd, and Au. In this regard, defect engineering has been presented as a solution to enable the use of other metals in these applications. Significant advances have been made in this area, including the use of transition metal oxides, which generally have oxygen vacancies in their structures. Suárez et al. studied the influence of oxygen vacancies on Ni_2CoO_4 spinels and cobaltite spinels supported on nitrogen-doped carbon nanotubes for ethylene oxidation. The introduction of defects can tailor electronic properties, electronic conductivity, electrocatalytic activity, and stability [12]. Electrochemical characterization showed that the spinel with a higher percentage of oxygen vacancies exhibited greater electrocatalytic activity for ethylene glycol oxidation as well as a decrease in charge transfer resistance. Additionally, when using $NiCo_2O_3$ EG_{33}/N-CNTG, a synergistic effect was obtained between the oxygen vacancies and the modification of the electronic interactions of the carbon atoms in the support due to nitrogen doping. This resulted in a reduction in the onset potential of the ethylene glycol oxidation reaction as well as a decrease in the apparent activation energy. Other studies have found that cobalt and iron oxide-based materials in an amorphous form exhibit superior performance compared to their crystalline counterparts for ORR and oxygen evolution reaction (OER). In this study, it was found that the spinel $CoFe_2O_4$ shows similar activity for ORR as Pt/Vulcan X-72, with an onset potential of 0.8 V vs. 0.85 V. The electrons transferred by this metal were analyzed using the Koutecky–Levich equation, resulting in a value close to $4e^-$ [13].

14.4 Carbon-Based Materials and Their Modification with Defects

Carbon-doped materials are widely used as a support or catalyst because this type of material can enhance the electrocatalytic activity for ORR, avoiding the use of noble metals, and it has been reported that the incorporation of atoms such as nitrogen into the carbon structure can create active sites for ORR with greater stability and four electrons mechanism [14]. In Figure 14.4, the HRTEM (high resolution transmission electron microscopy) images of nitrogen-doped carbon synthesized using melamine as a precursor are shown, along with a comparison of its electrocatalytic activity for ORR. The results indicate that nitrogen-doped carbon derived from melamine approaches the activity of platinum, which is considered the best material for this reaction, with a difference of 90 mV. A major advantage of this material is that it is metal-free. Additionally, panel (d) demonstrates that this material exhibits greater stability compared to the conventional platinum-based materials used for this reaction.

14.5 Conclusions and Future Perspectives

Defect engineering has revolutionized the design of catalysts for various material classes, including transition metals, carbon-based materials, and non-precious

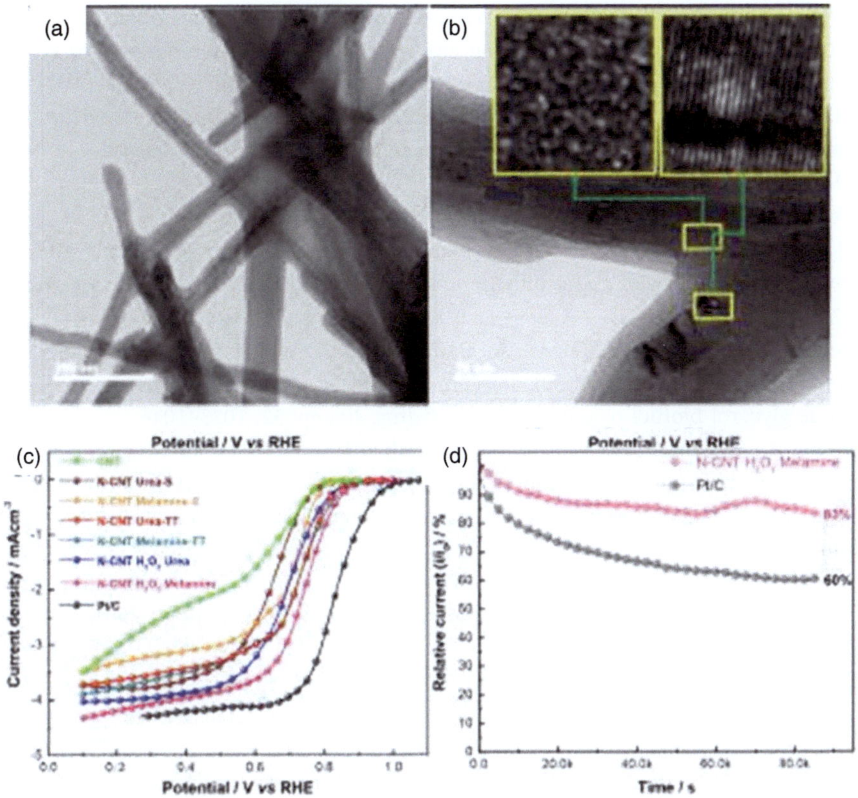

Figure 14.4 (a) and (b) HRTEM images of carbon doped with melamine as the precursor, (c) comparison of the oxygen reduction reaction ORR, and (d) comparison of stability of carbon doped with melamine as a precursor and Pt/C. Bejar et al. [14] / with permission of Elsevier.

metals. By strategically introducing and manipulating defects, unprecedented levels of catalytic activity, efficiency, and selectivity can be achieved. This approach not only addresses the limitations of platinum-group metals but also paves the way for cost-effective, scalable, and sustainable catalytic solutions. As advancements in nanotechnology and materials science continue, defect engineering will remain at the forefront of innovations in energy, the environment, and industrial catalysis. However, there are many obstacles to the extensive use of these materials, such as the precise control of defects, scalable synthesis of defective materials, physicochemical characterization combined with computational analysis such as DFT and other computational techniques.

References

1 Arjona, N., Guerra-Balcázar, M., and Álvarez-Contreras, L. (2022). Hierarchical Pd and pt structures obtained on 3D carbon electrode as electrocatalyst for ethylene glycol electro-oxidation, applied surface. *Science* 571: 151246.

2 Arjona, N., Torres–Pacheco, L.J., Álvarez-Contreras, L., and Guerra-Balcázar, M. (2022). Gold structures on 3D carbon electrodes as highly active nanomaterials for the clean energy conversion of crude glycerol. *Electrochim. Acta* 430: 141098.

3 Li, M., Zhao, Z., Cheng, T. et al. (2016). Ultrafine jagged platinum nanowires enable ultrahigh mass activity for the oxygen reduction reaction. *Science* 354: 6318.

4 Velazquez-Hernández, I., Lair, V., Cassir, M. et al. (2020). Ethanol electro–oxidation and spectroelectrochemical analysis of highly active sub<10 nm $PdFe_2O_3$, PdPt and PdAu bimetallic nanoparticles. *Int. J. Hydrog. Energy* 45: 9758–9772.

5 Velazquez-Hernández, I., Olivas, A., Bañuelos, J.A. et al. (2024). Noé Arjona; Defect engineering on metal oxides (Zn, Mo and Fe) and their impact on the crude glycerol biofuel oxidation. *Int. J. Hydrog. Energy* 52: 1033–1046.

6 Wang, X., Figueroa-Cosme, L., Yang, X. et al. (2016). Pt-based icosahedral nanocages: using a combination of {111} facets, twin defects, and ultrathin walls to greatly enhance their activity toward oxygen reduction. *Nano Lett.* 16 (2): 1467–1471.

7 Mao, J., Chen, Y., Pei, J. et al. (2016). Pt–M (M = Cu, Fe, Zn, etc.) bimetallic nanomaterials with abundant surface defects and robust catalytic properties. *Chem. Commun.* 52: 5985–5988.

8 Tian, X., Zhao, X., Su, Y.-Q. et al. (2019). Engineering bunched Pt-Ni alloy nanocages for efficient oxygen reduction in practical fuel cells. *Science* 366 (6467): 850–856.

9 Wang, M., Li, L., Wang, M., and Huang, X. (2022). Recent progress in palladium-nonmetal nanostructure development for fuel cell applications. *NPG Asia Mater.* 14: 78.

10 Wang, M., Li, L., Wang, M. et al. (2022). Recent progress in palladium-nonmetal nanostructure development for fuel cell applications. *NPG Asia Mater 14*, 78: https://doi.org/10.1038/s41427-022-00423-2.

11 Chen, L., Lilin, L., Zhu, H. et al. (2017). Improved ethanol electrooxidation performance by shortening Pd–Ni active site distance in Pd–Ni–P nanocatalysts. *Nat. Commun.* 8: 14136.

12 Alexander Suarez-Barajas, C.M., Ramos-Castillo, A.O., Guerra-Balcázar, M. et al. (2024). Oxygen vacancy-enriched $NiCo_2O_4$ spinels/N-doped carbon nanotubes-graphene composites for the ethylene glycol electro-oxidation. *Fuel* 360: 130371.

13 Indra, A., Menezes, P.W., Sahraie, N.R. et al. (2014). Unification of catalytic water oxidation and oxygen reduction reactions: amorphous beat crystalline cobalt iron oxides. *J. Am. Chem. Soc.* 136 (50): 17530–17536.

14 Béjar, J., Espinosa–Magaña, F., Avelar, J. et al. (2023). Rational design of nitrogen–doped carbon nanotubes by defect engineering for Zn–air batteries with high performance. *Carbon* 204: 411–426.

15

Defect and Interface Engineering in Electrolyzers

J.C. Cruz[1], *B. Pamplona Solis*[1], *K. García Uitz*[1], *and M.P. Gurrola*[2]

[1] *Tecnológico Nacional de México/ I.T. Chetumal, División de Estudios de Posgrado e Investigación, Av. Insurgentes 330, 77013, Chetumal, Quintana Roo, México*
[2] *IxM-SECIHTI-Tecnológico Nacional de México/ I.T. Chetumal, División de Estudios de Posgrado e Investigación, Av. Insurgentes 330, 77013, Chetumal, Quintana Roo, México*

15.1 Introduction to Electrolyzers

Electrolysis is an electrochemical process for producing the most developed, ecological, and technologically evolved hydrogen energy vector at a commercial level. In this process, electricity from an alternate energy source is used to separate the water molecule into hydrogen and oxygen. This process is carried out in a device called an electrolyzer.

Electrolyzers can be found from a kilowatt (kW) scale to gigawatts (GW), and these are interconnected with photovoltaic systems, wind systems, or hybrid systems. This renewable technology produces an environmentally friendly fuel by storing electrical energy in chemical energy.

Electrolyzers are electrochemical devices that are made up of an anode (positive) and a cathode (negative) separated by an electrolytic material. This last characteristic defines their classification in different electrolyzers (Figure 15.1), such as:

(i) Proton exchange membrane electrolyzer (PEMEL)
(ii) Alkaline solution electrolyzer (AEL)
(iii) Anion exchange membrane electrolyzer (AEMEL)
(iv) Solid oxide electrolyzer (SOEL)

Electrolyzers work under specific conditions to obtain their highest performance, with the aim of satisfying the needs of their transportation, use, and operation of the gases produced. Some of the conditions are shown in Table 15.1 [1].

This chapter addresses two of the electrolyzers: PEMEL and AEL, which constitute the majority of commercial technologies and have been on the market for a long time. As well as the introduction to the trends of AEMEL soon to its application. The materials used in the different interfaces of both systems are described and the problems generated in their interfaces in the different media (acid and base) are also

Surface and Interfacial Defects in Nanomaterials for Sustainable Energy Production and Storage, First Edition. Edited by Noé Arjona, Lorena Álvarez Contreras, and Minerva Guerra Balcázar.

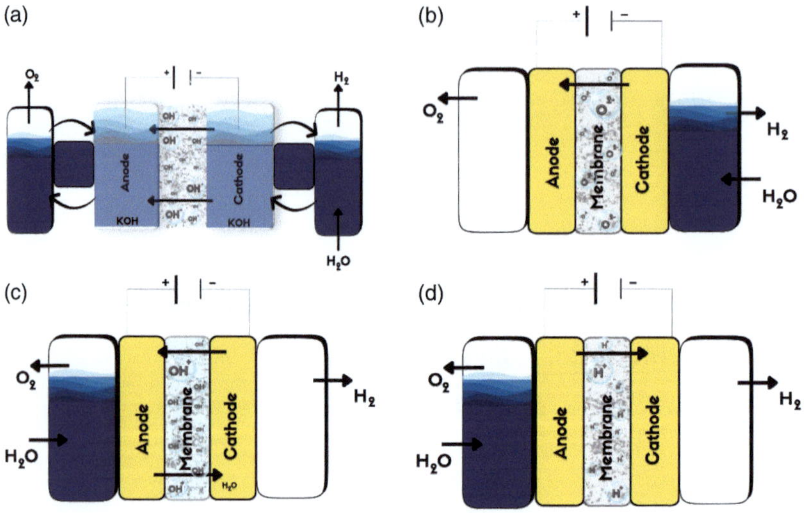

Figure 15.1 Classification of an electrolyzer with respect to the electrolyte used: (a) AEL, (b) PEMEL, (c) AEMEL, and (d) SOEL.

Table 15.1 Operating conditions of electrolyzers.

Technology	Alkaline large scale	Alkaline high pressure	Advanced alkaline	PEM	SOE
Status	Commercial	Commercial	Pre-commercial	Pre-commercial	Prototype
T (°C)	70–90	70–90	80–140	80–150	900–1000
P (bar)	1–25	Up to 690	Up to 120	Up to 400	Up to 30
kWh kg $-^1$ H_2	48–60	56–60	42–48	40–60	28–39

addressed. In addition to this, the components of both types of electrolyzers and the engineering problems are described; as well as those caused by the electrochemical phenomenon.

15.1.1 Proton Exchange Membrane Electrolyzer (PEMEL)

PEMEL are so named because they use a proton exchange polymeric membrane (PEM) as an electrolyte that separates the anode from the cathode. The parts that make up a mono-electrolyzer commonly are the following: housings for the anode and cathode, where the flow channels are located, then the seals, conductive meshes to evenly distribute the current from the source to the device, and the membrane electrode assembly (MEA) change membranes.

The reactions that occur in a PEM-type electrolyzer are described in the following equations:

$$\text{Anode: } 2\,H_2O \rightarrow \; O_2 + 4H^+ + 4e^- \qquad E^0 = +1.229\text{ V} \qquad (15.1)$$

$$\text{Cathode: } 4H^+ + 4e^- \rightarrow 2H_2 \qquad E^0 = 0 \qquad (15.2)$$

$$\text{Overall reaction: } 2H_2O \rightarrow 2H_2 + O_2 \qquad E^0 = -1.229\text{ V} \qquad (15.3)$$

PEMEL have the following advantages when compared to alkaline devices: they are less caustic and can operate at lower voltages, i.e. they can withstand higher temperatures, high current densities, and pressures, and they can achieve efficiencies of over 80% [2]. However, they also have high costs and corrosion in their elements due to the potential with which they operate [3].

The oxygen evolution reaction (OER) is the imitating one in an electrolyzer, since its kinetics are the slowest and it is the half-reaction of O_2 production, therefore, materials that resist corrosion caused in the environment are required [4].

15.1.2 Alkaline Electrolyzer

AEL are considered to be one of the first technologies used for electrolysis of water. This is why this technology has greater technological development, and its use has been promoted worldwide. However, it still presents different challenges for science. This type of electrolyzer uses aqueous solutions of sodium hydroxide (NaOH) or potassium hydroxide (KOH) with a variable concentration that can reach from 20% to 40% by weight to increase ionic mobility during the electrolysis process. These are separated by a diaphragm or membrane so as not to mix the gases produced during the reaction. However, it is risky to have high concentrations of the electrolyte since it can accelerate the corrosion process on the electrodes and their interfaces, which can affect the efficiency and useful life of the system.

During the operation of an electrolyzer, hydrogen is produced at the cathode, giving way to the hydrogen evolution reaction (HER), while oxygen is produced at the anode, carrying out the OER. In Eqs. (15.1)–(15.3), the half-reactions that occur during the electrolysis process in an AEL and the potential difference required under standard conditions (298.15 K and pH = 13.996) can be observed; however, the activation energies and the ohmic drops in the system are not considered [5].

$$\text{Cathode: } 2\,H_2O + 2\,e^- \rightarrow H_2 + 2\,OH^- \qquad E^0 = -0.8280\text{ V} \qquad (15.4)$$

$$\text{Anode: } 2\,OH^- \rightarrow 0.5\,O_2 + H_2O + 2e^- \qquad E^0 = +0.4011\text{ V} \qquad (15.5)$$

$$\text{Overall reaction: } H_2O \rightarrow H_2 + 0.5\,O_2 \qquad E^0 = -1.2291\text{ V} \qquad (15.6)$$

15.2 Materials Used as Catalysts in Electrolyzers

Electrocatalysts used in PEMELs and AELs are critical to improving the efficiency of HER and OER. Due to differences in the operating environment of each technology, the materials used in the electrodes vary significantly [6].

Volcano diagrams are used to analyze and predict the catalytic activity of different materials in electrochemical and heterogeneous reactions. These diagrams show the relationship between the adsorption-free energy of a reaction intermediate and the overpotential of the catalyzed reaction, forming a curve with an optimal region at the top. The "volcano" shape arises because too weak adsorption reduces reactivity by hindering the formation of intermediates, while too strong adsorption also limits it by hindering the release of final products. These diagrams are widely used in the development of catalysts for oxygen evolution and reduction, hydrogen production, and other key electrochemical processes in fuel cells and electrolyzers.

Figure 15.2 shows volcano diagrams showing the catalytic activity trends for oxygen evolution and reduction in different materials: (a) relationship between the negative theoretical overpotential and the standard free energy of the rate-limiting step in perovskites, considering a low coverage regime. The calculated values are used to evaluate the activity of each oxide, and the volcano curve is established from the scaling relationship between the reaction in alkaline media. (b) Comparison between the theoretical and experimental overpotential in acid media for rutile, anatase, Co_3O_4, and MnxOy oxides, considering the interaction with the neighboring site oxygen. For NiO_2, PbO_2, and SnO_2, the reaction occurs at bridging sites. (c) Relation between theoretical and experimental overpotentials in perovskites,

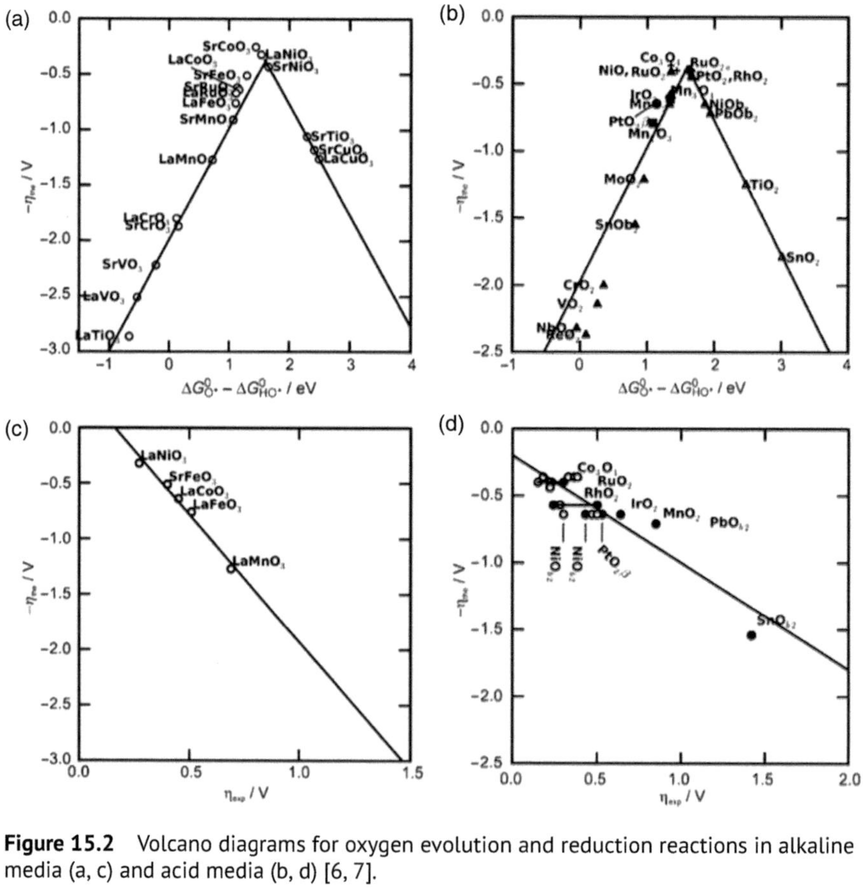

Figure 15.2 Volcano diagrams for oxygen evolution and reduction reactions in alkaline media (a, c) and acid media (b, d) [6, 7].

evaluated in alkaline medium at 10 mA cm^{-2}, room temperature, and pH 14. (d) Analogous comparison for transition metal oxides in acid media, with experimental data taken from the literature.

15.2.1 Electrocatalysts for PEMEL

PEMEL operate in an acidic environment and require highly stable and corrosion-resistant electrocatalysts. Due to the aggressiveness of the environment, the most efficient and durable materials are noble metals, although cheaper alternatives are being explored [8].
Anode (OER):

- Iridium (Ir) and its oxides (IrO_2) → standard material due to its high stability and catalytic activity.
- Ruthenium (Ru) and its oxides (RuO_2) → more active than Ir, but less stable in the long term.
- Ir–Ru, Ir–Ta, Ir–Sn alloys → optimization to reduce noble metal content without compromising performance.

Cathode (HER):

- Platinum (Pt) and Pt nanoparticles on carbon → more efficient catalyst due to its high activity and stability in acidic media.
- Pt alloys (Pt–Ir, Pt–Co, Pt–Ni) → reduce the amount of Pt while maintaining a high catalytic activity.
- Emerging alternatives such as carbides, nitrides, and materials doped with metal atoms (e.g. MoS_2, WC, VN), although still under development.

15.2.2 Electrocatalysts for AEL

Alkaline electrolyzers (AEL) operate in a less corrosive medium, allowing the use of more abundant and cheaper materials compared to PEMEL [9, 10].
Anode (OER):

- Nickel (Ni) and Ni oxides (NiO, $Ni[OH]_2$) → widely used material due to its good activity and stability.
- Cobalt (Co) and Co oxides (Co_3O_4, CoO) → efficient and economical alternative.
- Ferrites ($NiFe_2O_4$, $CoFe_2O_4$) → improves catalytic activity by combining Ni and Fe.
- Ni–Fe, Ni–Co, Ni–Mn alloys → increases efficiency by reducing the overpotential of the reaction.

Cathode (HER):

- Metallic nickel (Ni) → most widely used base material due to its low cost and stability.
- Nickel–molybdenum (NiMo) → improves catalytic activity compared to pure Ni.

15.2.3 Pourbaix Diagram for Water Electrolysis

The Pourbaix diagram, also known as the potential–pH diagram, is a graphical tool that shows the stability of the chemical species of an element as a function of the electric potential (E) and the pH of the aqueous solution. Its main uses are to predict the way in which a metal will behave in a given chemical environment, helping to identify whether it will be stable, whether it will corrode, or whether it will form a protective layer of oxides or hydroxides.

Pourbaix diagrams provide a large amount of thermodynamic information in an efficient and compact manner. This information is especially useful for controlling the corrosion of pure metals in aqueous media. By adjusting the pH and potential within the immunity and passive regions, it is possible to reduce corrosion.

For example, by increasing the pH of the environment, there are slightly alkaline values, which can mitigate water corrosion, something that can be achieved through water treatments. In a similar way, modifying the power potential to more negative values prevents corrosion, a technique known as cathodic protection. On the other hand, raising the potential has more positive values and reduces corrosion, which favors the formation of stable oxide films on the surface of transition metals.

A practical case is steel in reinforced concrete, which does not corrode as long as it remains in an alkaline environment. However, for aluminum, an alkaline environment can be harmful, as if the pH exceeds 8.0, corrosion accelerates significantly. These examples clearly demonstrate the usefulness of Pourbaix diagrams to predict and control corrosion in different materials.

On the vertical side of the diagram, the electrical potential (E) in volts (V) is represented with respect to the standard hydrogen electrode (SHE). This potential indicates the tendency of a species to oxidize or reduce. Positive potential values indicate oxidizing conditions, while negative values indicate reducing conditions. On the horizontal axis, the pH is measured, which indicates the acidity or basicity of the solution [12].

The Pourbaix diagram is divided into different regions that represent the stability of a metal as a function of electric potential (E) and pH. Each of these regions indicates whether the metal will remain in its pure state, whether it will dissolve in the solution, or whether it will form a protective layer of oxides or hydroxides. This classification is fundamental to understanding the corrosion and the immunity of metals in watery media. The Pourbaix diagram is divided into distinct regions that indicate the stability of the chemical species of a metal, briefly described in the following section.

The immunity region is the one where the metal remains in its pure state and does not experience corrosion. In this zone, the potential is sufficiently negative to prevent the metal from dissolving in solution, so no metallic ions are formed. This region is key in cathodic protection applications, where the potential of a metal is artificially reduced to prevent corrosion. An example of this is iron, which in the region of immunity remains in a solid state without undergoing oxidation.

In the region of corrosion, the metal oxidizes and dissolves in the aqueous solution in the form of metallic ions. This zone is generally at intermediate or high potentials and can be especially aggressive in acidic environments or in the presence of strong oxidants. When a metal is in this region, it suffers continued degradation, which makes it vulnerable to corrosion. An example of this is iron in acidic conditions, which dissolves to form Fe^{2+} or Fe^{3+}. To avoid corrosion in this region, techniques such as adjusting the pH or modifying the electrical potential can be used.

The passive region is the one where the metal oxidizes, but instead of dissolving, it forms a protective layer of oxides or hydroxides on its surface. This layer acts as a barrier that reduces the rate of corrosion, protecting the metal from further degradation. Passivation occurs at more positive potentials, where oxides become stable. A typical case is stainless steel, which forms a thin layer of chromium oxide (Cr_2O_3) that provides resistance to corrosion. This property is used in industrial applications and structures exposed to aggressive environmental conditions [13].

In addition to these regions, the Pourbaix diagram includes water stability lines (Figure 15.3), which mark the limits where water can decompose into hydrogen (H_2) or oxygen (O_2). Due to the low hydrogen evolution line, water is reduced to form gaseous hydrogen, a process that occurs at the cathode of an electrolyzer. Above the oxygen evolution line, water oxidizes and releases gaseous oxygen, which is relevant in processes such as water electrolysis or corrosion in aerobic media. A key aspect of the diagram is the water stability lines, which indicate the limits where the water can decompose into hydrogen (H_2) or oxygen (O_2). Below the hydrogen evolution line, water is reduced to H_2, while above the oxygen evolution line, water is oxidized to O_2. These zones are particularly important in processes such as water electrolysis, where electrodes operate at specific potentials to generate hydrogen and oxygen.

The regions of the Pourbaix diagram allow us to understand the electrochemical behavior of metals under different conditions. The identification of these zones is

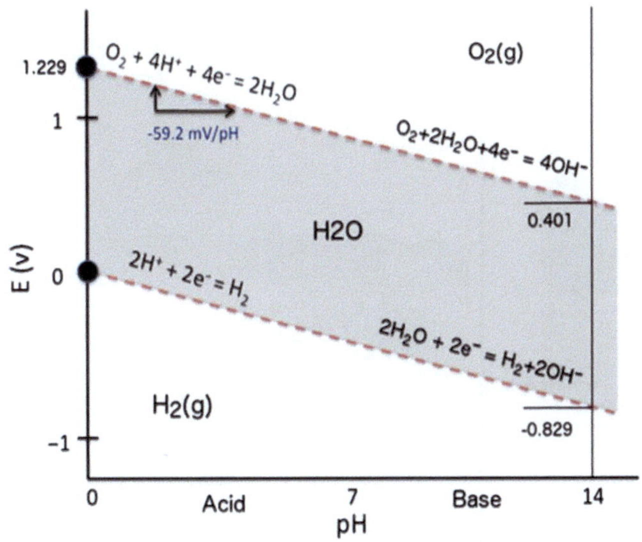

Figure 15.3 Pourbaix diagram showing the balance between water, oxygen, and hydrogen. Source: Adapted from Zhang et al. [11].

key to protection against corrosion, the design of resistant structures, and the optimization of industrial processes. By adjusting the pH and potential, it is possible to control the stability of a metal and prevent its degradation in practical applications.

15.3 Components of an Electrolysis System

As mentioned above, to achieve large-scale hydrogen production, an important strategy is through electrochemical splitting with electrolysis systems. In this regard, four types of technologies have been developed that can be categorized according to the charge carrier (OH^-, H^+, O^{2-}), operating conditions, and their electrolyte, as (a) AEL, (b) AEMEL, (c) PEMEL, and (d) solid oxide water electrolysis (SOEL), each with unique characteristics and progress prospects. However, the operating principles are the same for all systems. Currently, these technologies focus on using components that have affordable costs to increase hydrogen production capacity [14]. Here we will focus on AWE and PEMWE (Figure 15.4), since these techniques are commercially available and have been widely studied.

15.3.1 Components of an AEL-Type Electrolyzer

The main components of alkaline water electrolytic cells are two electrodes, diaphragm/separator as porous medium, current collectors (gas diffusion layer), separator plates (bipolar plates), and end plates. Generally, asbestos/zircon/nickel-coated perforated stainless steel diaphragms are used as separators in alkaline water electrolysis. Normally, nickel mesh is used as gas diffusion layer, and stainless steel/nickel-coated stainless steel separator plates are also used as bipolar and end plates, respectively. The operating temperature is around 40–90 °C, and the strong inert electrolyte is used as an aqueous solution of NaOH/KOH [14].

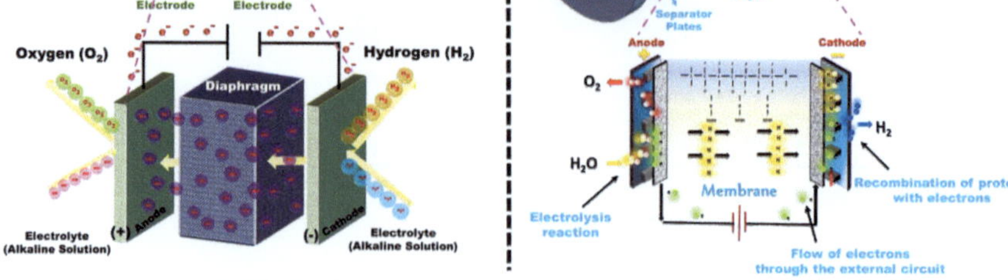

Figure 15.4 Principal components of (a) AEL and (b) PEMEL.

The electrode/catalyst interface for both hydrogen and oxygen is nickel-coated perforated stainless steel. On the other hand, the gas diffusion layer is a nickel mesh, while the bipolar plates are stainless steel/nickel-coated stainless steel. In this system, attention must be paid to the diaphragm and the current density used, since it is not completely impermeable and a gas from the end could pass through it, causing cross-contamination. Due to the characteristics of the equipment, it requires a long start-up preparation and has a slow response to changes in electrical energy, which makes it difficult to adapt to renewable energy sources such as solar and wind [15].

15.3.2 Components of a PEMEL-Type Electrolyzer

A stack is composed of hundreds of individual cells (Figure 15.4), while a single cell consists of a core of bipolar plates (BPs), porous transport layers (PTLs), and a membrane electrode assembly (MEA). BPs transfer mass and electrons and as a component contribute up to ~53% of the total stack cost and ~25% of the system costs (Figure 15.4a). It is known that the BPs of a PEMEL generally work in corrosive environments (typically 1.6–2 V, pH 2–4, 50–90 °C and O_2 saturation) for a lifetime of more than 80,000 hours and where pressures are high (MPa), temperatures around 50–80 °C, and pH in a range of 2–4. Therefore, anticorrosive properties are demanded and need to be examined from different aspects such as potentiodynamic/potentiostatic polarization, weight loss, and composition/structure characterizations. Therefore, the development of cheaper BPs should be a priority, thereby reducing the cost of PEMEL [16]. Gas diffusion layers are important because they control the diffusion of water into the catalyst layer and the diffusion of water, hydrogen, and oxygen out of the catalyst [17]. This type of system requires stable and highly acidic conditions, so expensive metal catalysts such as Pt, Ir, graphite, and among others, are used. Therefore, research is currently being carried out to obtain catalysts composed of accessible metals that present stability and high reactivity [18].

15.4 Common Problems in Materials Engineering

The description of the technologies for obtaining hydrogen through water electrolysis has been mentioned above; however, it is necessary to mention the challenges that each system presents with respect to the materials it uses and that directly impact the efficiency, durability, and cost.

15.4.1 AEL System

In this system, one disadvantage is the current density, which is low (200–500 mA cm^{-2}), the cell is larger than the PEMEL system (800–2500 mA cm^{-2}), this disadvantage is related to the poor ionic conductivity, which causes low production of green hydrogen and slow evolution of oxygen and also requires very efficient catalysts [19]. Another limitation is the diaphragm, which, being made of asbestos, zirconium, or steel, can cause cross-contamination of gases, which affects the quality of the hydrogen and the durability of the system and can spontaneously cause an explosion. Since high concentrations of KOH and NaOH electrolytes are required for its operation, oxidation occurs, causing the nickel-based electrodes and the nickel-coated stainless steel bipolar plates to degrade rapidly, leading to a loss of catalytic activity due to the corrosion they suffer in this alkaline environment. This is related to the temperatures used, which cannot be higher than 90 °C, since at higher temperatures, the corrosion resistance of the materials decreases, and this directly affects the purity of the hydrogen [20, 21]. Due to the materials used, it is a system that presents difficulties for integration with intermittent renewable energy sources due to slow start-up and shutdown times.

15.4.2 PEMEL System

Compared to the EAL system, this system requires expensive noble metal catalysts (Platinum in the cathode and Iridium in the anode), so today the use of non-noble metals that maintain an efficiency at least around 60% of hydrogen is being studied [16]. At the same time, the membrane used as Nafion is expensive, and it must be considered that they degrade due to mechanical stress, chemical attack, and the hydration/dehydration cycle [22]. The system uses titanium plates with protective coatings to avoid high anode corrosion and an acidic environment and the diffusion layers have to balance the passage of water and the transport of gas, avoiding flooding, which increases costs [20, 22]. In the long term, the iridium catalysts, as well as their migration and dissolution, degrade and reduce the efficiency of the system [22]. One of the problems is that this system uses temperatures of 79–90 °C with pressure ranges above 80 bars, which induces oxidation of the material, increasing its cost [19]. Today there is a need to investigate the reduction of the degradation of materials, which has given way to studies where the catalysts and the robustness of the cell are analyzed to reduce costs and increase efficiency in hydrogen production [23].

15.5 Future Trends of PEMEL, AEL, and AEMEL

As technology advances toward energy sustainability, hydrogen production through electrolysis is gaining ground over other methodologies used. Technologies that use PEMEL and AEL are the methods that are currently being widely studied and are in a trend that seeks to improve their efficiency [21, 24, 25], reduce costs [26, 27], and facilitate the transfer of technology on a large scale [15]. Currently, PEMEL and AEL have been studied to improve their efficiency in hydrogen production performance. Since it is still far from the theoretical limits, it is sought to increase their efficiency above 60%, which would benefit the sector since it would reduce energy consumption per unit of hydrogen produced, the above would be achieved if alternative electrode materials were used, as well as the development of stable electrolytes and good conductors for high pressure and temperature electrolysis [16, 27]. To respond to the sustainable development goals against climate change [28], it is important to use

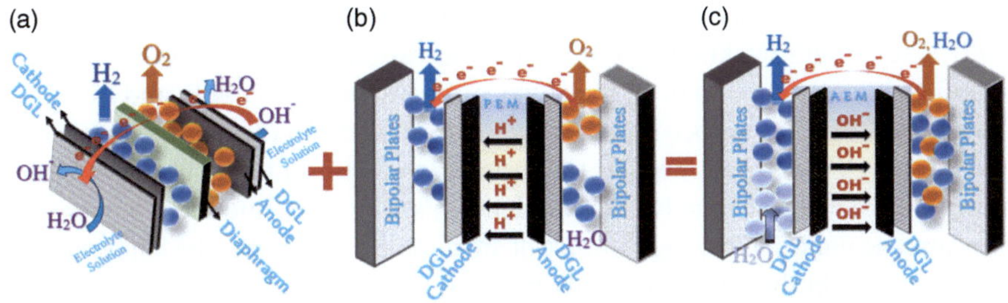

Figure 15.5 Trend for 2030 for green H_2 production. (a) AEL; (b) PEMEL; (c) AEMEL.

green hydrogen production sources, however, less than 1% comes from electrolysis [29]. Although current costs are very high, compared to fossil fuels, it is seen as a viable trend for 2050. Research shows mass-based costs of hydrogen considering the lower calorific value of hydrogen (120 MJ kg^{-1}, or 33.3 kWh kg^{-1}) and renewable hydrogen costs at USD 1.1–10.3/MMBtu and coal prices of USD 40–116/t [30].

As demand for green hydrogen increases, electrolyzer technology must scale to meet the needs of industrial applications in both the private sector and national governments, which is in line with the International Energy Agency (IEA) [31]. It is time to scale up these technologies and enable hydrogen to be widely used with installed capacity from MW to GW scale with the aim of having 40 GW of electrolysis capacity installed in the EU by 2030 [32]. Various industries have launched projects to build 50 MV electrolyzer that are intended to power a refinery in Germany, gas network operators, among others [27, 33]. The success of hydrogen production and a future economy that has a significant impact will depend on improvements in current technologies and a competitive cost, as well as on a transition of the energy system toward green alternatives that are integrated with other solutions without losing sight of the potential benefits in terms of reducing greenhouse gas emissions [34]. Regarding AEMEL, they represent a promising trend in green hydrogen production, combining advantages of PEMEL and AEL technologies. Unlike PEMEL, which require precious metals such as platinum and iridium, AEMEL use less expensive materials, which significantly reduces production costs [35]. Compared to traditional AEL, AEMEL offer higher efficiency and operate at lower current density, improving their integration with renewable sources such as solar and wind energy. Furthermore, advances in high conductivity and chemical stability membranes are increasing their durability, allowing longer lifetimes without the accelerated degradation faced by some alkaline technologies [36]. With a focus on sustainability and scalability, AEMEL (Figure 15.5) emerge as a viable alternative for decentralized hydrogen production, with applications in mobility, energy storage, and processes [37, 38].

15.6 Conclusions

In this chapter, the concepts of electrolyzers were defined, as well as their classification by the electrolyte used, which are AEL, PEMEL, AEMEL, and SOEL. The OER and the HER were addressed, and it was determined that the limiting reaction is the OER due to the slow kinetics and it favors corrosion at the interface of the catalyst, diffuser, and plates. Some of the engineering problems that AEL and PEMEL have suffered were defined because these types are already on the market, however, due to their high costs, AEL exceeds the installation at present. The trends of electrolyzers were addressed to solve the problems of sizing, cost, and more accessible electrocatalytic materials through the use of an anionic membrane. The latter, called AEMEL, are still in their experimental phase, but the aim is to exceed the installation capacity, mitigate the problems caused by corrosion, and achieve an efficiency very comparable to PEM using inexpensive materials.

References

1 Gunawardane, K. (2023). Evolution of hydrogen energy and its potential opportunities around the globe. In: *Hydrogen Energy Conversion and Management*, 3–33. Elsevier https://doi.org/10.1016/B978-0-443-15329-7.00007-7.

2 Fernández-Arias, P., Antón-Sancho, Á., Lampropoulos, G., and Vergara, D. (2024). On green hydrogen generation technologies: a bibliometric review. *Appl. Sci.* 14 (6). Multidisciplinary Digital Publishing Institute (MDPI). https://doi.org/10.3390/app14062524.

3 Cruz, J.C., Barbosa, R., Escobar, B. et al. (2020). Electrochemical and microstructural analysis of a modified gas diffusion layer for a PEM water electrolyzer. *Int. J. Electrochem. Sci.* 15: 5571–5584. https://doi.org/10.20964/2020.06.12.

4 F. M. Cuevas-Muñiz et al. (2015). Correlation between theoretical data and experimental selective properties of PtAg core-shell nanoparticles for oxygen reduction reactions, *Int. J. Hydrog. Energy*. pp. 17284–17290. doi: https://doi.org/10.1016/j.ijhydene.2015.06.096

5 Liu, Z., Sajjad, S.D., Gao, Y. et al. (2017). The effect of membrane on an alkaline water electrolyzer. *Int. J. Hydrog. Energy* 42 (50): 29661–29665. https://doi.org/10.1016/j.ijhydene.2017.10.050.

6 Quaino, P., Juarez, F., Santos, E., and Schmickler, W. (2014). Volcano plots in hydrogen electrocatalysis-uses and abuses. *Beilstein J. Nanotechnol.* 5 (1): 846–854. https://doi.org/10.3762/bjnano.5.96.

7 Man, I.C., Su, H.Y., Calle-Vallejo, F. et al. (2011). Universality in oxygen evolution electrocatalysis on oxide surfaces. *ChemCatChem* 3 (7): 1159–1165. https://doi.org/10.1002/cctc.201000397.

8 Araújo, H.F., Gómez, J.A., and Santos, D.M.F. (2024). Proton-exchange membrane electrolysis for green hydrogen production: fundamentals, cost breakdown, and strategies to minimize platinum-group metal content in hydrogen evolution reaction electrocatalysts. *Catalysts* 14 (12). Multidisciplinary Digital Publishing Institute (MDPI): 845. https://doi.org/10.3390/catal14120845.

9 Brauns, J. and Turek, T. (2020). Alkaline water electrolysis powered by renewable energy: a review. *Processes* 8 (2). MDPI AG: 248. https://doi.org/10.3390/pr8020248.
10 Şahin, M.E. (2024). An overview of different water electrolyzer types for hydrogen production. *Energies* 17 (19): 4944. https://doi.org/10.3390/en17194944.
11 Exner, K.S. (2024). Four generations of volcano plots for the oxygen evolution reaction: beyond proton-coupled electron transfer steps? *Acc. Chem. Res.* 57 (9): 1336–1345. https://doi.org/10.1021/acs.accounts.4c00048.
12 Pourbaix, M. (1966). *Atlas of Electrochemical Equilibria in-Aqueous Solutions.* NACE.
13 Bard, A.J. and Faulkner, L.R. (2001). *Electrochemical Methods: Fundamentals and Applications.* Wiley.
14 Shiva Kumar, S. and Himabindu, V. (2019). Hydrogen production by PEM water electrolysis – A review. *Mater. Sci. Energy Technol.* 2 (3): 442–454. https://doi.org/10.1016/j.mset.2019.03.002.
15 Zhang, K., Liang, X., Wang, L. et al. (2022). Status and perspectives of key materials for PEM electrolyzer. *Nano Res. Energy* 1: e9120032. https://doi.org/10.26599/NRE.2022.9120032.
16 Franco, A. and Giovannini, C. (2023). Recent and future advances in water electrolysis for green hydrogen generation: critical analysis and perspectives. *Sustainability* 15 (24): 16917. https://doi.org/10.3390/su152416917.
17 Mo, J., Steen, S., Kang, Z. et al. (2017). Study on corrosion migrations within catalyst-coated membranes of proton exchange membrane electrolyzer cells. *Int. J. Hydrog. Energy* 42 (44): 27343–27349. https://doi.org/10.1016/j.ijhydene.2017.09.020.
18 Ogawa, T., Takeuchi, M., and Kajikawa, Y. (2018). Analysis of trends and emerging technologies in water electrolysis research based on a computational method: a comparison with fuel cell research. *Sustainability* 10 (2): 478. https://doi.org/10.3390/su10020478.
19 Lange, H., Klose, A., Lippmann, W., and Urbas, L. (2023). Technical evaluation of the flexibility of water electrolysis systems to increase energy flexibility: a review. *Int. J. Hydrog. Energy* 48 (42): 15771–15783. https://doi.org/10.1016/j.ijhydene.2023.01.044.
20 Brauns, J. and Turek, T. (2022). Experimental evaluation of dynamic operating concepts for alkaline water electrolyzers powered by renewable energy. *Electrochim. Acta* 404: 139715. https://doi.org/10.1016/j.electacta.2021.139715.
21 Sakas, G., Ibáñez-Rioja, A., Ruuskanen, V. et al. (2022). Dynamic energy and mass balance model for an industrial alkaline water electrolyzer plant process. *Int. J. Hydrog. Energy* 47 (7): 4328–4345. https://doi.org/10.1016/j.ijhydene.2021.11.126.
22 Kiemel, S., Smolinka, T., Lehner, F. et al. (2021). Critical materials for water electrolysers at the example of the energy transition in Germany. *Int. J. Energy Res.* 45 (7): 9914–9935. https://doi.org/10.1002/er.6487.

23 Chatenet, M., Pollet, B.G., Dekel, D.R. et al. (2022). Water electrolysis: from textbook knowledge to the latest scientific strategies and industrial developments. *Chem. Soc. Rev.* 51 (11): 4583–4762. https://doi.org/10.1039/D0CS01079K.

24 Maier, M., Smith, K., Dodwell, J. et al. (2022). Mass transport in PEM water electrolysers: a review. *Int. J. Hydrog. Energy* 47 (1): 30–56. https://doi.org/10.1016/j.ijhydene.2021.10.013.

25 Roggi, A., Guazzelli, E., Resta, C. et al. (2023). Vinylbenzyl chloride/styrene-grafted SBS copolymers via TEMPO-mediated polymerization for the fabrication of anion exchange membranes for water electrolysis. *Polymers* 15 (8): 1826. https://doi.org/10.3390/polym15081826.

26 Millet, P. and Grigoriev, S. (2013). Water electrolysis technologies. In: *Renewable Hydrogen Technologies*, 19–41. Elsevier https://doi.org/10.1016/B978-0-444-56352-1.00002-7.

27 Noussan, M., Raimondi, P.P., Scita, R., and Hafner, M. (2020). The role of green and blue hydrogen in the energy transition—A technological and geopolitical perspective. *Sustainability* 13 (1): 298. https://doi.org/10.3390/su13010298.

28 Naciones Unidas. (2018). La Agenda 2030 y los Objetivos de Desarrollo Sostenible: una oportunidad para América Latina y el Caribe (LC/G. 2681-P/Re.

29 Corengia, M. and Torres, A.I. (2022). Coupling time varying power sources to production of green-hydrogen: a superstructure based approach for technology selection and optimal design. *Chem. Eng. Res. Des.* 183: 235–249. https://doi.org/10.1016/j.cherd.2022.05.007.

30 BloombergNEF. (2020). Hydrogen economy outlook. 2020. *Disponible en línea:* https://data.bloomberglp.com/professional/sites/24/BNEF-Hydrogen-Economy-Outlook-Key-Messages-30-Mar-2020.pdf. Consultado el 25 de enero del 2025.

31 Birol, F. (2019). El futuro del hidrógeno: aprovechar las oportunidades actuales. *Informe de La AIE,* 20.

32 Erbach, G., & Jensen, L. (2021). EU hydrogen policy: hydrogen as an energy carrier for a climate-neutral economy.

33 Franke, A., & Baratti, G. (2020). Hidrógeno verde: una guía para la formulación de políticas. *S&P Global Platts, 9 de Diciembre de 2020. Disponible En Línea*: Https://Www.Spglobal.Com/Platts/En/Market-Insights/Latest-News/Metals/120920-European-Groups-Join-Hydrogen-Project-Targeting-2kg-Production (Consultado El 15 de Enero de 2025).

34 Velazquez Abad, A. and Dodds, P.E. (2020). Green hydrogen characterisation initiatives: definitions, standards, guarantees of origin, and challenges. *Energy Policy* 138: 111300. https://doi.org/10.1016/j.enpol.2020.111300.

35 Chen, Y., Liu, C., Xu, J. et al. (2023). Key components and design strategy for a proton exchange membrane water electrolyzer. *Small Struct.* 4 (6): https://doi.org/10.1002/sstr.202200130.

36 Zakaria, Z. and Kamarudin, S.K. (2021). A review of alkaline solid polymer membrane in the application of AEM electrolyzer: materials and characterization. *Int. J. Energy Res.* 45 (13): 18337–18354. https://doi.org/10.1002/er.6983.

37 Gul, E., Baldinelli, G., Farooqui, A. et al. (2023). AEM-electrolyzer based hydrogen integrated renewable energy system optimisation model for distributed communities. *Energy Convers. Manag.* 285: 117025. https://doi.org/10.1016/j.enconman.2023.117025.

38 Hora, C., Dan, F.C., Rancov, N. et al. (2022). Main trends and research directions in hydrogen generation using low temperature electrolysis: a systematic literature review. *Energies* 15 (16): 6076. https://doi.org/10.3390/en15166076.

16

Defect and Interface Engineering for the Oxygen Reduction Reaction

Heriberto Cruz-Martínez, Lidia Santiago-Silva, Brenda García-Hilerio, and Víctor A. Franco-Luján

Tecnológico Nacional de México, Instituto Tecnológico del Valle de Etla, Abasolo S/N, Barrio del Agua Buena, Santiago Suchilquitongo, Oaxaca 68230, Mexico

16.1 Introduction

Proton exchange membrane fuel cells (PEMFCs) are vital for developing clean and sustainable energy technologies. Their ability to efficiently convert hydrogen into electricity without producing harmful emissions makes them a key solution for the transition to greener energy. These PEMFCs are renowned for their high-power density, low-temperature operation, and versatility in applications such as electric vehicles, residential energy systems, and portable devices [1–5]. In these technologies, the energy conversion begins at the anode (hydrogen oxidation reaction, HOR), where hydrogen (H_2) is split into protons (H^+) and electrons (e^-). The protons pass through the PEM, while the electrons are transferred to the cathode through an external circuit. At the cathode (oxygen reduction reaction, ORR), O_2 is reduced with the e^- and H^+, with water and heat as the only by-products [6, 7]. The ORR primarily determines the efficiency of the overall process, as it is several orders of magnitude more slowly than the HOR [6–8]. Therefore, catalysts are vital to accelerate the ORR kinetics in acidic media.

Pt nanoparticles deposited on Vulcan carbon (Pt/C) are typically used as catalysts for the ORR in PEMFCs. However, the main drawbacks of Pt include its high cost, limited availability, and relatively low stability [6, 7]. Therefore, investigations of more stable and cost-effective catalysts with high catalytic activity and stability are crucial for the PEMFCs widespread utilization. In this context, various defect and interface engineering strategies have been established to improve the activity, stability, and selectivity of ORR catalysts [9, 10]. In the first instance, different defect and interface engineering approaches have been studied and applied to nanoparticles used as catalysts such as forming bimetallic and trimetallic nanoparticles with varying atomic arrangements and defects. Implementing these strategies has resulted in the development of Pt-based alloy catalysts with catalytic activities comparable to or superior those of commercial Pt/C catalysts [9–14]. Unfortunately, Pt-based

Surface and Interfacial Defects in Nanomaterials for Sustainable Energy Production and Storage, First Edition. Edited by Noé Arjona, Lorena Álvarez Contreras, and Minerva Guerra Balcázar.

nanoparticles are prone to agglomeration, hindering their catalytic efficiency. Therefore, it is essential to employ support materials with a high surface area that can act as anchoring sites, reducing agglomeration and enhancing catalytic performance [15, 16]. In addition to enhanced dispersion, the support material plays a crucial role in the catalytic activities, stabilities, and overall performances of Pt-based catalysts. Therefore, diverse defect and interface engineering approaches have been implemented for the support materials to improve the catalytic activity, stability, and selectivity of the catalysts [17, 18].

Carbon-based materials are attractive as support materials due to their high surface area, excellent conductivity, and cost-effectiveness [17–22]. Carbon exists in various allotropes, each with unique shapes and microstructures. Owing to their diverse morphologies and characteristics, carbon materials are widely utilized as catalyst support materials, highlighting the use of graphene. This allotrope has proven to be an effective support material due to its featured properties such as good conductivity, high surface area, and good mechanical properties [23–25]. Despite these advantages, graphene is known to exhibit low chemical reactivity. To address this, different defect and interface engineering strategies have been developed to enhance the properties of graphene, making it a more effective support material for Pt-based catalysts in the ORR. Therefore, this chapter examines representative studies on modifying graphene through defects and their impact on graphene properties as a catalyst support material.

16.2 Types and Effects of Defects in Graphene for ORR

Defects in graphene have been widely studied to modify its chemical reactivity and thus be a more suitable support material for Pt-based nanoparticles that are used as catalysts for ORR [26–28]. Some of the defects that have been investigated and implemented in graphene are illustrated in Figure 16.1. These defects are very important for improving the properties of graphene, resulting in a more suitable support material for Pt-based nanoparticles that are used as catalysts for ORR. When defects are present in graphene, this structure exhibits various changes in its chemical, electrical, structural, and electronic properties, which brings with it an improvement in catalytic activity, stability, and selectivity of Pt-based nanoparticles for ORR.

16.3 Roles of Vacancies in Graphene for ORR

Vacancies in graphene are point defects where one or more carbon atoms are missing from the structure, creating vacant or disordered sites. These vacancies play a crucial role when graphene is employed as a support material for Pt-based catalysts in the ORR, as they enhance the properties of graphene, making it a more suitable support material. Several theoretical studies are exploring the stability and catalytic properties of Pt-based nanoparticles supported on graphene with vacancies [29–37]. In the first instance, various density functional theory (DFT) studies have explored

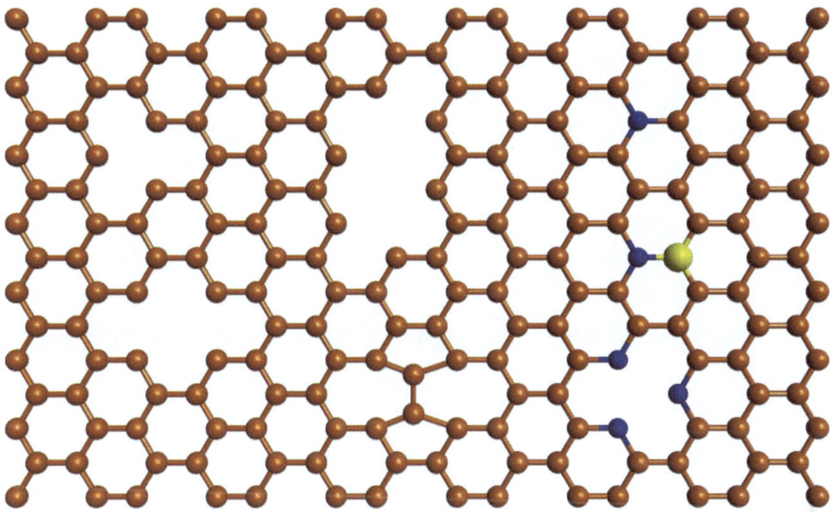

Figure 16.1 Types of defects in graphene.

the stability and reactivity of Pt-based nanoparticles deposited on graphene with different vacancies [29–34]. For instance, the stability of small Pt clusters deposited on graphene with vacancies (single vacancy, double vacancy, 585-type vacancy, and 555–777-type vacancy) was studied employing DFT calculations (Figure 16.2). It is observed that when the Pt_{13} nanoclusters are deposited on graphene with vacancies, the Pt_{13} nanoclusters show a structural change. It is also observed that the stability of the Pt_{13} clusters on graphene with vacancies was higher than that of pristine graphene, which may be associated with a more significant charge transfer from the Pt_{13} nanoclusters to the defective graphene [30].

More recently, stabilities of Pt_n ($n = 4$, 6, 10, 13, 16, and 24) clusters supported on single vacancy were studied, employing DFT calculations [33]. The interaction energy of Pt clusters on single-vacancy graphene was higher than −6 eV, which can be associated with good stability of the Pt clusters on single-vacancy graphene. According to the charge analysis, it was observed that the Pt clusters donated charge to single-vacancy graphene, which is consistent with previous results [29–34]. It was observed that the charge transferred from Pt clusters is located mainly at the interface between the Pt clusters and the defective graphene.

Interestingly, several studies investigated the ORR mechanism or the adsorption of ORR species on Pt-based nanoparticles supported on graphene with vacancies [35–37]. For instance, DFT computations were carried out to investigate the electronic and reactivity properties of Pt_{13} nanoclusters adsorbed on single-vacancy graphene [35]. It was demonstrated that the single-vacancy graphene plays a crucial role as an anchoring point for Pt_{13} nanoclusters and suggests an enhanced catalytic activity in their interaction with O_2. In another study, DFT calculations were used to investigate the stability and reactivity of Fe_nPt_{13-n} ($n = 0$, 1, 2, and 3) nanoclusters supported on defective graphene for the ORR [37]. The interaction

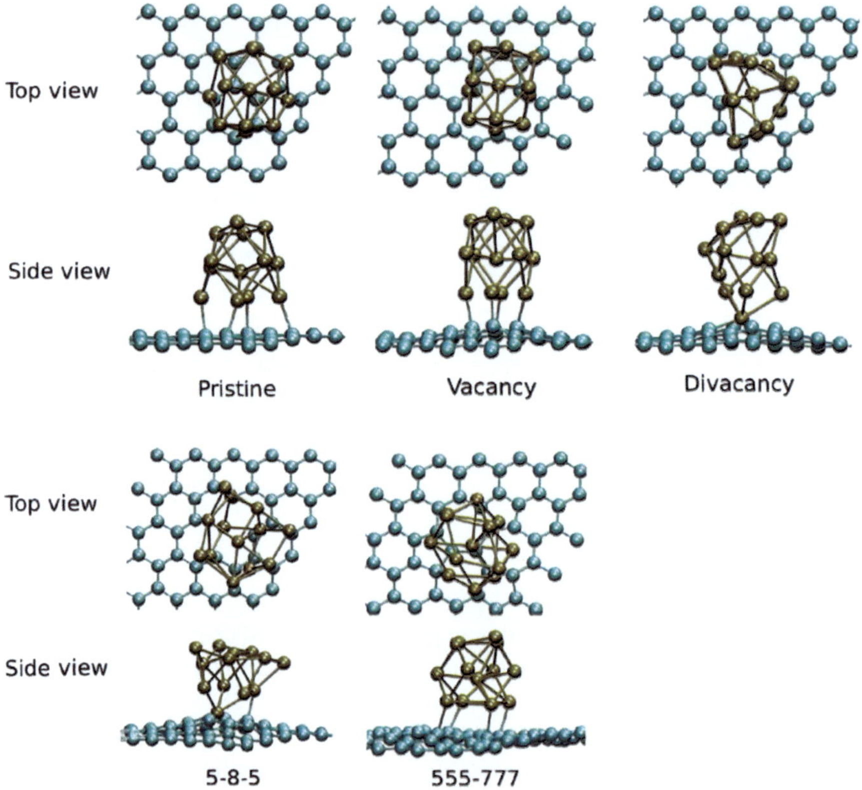

Figure 16.2 Low-energy structures of Pt_{13} nanoclusters on graphene with vacancies. Cyan and gold spheres represent C and Pt atoms, respectively. Source: Fampiou and Ramasubramaniam [30] / with permission of American Chemical Society.

energy of Fe_nPt_{13-n} nanoclusters on single-vacancy graphene was higher than −5 eV, which can be associated with good stability of these nanoclusters on defective graphene. Also, according to the O_2 adsorption energy and the *d*-band center, FePt nanoclusters are promising candidates for the ORR [37]. Also, the ORR mechanisms on Pt_{13} nanoclusters supported on single-vacancy graphene have been studied to understand the impact of the support material on the ORR [36]. It was demonstrated that the single-vacancy graphene helps balance the binding of ORR intermediates on Pt_{13} nanoclusters. Figure 16.3 compares the ORR mechanism through a four-electron pathway on Pt_{13} nanocluster supported on single-vacancy graphene and on unsupported Pt_{13} nanocluster. Both exhibit strong exothermic behavior at zero potential ($U = 0\,V$) except for the final step related to H_2O desorption. For the supported and unsupported Pt_{13} nanocluster, at the equilibrium potential ($U = 1.23\,V$), the rate-limiting step is in the last proton/electron transfer and H_2O desorption, with energy barriers of 1.68 eV for supported Pt_{13} nanocluster and of 2.71 eV for free Pt_{13} nanocluster. The defective graphene support plays a significant role in lowering the energy barrier, offering a potential way to tune

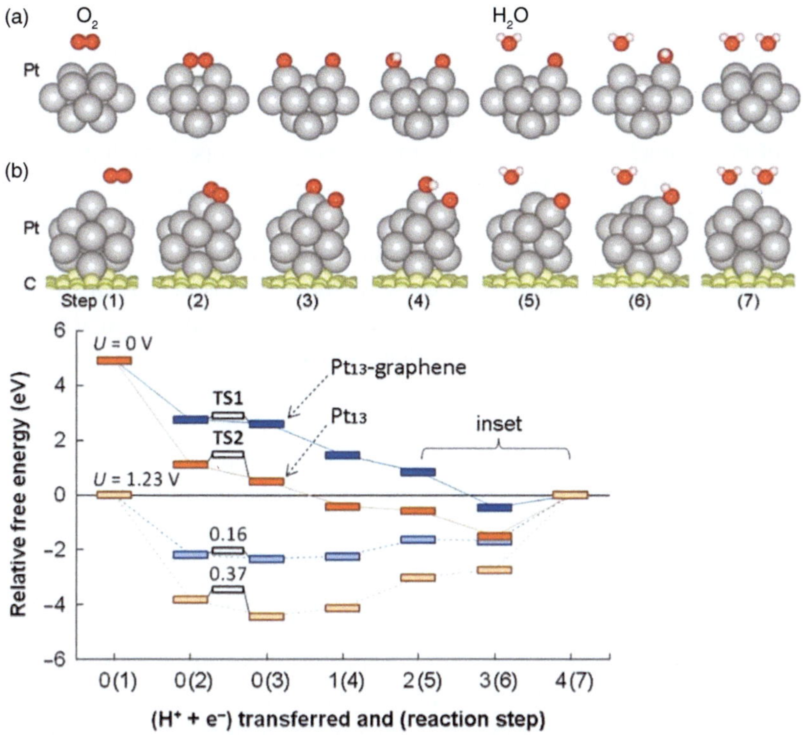

Figure 16.3 The ORR mechanism through a direct four-electron pathway on free Pt_{13} nanocluster (A and markers in orange) and on Pt_{13} nanocluster supported on single-vacancy graphene (B and markers in blue). Source: Lim and Wilcox [36] / with permission of American Chemical Society.

the activity of Pt nanoparticles through the graphene. The reduced stability of the ORR intermediates may result from charge redistribution induced by the defective graphene support [36].

16.4 Roles of Doping in Graphene for ORR

Another widely used approach to enhance the reactivity of pristine graphene is doping. Doping atoms have been shown to significantly alter the electronic, chemical, and structural properties of graphene [38–40]. These modifications are crucial when doped graphene is employed as a support material for Pt-based nanoparticles in the ORR, making it a more effective and suitable support material. To date, various doping (e.g. N, S, Fe, P, O) atoms have been studied to enhance the reactivity of graphene as support materials for Pt-based nanoparticles for ORR [41–60], being N the most studied as a doping element [41–56]. For instance, the catalytic activity and the stability for the ORR of two types of Pt nanotubes supported on N-doped graphene synthesized under different conditions were investigated [41]. Both catalysts demonstrated higher catalytic activity than commercial Pt/C.

Additionally, these catalysts were subjected to accelerated durability tests (ADTs) involving cyclic potential sweeps for 10,000 cycles, exhibiting higher stability than Pt/C, which is associated with the strong interaction between the Pt nanocubes and the N-doped graphene [41]. In another study, the electronic and reactivity properties of Pt_{13} nanocluster supported on three types of N-doped graphene were investigated using DFT computations [42]. The interface between the N-doped graphene and the Pt_{13} nanoclusters changed substantially due to the presence of N atoms in the graphene and produced a strong interaction between the nanocluster and the support material. Additionally, a decrease in the O adsorption energy on the Pt_{13} nanocluster supported on N-doped graphene was observed compared to the free Pt_{13} nanocluster [42]. Also, Pt nanoparticles supported on N-doped graphene and on nanocarbon-wedged N-doped graphene were evaluated for the ORR [49]. These catalysts exhibited higher catalytic activities than commercial Pt/C catalysts. Also, Pt nanoparticles deposited on nanocarbon-wedged N-doped graphene and commercial Pt/C were subjected to ADTs involving cyclic potential sweeps for 5000 cycles, presenting a higher stability than Pt/C [49].

Recently, Pt nanoparticles deposited on N-doped graphene were investigated for the ORR, where N-doped graphene was produced by using an electrochemical exfoliation method varying the concentration of ammonium sulfate as an electrolyte and urea as a nitrogen precursor (ENG x–y = exfoliated N-doped graphene, where x = molarity of urea and y = molarity of ammonium sulfate) [52]. It was observed that by changing the synthesis parameters of the support material, a change in the concentration of N in graphene was observed, where N content was 5.7, 7.32, and 10.05% for ENG 1-1, ENG 3-1, and ENG 5-1, respectively. Higher electrochemically active surface areas (ECSAs) and mass activity were observed for Pt nanoparticles deposited on N-doped graphene compared to those deposited on undoped support, suggesting that the N doping played a crucial role in enhancing the performance toward the ORR. After, Pt/ENG 3-1 catalyst stability was analyzed by employing ADTs considering cyclic potential sweeps for 30,000 cycles. The cyclic voltammograms recorded before and after the ADTs tests showed that the ECSA drastically reduced for the commercial Pt/C catalyst (Figure 16.4a), with only 60.7% retention in ECSA after 5000 cycles. In contrast, the retention was better for Pt/ENG 3-1 (Figure 16.4b), with a retention of 78.4% in ECSA. Pt/C recorded an 11 mV change in half-wave potential (Figure 16.4c) concerning the 10 mV shift for Pt nanoparticles deposited on ENG 3-1 (Figure 16.4d). The Pt/ENG 3-1 catalyst retained 62.8% of its ECSA after 30,000 cycles [52], which shows the critical role of the support materials on the stability and catalytic activity. Interestingly, some studies have explored the role of temperature in the production of Pt nanoparticles deposited on N-doped graphene [46, 47]. Pt nanoparticles deposited on N-doped graphene were studied for ORR, where N-doped graphene was produced at different temperatures (150, 300, 450, 600, 750, and 900 °C). It was observed that increasing the synthesis temperature of N-doped graphene modified the N concentration in the graphene. Additionally, the incorporation of pyridinic N into the graphene was found to influence the ORR activity [47].

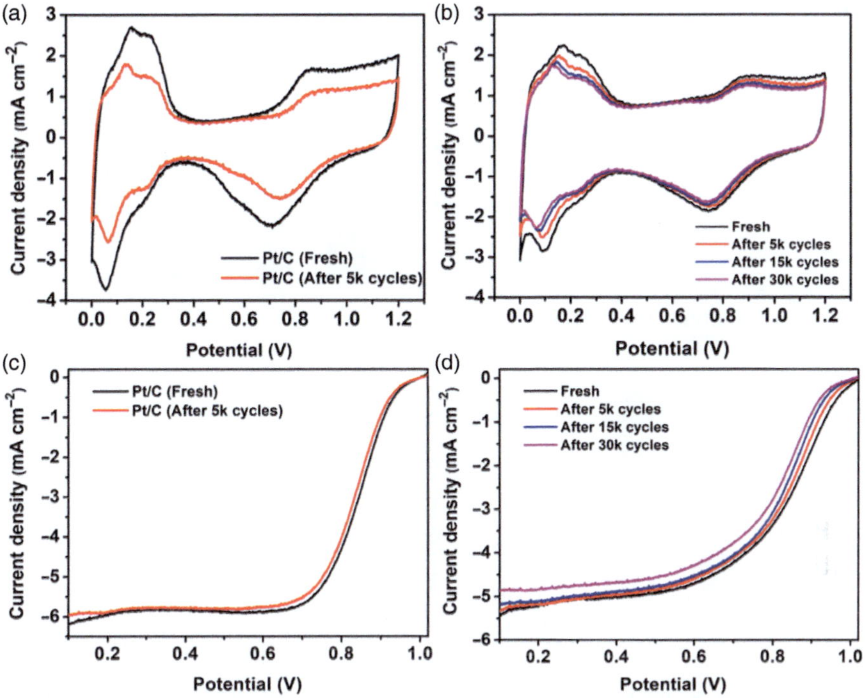

Figure 16.4 (a, b) Cyclic voltammograms and (c, d) linear sweep voltammetry of (a, c) commercial Pt/C, fresh catalyst, and after 5000 cycles and (b, d) Pt/ENG 3-1, fresh catalyst, and after 5000 (5 k), 15,000 (15 k), and 30,000 (30 k) cycles. Source: Bhaskaran and Chetty [52] / with permission of American Chemical Society.

The co-doping is another strategy that has been implemented to improve the properties of doped graphene as support materials of Pt-based nanoparticles for the ORR [53–56]. For instance, Pt nanoparticles supported on Fe and N co-doped holey graphene were evaluated for ORR [54]. The catalytic performance was more significant than N-doped holey graphene and commercial Pt/C (Figure 16.5), demonstrating the importance of co-doping as a good strategy to improve catalytic activity for ORR. In another study, Pt nanoparticles supported on co-doped (N – B and N – P) graphene were studied for the ORR [55]. Pt nanoparticles supported on co-doped graphene exhibited higher mass activities than their counterparts of single-doping. After, the stability of Pt nanoparticles deposited on N and B co-doped were studied employing ADTs considering cyclic potential sweep for 5000 cycles. Pt nanoparticles supported on N- and B-doped graphene lost only 6.7% of their ECSA, followed by Pt nanoparticles supported on B-doped graphene Gr (26%), while Pt nanoparticles supported on N-doped graphene lost the most (53%). These studies demonstrate the importance of co-doping as a good strategy to improve catalytic activities and stabilities of Pt-based nanoparticles supported on co-doped graphene for the ORR [53–56].

Finally, various theoretical studies have analyzed the catalytic activity and stability of Pt nanoparticles supported by oxygen-doped graphene for

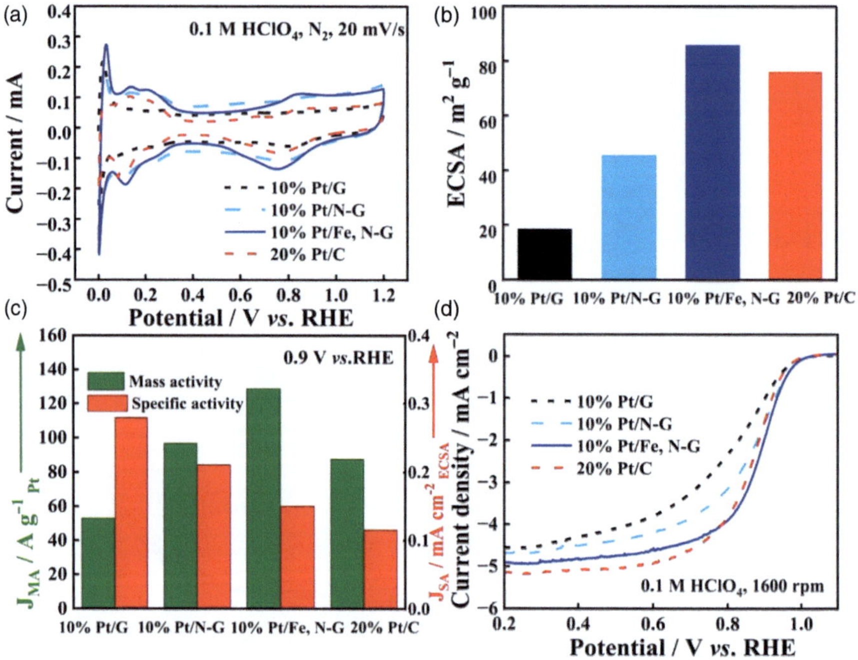

Figure 16.5 Plots of 10% Pt/G, 10% Pt/N-G, 10% Pt/Fe, N-G, and 20% Pt/C (JM) catalysts. (a) Cyclic voltammogram curves, (b) ECSA, (c) mass and specific activity (0.9 V vs RHE), and (d) linear sweep voltammetry curves. Source: Fu et al. [54] / with permission of Elsevier.

ORR [57–59]. For example, Pt_3M (M = 3*d* transition metals) clusters supported on oxygen-doped graphene were investigated as catalysts for the ORR employing DFT calculations [59]. First, the O_2 adsorption was studied on different Pt_3M-supported O-doped graphene. The O_2 was adsorbed in a bridge-type mode with one oxygen atom adsorbed on a Pt atom and the other on M. For most of the systems studied, the O–O bond length exhibited an elongation when the O_2 molecule was adsorbed on these composites (Figure 16.6). Meanwhile, a dissociation of molecular oxygen was observed for Pt_3Ti and Pt_3V supported on O-doped graphene (Figure 16.6). After, the ORR mechanism was studied on Pt_4, Pt_3Ti, and Pt_3Ni supported on O-doped graphene for the ORR. According to the free energy profiles, Pt_3Ni supported on O-doped graphene could be a better catalyst than supported Pt_4 and Pt_3Ti clusters [59].

16.5 Conclusions

In this chapter, we examine the various modifications made to graphene to alter the interface between Pt-based nanoparticles and the modified graphene to enhance the stability and catalytic activity of Pt-based catalysts for the ORR. Initially, different types of vacancies have been introduced in graphene. In most cases, the Pt-based nanoparticles supported on graphene demonstrated superior catalytic properties,

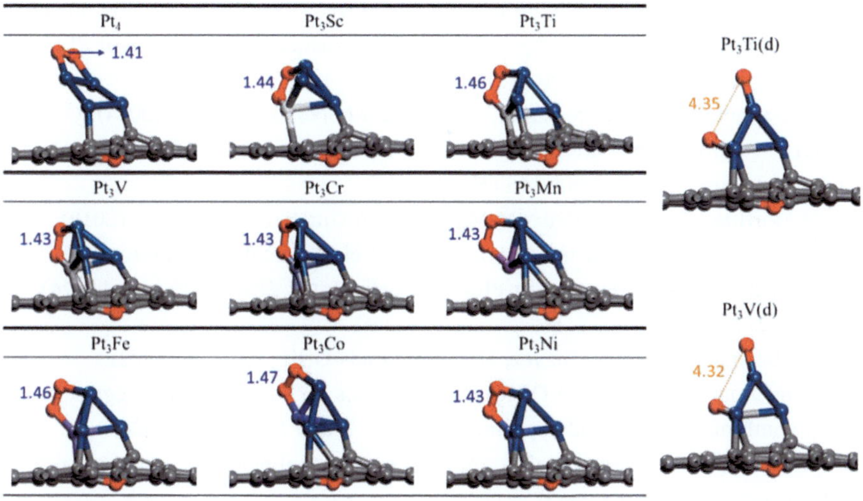

Figure 16.6 Adsorption mode and O−O bond length (in Å) of O_2 on the Pt_3M supported on O-doped graphene. The left three columns show O_2 adsorbed as di-oxygen species, while the two structures on the right are the O adatoms on supported Pt_3Ti and Pt_3V. Red, dark blue, and grey spheres represent O, Pt, and C, respectively. The color of M is different and depends on the element. Source: Cui et al. [59] / MDPI / CC BY 4.0.

stability, and activity compared to commercial Pt/C. Furthermore, Pt-based catalysts supported on doped graphene have been extensively studied, with various doping elements (e.g. N, S, Fe, P, O) explored to improve the reactivity of graphene as a support material for Pt-based nanoparticles in ORR. Among these, N is the most widely studied dopant. Additionally, co-doping has emerged as another effective strategy to enhance the properties of doped graphene as a support material for Pt-based nanoparticles in ORR.

Acknowledgments

This research was funded by the Tecnológico Nacional de México, Grant number 22639.25-P.

References

1 Mo, S., Du, L., Huang, Z. et al. (2023). Recent advances on PEM fuel cells: from key materials to membrane electrode assembly. *Electrochem. Energy Rev.* 6 (1): 28.

2 Tellez-Cruz, M.M., Escorihuela, J., Solorza-Feria, O., and Compañ, V. (2021). Proton exchange membrane fuel cells (PEMFCs): advances and challenges. *Polymers* 13 (18): 3064.

3 Tawalbeh, M., Alarab, S., Al-Othman, A., and Javed, R.M.N. (2022). The operating parameters, structural composition, and fuel sustainability aspects of PEM fuel cells: a mini review. *Fuel* 3 (3): 449–474.

4 Gomez-Sanchez, A., Franco-Luján, V.A., Alfaro-López, H.M. et al. (2024). Carbon material-reinforced polymer composites for bipolar plates in polymer electrolyte membrane fuel cells. *Polymers* 16 (5): 671.

5 Cruz-Martínez, H., Tellez-Cruz, M.M., Guerrero-Gutiérrez, O.X. et al. (2019). Mexican contributions for the improvement of electrocatalytic properties for the oxygen reduction reaction in PEM fuel cells. *Int. J. Hydrog. Energy* 44 (24): 12477–12491.

6 Cruz-Martínez, H., Rojas-Chávez, H., Matadamas-Ortiz, P.T. et al. (2021). Current progress of Pt-based ORR electrocatalysts for PEMFCs: an integrated view combining theory and experiment. *Mater. Today Phys.* 19: 100406.

7 Debe, M.K. (2012). Electrocatalyst approaches and challenges for automotive fuel cells. *Nature* 486 (7401): 43–51.

8 Jacob, T. (2006). The mechanism of forming H_2O from H_2 and O_2 over a Pt catalyst via direct oxygen reduction. *Fuel Cells* 6 (3–4): 159–181.

9 Shao, Q., Wang, P., and Huang, X. (2019). Opportunities and challenges of interface engineering in bimetallic nanostructure for enhanced electrocatalysis. *Adv. Funct. Mater.* 29 (3): 1806419.

10 Zhao, R., Li, Q., Jiang, X. et al. (2021). Interface engineering in transition metal-based heterostructures for oxygen electrocatalysis. *Mater. Chem. Front.* 5 (3): 1033–1059.

11 Zaman, S., Huang, L., Douka, A.I. et al. (2021). Oxygen reduction electrocatalysts toward practical fuel cells: progress and perspectives. *Angew. Chem.* 133 (33): 17976–17996.

12 Cruz-Martínez, H., Tellez-Cruz, M.M., Rojas-Chávez, H. et al. (2019). NiPdPt trimetallic nanoparticles as efficient electrocatalysts towards the oxygen reduction reaction. *Int. J. Hydrog. Energy* 44 (24): 12463–12469.

13 Hussain, S., Erikson, H., Kongi, N. et al. (2020). Oxygen reduction reaction on nanostructured Pt-based electrocatalysts: a review. *Int. J. Hydrog. Energy* 45 (56): 31775–31797.

14 Huang, L., Zaman, S., Tian, X. et al. (2021). Advanced platinum-based oxygen reduction electrocatalysts for fuel cells. *Acc. Chem. Res.* 54 (2): 311–322.

15 Jayabal, S., Saranya, G., Geng, D. et al. (2020). Insight into the correlation of Pt–support interactions with electrocatalytic activity and durability in fuel cells. *J. Mater. Chem. A* 8 (19): 9420–9446.

16 Zaman, S., Wang, M., Liu, H. et al. (2022). Carbon-based catalyst supports for oxygen reduction in proton-exchange membrane fuel cells. *Trends Chem.* 4 (10): 886–906.

17 Zhang, N. and Jiang, R. (2021). Interfacial engineering of metal/metal oxide heterojunctions toward oxygen reduction and evolution reactions. *ChemPlusChem* 86 (12): 1586–1601.

18 Shao, Q., Wang, P., and Huang, X. (2019). Opportunities and challenges of interface engineering in bimetallic nanostructure for enhanced electrocatalysis. *Adv. Funct. Mater.* 29 (3): 1806419.

19 Ortiz-Herrera, J.C., Cruz-Martínez, H., Solorza-Feria, O., and Medina, D.I. (2022). Recent progress in carbon nanotubes support materials for Pt-based cathode catalysts in PEM fuel cells. *Int. J. Hydrog. Energy* 47 (70): 30213–30224.

20 Rajalakshmi, N., Ryu, H., Shaijumon, M.M., and Ramaprabhu, S. (2005). Performance of polymer electrolyte membrane fuel cells with carbon nanotubes as oxygen reduction catalyst support material. *J. Power Sources* 140 (2): 250–257.

21 Bhuvanendran, N., Ravichandran, S., Lee, S. et al. (2024). Recent progress in Pt-based electrocatalysts: a comprehensive review of supported and support-free systems for oxygen reduction. *Coord. Chem. Rev.* 521: 216191.

22 Peera, S.G., Koutavarapu, R., Akula, S. et al. (2021). Carbon nanofibers as potential catalyst support for fuel cell cathodes: a review. *Energy Fuel* 35 (15): 11761–11799.

23 Sánchez-Rodríguez, E.P., Vargas-Hernández, C.N., Cruz-Martínez, H., and Medina, D.I. (2021). Stability, magnetic, energetic, and reactivity properties of icosahedral $M@Pd_{12}$ (M=Fe, Co, Ni, and Cu) core-shell nanoparticles supported on pyridinic N_3-doped graphene. *Solid State Sci.* 112: 106483.

24 Soldano, C., Mahmood, A., and Dujardin, E. (2010). Production, properties and potential of graphene. *Carbon* 48 (8): 2127–2150.

25 Castro Neto, A.H., Guinea, F., Peres, N.M. et al. (2009). The electronic properties of graphene. *Rev. Mod. Phys.* 81 (1): 109–162.

26 Gerber, I.C. and Serp, P. (2019). A theory/experience description of support effects in carbon-supported catalysts. *Chem. Rev.* 120 (2): 1250–1349.

27 Zhou, X., Qiao, J., Yang, L., and Zhang, J. (2014). A review of graphene-based nanostructural materials for both catalyst supports and metal-free catalysts in PEM fuel cell oxygen reduction reactions. *Adv. Energy Mater.* 4 (8): 1301523.

28 Samad, S., Loh, K.S., Wong, W.Y. et al. (2018). Carbon and non-carbon support materials for platinum-based catalysts in fuel cells. *Int. J. Hydrog. Energy* 43 (16): 7823–7854.

29 Yumura, T., Awano, T., Kobayashi, H., and Yamabe, T. (2012). Platinum clusters on vacancy-type defects of nanometer-sized graphene patches. *Molecules* 17 (7): 7941–7960.

30 Fampiou, I. and Ramasubramaniam, A. (2012). Binding of Pt nanoclusters to point defects in graphene: adsorption, morphology, and electronic structure. *J. Phys. Chem. C* 116 (11): 6543–6555.

31 Xu, D., Zhao, J., and Wang, X. (2013). A density functional theory study of the adsorption of bimetallic Fe_n Pt_m clusters on defective graphene: structural, electronic, and magnetic properties. *J. Nanopart. Res.* 15: 1–14.

32 Tang, Y.N., Li, C.G., Chen, W.G., and Dai, X.Q. (2015). Geometric stability, electronic structure and reactivity of Pt_4 cluster supported on defective graphene. *Integr. Ferroelectr.* 159 (1): 57–65.

33 Yang, G.M., Fan, X.F., Shi, S. et al. (2017). Stability of Ptn cluster on free/defective graphene: a first-principles study. *Appl. Surf. Sci.* 392: 936–941.

34 Bord, J., Kirchhoff, B., Baldofski, M. et al. (2023). An atomistic view of platinum cluster growth on pristine and defective graphene supports. *Small* 19 (10): 2207484.

35 Lim, D.H. and Wilcox, J. (2011). DFT-based study on oxygen adsorption on defective graphene-supported Pt nanoparticles. *J. Phys. Chem. C* 115 (46): 22742–22747.

36 Lim, D.H. and Wilcox, J. (2012). Mechanisms of the oxygen reduction reaction on defective graphene-supported Pt nanoparticles from first-principles. *J. Phys. Chem. C* 116 (5): 3653–3660.

37 Xu, D., Tian, Y., Zhao, J., and Wang, X. (2015). High stability and reactivity of defective graphene-supported Fe_nPt_{13-n} ($n = 1, 2$, and 3) nanoparticles for oxygen reduction reaction: a theoretical study. *J. Nanopart. Res.* 17: 1–12.

38 Ruiz-Villalobos, D., López-Sosa, L., García-Hilerio, B. et al. (2023). Stability and activity of PdCu clusters embedded on pyridinic N-doped graphene: a density functional theory investigation. *Mol. Phys.* 121 (5): e2192826.

39 Wang, Q., Tian, Y., Chen, G., and Zhao, J. (2017). Theoretical insights into the energetics and electronic properties of MPt_{12} (M=Fe, Co, Ni, Cu, and Pd) nanoparticles supported by N-doped defective graphene. *Appl. Surf. Sci.* 397: 199–205.

40 Montejo-Alvaro, F., Martínez-Espinosa, J.A., Rojas-Chávez, H. et al. (2022). CO_2 Adsorption over 3d transition-metal nanoclusters supported on pyridinic N_3-doped graphene: a DFT investigation. *Materials* 15 (17): 6136.

41 Zhu, J., Xiao, M., Zhao, X. et al. (2015). Strongly coupled Pt nanotubes/N-doped graphene as highly active and durable electrocatalysts for oxygen reduction reaction. *Nano Energy* 13: 318–326.

42 Tian, Y., Liu, Y.J., Zhao, J.X., and Ding, Y.H. (2015). High stability and superior catalytic reactivity of nitrogen-doped graphene supporting Pt nanoparticles as a catalyst for the oxygen reduction reaction: a density functional theory study. *RSC Adv.* 5 (43): 34070–34077.

43 Ma, J., Habrioux, A., Luo, Y. et al. (2015). Electronic interaction between platinum nanoparticles and nitrogen-doped reduced graphene oxide: effect on the oxygen reduction reaction. *J. Mater. Chem. A* 3 (22): 11891–11904.

44 Jukk, K., Kongi, N., Rauwel, P. et al. (2016). Platinum nanoparticles supported on nitrogen-doped graphene nanosheets as electrocatalysts for oxygen reduction reaction. *Electrocatalysis* 7: 428–440.

45 Rethinasabapathy, M., Kang, S.M., Haldorai, Y. et al. (2017). Ternary PtRuFe nanoparticles supported N-doped graphene as an efficient bifunctional catalyst for methanol oxidation and oxygen reduction reactions. *Int. J. Hydrog. Energy* 42 (52): 30738–30749.

46 Varga, T., Varga, Á.T., Ballai, G. et al. (2018). One step synthesis of chlorine-free Pt/Nitrogen-doped graphene composite for oxygen reduction reaction. *Carbon* 133: 90–100.

47 Siburian, R., Sebayang, K., Supeno, M., and Marpaung, H. (2017). Effect of N-doped graphene for properties of Pt/N-doped graphene catalyst. *ChemistrySelect* 2 (3): 1188–1195.

48 Rethinasabapathy, M., Kang, S.M., Haldorai, Y. et al. (2019). Quaternary PtRuFeCo nanoparticles supported N-doped graphene as an efficient bifunctional electrocatalyst for low-temperature fuel cells. *J. Ind. Eng. Chem.* 69: 285–294.

49 Xiong, Y., You, M., Liu, F. et al. (2020). Pt-decorated, nanocarbon-intercalated, and N-doped graphene with enhanced activity and stability for oxygen reduction reaction. *ACS Appl. Energy Mater.* 3 (3): 490–2495.

50 Phan, P.Q., Naraprawatphong, R., Pornaroontham, P. et al. (2021). N-Doped few-layer graphene encapsulated Pt-based bimetallic nanoparticles via solution plasma as an efficient oxygen catalyst for the oxygen reduction reaction. *Mater. Adv.* 2 (1): 322–335.

51 Hussain, S., Erikson, H., Kozlova, J. et al. (2024). Comparing Pt electrodeposition processes on nitrogen-doped graphene nanosheets for the electroreduction of oxygen. *J. Solid State Electrochem.* 28 (5): 1681–1690.

52 Bhaskaran, R. and Chetty, R. (2024). One-pot room temperature synthesis of nitrogen-doped graphene and its application as catalyst support for ORR in PEMFCs. *ACS Appl. Energy Mater.* 7 (2): 390–402.

53 Gao, Z., Chen, Z., Zhan, X. et al. (2023). Pt nanoparticles supported on iron and nitrogen-doped holey graphene for boosting ORR performance. *ACS Appl. Nano Mater.* 6 (12): 10521–10530.

54 Fu, L., Li, M., Pan, T. et al. (2024). Enhanced stability and activity of platinum-based catalyst using iron-nitrogen co-doped graphene as support for oxygen reduction reaction. *Int. J. Hydrog. Energy* 58: 1204–1213.

55 Chambers, R., Hussain, S., Kozlova, J. et al. (2024). Pt nanoparticles electrochemically deposited onto heteroatom-doped graphene supports as electrocatalysts for ORR in acid media. *J. Electrochem. Soc.* 171 (9): 096506.

56 Chambers, R., Piirsoo, H.M., Tamm, A. et al. (2024). Heteroatom-doped graphene-supported Pt nanoparticles as electrocatalysts for proton exchange membrane fuel cells. *ACS Appl. Nano Mater.* 7 (6): 5943–5955.

57 Groves, M.N., Malardier-Jugroot, C., and Jugroot, M. (2012). Improving platinum catalyst durability with a doped graphene support. *J. Phys. Chem. C* 116 (19): 10548–10556.

58 Jin, N., Han, J., Wang, H. et al. (2015). A DFT study of oxygen reduction reaction mechanism over O-doped graphene-supported Pt_4, Pt_3Fe and Pt_3V alloy catalysts. *Int. J. Hydrog. Energy* 40 (15): 5126–5134.

59 Cui, C., Sun, M., Zhu, X. et al. (2020). Oxygen reduction reaction catalyzed by Pt_3M (M=3d Transition Metals) supported on o-doped graphene. *Catalysts* 10 (2): 156.

60 Higgins, D., Hoque, M.A., Seo, M.H. et al. (2014). Development and simulation of sulfur-doped graphene supported platinum with exemplary stability and activity towards oxygen reduction. *Adv. Funct. Mater.* 24 (27): 4325–4336.

17

Defect and Interface Engineering in Li-Ion Batteries

Jesús Adrián Díaz-Real

Centro de Investigación y Desarrollo Tecnológico en Electroquímica, Dirección de Ciencia, Parque Tecnológico Querétaro S/N, Sanfandila, Pedro Escobedo, Querétaro, C. P. 76703, México

17.1 Introduction

Lithium-ion batteries (LIBs) have transformed energy storage technology and are thoroughly used across a number of modern-life applications like consumer electronics, electric vehicles (EV), and grid energy storage due to their high energy density, lifespan, and coulombic efficiency [1, 2]. However, their performance is limited by several aspects such as loss of capacity, sluggish charge transport, and safety concerns, which are derived from both material properties and electrode architecture [3, 4].

To overcome these challenges novel strategies are required, where defect and interface engineering stand as innovative approaches. Defect engineering involves the intentional manipulation of crystallographic and chemical impurification to modulate and optimize material properties. Point defects are a type of individual defect which modifies ionic and electronic transport like oxygen vacancies. Other defects are extended defects, which is a group that include dislocations and stacking faults, that affect structural stability and charge storage dynamics [5, 6]. Interface engineering is a complementary approach that focuses on optimizing the interfacial properties between battery components such as the electrode–electrolyte interface, minimizing resistances, and improving the overall battery performance [7, 8].

Recent literature report that synergistic benefits can be obtained from both approaches, thus tackling intrinsic limitations in LIBs. The introduction of oxygen vacancies along with the regulation of interfacial chemistries, as reported by Yang et al. [9] improved ionic conductivity and structural stability, leading to longer cycle life and higher energy densities [10]. While it is most common to work on the cathode side, several other works report physicochemical treatments on the anode side, where the defect concentration and interface quality are modulated through doping and controlled annealing [11].

In this chapter, we explore the critical role of defect and interface engineering in advancing LIB technology. Several approaches are presented, and their impact is discussed on electrochemical performance, focusing on charge transport, capacity

Surface and Interfacial Defects in Nanomaterials for Sustainable Energy Production and Storage, First Edition. Edited by Noé Arjona, Lorena Álvarez Contreras, and Minerva Guerra Balcázar.

enhancement, and durability. Through a combination of experimental insights and theoretical frameworks, we aim to provide a comprehensive overview of current progress and future directions in LIB. Finally, we highlight the interplay between defect structures and interfacial properties for designing next-generation LIBs with superior performance and reliability.

17.2 Defect Engineering in Li-Ion Batteries

Defect engineering is a complex approach in the field LIBs, wherein the crystalline lattice of a particular material is altered with the intention of modifying the structural characteristics to improve electrochemical performance. In spite of receiving a general pejorative denomination, controlled defects allow the tailoring of particular properties such as ionic conductivity, electronic transport, structural stability, and reaction kinetics. In this section, we will review the types of defects, their effects on LIB performance, and the methodologies used in their engineering.

17.2.1 Types of Defects and Their Mechanisms

Crystalline materials exhibit defects that are typically categorized as point defects, extended defects, and planar defects. Each of those defects has a particular impact on LIB performance and will be described.

17.2.1.1 Point Defects

This type of defect includes vacancies and substitutions, which owe their name to their localized nature, namely a disruption in the atomic arrangement. Oxygen vacancies (OV) have been widely investigated due to their impact on ionic transport. In general, vacancies are a type of defect that originated from the absence of an atom in a particular lattice position. In the case of LIB cathodes, OVs enhanced ionic conductivity as they facilitate Li-ion diffusion by generating specific pathways. An example has been described in the work of Li et al., where periodic native oxygen vacancies induce structural flexibility for ionic extraction in Na-ion batteries [12].

In NASICON-based solid electrolytes, OVs were promoted by Al^{3+} doping in an Al-doped $NaTi_2(PO_4)_3$ cathode as described by Tian et al. [7]. These OVs had an impact on the Li^+ ionic transport and also improved stability as a consequence of a broadened the electrolyte polarization range [7]. Similarly, an OVs-rich TiO_2/C composite described by Huo et al. exhibited enhanced cycling stability and remarkable capacity retention due to improved kinetics [13].

17.2.1.2 Extended Defects

Extended defects include dislocations and stacking faults, which are disruptions in the crystalline lattice extending over multiple atomic planes modifying the structural, mechanical, and electrochemical properties of cathodic materials. The work of Zeng et al. in lithium-rich layered oxides (LLOs) noted that stacking faults in layered oxides was detrimental for ionic mobility [5]. However, strategies to reduce stacking fault densities through synthesis optimization have been shown to improve the performance of Li-rich layered oxides by enhancing Li-ion diffusion pathways [5].

Dislocations are known for being associated with material instability. However, careful control of dislocation can lead to increased ionic diffusion. As an example, Yang et al. reported on the regulated density of dislocations in high-entropy oxides, which improved structural stability and ionic conductivity [9].

17.2.1.3 Planar Defects

This type of defect considers grain boundaries and interface defects and is often found in polycrystalline materials. The latter, grain boundaries, are commonly associated with increased charge transfer resistance; however, they can be engineered to facilitate ionic conductivity by doping or thermal treatments, as Kireeva et al. reported in garnet-structured solid electrolytes [14]. They pointed out that there is a significant impact on the total conductivity due to the composition of grain boundaries.

Interfacial defects also play a critical role in structural stability during charge/discharge cycles. Several strategies that have been reported to improve interfacial charge transfer include surface doping [7], coatings, and thermal treatments [3] to promote stability and enhanced performance of cathode materials.

17.2.2 Effects on Electrochemical Performance

17.2.2.1 Charge Transport

Ion mobility is directly related to charge transport and defects that can enhance both ionic and electronic transport. In TiO_2-based anodes, OVs significantly improved ionic transport by minimizing migration barriers [13]. Pulsed laser annealing was used in carbon-based anodes to promote CVs, which in addition created lithium intercalation sites, increased charging rate, and reduced interfacial resistance [6]. Dislocations in high-entropy materials created a network of pathways that enhanced ionic mobility and are highly desirable in high-power output demanding application [9].

17.2.2.2 Structural Stability

Structural stability is one of the most affected properties when tailoring defects in LIB materials. Excessive defects lead to poor stability, while controlled defect densities render improved mechanical stability and affect other properties. To exemplify this, the work of Gao et al. on single-crystallization and chemical treatment with boric acid on Li-rich cathodes induced lattice defects, improved mechanical, cyclic stability, and higher specific capacity and led to grain boundary-free surfaces [11].

In NASICON-based solid electrolytes, these defects demonstrated resilient structures to volume changes during cycling and also prevented structural collapse and extending battery life [7].

17.2.2.3 Energy Density and Capacity

These two important properties are extremely dependent on defect induction in LIB materials which can provide extra active sites for ionic allocation. Defect-rich hexagonal boron nitride (hBN) anodes in Lei et al. work [2] showed an increase in specific capacity due to the introduction of nitrogen antisite defects. Such types of defects were induced via cryomilling, which worked as active sites for Li-ion adsorption.

Analogously, perovskite-type materials ($Sr_xVO_{3-\delta}$) with controlled defect densities through thermal treatments were reported by Li et al., where higher capacities and enhanced cycling stability were obtained [15].

17.2.3 Methods for Defect Creation and Control

17.2.3.1 Chemical Doping

The chemical nature of LIB materials can be modified by replacing lattice-forming elements with suitable substitutes, i.e. doping, creating point defects such as vacancies. The process of doping usually takes place during the synthesis step, especially when the lattice-forming elements are reacting. Secondary doping processes can also work, with materials that require thermal treatment post-processing step under a particular chemical (gas) environment. In Al-doped NASICON electrolytes, OVs were obtained as part of the doping process which also enhanced ionic conductivity [7]. In a similar approach, Zhao et al. used first-principle calculations to study O-doping in VS_2 with the intention to induce S-vacancies resulting in a reduced diffusion barrier for Li ion and favoring the intercalation-deintercalation process [16].

17.2.3.2 Thermal and Mechanical Treatments

Both, thermal (annealing) and mechanical (milling) methods, are common strategies used to introduce extended defects in LIB materials. More advanced techniques involve complex experimental techniques to control such defects, as is the case for pulsed laser annealing which allows for tailoring carbon vacancies in graphite anodes [6]. Other methods imply high-energy strategies like ball milling which introduce large dislocations despite lesser control [8].

17.2.3.3 Non-equilibrium Synthesis

Other synthesis techniques, such as solution combustion and cryomilling, allow for advanced and precise control over defect induction. These methods typically rely on the control between crystalline phases and defect densities [15].

17.2.4 Challenges and Future Directions

Defect engineering is a field currently active which, while important progress has been reported, several challenges remain. The main challenge relies upon defect density control, as excessive densities lead to structural instability and material decay. These are not minor issues, as they are related to lithium plating and dendrite formation, which compromise battery safety with performance and efficiency [3]. Therefore, future research efforts should focus on optimizing defect densities control, and the development of specialized techniques to induce selectively a particular type of defect [14]. Modern techniques, like simulation through computational methods and machine learning, should be considered to accelerate the discovery of optimized defect configurations.

17.3 Interface Engineering in Li-Ion Batteries

A recent approach used in the optimization of LIB is interface engineering and it focuses on controlling the interfacial properties between battery components to enhance their performance. The interface between the electrode and the electrolyte results critical in determining ionic and electronic transport, structural stability, and overall energy efficiency of LIB.

17.3.1 Role of Interfaces in Battery Performance

Important electrochemical processes such as charge transfer, ionic diffusion, and the formation of solid–electrolyte interphases (SEIs) are governed by interfacial properties. The electrochemical performance of a LIB is highly sensitive to the chemical and structural characteristics of these interfaces, making their optimization essential.

Electrode–Electrolyte Interface: The electrode–electrolyte interface plays a vital role in the cycling stability and efficiency of LIBs. Side reactions often take place at the interface resulting in the formation of SEI layers, which passivates the surface and increases impedance over time [3]. Solid-state batteries can present important challenges for the SEI, compromising mechanical stability and charge transfer resistance, thus requiring precise engineering to ensure seamless ionic transport [7].

Cathode–Anode Interfaces (CEIs): Another critical consideration comes from the compatibility between cathode and anode materials. In high-voltage cathodes such as NCM811, the electrode surface will promote electrolyte degradation due to increased reactivity from metallic Li formed at high potentials, requiring protective coatings or modifications to stabilize the CEI [17].

17.3.2 Strategies for Interface Optimization

17.3.2.1 Surface Coatings and Modifications

One effective method in interface engineering is to create coating at the surface of the electrodes to modify its properties, serving as protective barriers, and preventing undesired reactions while enhancing interfacial stability.

Inorganic Coatings: LLO materials such as $Li_2B_4O_7$, Al_2O_3, and ZrO_2 have been used as coatings on cathode materials to reduce the effects of high-voltage operation and avoid undesired reactions. A spinel-type $Li_2B_4O_7$ coating on single-crystal LLO reduced the irreversible oxygen loss and also helped the stabilization of the lattice, which resulted in enhanced cycling stability [11].

Organic–Inorganic Hybrids: Hybrid NATP-PAN composite coatings have been used to combine the flexibility of polymers with the mechanical strength of inorganic particles, as demonstrated by Tian et al. exhibiting excellent interfacial stability and ionic transport properties in solid-state batteries [7].

17.3.2.2 Interface Doping

Doping at the electrode or electrolyte surface can significantly modify interfacial properties by introducing defects. Tian et al. doped NASICON-based electrolytes with Al^{3+}, creating OVs which in turn reduced the activation energy for lithium-ion diffusion by an improved ionic conductivity [7]. Kireeva et al. analyzed the parameters that influenced solid-state garnet-based electrolytes and showed that interfacial resistance is decreased as a consequence of oxygen vacancies doping [14].

17.3.2.3 Solid–Electrolyte Interface (SEI) Engineering

The SEI layer is a key component in LIBs that passivates the electrode surface, thus preventing electrolyte decomposition. The main challenge in SEI is to increase its stability without compromising an increased resistance. SEI layers chemically modified with non-fluorinated ether-based electrolytes have shown an increase in their lifespan from 100% to 500% in high-voltage NMC811 cathodes. This extended durability was explained in terms of a controlled surface reactivity which improved coulombic efficiency [17].

17.3.2.4 Advanced Surface Engineering Techniques

Some techniques such as pulsed laser deposition (PLD) allow precise control in surface modifications. This method created surface texturization improving Li ion intercalation rate, tailor control of surface chemistry by removing inactive PVDF binder, and inducing carbon vacancies [6].

17.3.3 Case Studies in Interface Engineering

Recent studies have highlighted the potential of interface engineering in improving LIB performance which are described as follows. The first example comprises an oxygen-vacancy-enriched $LiFePO_4$ cathode, which demonstrated enhanced electronic conductivity and lithium-ion diffusion and was achieved through controlled defect introduction at the interface [1]. Another example is in the use of high-entropy oxides, where a dual-phase high-entropy oxide with intergrown interfaces exhibited improved structural stability and ionic transport due to phase-selective defect engineering [9].

17.3.4 Challenges and Future Directions

Interface engineering has observed significant progress, however, several challenges remain to be addressed. Chemical instability leads to interfacial degradation and mechanical failure after extensive cycling, along with the elevated fabrication costs, and hinders the applicability of interface-engineered materials. The perspectives that should be followed include the integration of computational methods through which simulations of the optimal interface designs can be predicted, thus accelerating material discovery [14]. As aforementioned, the development of affordable and scalable methods compatible with current manufacturing technologies (roll-to-roll coatings) is imperative for industrial adoption [3].

The accelerated discovery of new multifunctional materials for interface stabilization is a promising approach expected in next-generation LIBs [8]. The combination of the several innovative approximations described, theoretical and experimental, offers an opportunity for the improvement of safety, efficiency, and durability of next-gen batteries.

17.4 Experimental Techniques and Analytical Methods

Progress in the research for defect and interface engineering for LIBs strongly depends on sophisticated experimental techniques. This is valid for the synthesis, properties modulation, and characterization methods to determine their structure, chemistry, and electrochemical performance. In this section, the state-of-the-art analytical methods for the study of defect and interface engineering are presented.

17.4.1 Characterization of Defect Analysis

A number of these defects can be studied using microscopic, spectroscopic, and diffractometric techniques, which are described as follows.

17.4.1.1 Microscopy-Based Techniques

Scanning Electron Microscopy (SEM): When combined with energy-dispersive X-ray spectroscopy (EDX), SEM is enabled for the identification of defects at a large scale with composition analysis. This is often used to relate the morphological changes and evolution of electrodes over extensive cycling, revealing surface erosion, fractures, or voids [9].

Transmission Electron Microscopy (TEM): High-resolution TEM (HR-TEM) provides atomic-scale imaging of dislocations, stacking faults, and grain boundaries. In scanning TEM (STEM), a module can be coupled which enables the so-called electron energy loss spectroscopy (EELS), enabling the chemical analysis of defects like OVs [3, 11].

17.4.1.2 Spectroscopy-Based Techniques

Raman Spectroscopy: It is often seen that cathodic materials are deposited on carbonaceous supports, and the performance is greatly dependent of the interface quality. Here is where Raman spectroscopy becomes important, as it is employed to identify structural disorders, particularly in carbon materials. Defects, including vacancies and substitutions, are reflected as shifts in Raman peaks, providing insights into their density and nature [6].

X-Ray Photoelectron Spectroscopy (XPS): XPS is a robust and powerful characterization technique widely used in the study of point defects, OVs, and changes in the chemical states in metals. While XPS is limited to a depth of analysis up to 10 nm in the surface, an accessory ion bombardment method enables XPS to perform depth profiling, particularly useful in the study of defect distributions across interfaces [7].

17.4.1.3 Crystallographic Techniques

X-Ray Diffraction (XRD): XRD is an essential technique in the study of crystal structures. From the XRD diffractograms, several defects can be identified such as lattice distortions or stacking faults. A more in depth-analysis can be performed on the diffractograms; Rietveld refinement, enabling the quantification of defect density and phase composition in multi-phase materials [11].

Neutron Diffraction: Complementary to XRD, neutron diffraction allows sensitivity to lighter elements, like lithium (Li). This technique has been used to track Li positions in defect-rich cathodes, offering insights into ionic Li transport phenomena [8].

17.4.2 Interfacial Characterization Techniques

The modified cathodic battery materials demonstrate their characteristics during the charging and discharging processes which take place at the electrode–electrolyte interface. The insights at this boundary are studied while the process is occurring (in situ/operando) using electrochemical-based techniques, which happen to be the most common and widely accepted methodology to study the performance of both materials and fully assembled cells. However, electrochemical characterization is out of the scope of the current chapter. Alternatively, the materials can be studied prior to their assembly or ex situ to evaluate the possible causes of failure (surface techniques). A few of the most used techniques are described as follows.

17.4.2.1 Ex Situ Techniques

The analysis of the surface topography of the interfaces can be done with atomic force microscopy (AFM). This technique has several available operation modes, while mechanical AFM is the most widely used. An example is conductive AFM which maps the electronic conductivity across interfacial regions, revealing the impact of engineering strategies [3].

When information regarding the chemical composition profile along the depth of an interface, secondary ion mass spectrometry (SIMS) provides a compositional mapping of elements, like lithium and oxygen, which cannot be quantified by other techniques. For example, durability tests that evaluate the interfacial degradation extent over prolonged cycling can be studied with SIMS [7].

17.4.2.2 In Situ and Operando Techniques

In situ microscopy techniques enable real-time monitoring of the morphological changes that materials undergo during a particular event. For example, in situ SEM has been used to study the development of crack propagation or dendrite growth, at interfaces during cycling. With this technique, a better understanding of the mechanical stability of engineered interfaces can be obtained [6]. A higher level of detail is obtained with in situ TEM, where real-time observations of defect evolution are possible, such as dislocation propagation or phase transformations, under applied electrochemical conditions. This technique provides direct evidence of structural changes caused by lithium insertion and desorption [11].

Compared to in situ techniques, operando techniques are measured not in a representative environment of the operation of the battery, but during actual use. An example is the use of Raman spectroscopy under operating conditions to monitor structural changes in real time such as the formation of intermediate phases in cathode materials [9]. Another example is operando XPS which tracks the chemical evolution of surfaces and interfaces during battery operation, revealing the dynamic interactions that occur at SEI layers [17].

17.4.3 Emerging Techniques

As mentioned before, the progress in material modification also requires more demanding and advanced techniques for its characterization. A few of the most recent available characterization methods are the following.

Four-Dimensional STEM (4D-STEM): It represents an extension of the capabilities of scanning electron microscopy (STEM) which combines spatial and diffraction data. This modern technique provides detailed information about strain, defects, and interfacial structuring with atomic resolution. 4D-STEM has proven to be remarkably useful in the analysis of complex materials like high-entropy oxides [9].

Machine Learning-Assisted Analysis: With the advent of machine learning, the integration of these advanced algorithms into microscopy and spectroscopy-based techniques can accelerate defect identification and interface optimization. Experimental data obtained with this innovative method can generate predictive models useful in designing next-generation materials [14].

Advanced Synchrotron Methods: High-energy spectrometry like synchrotron-based techniques have unparalleled resolution in the study of atomic vicinity. X-ray absorption spectroscopy (XAS) and tomography have been used to understand the electronic structure changes in defect-engineered cathodes and the 3D visualization of interface morphologies [3].

17.5 Challenges and Future Directions

Important progress has been discussed regarding the fields of defect and interface engineering bringing significant benefits to LIB technology. Nevertheless, a careful balance has to be kept when modulating the density of defects as properties are intrinsically dependent on them, like stability. For such reasons, it is important to remark the current challenges and research directions that require further investigation to overcome these limitations.

17.5.1 Challenges in Defect and Interface Engineering

17.5.1.1 Defect Density vs Material Stability

A fundamental challenge in defect engineering is to balance the density of defects without compromising the integrity of the overall structure. As discussed in previous sections, while it is desirable to promote vacancies and dislocations to enhance lithium-ion transport, the excess of defects leads to chemical instability. Zheng et al. reported on the effect of excess OVs which increased the probability to promote oxygen evolution, and eventually led to structure collapse [3]. In anodes, the lack of control in graphite defects was associated with Li deposition and consequently dendrite growth, resulting in a compromise to battery safety [6]. Future investigations should consider the scalability of processes for material synthesis with precise control of defect density.

17.5.1.2 Interfacial Degradation

The electrode–electrolyte interface is a key element that ultimately reflects the properties of both phases. In the case of optimized electrodes like high-voltage NMC811, Wang et al. observed that electrolyte decomposes at the interface, and consequently, an unstable SEI is formed [17]. In solid-state batteries, a situation arises from poor mechanical compatibility between the electrodes and the solid electrolyte that tends to increase the interfacial resistance leading to complete loss of contact over cycling [14]. Therefore, more robust materials are required for the assembly of stable interfaces as well as deeper understanding of chemical dynamics during the intercalation-deintercalation process.

17.5.1.3 Manufacturability and Scalability of Processes

Many of the techniques reported in the literature used in the advanced control and modification of electrodic materials are still not available to be reproduced at larger scales. This is the case of atomic layer deposition (ALD) or pulsed laser methods, which are far from being implemented in manufacturing lines [3]. Such methods are expensive, time-consuming, and present high variability when modifying their parameters, while at the same time being too complex to implement in industrial conditions. Aside from these situations, other aspects should also be taken into consideration like reproducibility and uniformity over larger electrode areas that are posing a significant technical challenge.

17.5.2 Future Directions

Several challenges have been underlined to this point and further investigation should prioritize certain areas to make more readily available the advantages of defect and interface engineering in next-gen batteries.

As aforementioned, the development of novel cathodes is it is an important goal, which integrates, in a single architecture, defect and interface engineering. Therefore, these multifunctional materials could benefit from both types of modifications without stressing or compromising the rest of their properties. Promising materials like dual-phase high-entropy oxides with intergrown interfaces could balance an increased capacity without sacrificing structural stability [9]. To accelerate the discovery of such materials, the integration of modern computational modeling tools should be considered in future research. The uses of density functional theory (DFT) and molecular dynamics (MD) are important assets to expedite and rationalize experimental methods. Newer algorithms, machine learning, and artificial intelligence (AI) also show promising effects in the prediction of material properties, interface configurations, and material discovery for the design of next-gen batteries [14]. Along with the prediction and optimization of their properties, advanced and complex characterization techniques are necessary to study and validate the modification in materials. While there is a standardized methodology to evaluate the performance of materials in half-cell and full-cell, the evolution of the interface, preservation of defects, and structural stability at atomic scale is still complicated to measure. Advances in high-energy methods like synchrotron light will also provide further understanding of these intricate material dynamics.

Once the material is optimized, synthetic methods to introduce them into manufacturing lines are required. While many of the current technologies are available at the laboratory level, the industrialization of these modified cathodes requires scalable processes which comply with uniformity, repeatability, and precise control without increasing significantly the manufacturing cost. This is one of the most complicated challenges to overcome but common in advanced material production. Finally, but no less important are the topics of sustainability and circularity of these processes. There are many demands, and international commitments made by individual countries and alliances toward the adoption of cleaner manufacturing practices, emphasizing the recyclability of materials and minimization of toxic components. The development of environmentally benign materials, such as bio-inspired interfaces or low-toxicity dopants, will be critical for sustainable battery systems [8].

17.6 Conclusions

Defect and interface engineering have emerged as recent strategies to overcome critical limitations LIB technology, offering pathways to enhance energy density, prolonged cyclability, and enhanced performance, while improving safety. The roles of defect manipulation, interface optimization, and their synergistic integration in advancing LIB performance were explored in this chapter. Defect engineering enables the precise tailoring of material properties through controlled modifications such as OVs, dislocations, and grain boundaries. Such characteristics facilitate enhanced ionic transport, improved electronic conductivity, and promoted structural stability. Interface engineering complements these improvements by optimizing critical boundaries, such as the electrode–electrolyte interface, by introducing protective coatings, employing hybrid materials, or SEI layers to mitigate degradation and reduce resistance. A concomitant approach of both defect and interface engineering provides synergistic effects addressing multifactorial challenges in LIBs. Integrated attempts of both modifications showed significant improvements in ionic and electronic transport as well as mechanical and interfacial stability. Despite significant progress, several challenges remain in balancing defect densities, addressing interfacial degradation, and scaling up fabrication methods for commercial applications. To overcome these situations, the use of advanced algorithms, such as machine learning, simulation, and AI tools, in combination with in situ and operando experimental techniques are resourceful strategies to follow. Surpassing the aforementioned limitations will require the collaboration of these approaches to develop safer, more efficient, and more sustainable next-gen LIBs.

References

1 Wang, G., Xu, P., Desta, H.G. et al. (2024). Water effect on the electronic properties and lithium-ion conduction in a defect-engineered $LiFePO_4$ electrode. *Batteries* 10 (8): 281.

2 Lei, Y., Avvaru, V.S., Ward, Z. et al. (2024). Defect-driven ion storage on hexagonal boron nitride for fire-safe and high-performance lithium-ion batteries. *Chem. Eng. J.* 498: 155347.

3 Zheng, Z., Wang, X., Wang, K. et al. (2025). Achieving ultra-low voltage fading in Co-free Li-rich layered oxides via effortless surface defect engineering. *J. Colloid Interface Sci.* 678: 572–582.

4 Liu, X., Guo, Q., Li, Y. et al. (2023). The oxygen-doped self-supporting functional carbon nanotube sponge realized the synergistic enhancement of lithium ion transport and lithium polysulfide reaction. *J. Alloys Compd.* 969: 172461.

5 Zeng, W., Shu, W., Zhu, J. et al. (2024). Stacking fault slows down ionic transport kinetics in lithium-rich layered oxides. *ACS Energy Lett.* 9: 346–354.

6 Khosla, N., Narayan, J., Narayan, R. et al. (2023). Microstructure and defect engineering of graphite anodes by pulsed laser annealing for enhanced performance of lithium-ion batteries. *Carbon* 205: 214–225.

7 Tian, H., Gu, X., Wu, Q. et al. (2024). Defect engineering build organic-inorganic composite electrolytes with stable electrochemical window and interface for all-solid-state lithium-ion batteries. *Electrochim. Acta* 492: 144383.

8 Shen, Q., Shi, Y., He, Y., and Wang, J. (2024). Defect engineering of hexagonal MAB phase Ti_2InB_2 as anode of lithium-ion battery with excellent cycling stability. *Adv. Sci.* 11: 2308589.

9 Yang, H., He, L., Chen, Q. et al. (2024). Phase-selective defects engineering in dual-phase high entropy oxide for Li-ion storage. *Chem. Eng. J.* 488: 151113.

10 Zhang, Y., Ugata, Y., Campéon, B.L., and Yabuuchi, N. (2024). Unlocking electrode performance of disordered rocksalt oxides through structural defect engineering and surface stabilization with concentrated electrolyte. *Adv. Energy Mater.* 14: 2304074.

11 Gao, X., Wang, L., Guo, J. et al. (2024). Lattice engineering toward extraordinary structural stability of high-performance single-crystal Li-rich layered oxides cathodes. *Adv. Funct. Mater.* 34: 2407692.

12 Li, Y., Wang, X., Gao, Y. et al. (2019). Native vacancy enhanced oxygen redox reversibility and structural robustness. *Adv. Energy Mater.* 9: 1803087.

13 Huo, J., Xiao, Y., Yang, H. et al. (2024). Ultrasmall TiO_2/C nanoparticles with oxygen vacancy-enriched as an anode material for advanced Li-ion hybrid capacitors. *J. Energy Storage* 89: 111586.

14 Kireeva, N.V., Tsivadze, A.Y., and Pervov, V.S. (2023). Modeling ionic conductivity and activation energy in garnet-structured solid electrolytes: the role of composition, grain boundaries and processing. *Solid State Ionics* 399: 116293.

15 Li, X., Lin, Z., Jin, N. et al. (2023). High-capacity anode materials for lithium-ion batteries: perovskite strontium vanadate with crystalline-amorphous dual-phase structure. *Mater. Today Chem.* 30: 101542.

16 Zhao, R., Wan, C., Qian, P., and Ju, X. (2023). Insight into the effects of S-vacancy and O-doping in monolayer VS_2 as lithium-ion battery anodes from first-principles calculations. *Surf. Interfaces* 38: 102851.

17 Wang, Z., Che, X., Wang, D. et al. (2024). Non-fluorinated ethers to mitigate electrode surface reactivity in high-voltage NCM811-Li batteries. *Angew. Chem. Int. Ed.* 63: e202404109.

18

Defects and Interface Engineering in Na-Ion Batteries

Zhen-Yi Gu, Xiao-Tong Wang, Xin-Xin Zhao, and Xing-Long Wu

State Key Laboratory of Integrated Optoelectronics, MOE Key Laboratory for UV Light-Emitting Materials and Technology, School of Physics, Changchun, 130024, China

The mature experience and technology in the development of Li-ion batteries (LIBs) provide a rich reference for Na-ion batteries (NIBs). Nevertheless, compared with Li (0.76 Å and 3.02 V vs standard hydrogen electrode, SHE), Na^+ has a larger radius and lower redox potential (1.02 Å and 2.71 V vs SHE), coupled with the complexity of the battery composition, the development of NIBs still needs to explore the appropriate model [1, 2]. The lifetime and electrochemical performance of NIBs are affected by many aspects, among which the most influential factors come from the defects of the electrode material itself and the unstable construction of the electrode interface [3, 4].

18.1 Defects in Electrode Materials

18.1.1 Defect Type

Due to the similarity of elemental radii, the transition metal ions are prone to migrate to the carrier sites during material synthesis, which belong to the inherent anti-site defects (IASD) of the material [5, 6]. For example, in the Na superionic conductor (NASICON)-structured polyanionic cathode material for NIBs, there are only 18 Na^+ in the 24 sites of Na. Thus, the incompletely occupied Na1 and Na2 sites provide vacancies for disordered transition metal ions. As in $Na_3MnTi(PO_4)_3$, Mn migrates to the Na2 site, blocking the Na^+ diffusion channel and inhibiting the $Mn^{2+/3+/4+}$ redox reaction, leading to voltage polarization and capacity loss. Hu et al. employed Mo-doping strategy to enhance the formation energy of IASDs, to reduce the defect concentration, and to improve the initial Coulomb efficiency from 76.2% to 85.9%.

Again, as in the case of $Na_3VCr(PO_4)_3$, the migration of V^{5+} to the Na1 site with a Wyckoff site of 6b occurs during the Na^+ de-embedding reaction at room temperature, leading to structural degradation during cycling. This transition metal migration during charging and discharging anyway is called the derived anti-site defect (DASD) [7]. The difference between IASD and DASD is shown in Figure 18.1.

Surface and Interfacial Defects in Nanomaterials for Sustainable Energy Production and Storage, First Edition. Edited by Noé Arjona, Lorena Álvarez Contreras, and Minerva Guerra Balcázar.

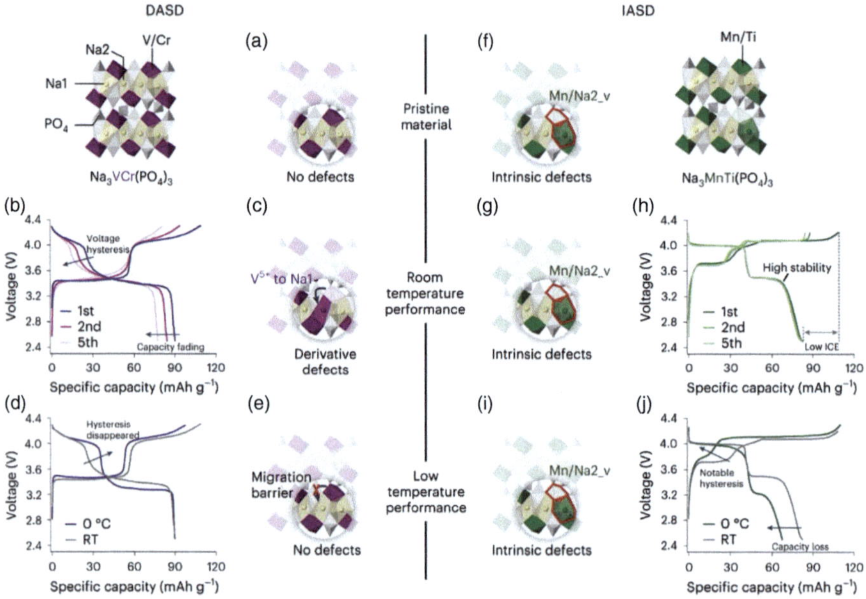

Figure 18.1 Schematic representation of DASDs vs the IASDs. (b, d, h, j) galvanostatic charge and discharge curves of $Na_3VCr(PO_4)_3$ (left panel) and $Na_3MnTi(PO_4)_3$ (right panel) at room temperature (b, h) and 0 °C (d, j) for 0.1C (1C = 117 mA h g^{-1}), respectively. (a) Na atoms have two Wyckoff positions, i.e. 6b (denoted by Na1) and 18e (denoted by Na2). (b–e) After cycling (b), DASD is generated during the denaturation process, and V^{5+} migrates to the Na1 vacancy (c) and induces a voltage hysteresis. However, at low temperatures, the apparent migration barrier of V^{5+} (e) hinders the above process, and thus the voltage hysteresis disappears (d). (f–j) Occupation of Na2 vacancies by Mn during synthesis (f) leads to the formation of IASDs, and the structure stays unchanged in subsequent cycles (f, g, i). RT: room temperature. Source: Liu, Y et al. [6] / with permission of Springer Nature.

Whereas, during electrochemical reactions at low temperatures, the V atoms are fixed in their original positions in the lattice, and V^{5+} migration is suppressed. Therefore accordingly, the researchers proposed to restore the degraded structure by room temperature deep sodiation (<1.7 V vs Na^+/Na, overdischarge) to force the dislocated V atoms to return to their original lattice. Specifically, the batteries were first discharged to 1.4 V, followed by a second cycle in the voltage interval of 1.4–2.5 V, followed by a cycle in the voltage window of 1.4–4.3 V.

This phenomenon is also observed in oxide cathode materials for NIBs. For example, the presence of DASD is clearly observed in O3-$NaFeO_2$ [8]. Fe irreversibly migrates from the Fe layer to the Na layer, resulting in the collapse of the material structure. In oxide-type cathode materials for LIBs, the change in the length of the transition metal-oxygen (TM-O) bond leads to strain in the octahedral or O-TM-O layer and provides the impetus for TM migration, making the material susceptible to the phase transition from the layered to the rock salt phase. In the oxide-type cathode materials for NIBs, on the other hand, due to the large radius of Na^+, such a phase transition behavior does not occur readily, and the lamellar structure, although preserved, results in phase transitions in the O3 and P3 structures. In addition, the layered transition metal oxides are prone to structural defects of dislocations and stacked laminar dislocations during charging and discharging, which are fatigued by the repeated de-embedding of Na^+, leading to lattice distortion, and thus the gradual accumulation of microstrain within the cathode particles, which significantly affects the mechanical strength of the energy materials. This accumulation of microstrain not only leads to internal structural fatigue but also exacerbates the capacity degradation problem of the material during cycling [9–12].

18.1.2 Defect-Modulated Pure Phase

The preparation of materials is often inevitably accompanied by the generation of thermodynamically favorable impurity phases that limit the capacity of the material. Defect modulation can be used to suppress the generation of impurity phases and thus enhance the electrochemical performance of electrode materials. For example, $Na_4Fe_3(PO_4)_2P_2O_7$ (NFPP) has become a sought-after material in the market because of its low cost, safety, non-toxicity, moderate voltage, and specific capacity, but it is easy to coexist with $NaFePO_4$ (NFP) impurity phases in the material preparation. Researchers have discovered that in the $Na_4F_xP_4O_{12+x}$ system, the $Na_2FeP_2O_7$ (NFPO, $x = 2.00$), NFPP ($x = 2.91$), and NFP ($x = 4.00$) phases coexist. However, when x takes other values, a two-phase mixture forms instead [13]. It is further found that the NFPP lattice tends to form an iron defect structure at the Fe2 site, and by precisely adjusting the defect content, the highly pure phase $Na_4F_{2.91}(PO_4)_2P_2O_7$ ($NF_{2.91}PP$) can be obtained. The presence of iron defects in $NF_{2.91}PP$ results in the material having a lower bandgap and Na^+ migration barriers, which further enhances the electronic and ion-conducting properties of the material. In the synthesis of $Na_2CoP_2O_7$ (NCPO), the pure Co rose phase is preferred, but the Co blue phase with orthorhombic polycrystalline structure (Pna21) is more likely to be produced [14]. Taking temperature as a node, below the melting point (700 °C), the Co blue phase is preferentially formed, and above 800 °C, the Co rose phase solidifies slowly ($10\,°C\,h^{-1}$). High temperatures tend to form the desired Co rose phase, but the boiling point of Na (883 °C) leads to easy evaporation of Na at this temperature, resulting in losses. The researchers found that the Co rose phase was more favorable for the formation of Na defects compared to the Co blue phase. By adjusting the Na defect ($Na_{2-x}CoP_2O_7$), then sintering at low temperatures would result in the stabilization of the Co rose phase, yielding pure phase NCPO. The researchers constructed phase transition-free, sustainable P2-NaFeMnO cathodes by lattice defect engineering [15]. Specifically, a P2-$Na_{0.7}Fe_{0.1}Mn_{0.75}\square_{0.15}O_2$ ("□" stands for intrinsic TM vacancy) cathode containing intrinsic TM vacancies was designed by introducing intrinsic vacancies into the TM layer, which does not undergo a P-O conformational transition in the high charge state. The uniformly dispersed intrinsic TM vacancies significantly increase the migration energy barriers of the Na^+ adjacent to them and immobilize the Na^+ within each TM layer, allowing the P-type layer to be maintained even in the Na-poor state. Compared with the O/P symbiotic structures obtained in typical vacancy-free P2-$Na_{0.7}Fe_{0.33}Mn_{0.67}O_2$

cathodes, the phase-change-free P-type structures in the Na-poor state have higher TM-O local structural stability and lower risk of thermal runaway. In addition, the specific Na-O-□ configuration can stimulate reversible anion redox activity, thus increasing the energy density of the batteries.

18.2 Interface Engineering

18.2.1 Air Stability

Due to its sensitivity to air, when NIB cathode materials are exposed to air, they will deteriorate rapidly in just a few hours, especially oxide-type cathodes, resulting in rapid capacity degradation, difficulty in making electrodes, and poor electrochemical performance.

During the synthesis process, the electrode material reacts with CO_2 and H_2O in the air tends to generate insulating Na_2CO_3 and NaOH, which becomes an insulating layer wrapped around the surface of the cathode. This makes the material less conductive and the surface stability of the material deteriorates, reducing the output and capacity of the battery. In terms of deterioration, the combination of CO_2 and water vapor proved to be the most harmful atmosphere, followed by the combination of O_2 and water vapor [16, 17]. Sufficient protons generated by the combination of water vapor and CO_2 to exchange with water vapor and inserted into the Na layer can trigger acid degradation. As observed for O3-$NaNi_{1/3}Fe_{1/3}Mn_{1/3}O_2$ (NFM111), the surface becomes rough within 1 hour of exposure to CO_2 and water vapor and is covered with a large amount of $NaHCO_3$ with a thickness greater than 15 nm after 2 hours. Severe surface degradation including crack extension, roughening, and growth of secondary phases was also traced, confirming the damaging effect of the water vapor and CO_2 combination.

During the mass production of electrodes, the material initially undergoes processes including cooling, crushing, sieving, and decontamination. During this period, close contact with air is often unavoidable, which can lead to rapid degradation of the material. Once processing is complete, the cathode material is packaged, stored, and transported for pending electrode preparation. Ideally, airtight packaging prevents interaction between the material and air. However, prolonged storage and transport increases the risk of air leakage and also leads to slow degradation of the material. The accumulation of Na residues such as NaOH, Na_2CO_3, and $NaHCO_3$ increases the alkalinity of the cathode powder, polyvinylidene fluoride (PVDF), and N-methyl pyrrolidone (NMP) mixtures, which can interfere with electrode fabrication. These alkalis break down the PVDF and cause slurry coagulation and particle aggregation, which prevents the slurry from being applied to the current collector. In addition, these alkalis corrode the current collector. As a result, the loss of Na^+ during air exposure and the resulting serious defects in electrode preparation can cause irreversible capacity loss, increased internal resistance, outgassing, swelling, and additional safety issues in the assembled cells. The inert gas protection, ambient humidity control, and surface encapsulation

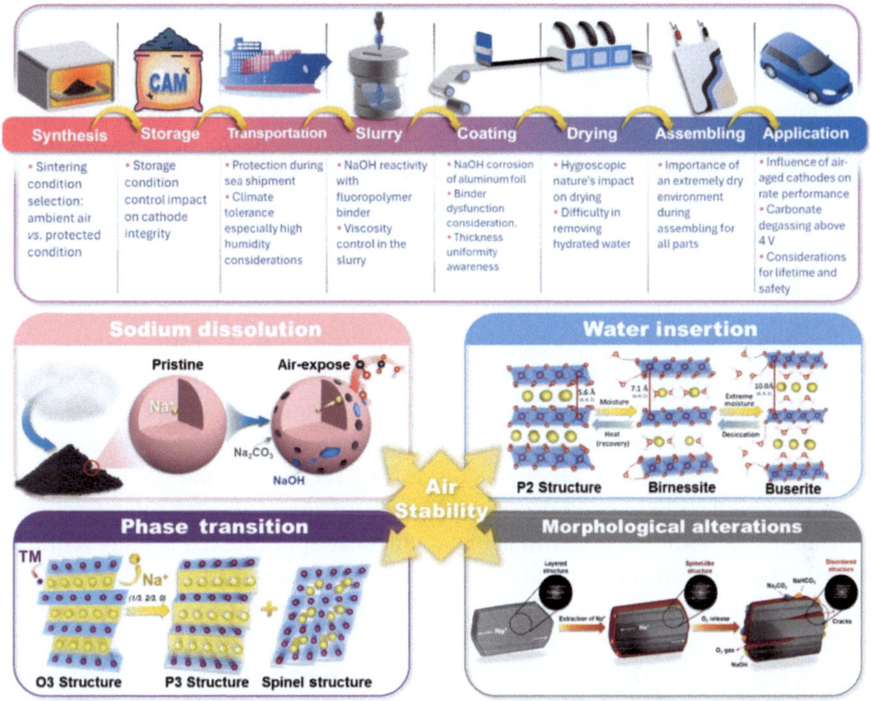

Figure 18.2 Schematic illustration of the key role of air stability during battery manufacturing, with the four main degradation pathways of layered oxide materials in air/water in NIBs is shown. Royal Society of Chemistry/Source: Jia, S et al. [18] / with permission of Royal Society of Chemistry.

strategies can be used during preparation. After preparation, re-sintering, material drying, vacuum/inert gas encapsulation with desiccants or CO_2 adsorbents can be used to effectively mitigate the adverse effects of air [17, 18]. Figure 18.2 shows a schematic diagram of the key role of air stability during battery manufacturing and the four main degradation pathways of layered oxide materials in air/water in NIBs.

After a long time of cycling of the positive electrode material, the H_2O and CO_2 in the surface of residual alkali will have a side reaction with the electrolyte to generate acid. The acid will corrode the layered cathode material and react with Na_2CO_3 and NaOH to generate H_2O and CO_2, introducing water and gas into the battery cell, which not only affects the performance of the battery cell but also expands the volume of the battery, leading to safety issues.

Among different coating modification materials, metal oxides such as Al_2O_3, MgO, ZrO, SnO, ZnO, and TiO_2 [19] are widely used for the surface coating of layered cathodes. Amphoteric oxides such as Al_2O_3, and ZnO are aimed at the problem of residual alkali on the surface. The researchers improved the capacity retention and cycling ability of layered P2-type $Na_{2/3}[Ni_{1/3}Mn_{2/3}]O_2$ by obtaining a thin coating of Al_2O_3, which suppressed the phenomenon of positive electrode particle exfoliation. The academics applied an Al_2O_3 coating on a layered P2-$Na_{2/3}(Mn_{0.54}Ni_{0.13}Co_{0.13})O_2$

cathode by applying the atomic layer deposition (ALD) technique, which offers a number of advantages including more precise control of the deposited film at the atomic level as well as consistency and uniformity of the deposited film. In another study, layered $Na_{0.44}MnO_2$ encapsulated with an Al_2O_3 coating (2 wt.%) obtained by wet ball milling initially provided a high discharge capacity of 109.8 mA h g^{-1}, with a capacity retention of 93.2% after 200 cycles at 0.4 °C. The XRD pattern of the Al_2O_3-coated $Na_{0.44}MnO_2$ showed that the Al_2O_3 coating prevented the volume expansion of $Na_{0.44}MnO_2$ during the electrochemical process of complex multiphase transition and stabilized the structure of $Na_{0.44}MnO_2$.

For polyanionic cathode materials, although they are relatively less affected by air stability, the erosion of the materials by airborne water and CO_2 cannot be ignored. Improving the air stability of the material is essential to facilitate the development of its all-weather performance and storage resistance. In the case of $Na_3TiV(PO_4)_3$, for example, after 30 days of storage in air, it will spontaneously undergo a de-sodiation reaction and degrade into by-products such as $Na_2TiV(PO_4)_3$, leading to the destruction of its crystal structure and decrease in its electrochemical performance storage of unstable electrode materials to achieve practical applications, usually requires harsh storage and transport conditions, which undoubtedly increase the cost of use and limit the breadth of its application and lifespan. The researchers successfully developed a storage-stable cathode material, $Na_3V_2(PO_4)O_2F_{0.99}Cl_{0.01}$ (NVPFCl), based on $Na_3V_2(PO_4)_2O_2F$ (NVPOF) [20]. The prepared NVPFCl cathode exhibited extremely high storage stability. No significant capacity degradation was observed after 7 days of immersion in water, 24 hours of storage at 50 °C or 25 °C, and up to 26 months of exposure to air at room temperature.

18.2.2 Electrode–Electrolyte Interphase

A key characteristic of NIBs that ensures stable operation over multiple charge/discharge cycles is the formation of a stable electrode–electrolyte interphase (EEI) at the surface of the anode and the cathode. During charge/discharge cycling, the highly reducing environment at the anode and oxidizing conditions at the cathode will cause the electrolyte to degrade, resulting in a complex of organic and inorganic decomposition products being deposited on the respective electrode surfaces to form EEI, which acts as the passivation layer, preventing further degradation of the electrolyte and providing NIBs with sustained cycling stability. The ideal EEI should have high ion mobility, electron resistivity, uniform coverage, and interfacial stability [21]. The EEI formed on the cathode side can be called cathode electrolyte interphase (CEI) or solid electrolyte interphase (SEI), and the one formed on the anode side is called SEI (more researched), which forms a thin but robust SEI layer on the electrode surface, which prevents the electron flow to the electrolyte and maintains reversible Na^+ de–/intercalation in subsequent cycles. The thickness of EEI affects the carrier transport and hence the electronic performance of NIBs [22]. When the carrier passes through the EEI film and comes into contact with the cathode surface, an electron transfer reaction occurs and the additional SEI layer is created beyond the initial SEI layer. Repeated cycling will increase the internal resistance

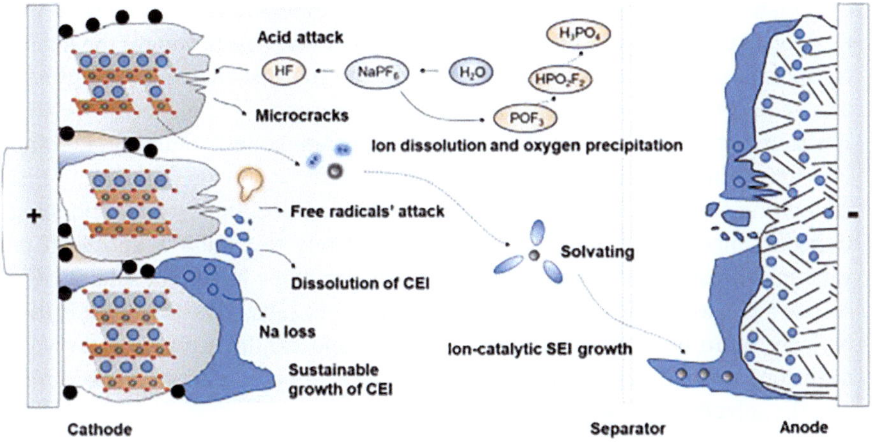

Figure 18.3 Summary schematic of typical electrochemical reactions and side reactions at the EEI in NIBs. The left and right sides show the damage reactions and consequences on the CEI and SEI, respectively. Source: Zhang J et al. [28] / with permission of Elsevier.

of the batteries, potentially affecting the performance, lifespan, and thermal stability of NIBs.

The composition of the electrolyte, salt concentration, and dosage strongly affect the composition, thickness, and stability of EEI. The electrochemical reactions and side reactions that CEI and SEI intermediates in NIBs may suffer during the electrochemical reaction are shown in Figure 18.3. Compared with LIBs, EEI in NIBs has the following characteristics:

i) Solvated structure of the electrolyte: Na^+ has a larger radius (1.02 Å) and lower charge distribution density than Li^+ (0.76 Å). As a result, the solvation capacity between Na^+ and polar solvent molecules in the electrolyte is less than that of Li^+, which can lead to greater participation of anions in the solvated structure of Na^+ and more decomposition at the interface. As a result, SEI/CEI generated in Na-based systems may contain more inorganic components such as NaCl, NaF, Na_2O, and Na_2CO_3.
ii) Electrochemical potential: With reference to the potential of SHE, the electrochemical potential of Na^+/Na is higher than that of Li^+/Li ($E^0 = -2.71$ V vs -3.04 V). Therefore, the potential strength at the electrode surface should be greater in the Na-based system than in the Li system. This poses a greater challenge to the electrochemical stability of the electrolyte for NIBs, especially on the high potential positive side, which may promote excessive oxidative decomposition of the electrolyte and the formation of thicker CEI on the surface of the Na-based cathode.
iii) Chemical stability: Since Na^+ has a lower charge density than Li^+, it is easier to establish weak and long covalent bonds with other elements. In LIBs and NIBs, there are differences in the physicochemical properties of the component substances within the intermediate phase layer (SEI/CEI). Typically, Na-based

carbonate materials are less electrochemically stable and more soluble in the electrolyte compared to Li-based carbonate materials. This promotes the loss of Na species and the failure of EEI and leads to an increase in self-discharge and a rapid decrease in the capacity of Na-based batteries.

iv) Ion transfer: Depending on the effect of ionic radius and solubility, Li^+ and Na^+ may exhibit different diffusion modes and kinetic processes in EEI. In EEI containing Li and Na components, Li^+ exhibits a knock-off or direct hopping transport mechanism, whereas Na^+ is more likely to exhibit a knock-off or vacuum-diffusion transport mechanism. On the other hand, Na salts have a somewhat lower ionic conductivity compared to Li salts. The difference is that Na^+ are more likely to de-solvate at the EEI and exhibit stronger interfacial diffusion in the same electrolyte due to the weaker coordination between Na^+ and the solvent in the solvated structure.

Side reactions: The salts such as NaTFSI and NaFSI are more corrosive to Al current collectors and cell casings than Li salts and are therefore rarely used as a single electrolyte salt in NIBs. In addition, some organic electrolytes also produce more serious side reactions on the Na metal anode, such as the continuous corrosion of Na metal by the electrolyte (1 M $NaClO_4$ in diglyme). Their derivatives may increase the complexity and instability of the CEI and further hinder the transport of Na^+ at the interface.

Decomposition products of the electrolyte are the main components of EEI on the electrode surface. Unusually, for oxide-typed cathodes, the CEI formed may contain additional natural passivation layers (e.g. NaOH and Na_2CO_3) and side reaction products (metal fluorides). For polyanionic compounds, decomposition products of the electrolyte dominate the formation of CEI. Whereas for Prussian blue analogues, the CEI formed may contain more carbonate material.

Optimization of the electrolyte is the key to constructing highly stable EEI membranes for NIBs in the high voltage regime. The stable EEI membranes are usually constructed through electrolyte composition modulation, introduction of additives, and design of artificial interfacial layers [23]. The researchers used a film formation modulation strategy to optimize the formation of the interface through the phosphorus/silicon intermediates formed by the additives (1 M $NaClO_4$ in PC +10 vol% FEC +1 vol% TMSPi) [24]. The improved CEI/SEI with excellent Na^+ transport capacity and structural/thermal stability enables stable operation of NIBs over a wide temperature range. In the half-cell, NVPOF exhibits excellent performance over a wide temperature range of −25 to 75 °C, with a capacity retention of 71% at 60 °C for 1700 cycles. In addition, the dynamic all-weather testing further demonstrated the reliability of the battery system's smooth operation supported by this electrolyte formulation. By introducing sodium difluoro(oxalate)borate (NaDFOB) as an additive in the ether-based electrolyte, a robust additive-induced borate and fluorine-rich interface was constructed to effectively inhibit the growth of sodium dendrites and the dissolution of transition metal ions [25]. The capacity retention rate of the constructed FeMn-based Prussian blue (FeMnHCF)||HC full cell reaches 82.1%.

18.2.3 Solid-State Electrolyte

The solid-state NIBs (SSBs) use solid electrolytes (SEs) instead of traditional organic electrolytes, which solves the safety problems of combustion and explosion caused by electrolyte volatilization and leakage. At the same time, the superior mechanical properties and thermal/chemical stability of SEs can improve battery life and stability, realizing the matching use of high-voltage cathode and Na metal anode in SSBs [26]. In addition, SEs can simplify the design of NIBs without the need for additional electrolyte containers or diaphragm components, thus improving the energy density of NIBs. The SSBs combine high energy density with high safety, which is of great significance for the development of next-generation high-energy-density NIBs and the resolution of the energy crisis [27, 28]. The difference between the configurations of conventional NIBs and SSBs is shown in Figure 18.4. Nonetheless, SEs, as the core material of SSBs, are still facing problems such as low room temperature ionic conductivity, narrow electrochemical window, poor compatibility with electrode interface, and poor contact. Therefore, the enhancement of Na^+ conductivity, interfacial compatibility, interfacial stability, and reduction of interfacial impedance of SEs are the keys to the performance enhancement and commercial application of SSBs. Among them, the interfacial impedance between the cathodes and SEs mainly originates from the incomplete contact, the electrode volume effect, and the space charge layer during the charging and discharging process, and the interfacial stability problem is mainly manifested by the chemical reaction and the interfacial elements diffusing into each other [29]. The interface problems between the anodes and SEs are mainly due to dendrite growth, poor interfacial contact, and poor interfacial compatibility. In response to the above problems, currently, surface coating modification, electrode composition modulation, introduction of interfacial intercalation, and design of new cell structures are generally used to reduce the interfacial impedance and improve the interfacial stability of SSBs. In summary, constructing electrode/SEs interfaces with good wettability, compatibility, high ionic conductivity, and electronic insulation is crucial for obtaining SSBs with excellent multiplicity performance, charge/discharge efficiency, and cycling stability.

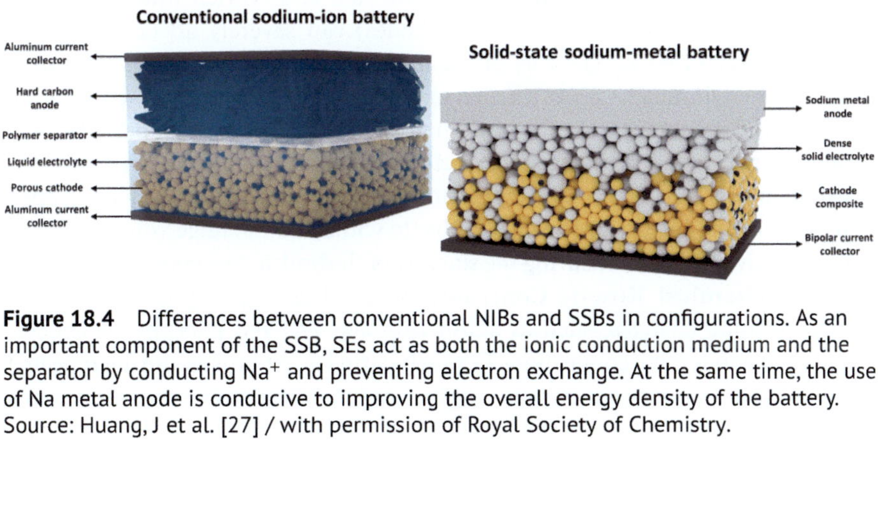

Figure 18.4 Differences between conventional NIBs and SSBs in configurations. As an important component of the SSB, SEs act as both the ionic conduction medium and the separator by conducting Na^+ and preventing electron exchange. At the same time, the use of Na metal anode is conducive to improving the overall energy density of the battery. Source: Huang, J et al. [27] / with permission of Royal Society of Chemistry.

18.2.4 Binder

The electrode consists of the active substance, the conductive carbon black, and the binder. A properly designed binder can improve the electrochemical properties of NIBs [30]. At present, the most commonly used binder for the cathode in NIBs is PVDF, and in addition, water-soluble binders such as carboxymethylcellulose sodium (CMC) and sodium alginate (SA) are used in some polyanionic cathode materials that are slightly passive to water. A mixed binder of CMC and styrene (SBR) is commonly used for anode [31, 32]. The ideal binder should have the following features:

i) **Bonding:** Bonding is the most basic and important working function of a binder. An ideal binder should be able to cope with repeated electrochemical cycling, firmly bonding the electrode components (active material, conductive additive, and current collector) together, thereby maintaining the mechanical and conductive integrity of the electrode.
ii) **Interfacial Stability:** During cycling, when the operating potential of the electrode exceeds the electrochemical stability limit of the electrolyte, the electrolyte decomposes, creating an interface (EEI) at the anode and/or cathode. The stability of the EEI as a passivation layer is critical to the stability of the electrode. A stable interfacial film inhibits further decomposition of the electrolyte and achieves high Na^+ selectivity and conductivity. The unstable intermediate layers can break down in subsequent cycles, exposing new electrode surfaces. The electrolyte breaks down on these surfaces and the interface is re-formed. This process is repeated during cycling, with a continuous depletion of Na^+. The use of novel binders is an effective strategy for stabilizing the interface. The functional binders can uniformly cover the electrode surfaces and act as stable artificial interfaces, minimizing self-discharge and mitigating side reactions between the electrode and the electrolyte.
iii) **Material Protection:** The binder coated on the surface of the active material provides protective isolation and mechanical cushioning to maintain the integrity of the electrode. On the one hand, the dissolution of certain active materials or intermediates (e.g. Mn-rich and Ni-rich transition metal compounds, polysulfides in sulfur cathodes) can severely affect the cycling performance of NIBs. The role of the binder is to act as a coating to provide a barrier environment for the active materials and to prevent the degradation of harmful substances. On the other hand, some active material particles (e.g. alloy anodes) undergo significant volume changes with repeated insertion/extraction of Na^+. The three-dimensional cross-linked polar binder wrapped tightly around the active material is able to withstand the huge volume change of the active material, ensuring the stability of the bulk active material structure.
iv) **Electrochemical Kinetic Contribution:** For high-capacity electrodes with large volume changes, such as sulfur cathodes and red phosphorus anodes, the rigid conductive connections between particles are broken during repeated cycling due to segregation and agglomeration (discontinuous connections) of the conducting carbon additives. Many active materials with significant volume effects have poor electrical conductivity because when the conductive connections formed by the conductive carbon black fail, electron transport in

the electrode shuts down, leading to a fading of the electrode's electrochemical performance. During the cycling process, the functionalized conjugated conductive binder forms a solid and continuous conductive network in the electrode while tightly binding the active material to ensure a stable charge transfer process. Conductive carbon typically has an electronic conductivity of 1–100 S cm^{-1}, while advanced conductive binders can be as high as 1000 S cm^{-1}. The advanced conductive binders greatly increase the overall conductivity of the electrodes, which is critical for improving the fast-charging performance and power density of NIBs. In addition, ionic conductivity is an important factor affecting the kinetics of electrochemical reactions. Especially at high current rates, efficient Na^+ transport in the electrode becomes particularly important. Binders with ionic conductive groups allow for fast Na^+ transport between the binder and the active material surface, thus improving the electrochemical performance of NIBs.

v) **Enabling Electrochemical Performance:** As a typical cathode material for NIBs, NVPOF is favored for its high operating voltage and specific capacity. Nevertheless, NVPOF can only develop its capacity when used with water-soluble binders such as CMC, which is not compatible with PVDF binders. Due to the large surface tension of aqueous binders, high loaded electrodes cannot be obtained due to the large surface tension during electrode preparation, which tends to cause cracking of the electrode or even peeling off of the active material from the current collector. In this case, the researchers modified the commercial binder. By mixing polyacrylic acid (PAA) and PVDF in specific ratios to form intermolecular hydrogen bonds, the resulting hybrid binder can effectively bring out the electrochemical performance of NVPOF, and at the same time facilitate the possibility of highly loaded electrodes, which greatly advances the commercialization of NVPOF (Figure 18.5) [33].

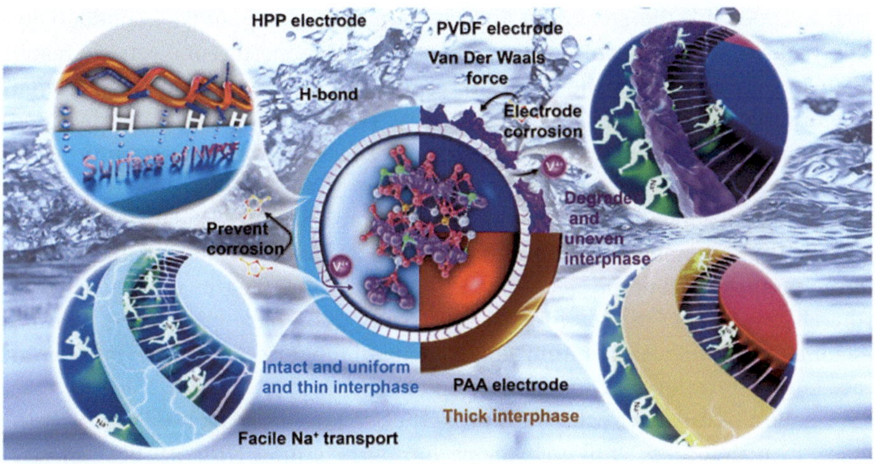

Figure 18.5 Schematic representation of the characteristics and functions of the hybrid binders. The uniformly coated hybrid binder and strong hydrogen bonding forces protect NVPOF electrodes from side reactions and improve their cycling stability. Source: Gu, Z.Y et al. [33] / with permission of American Chemical Society.

18.3 Summary

The strong deployment of renewable energy technologies has led to a significant increase in the capacity of their supporting electrochemical energy storage devices and a new level of efficiency and reliability in energy storage and conversion. The performance of NIBs, as a promising energy storage technology, not only depends on the intrinsic properties of the electrode materials but also significantly affects the interaction of the electrode–air and EEI membranes. The quality and stability of these defects and interfaces are directly related to the performance and lifetime of NIBs during the charging and discharging process. In this regard, the configuration design of NIBs should consider the following points:

a) Advanced in situ and synchrotron characterization tools are required to characterize the bulk phase and surface of the electrode materials at the atomic level (i.e. cryo-electron microscopy, spherical difference electron microscopy, in situ optical microscopy, atomic force microscopy, in situ transmission electron microscopy, in situ X-ray photoelectron spectroscopy, secondary ion mass spectroscopy, nuclear magnetic resonance imaging, X-ray computed tomography, transmission X-ray microscopy, etc.), with a view to tracing the chemical and structural evolution of the electrodes during the electrochemical reactions and identifying the root cause of cell polarization, capacity loss, and structural collapse.
b) Combine the theoretical calculations and intelligent simulation tools (machine learning, artificial intelligence, multi-physics field simulation, and density functional theory) to establish electrode failure model and simulate the battery failure process, focusing on elucidating the battery failure mechanisms and managing to put forward empirical prevention and solution methods for experimental reference.
c) Design and screening of surface robust composite structures to enable the material to cope with the erosion of air layers. In-depth study of Na^+ conductivity mechanism, material composition, and structural properties of electrode materials for NIBs improves the carrier transport and electron conduction through the intersection of physics, electrochemistry, and materials science. At the same time, the air stability and storage resistance of the materials need to improve to enhance the durability of electrode materials for NIBs.
d) Adaptation of appropriate electrolyte formulations and functional binders to construct strong, homogeneous, and inorganic-rich EEI films that inhibit successive decomposition of the electrolyte and maintain the integrity of the electrode, thereby matching the use of high-voltage cathodes. Flame retardancy, thermal stability, and wide temperature range compatibility also need to be taken into account for safe and all-climate operation for NIBs.
e) Evaluation of the electrochemical properties of the materials used using high-capacity pouch cells. Until now, the performance of most NIBs has been mostly evaluated with small capacity button cells. The pressure applied in pouch cells is low and nonuniform due to the different cell structures. In addition, the loading

of the electrode material and the amount of electrolyte are several orders of magnitude higher in pouch cells than in button cells, resulting in a nonuniform distribution of the electrode material and electrolyte. Therefore, these large differences between laboratory-level button cells and industrial-level pouch cells make the practical feasibility of many structure optimization strategies questionable to the extent that they are difficult to generalize. Therefore, validating the effectiveness of the designed materials in high-capacity batteries is important to improve the energy density and cycle life of NIBs at the industrial level.

References

1 Zhang, F., He, B., Xin, Y. et al. (2024). Emerging chemistry for wide-temperature sodium-ion batteries. *Chem. Rev.* 124: 4778–4821.

2 Guo, Y.J., Jin, R.X., Fan, M. et al. (2024). Sodium layered oxide cathodes: properties, practicality and prospects. *Chem. Soc. Rev.* 53: 7828–7874.

3 Huang, F., Xu, P., Fang, G. et al. (2024). In-depth understanding of interfacial Na^+ behaviors in sodium metal anode: migration, desolvation, and deposition. *Adv. Mater.* 36: e2405310. https://www.ncbi.nlm.nih.gov/pubmed/39152941

4 Gu, Z.Y., Wang, X.T., Heng, Y.L. et al. (2023). Prospects and perspectives on advanced materials for sodium-ion batteries. *Sci. Bull.* 68: 2302–2306.

5 Yu, H., Gao, A., Rong, X. et al. (2024). A prismatic alkali-ion environment suppresses plateau hysteresis in lattice oxygen redox reactions. *Energy Environ. Sci.* 17: 3855–3867.

6 Liu, Y., Rong, X., Bai, R. et al. (2023). Identifying the intrinsic anti-site defect in manganese-rich NASICON-type cathodes. *Nat. Energy* 8: 1088–1096.

7 Liu, R., Zheng, S., Yuan, Y. et al. (2020). Counter-intuitive structural instability aroused by transition metal migration in polyanionic sodium ion host. *Adv. Energy Mater.* 11: 2003256.

8 Li, Y., Gao, Y., Wang, X. et al. (2018). Iron migration and oxygen oxidation during sodium extraction from $NaFeO_2$. *Nano Energy* 47: 519–526.

9 Li, M., Zhuo, H., Xu, Y. et al. (2024). Thermodynamically stable low-Na O_3 cathode materials driven by intrinsically high ionic potential discrepancy. *Energy Environ. Sci.* 17: 7058–7068. https://pubs.rsc.org/en/content/articlelanding/2024/ee/d4ee02359e

10 Zuo, W., Gim, J., Li, T. et al. (2024). Microstrain screening towards defect-less layered transition metal oxide cathodes. *Nat. Nanotechnol.* **19**: 1–10. https://www.ncbi.nlm.nih.gov/pubmed/39164411

11 Ding, F., Ji, P., Han, Z. et al. (2024). Tailoring planar strain for robust structural stability in high-entropy layered sodium oxide cathode materials. *Nat. Energy* **9**: 1–11. https://doi.org/10.1038/s41560-024-01616-5

12 Wang, X., Zhang, Q., Zhao, C. et al. (2024). Achieving a high-performance sodium-ion pouch cell by regulating intergrowth structures in a layered oxide cathode with anionic redox. *Nat. Energy* 9: 184–196.

13 Zhao, A., Liu, C., Ji, F. et al. (2022). Revealing the phase evolution in $Na_4Fe_xP_4O_{12+x}$ ($2 \le x \le 4$) cathode materials. *ACS Energy Lett.* 8: 753–761.

14 Kim, H., Park, C.S., Choi, J.W. et al. (2016). Defect-controlled formation of triclinic $Na_2CoP_2O_7$ for 4 V sodium-ion batteries. *Angew. Chem. Int. Ed.* 55: 6662–6666.

15 Tang, Y., Zhang, Q., Zuo, W. et al. (2024). Sustainable layered cathode with suppressed phase transition for long-life sodium-ion batteries. *Nat. Sustainability* 7: 348–359.

16 Zhan, J., Huang, J., Li, Z. et al. (2024). Air-stable high-entropy layered oxide cathode with enhanced cycling stability for sodium-ion batteries. *Nano Lett.* 24: 9793–9800.

17 Yang, Y., Wang, Z., Du, C. et al. (2024). Decoupling the air sensitivity of Na-layered oxides. *Science* 385: 744–752.

18 Jia, S., Kumakura, S., and McCalla, E. (2024). Unravelling air/moisture stability of cathode materials in sodium ion batteries: characterization, rational design, and perspectives. *Energy Environ. Sci.* 17: 4343–4389.

19 Wang, J., Zhu, Y.F., Su, Y. et al. (2024). Routes to high-performance layered oxide cathodes for sodium-ion batteries. *Chem. Soc. Rev.* 53: 4230–4301.

20 Gu, Z.-Y., Guo, J.-Z., Sun, Z.-H. et al. (2021). Air/water/temperature-stable cathode for all-climate sodium-ion batteries. *Cell Rep. Phys. Sci.* 2: 100665.

21 Dinda, S., Trivedi, S., Roy, A. et al. (2023). Capturing nano-scale inhomogeneity of the electrode electrolyte interface in sodium-ion batteries through tip-enhanced raman spectroscopy. *Adv. Energy Mater.* 13: 2302176.

22 Chen, Y., Xue, T., Chen, C. et al. (2024). Helical peptide structure improves conductivity and stability of solid electrolytes. *Nat. Mater.* **23**: 1–8. https://www.ncbi.nlm.nih.gov/pubmed/39107570

23 Chen, H., Chen, K., Yang, J. et al. (2024). Designing advanced electrolytes for high-safety and long-lifetime sodium-ion batteries via anion-cation interaction modulation. *J. Am. Chem. Soc.* 146: 15751–15760.

24 Liang, H.J., Liu, H.H., Zhao, X.X. et al. (2024). Electrolyte chemistry toward ultrawide-temperature (−25 to 75 °C) sodium-ion batteries achieved by phosphorus/silicon-synergistic interphase manipulation. *J. Am. Chem. Soc.* 146: 7295–7304.

25 Liu, X., Zhao, J., Dong, H. et al. (2024). Sodium difluoro(oxalato)borate additive-induced robust SEI and CEI layers enable dendrite-free and long-cycling sodium-ion batteries. *Adv. Funct. Mater.* 34: 2402310.

26 Hou, J., Zhu, T., Wang, G. et al. (2024). Composite electrolytes and interface designs for progressive solid-state sodium batteries. *Carbon Energy* 6: e628. https://doi.org/10.1002/cey2.628, e628

27 Huang, J., Wu, K., Xu, G. et al. (2023). Recent progress and strategic perspectives of inorganic solid electrolytes: fundamentals, modifications, and applications in sodium metal batteries. *Chem. Soc. Rev.* 52: 4933–4995.

28 Zhang, J., Ma, S., Zhang, J. et al. (2024). Critical review on cathode electrolyte interphase towards stabilization for sodium-ion batteries. *Nano Energy* 128: 109814.

29 Li, J., Hu, Z., Zhang, S. et al. (2024). Molecular engineering of renewable cellulose biopolymers for solid-state battery electrolytes. *Nat. Sustain.* 7: 1481–1491. https://doi.org/10.1038/s41893-024-01414-7

30 He, Q., Ning, J., Chen, H. et al. (2024). Achievements, challenges, and perspectives in the design of polymer binders for advanced lithium-ion batteries. *Chem. Soc. Rev.* 53: 7091–7157.

31 Gu, Z.Y., Sun, Z.H., Guo, J.Z. et al. (2020). High-rate and long-cycle cathode for sodium-ion batteries: enhanced electrode stability and kinetics via binder adjustment. *ACS Appl. Mater. Interfaces* 12: 47580–47589.

32 Li, Z., Wu, Z., Wu, S. et al. (2023). Designing advanced polymeric binders for high-performance rechargeable sodium batteries. *Adv. Funct. Mater.* 34: 2307261.

33 Gu, Z.Y., Cao, J.M., Guo, J.Z. et al. (2024). Hybrid binder chemistry with hydrogen-bond helix for high-voltage cathode of sodium-ion batteries. *J. Am. Chem. Soc.* 146: 4652–4664.

19

Defect and Interface Engineering in K-Ion Batteries

Yahreli Audeves-Audeves[1,], Raúl Castellanos-Espinoza[2,*], and Minerva Guerra Balcázar[2]*

[1] *Centro de Investigación y Desarrollo Tecnológico en Electroquímica S.C., Pedro Escobedo, Querétaro, C.P. 76703, México*
[2] *Universidad Autónoma de Querétaro, Facultad de Ingeniería, División de Investigación y Posgrado, Querétaro, C.P. 76010, México*

19.1 Introduction to Potassium-Ion Batteries

Potassium-ion batteries (PIBs) are energy storage devices that use potassium ions (K^+) as charge carriers. PIBs have a more negative standard electrode potential compared to the standard hydrogen electrode (SHE) (−2.93 V for K^+/K), ensuring a higher operating voltage and greater energy density due to the standard redox potential of the K^+/K couple [1]. The K^+ exhibits noteworthy characteristics when compared to sodium ions (Na^+) and lithium ions (Li^+). These include a reduced Lewis acidity and a diminished ΔGads, which consequently leads to a reduced number of solvated ions. This reduction in solvated ions facilitates their transportation, thereby enhancing ionic conductivity [2]. Furthermore, PIBs have garnered significant attention due to the substantial abundance and distribution of potassium (2.09 wt.%) in the Earth's crust, which is considerably higher than that of lithium (0.0017 wt.%) [3].

PIBs comprise an anode, which typically utilizes carbon-based materials; a cathode, which often employs metal oxides or phosphate compounds; and a separator, which serves the primary function of facilitating effective ionic transport between the anode and cathode while impeding short circuits between the electrodes [1, 4]. PIBs operate like a pendulum, where K^+ migrate between the electrodes during charge/discharge processes (Figure 19.1a), giving a reversible intercalation reaction [5]. During the charging process, potassium atoms in the cathode transform into K^+ by releasing electrons into the external circuit. The formed K^+ travels toward the anode side through the separator and electrolyte. When the K^+ reaches the anode, they combine with the electrons from the external circuit. This process of conversion and storage of electrical energy from an external power source into the

* These authors contributed equally to this work.

Surface and Interfacial Defects in Nanomaterials for Sustainable Energy Production and Storage, First Edition. Edited by Noé Arjona, Lorena Álvarez Contreras, and Minerva Guerra Balcázar.

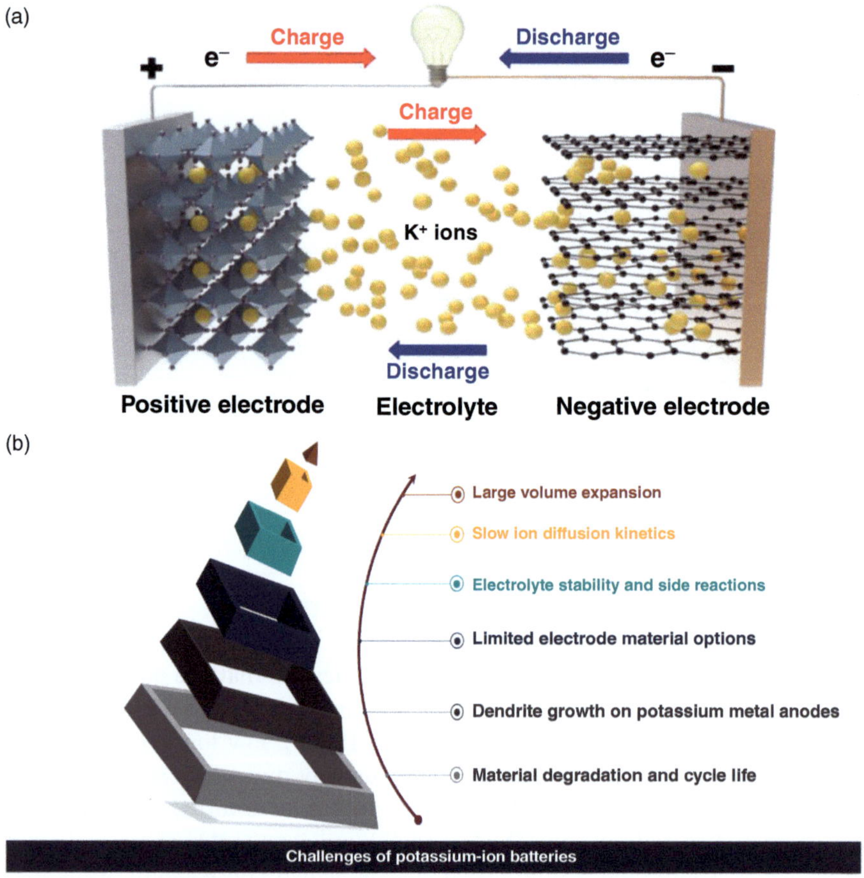

Figure 19.1 (a) Scheme of intercalation/deintercalation in PIBs during the charge/discharge process. Source: Adapted from Min et al. [5]. (b) Challenges of PIBs.

battery is known as a charging process. Conversely, during discharge, the process is reversed. When the battery discharges through an external load, K^+ deintercalated from the anode side and moves toward the cathode through the separator and electrolyte. Concurrently, electrons migrate from the anode to the external load [6].

The development of PIBs is hindered by the challenge of identifying suitable host materials that exhibit acceptable electrochemical performance. This difficulty is primarily attributed to the large atomic radius (1.38 Å) of K^+, which significantly reduces the insertion sites and diffusion channels available for K^+ in the lattices of host materials [7]. Additionally, electrodes frequently undergo volume expansion, leading to suboptimal electrochemical performance during potassium deposition/dissolution processes, as well as structural degradation of the electrode materials [8]. As outlined above, a number of studies have concentrated on electrode stabilization for K^+ storage and diffusion in PIBs [9]. Defect engineering has emerged as a potent method for manipulating the electrode structure. It has been demonstrated that defect engineering possesses the capacity to damage and reconstruct periodic crystal structures, thereby conferring electrode materials with a greater number of exposed active sites, enhancing K^+ adsorption energy and promoting ion diffusion and electron transfer properties [9, 10]. This augmentation in electrochemical performance in PBIs signifies a significant advancement in the field.

PIBs face several challenges (Figure. 19.1b): first, the insertion and extraction of K^+ cause significant volume expansion, leading to the structural degradation of the electrodes. Additionally, the ionic diffusion kinetics are slow, the selection of electrode materials is limited, dendrite formation occurs, energy density is low, and material degradation and limited lifespan reduce the efficiency of charge and discharge cycles. Next, recent strategies and advances in defect engineering are explored in detail, focusing on the relationship between material characteristics (composition, structure, and morphology) and their electrochemical performance (capacity, cycling stability, and rate performance).

19.2 Defect Engineering in Materials of Potassium-Ion Batteries

The inherent complexities of defects in the materials of the PIBs, as evidenced by their diverse manifestations and intricate structures, underscore the imperative for a systematic and comprehensive classification system. This classification constitutes a foundational element in the formulation and execution of defect engineering strategies, thereby ensuring the efficacy and efficiency of defect management methodologies [11].

The classification of defects in solid materials is predicated on their dimensionality, which can be categorized into three primary types: zero-dimensional (0D), one-dimensional (1D), and two-dimensional (2D). 0D defects are point defects, including vacancies, interstitials, substitutions, and heteroatom doping defects. Edge defects (line-like) constitute the category of 1D defects, involving steps and dislocations. 2D defects manifest as planar defects, primarily grain boundaries and phase boundaries [4]. The introduction of defects (0D, 1D, and 2D) has been demonstrated to exert a substantial influence on the electrochemical behavior of electrode materials. For instance, it has been observed that potassium variation in metal oxides (Mn/Ni) results in a disordered K^+/vacancy structure, leading to the formation of continuous interconnected channels for K^+ diffusion and enhanced active storage, leading to improved insertion and extraction kinetics [12]. In contrast, oxygen vacancies in ligand-doped transition metal oxides (δ-MnO_2) have been demonstrated to enhance their specific capacitance [13]. Conversely, edge defects in monolayer MoS_2 have been observed to decrease the overpotential [14], thereby facilitating electron transfer and mitigating the Ohmic drop within batteries [15]. This enhancement in the system's efficiency is a notable outcome. While defects in the basal plane generate additional sites for ion intercalation, thereby facilitating augmented K^+ storage and diffusion [16]. It is noteworthy to mention that point defects can be classified into two categories: intrinsic and extrinsic defects [11]. This underscores the significance of defect engineering in this context. Another salient parameter in defect engineering is the material from which the electrodes are fabricated.

Nanostructured designs have been successfully applied to most anode and cathode materials, including carbon-based materials, hollow carbon graphite, and graphene, as well as alloy-based electrodes, oxides, sulfides, selenides, and Prussian blue and its analogs. Various nanostructures have been studied for graphite, including a three-dimensional (3D) interconnected network structure, which can offer rapid electron transfer capability due to its naturally high electrical conductivity and provide sufficient channels for ion transport [17].

19.3 Defects in Anode Materials Used in PIBs

In the field of PIB technology, the use of carbon materials as an anode has been contemplated, taking into consideration their favorable physical and chemical properties, such as high electrical conductivity and environmental compatibility [18]. Potassium storage in carbon materials is governed by a mechanism comprising two main components: (i) the adsorption of K^+ on the surface and pores and (ii) intercalation of K^+ into the carbon layer to form an intercalation compound [19].

However, a significant limitation in the use of carbon materials has been identified, which is attributed to the ionic radius of K^+. This intrinsic characteristic results in considerable volumetric expansion (approximately 61%) and collapse of the structure during repeated loading and unloading cycles. This phenomenon, in turn, results in a substantial reduction in capacity over the long-term cycle [20]. Recent studies have focused on enhancing the efficiency of carbon materials as an anode in PIBs, exploring strategies to avert the structural failure of the material. In this regard, research has been conducted in the field of defect engineering, including the creation of intrinsic and extrinsic defects (point defects).

Defects in carbon-based materials (graphite and hard/soft carbon) consist of intrinsic and extrinsic defects [11]. Intrinsic defects can be divided into vacancies, edges, and holes that are associated with the microstructure and regulation in material preparation [21]. These defects are associated with the microstructure and regulation during material preparation. The synthesis methods employed to create these defects are crucial, as they significantly influence the total number of defects present. Furthermore, it is essential to meticulously control the synthesis conditions to effectively enhance the number of well-defined and available active sites within the material structure [22]. Intrinsic defects have been shown to promote efficient K^+ diffusion, while edge sites and nanopores have been observed to enhance capacity by adsorbing K^+ [9].

The extrinsic defects, primarily attributable to the incorporation of heteroatoms (metallic and nonmetallic), can be mitigated to a certain extent through the meticulous regulation of synthesis conditions and processes. As the promotion of electrocatalytic activity in heteroatom-doped carbons can be predominantly ascribed to the redistribution of charge density within the carbon structure, resulting from the incorporation of heteroatoms with varying electronegativity [23]. Specifically, the heteroatoms disrupt the integrity of the π-conjugation of the carbon plane, thereby sequentially optimizing the charge state of neighboring carbon atoms and converting them into potential active sites. Additionally, transition metal–nitrogen–carbon (M – N – C) compounds, including single-atom transition metal-doped catalysts (Fe, Co, Mn, etc.), have been developed. Conversely, self-doped defects are formed in materials without external dopants, thereby creating well-defined active sites [24]. The superior electrocatalytic performance of these materials is commonly ascribed to the optimal adsorption and desorption behavior at the M_x–N_y active sites, which increases the number of ion storage sites, adsorption, and activity.

19.3.1 Intrinsic Defect Carbon-Based Materials

19.3.1.1 Vacancies in Carbon-Based Anodes

Vacancies, also known as point defects, are defined as the absence of carbon atoms within the graphitic lattice. These absences give rise to the presence of additional active sites, thereby facilitating the intercalation of potassium ions, which in turn increases the energy storage capacity and the diffusion of K^+. Recent studies have introduced surface vacancies in graphite by employing ball milling, where the defects exhibit high discharge capacity compared to graphite (without defects). Specifically, the discharge capacity increased from 183.5 to 286.2 mAh g^{-1} at a current density of 0.025 A g^{-1}, thereby demonstrating the importance of mechanical defect generation in PIBs [19]. Also, the generation of vacancies (point defects) in the graphitic plane is attributed to the preferential decomposition of anions in the graphitic lattice (N and S doping). The results obtained indicate that the formation of such vacancies leads to high potassium ionic conductivity, lower desolvation, and higher storage of K^+. Notably, the study reports a high discharge capacity of 504 mAh g^{-1} at 0.1 A g^{-1}, underscoring the significance of this material for energy storage applications [25].

19.3.2 Edge-Defect in Carbon-Based Anodes

Edge defects in graphite correspond to imperfections or irregularities in the crystalline structure that manifest themselves at the edges of the constituent sheets. These defects may include the presence of missing carbon atoms, the formation of non-hexagonal bonds (such as pentagons or heptagons), or the appearance of chemical impurities. These modifications in the atomic organization result in the alterations of the electronic, mechanical, and chemical properties of the material, influencing its reactivity and conductivity [26]. The implementation of a graphitic structure with a high concentration of edge defects has proven advantageous in the diffusion of K^+ and the formation of additional active sites, facilitating the storage of K^+. This approach has yielded noteworthy performance in terms of speed and cycling stability of PIBs [27]. A notable example is the study by Quan et al. [27] in which an expanded pitch/graphite-derived carbon anode was fabricated, characterized by a high concentration of edge defects. These defects were introduced by melt impregnation and confined pyrolysis process. The reversible capacity of the material with the introduction of edge defects was significantly higher than the reversible capacity values of 338.5 mAh g^{-1} at 0.1 A g^{-1} and 307 mAh g^{-1} at 0.1 A g^{-1}, respectively.

19.3.3 Extrinsic Defects Carbon-Based Materials

19.3.3.1 Vacancies/Doping in Carbon-Based Anodes

The generation of vacancies/doping, through the introduction of heteroatoms such as N (nitrogen) and S (sulfur), modifies the electronic properties of graphite [28]. The incorporation of N atoms into the graphite structure results in an increase in charge carrier density due to the higher electronegativity of nitrogen compared to carbon [29]. This property has been shown to lead to the generation of a larger number of active sites for the storage and adsorption of K^+ [28]. The discharge capacity efficiency of nitrogen-doped graphite is influenced by various factors, such as doping percentages, composition, distribution of N-doping species, and different morphologies [29, 30]. Previous studies have synthesized defect-rich nitrogen-doped graphitic nanocarbons with controllable defect densities by annealing a coordination compound of nickel and ethylenediaminetetraacetic acid at different temperatures (600 °C, 700 °C and 800 °C), where superior performance of the nanocarbons is observed. Notably, at 600 °C, the discharge capacity of 280 mAh g^{-1} at 0.05 A g^{-1} has been achieved, underscoring the impact of temperature on the material's properties. This capacity is attributed to the percentage of pyrrolic and pyridinic nitrogen, which has been shown to be present at the trailing edges of the graphitic sheet and to possess high electrochemical activity [18, 29].

S-doping has been shown to enhance the reversible capacity of carbon materials in PIBs, which is attributed to the formation of C–S–K bonds. Concurrently, due to the substantial covalent radius of S atoms, S-doped carbon materials manifest augmented carbon interlayer spacing, which not only accommodates volume changes but also promotes charge transfer and ion diffusion kinetics, thereby ensuring the optimal performance of PIBs [31]. The doping of amorphous carbon microspheres with S by carbonization at varying temperatures (500 °C, 700 °C, and 900 °C) using $(CH_3C_6H_4SO_3)_3$ as an S source has been documented [8]. It has been observed that the modulation of S-doping in the graphitic network depends on the carbonization times, resulting in a higher percentage of inherent defects X-ray photoelectron spectroscopy (XPS) at a temperature of 700 °C. Moreover, it has been found that this temperature exhibits optimum performance and obtains a high reversible capacity of 199.6 mAh g^{-1} at 0.1 A g^{-1}.

19.3.3.2 Vacancies/Doping with Order-in-Disorder Engineering in Carbon-Based Anodes

The doping of a graphitic lattice with heteroatoms (N and S), in conjunction with the utilization of templates for the purpose of controlling the morphology of graphite-based materials for PIB anodes, results in the creation of order-in-disorder engineered materials [32]. The implementation of templates has been demonstrated to have a significant impact on the modification of the material structure, resulting in the formation of a larger number of active sites, the generation of more complex carbon structures, the production of numerous defects, structural vacancies, and disorder in the material [18]. These changes contribute to improved efficiency in PIBs. In this regard, Zhou et al. [20] have investigated the synthesis of N-doped carbon nanowires derived from carbon nanotubes. The combination of a tubular morphology with doping results in the generation of fast channels for charge transport and numerous active sites for K^+ storage. The nanowires demonstrated a maximum capacity of 441.4 $mA\,h\,g^{-1}$ at a current density of 0.2 $A\,g^{-1}$ after 200 cycles. In contrast, Huang et al. [18] utilized a MgO template for the formation of 3D nanolayers, yielding a reversible specific capacity of 321 $mAh\,g^{-1}$ at 5 $A\,g^{-1}$. In a similar vein, Chen et al. [32] have employed a synergistic order-in-disorder engineering approach in carbon materials via self-templating, thereby ensuring a high concentration of reversible carbon vacancies and expedited charge transfer kinetics, culminating in an impressive reversible capacity of 425 $mAh\,g^{-1}$ at 0.05 $A\,g^{-1}$.

19.3.4 Intercalation and Storage of Metal Ions in Graphite Anodes

A high-performance anode material must possess two essential qualities: ample storage capacity and a multitude of available storage sites for metal ions [23]. In the context of PIBs, the prevalence of defects and edge sites gives rise to surface-driven capacitance, which invariably results in a suboptimal intercalation capacity (<100 $mAh\,g^{-1}$ below 0.5 V) [24]. The large ionic radius of K^+ (1.38 Å vs. 0.76 Å for Li^+) results in slow K^+ diffusion and low cycle performance [33]. However, a defect-rich material can enhance this process by increasing the intercalation sites for K^+ and shortening migration paths, leading to improved ion diffusion kinetics and intercalation capacity [24]. Interfacial transfer involves the desolvation of potassium ions on the surface of a solid electrolyte interphase (SEI) film and the transfer of desolvated potassium ions through the SEI film [25]. The kinetics of interfacial transfer and bulk diffusion of potassium directly influence the performance rate of PIBs, making it essential to explore rational strategies for accelerating potassium transport. Cycle capacity and charge/discharge rate are also key indicators of battery performance. Rapid metal diffusion/mobility is required to enable fast charging/discharging, achieve high reversible capacity, and limit dendritic growth. Therefore, it is also important to consider metal migration and diffusion in anode materials [33]. Consequently, exploring interfacial transmission mechanisms and controlling interfacial transport to improve diffusion/reaction kinetics are highly desirable, as they are crucial to enhancing electrochemical performance in terms of cycle stability and rate capacity [34].

19.4 Defects in Cathode Materials Used in PIBs

Cathode materials play a crucial role in the performance of PIBs. Several types of cathode materials have been developed for PIBs, including Prussian blue analogs,

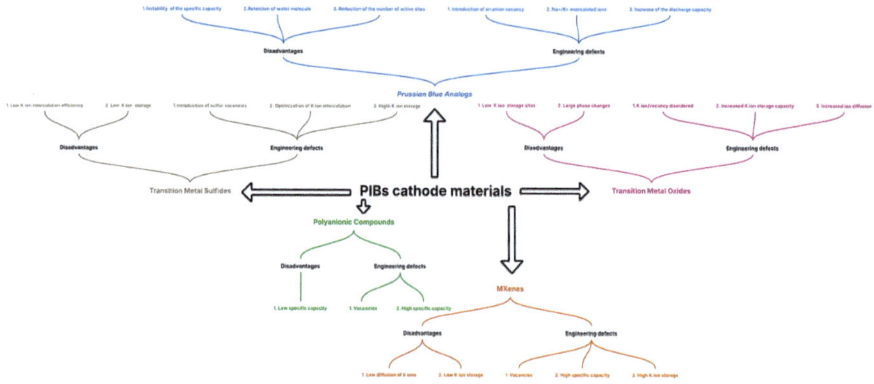

Figure 19.2 Main cathodic materials for PIB.

layered transition metal oxides, organic materials, and polyanionic compounds. The most used cathodic materials are presented in Figure 19.2.

19.4.1 Prussian Blue Analogs (PBAs)

Prussian Blue Analogs (PBAs) are considered one of the most promising cathode materials because they exhibit a high theoretical capacity. The PBAs are cost-effective, easy to synthesize, chemically tunable, and structurally controllable [35].

PBAs have emerged as a particularly noteworthy class of cyanide-bridged materials. These materials are formed through reactions involving metal ions and transition metal cyanometalates. These materials exhibit cubic, face-centered structures (space group *Fm 3-m*) and possess a general chemical formula represented as $A_xM[M'(CN)_6]_{1-y}\cdot zH_2O$ $(0 \leq x \leq 2;\ 0 \leq y < 1)$, where A denotes an alkali metal cation (e.g. Li, Na, or K), M and M′ represent transition metal ions (e.g. Mn, Fe, or Co) coordinated to the cyano groups, both M and M′ sites are redox-active centers, and y is the fraction of $[M'(CN)_6]$ vacancies [36]. The organization of the structure in question gives rise to a 3D open framework that confers on the structure of the K^+ ion transport channels. The design in question inherently accelerates the entry and exit of K^+, ensuring superior ionic diffusion kinetics [37]. Additionally, the robust 3D framework structure reduces lattice volume changes during charge and discharge cycles [35]. However, the presence of $[M'(CN)_6]$ vacancies that are difficult to remove, as well as the presence of crystalline water, resulting in the structural instability and deterioration of the specific capacity [38]. This phenomenon can be attributed to the presence of water molecules, which occupy the sites intended to host K^+. Consequently, this results in a reduction in the number of redox-active sites [39]. In order to enhance the specific capacitance efficiency of PBAs, the intercalation of Na^+ and K^+ ions has been employed. These ions replace the vacancies of $[Fe(CN)_6]$, leading to a decrease in the water content within the PBA structure [40]. In a recent study, Li et al. [40] synthesized a PBA-based material, in which the vacancies were occupied by Na^+/K^+ intercalated ions. This process resulted in a water content reduction of 21.25% for the Na^+/K^+ intercalated ion

(KNaFeHCF) structure, which resulted in an increase in the discharge capacity from 108.1 mA h g^{-1} at 0.05 A g^{-1} to 130.3 mA h g^{-1} at 0.05 A g^{-1} for the materials with Na^+ (NaFeHCF) and Na^+/K^+ intercalated (KNaFeHCF), respectively. In this regard, Wei et al. [41] have developed a hybrid cathode based on PBA by Na^+/K^+ ion intercalation. It was observed that the cation intercalation and electrochemical performance of the cathode can be significantly affected by anion vacancies. It was further demonstrated that the presence of $[Fe(CN)_6]_{4-}$ anion vacancies can significantly influence these processes. The study's findings revealed that the introduction of an anion vacancy level can effectively maximize the number of K^+ ion intercalation and diffusion sites within the PBA structure. This phenomenon has been shown to facilitate K^+ ion intercalation, while concurrently suppressing Na^+ ion intercalation. Additionally, it has been observed to reduce the K^+ migration energy barrier and enhance ion diffusion kinetics. Consequently, a high discharge voltage and K^+ ion-dominated ion storage process have been observed in the hybrid PIB cell. It has been observed that the PBA with 20% anion vacancy delivers 128 mA h g^{-1} at 0.025 A g^{-1} and 67 mA h g^{-1} at 1 A g^{-1}, and also retains a capacity of 89% and 81% after 100 and 300 cycles, respectively. These findings are particularly noteworthy when compared to those achieved with a system exhibiting an anion vacancy of 7%, underscoring the efficacy of these systems in PIBs.

19.4.2 MXenes

MXenes are a new type of 2D material with a typical 2D structure. They exhibit high electrical conductivity, high thermal conductivity, and excellent mechanical strength. The distance between the two adjacent intermediate layers in the MXene family is approximately 10 Å, which is much larger than the radius of K^+ (0.38 Å). This results in very small volume changes in MXene materials during charge/discharge processes, making them beneficial for the fast transport of K^+ [35].

19.4.3 Transition Metal Oxides

Layered transition metal oxides can accommodate the larger radius of K^+ and have a higher energy density due to their unique open-framework structure, which is expected to enable efficient K^+ storage. Among these, manganese-based oxide cathode electrode materials are low-cost and have thus become the most practical cathode materials due to their multivalence potassium storage activity and high energy density from multi-electron transfer processes [42]. Unfortunately, the practical application of manganese-based oxides is hindered by severe electrode surface erosion caused by electrolyte decomposition products, as well as the expansion and collapse of the material during potassium ion intercalation [43]. To optimize the critical interface structure, strategies such as element substitution and surface coating are widely employed to stabilize the crystalline structure and modify the interface between the active materials and the electrolyte [44].

However, these materials exhibit slow kinetics of K^+ ion transport and a low number of K^+ storage sites leading to the low capacity of PIBs [12]. To increase the efficiency in PIBs, defects have been introduced in layered transition metal oxides, resulting in an improvement in K^+ transfer rate and ion storage capacity [39], which has seen an increase in the PIB efficiency. In the area of layered transition metal oxides, particular emphasis has been placed on those based on Mn/Ni, whose synthesis involves the modulation of the K^+ content. The results obtained demonstrate that an elevated K^+ content gives rise to a more K^+/vacancy disordered structure. Furthermore, it has been observed that these materials exhibit a substantially higher yield and discharge capacity in comparison to their ordered counterparts, which possess a lower K^+ content. The $K_{0.7}Mn_{0.7}Ni_{0.3}O_2$ material, which exhibits a higher K^+ content, demonstrates a higher discharge specific capacity of 125.4 mA h g^{-1} at 0.1 A g^{-1}, while the $K_{0.4}Mn_{0.7}Ni_{0.3}O_2$ material, which exhibits a lower K+ content, exhibits a discharge specific capacity of 93.9 mA h g^{-1} at 0.1 A g^{-1} [12]. In addition, oxygen vacancies in layered transition metal oxides have been generated and modified. In this context, oxygen vacancies in δ-MnO_2 are occupied or stabilized by K^+ ions embedded by 2-methylimidazole (2-MI) ligands. The abundance of oxygen vacancies stabilized with 2-MI can improve the structural stability and tune the electronic structure of δ-MnO_2, thus further facilitating the electron/ion diffusion efficiency. The enhanced electrical conductivity and the augmented number of active sites result in a high specific area capacitance of 721.2 mF cm^{-2}, accompanied by excellent cycle durability maintained at 93.4% after 10,000 cycles [13].

19.4.4 Polyanionic Compounds

Polyanionic compounds, defined by their stable structure and thermal stability resulting from robust covalent bonds, have demonstrated considerable promise. These compounds possess an open 3D structure with extensive channels that facilitate reversible volume changes during the extraction and insertion of K^+ [45]. Vanadium-based polyanionic compounds, distinguished by their variable valence and high electrochemical activity, have undergone substantial advancements. The presence of PO_4^{3-} in these compounds has been shown to enhance thermal stability, attributed to its role in stabilizing the structure and facilitating electron transfer. Vanadium-based polyanionic cathodes that have been explored for PIBs include $K_3V_2(PO_4)_3/C$, $K_3V(PO_4)_2/C$, $KVOPO_4$, $K_2[(VO)_2(HPO_4)_2(C_2O_4)]$, and $K_3V_3(PO_4)_4{\cdot}H_2O$ [46].

19.4.5 Transition Metal Sulfides

Transition metal sulfides (MSs) have attracted significant interest in the field of PIBs due to a number of properties [47]. It has been observed that they possess a shallow valence band (VB), resulting in enhanced stability and reactivity in certain applications. In addition, the presence of broad active sites, a suitable electronic band structure, high specific capacitance, excellent mechanical and thermal stability, as well as electrical conductivity have been highlighted [48]. The promotion of faster migration, the requirement of lower energy for intercalation, and the display of commendable chemical stability render them particularly attractive for PIB cathode material applications [10]. The implementation of defect engineering in MSs has been shown to result in the generation of a larger number of active sites. This, in turn, has been demonstrated to translate into an optimization of K^+ intercalation, transport, and adsorption processes [10, 16]. This advance, in turn, has been shown to increase the efficiency of PIBs. For instance, Lin et al. [10] synthesized a vanadium disulfide cathode material with sulfur vacancies integrated with reduced graphene oxide (VS_2-x@rGO). The introduction of sulfur vacancies increases the active sites for K^+ storage, while the rGO ensures stable electronic conductivity and structural integrity. The VS_2-x@rGO cathode exhibited a reversible capacity of 116.2 mA h g^{-1} at 0.02 A g^{-1}, 77.5 mA h g^{-1} at 0.5 A g^{-1} and demonstrated exceptional cyclic stability with 94.5% capacity retention after 100 cycles at 0.1 A g^{-1}. In this regard, Xu et al. [16] introduced defects in the basal plane of MoS_2, which enlarged the MoS_2 interlayers and increased storage and interlayer K^+ diffusion. MoS_2 with interlayer spacing (defects in its basal planes) showed 40% higher reversible capacity and 120% higher rate capability than its defect-free counterpart in a controlled voltage window for intercalation reactions.

19.5 Recent Advances in PIBs Through Defect/Interface Engineering

The performance and rechargeability of the battery in the PIBs are contingent upon various factors, including mobility in the solid electrode, interfacial chemistry, and electrolyte functions derived from low ion diffusivity, large volume changes during potassiation/depotassiation, side reactions, dendrite formation, and electrolyte consumption [17]. In this regard, defect and interface engineering have a significant impact on ion diffusivity through the solid electrode and interface engineering on the interfacial chemistry. This subsection will focus on the most significant results regarding the enhancement of power density and cyclability.

19.6 Applications and Future Perspectives

Defect engineering is of great importance in improving the performance of PIBs, with the potential to become a viable alternative to lithium technology. One of the substantial challenges in developing these devices is scalability and achieving a good balance between cost and efficiency. It is essential to find an equilibrium between electrochemical performance, stability, lifespan, and cost, which can be achieved through defect engineering combined with computational modeling. Through defect control, it is possible to enhance efficacy, making these batteries suitable for a varied range of applications, such as portable and wearable devices, and electric vehicles. In brief, defect engineering has the potential to transform PIB technology into a viable option for efficient energy storage.

References

1 Kubota, K., Dahbi, M., Hosaka, T. et al. (2018). Towards K-ion and Na-ion batteries as "beyond Li-Ion." *Chem. Rec.* 18 (4): 459–479.

2 Zhou, M., Bai, P., Ji, X. et al. (2021). Electrolytes and interphases in potassium ion batteries. *Adv. Mater.* 33 (7): 2003741.

3 He, X., Iqbal, N., Ghani, U., and Li, T. (2024). Potassium ion batteries: recent advancements in anodic, cathodic, and electrolytic materials. *J. Alloys Compd.* 981: 173680.

4 Wei, X., Liu, B., Chen, Z. et al. (2022). Recent advances in modulation engineering-enabled metal compounds for potassium-ion storage. *Energy Storage Mater.* 51: 815–839.

5 Min, X., Xiao, J., Fang, M. et al. (2021). Potassium-ion batteries: outlook on present and future technologies. *Energy Environ. Sci.* 14 (4): 2186–2243.

6 Rajagopalan, R., Tang, Y., Ji, X. et al. (2020). Advancements and challenges in potassium ion batteries: a comprehensive review. *Adv. Funct. Mater.* 30 (12): 1909486.

7 Liu, S., Kang, L., and Jun, S.C. (2021). Challenges and strategies toward cathode materials for rechargeable potassium-ion batteries. *Adv. Mater.* 33 (47): 2004689.

8 Wu, X.-F., Li, Z.-J., Liu, J.-X. et al. (2022). Defect engineering of hierarchical porous carbon microspheres for potassium-ion storage. *Rare Metals* 41 (10): 3446–3455.

9 Dong, S., Li, Y., Zhang, M. et al. (2023). Controllable engineering of intrinsic defects on carbon nanosheets enables fast and stable potassium storage performance. *Carbon* 204: 507–515.

10 Lin, H., Wang, Z., Quan, J., and Li, H. (2024). Defect engineering of sulfur vacancies in vanadium disulfide integrated with graphene as advanced potassium-ion battery cathode. *Chem. Eng. J.* 498: 155644.

11 Zhao, Z., Chen, H., Zhang, W. et al. (2023). Defect engineering in carbon materials for electrochemical energy storage and catalytic conversion. *Mater. Adv.* 4 (3): 835–867.

12 Xiao, Z., Meng, J., Xia, F. et al. (2020). K+ modulated K+/vacancy disordered layered oxide for high-rate and high-capacity potassium-ion batteries. *Energy Environ. Sci.* 13 (9): 3129–3137.

13 Zhao, J., Liu, X., Liu, P. et al. (2023). Oxygen vacancies refilling and potassium ions intercalation of δ-manganese dioxide with high structural stability toward 2.3 V high voltage asymmetric supercapacitors. *J. Colloid Interface Sci.* 629: 1039–1048.

14 Florindo, B.R., Hasimoto, L.H., de Freitas, N. et al. (2023). Patterning edge-like defects and tuning defective areas on the basal plane of ultra-large MoS2 monolayers toward the hydrogen evolution reaction. *J. Mater. Chem. A* 11 (37): 19890–19899.

15 Appiah, W.A., Rieger, L.H., Flores, E. et al. (2024). Unravelling degradation mechanisms and overpotential sources in aged and non-aged batteries: a non-invasive diagnosis. *J. Energy Storage* 84: 111000.

16 Xu, Y., Bahmani, F., Zhou, M. et al. (2019). Enhancing potassium-ion battery performance by defect and interlayer engineering. *Nanoscale Horiz.* 4 (1): 202–207.

17 Zhang, W., Liu, Y., and Guo, Z. (2025). Approaching high-performance potassium-ion batteries via advanced design strategies and engineering. *Sci. Adv.* 5 (5): eaav7412.

18 Huang, H., Xu, R., Feng, Y. et al. (2020). Sodium/potassium-ion batteries: boosting the rate capability and cycle life by combining morphology, defect and structure engineering. *Adv. Mater.* 32 (8): 1904320.

19 Zhang, K.-Y., Liu, H.-H., Su, M.-Y. et al. (2024). Defect engineering unveiled: enhancing potassium storage in expanded graphite anode. *J. Colloid Interface Sci.* 664: 607–616.

20 Zhou, Y., Zhu, Y., Liu, Y., and Xu, B. (2025). Topological defect-regulated porous carbon nanoribbon for high-performance potassium-ion batteries. *Small* 21 (4): 2408974.

21 Guo, K., Li, N., Bao, L. et al. (2023). Intrinsic carbon structural imperfections for enhancing energy conversion electrocatalysts. *Chem. Eng. J.* 466: 143060.

22 Xue, D., Xia, H., Yan, W. et al. (2020). Defect engineering on carbon-based catalysts for electrocatalytic CO_2 reduction. *Nano Micro Lett.* 13 (1): 5.

23 Zhu, J. and Mu, S. (2020). Defect engineering in carbon-based electrocatalysts: insight into intrinsic carbon defects. *Adv. Funct. Mater.* 30 (25): 2001097.

24 Dong, S., Gu, X., Li, Y. et al. (2025). Mechanically induced surface defect engineering in expanded graphite to boost the low-voltage intercalation kinetics for advanced potassium-ion batteries. *Carbon* 232: 119791.

25 Huang, J., Chen, Y., Cen, Z. et al. (2024). Topological defect-regulated porous carbon anodes with fast interfacial and bulk kinetics for high-rate and high-energy-density potassium-ion batteries. *Adv. Mater.* 36 (30): 2403033.

26 Castro Neto, A.H., Guinea, F., Peres, N.M.R. et al. (2009). The electronic properties of graphene. *Rev. Mod. Phys.* 81 (1): 109–162.

27 Quan, Z., Wang, F., Wang, Y. et al. (2024). Robust micro-sized and defect-rich carbon–carbon composites as advanced anodes for potassium-ion batteries. *Small* 20 (4): 2305841.

28 Fu, Y., Dong, Y., Shen, Y. et al. (2024). Recent advances in developing high-performance anode for potassium-ion batteries based on nitrogen-doped carbon materials. *Small* 20 (51): 2406630.

29 Zhang, W., Ming, J., Zhao, W. et al. (2019). Graphitic nanocarbon with engineered defects for high-performance potassium-ion battery anodes. *Adv. Funct. Mater.* 29 (35): 1903641.

30 Wang, B., Gu, L., Yuan, F. et al. (2022). Edge-enrich N-doped graphitic carbon: boosting rate capability and cyclability for potassium ion battery. *Chem. Eng. J.* 432: 134321.

31 Wang, X. and Wang, H. (2022). Designing carbon anodes for advanced potassium-ion batteries: materials, modifications, and mechanisms. *Adv. Powder Mater.* 1 (4): 100057.

32 Chen, Y., Xi, B., Huang, M. et al. (2022). Defect-selectivity and "order-in-disorder" engineering in carbon for durable and fast potassium storage. *Adv. Mater.* 34 (7): 2108621.

33 Liu, Y., Gu, Z.-Y., Heng, Y.-L. et al. (2024). Interface defect induced upgrade of K-storage properties in KFeSO4F cathode: from lowered Fe-3d orbital energy level to advanced potassium-ion batteries. *Green Energy Environ.* 9 (11): 1724–1733.

34 Zhang, J., Shi, M., Gao, H. et al. (2024). Engineering interfaces of zinc metal anode for stable batteries. *Chem. Eng. J.* 491: 152050.

35 Chen, X., Zhong, Z., Chen, X., and Li, H. (2025). Advancements in cathode materials for aqueous potassium-ion batteries. *Energy Storage Mater.* 74: 103948.

36 Singh, B. and Indra, A. (2020). Prussian blue- and Prussian blue analogue-derived materials: progress and prospects for electrochemical energy conversion. *Mater. Today Energy* 16: 100404.

37 Shu, W., Han, C., and Wang, X. (2024). Prussian blue analogues cathodes for nonaqueous potassium-ion batteries: past, present, and future. *Adv. Funct. Mater.* 34 (1): 2309636.

38 Liao, J., Yuan, Z., Hu, Q. et al. (2024). Heat-resistant carbon-coated potassium magnesium hexacyanoferrate nanoplates for high-performance potassium-ion batteries. *Angew. Chem. Int. Ed.* 63 (35): e202409145.

39 Liu, Z., Su, H., Yang, Y. et al. (2021). Advances and perspectives on transitional metal layered oxides for potassium-ion battery. *Energy Storage Mater.* 34: 211–228.

40 Li, J., Liu, L., Gao, Y. et al. (2025). Ion-exchange synthesis of low-water prussian blue analogs for enhanced sodium storage. *EcoMat* 7 (2): e70000.

41 Wei, R., Zhai, X., Tinker, H.R. et al. (2023). Anion vacancy regulated sodium/potassium intercalation in potassium prussian blue analog cathodes for hybrid sodium-ion batteries. *Adv. Funct. Mater.* 33 (52): 2308227.

42 Cong, J., Luo, S., Lin, Y. et al. (2024). Stable electrochemical properties of trace titanium doping layered P3-type $K_{0.5}Mn_{0.92}Ti_{0.08}O_2$ cathode material for potassium ion batteries. *J. Energy Storage* 102: 114017.

43 Chen, X., Hao, C., Shang, M. et al. (2024). Enhanced surface stability of $K_{0.27}MnO_2{\cdot}0.54\,H_2O$ cathode materials with $LiF/Li_xPF_yO_z$ coating for potassium-ion batteries. *Colloids Surf. A Physicochem. Eng. Asp.* 700: 134772.

44 Liu, Z., Li, S., Mu, J. et al. (2024). Element-tailored quenching methods: phase-defective $K_{0.5}Mn_1$-xCrxO_2 cathode materials for potassium ion batteries. *Mater. Today Chem.* 40: 102251.

45 Li, S., Li, M., Peng, L. et al. (2025). Enhancement of K storage performance in K3V3-xLax(PO4)4/C cathode materials for potassium-ion batteries via La3+ gradient doping. *J. Power Sources* 627: 235802.

46 Li, S., Li, M., Shen, R. et al. (2024). Research on Fe doping regulation of vanadium-based phosphate material K3V3-xFex(PO4)4/C for potassium-ion battery cathodes. *J. Alloys Compd.* 1008: 176692.
47 Wang, L., Zou, J., Chen, S. et al. (2018). TiS2 as a high performance potassium ion battery cathode in ether-based electrolyte. *Energy Storage Mater.* 12: 216–222.
48 Zhang, Y., Zhang, Y., Zhang, H. et al. (2021). Defect engineering in metal sulfides for energy conversion and storage. *Coord. Chem. Rev.* 448: 214147.

20

Defect and Interface Engineering in Lithium-Air Batteries

Lorena Álvarez Contreras and J. Antonio Cruz-Navarro

Centro de Investigación en Materiales Avanzados S.C., Miguel de Cervantes 120, Chihuahua, C.P. 31136, México

20.1 Electrochemical Dynamics of Li-Air Systems

Since their introduction in 1996 [1], LABs emerged as promising candidates for high-energy-density storage, with theoretical energy densities approaching 3500 Wh/kg, surpassing those of Li-ion systems [2]. LABs are classified into nonaqueous, aqueous, hybrid aqueous/nonaqueous, and solid-state configurations, with nonaqueous systems being the most studied (Figure 20.1).

The fundamental chemistry of LABs relies on oxygen (O_2) from the air reacting with lithium at the cathode to form lithium peroxide (Li_2O_2) or lithium oxide (Li_2O) [3]. Their architecture includes a lithium metal anode, a porous cathode, and a nonaqueous electrolyte containing lithium salts in an aprotic solvent. A polypropylene or glass-fiber membrane separates the electrodes to prevent short circuits [4].

During discharge, the lithium metal at the anode oxidizes, releasing lithium ions that migrate through the electrolyte. At the cathode, oxygen undergoes reduction, forming lithium-containing discharge products, mainly Li_2O_2. The porous electrode facilitates the oxygen reduction reaction (ORR) and oxygen evolution reaction (OER) [5].

$$2\text{Li} + \tfrac{1}{2}\,\text{O}_2 \leftrightarrow \text{Li}_2\text{O}\ (E_{\text{eq}} = 2.91\ \text{V}) \tag{20.1}$$

$$2\text{Li} + \text{O}_2 \leftrightarrow \text{Li}_2\text{O}_2\ (E_{\text{eq}} = 2.96\ \text{V}) \tag{20.2}$$

While Li_2O provides a higher theoretical energy density (10.49 kWh/L), Li_2O_2 forms preferentially due to kinetic favorability [2, 6]. LiO_2 acts as an intermediate before further reduction to Li_2O_2. However, Li_2O_2 exhibits poor solubility, leading to cathode surface deposition, blocking pores, and hindering oxygen and electron transport [7]. Nanomaterials such as graphene oxide, carbon nanotubes (CNTs), and metal nanoparticles have been used to stabilize intermediates and prevent Li_2O_2 accumulation, enhancing cycle life and current density [8, 9].

Surface and Interfacial Defects in Nanomaterials for Sustainable Energy Production and Storage, First Edition. Edited by Noé Arjona, Lorena Álvarez Contreras, and Minerva Guerra Balcázar.

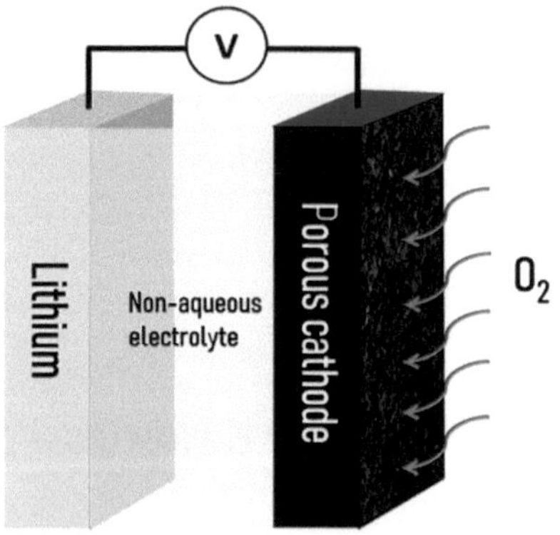

Figure 20.1 Basic architecture of a nonaqueous Li-air battery.

Charging reverses the reaction, decomposing Li_2O_2 back into lithium ions and oxygen (reaction 20.3). The OER requires a high voltage due to the insulating nature of Li_2O_2, resulting in overpotentials exceeding 1.0 V [10]. This overpotential accelerates electrode and electrolyte degradation, forming irreversible byproducts that limit the battery cycle life.

$$Li_2O_2 \rightarrow O_2 + 2e^- \tag{20.3}$$

OER kinetics depend on the electrode material and Li_2O_2 morphology and composition, which influence reaction dynamics. Lithium ions migrate back to the anode, where they are reduced and deposited as lithium metal (Eq. (20.4)). The reversibility of this plating process is crucial, as dendrite formation and electrolyte decomposition lead to capacity loss and safety hazards.

$$Li^+ + e^- \rightarrow Li \tag{20.4}$$

A primary limitation of LABs is their sluggish electrochemical kinetics, impacting power density due to the slow ORR and OER. These reactions involve multiple electron transfer steps and the formation/decomposition of Li_2O_2, which has poor conductivity and solubility. Consequently, high overpotentials lead to energy inefficiencies and degradation. The limitations arise from the wide-band-gap nature of Li_2O_2 and electrode passivation by parasitic byproducts [11].

Further challenges stem from electrolyte decomposition. The oxygen anion is highly reactive, degrading conventional nonaqueous electrolytes, such as alkyl carbonates. Alternative solvent systems, including acetonitrile, dimethyl sulfoxide, and ionic liquids, improve stability and performance [5].

20.2 Defect-Driven Modulation of Lithium Reactivity

The reactivity of lithium species in LABs critically influences charge–discharge efficiency and stability [12]. This behavior depends on electrolyte composition, electrode materials, and structural defects in lithium metal and electrode surfaces [13]. Defects, including vacancies, dislocations, grain boundaries, and surface irregularities, significantly impact lithium-ion transport and electrochemical performance [14]. Additionally, these defects act as active sites for ORR and OER reactions. In lithium metal anodes, surface defects serve as nucleation sites for lithium-ion deposition, influencing dendrite formation. While dendrites pose safety risks, controlling defect density can mitigate their growth, enhancing battery cycle life and safety. On the cathode, defects improve electrocatalytic activity, increasing discharge capacity and lowering overpotentials. Strategies such as heteroatom or metal doping further optimize performance by modifying electronic conductivity and reaction energetics.

Li et al. [15] categorized defect modulation in carbonaceous cathodes into three main effects: (i) enhancing electron conductivity, enabling the cathode to behave as a zero band-gap semiconductor or metallic material; (ii) reducing Li_2O_2 adsorption energies, ensuring homogeneous deposition; and (iii) engineering Li_2O_2 morphology through defect chemistry to enhance electronic conductivity and decrease charging overpotential. Specific catalysts (e.g. graphene oxide and CNTs) and electrolyte additives induce Li_2O_2 defects such as lithium vacancies ($Li_{2-x}O_2$), doped Li_2O_2, and amorphous phases [14]. These defects increase local electron density, facilitating oxygen adsorption and activation, which lowers ORR overpotentials and improves discharge efficiency. Optimizing defect concentration and distribution in anode and cathode materials enhances overall reversibility.

Defect modulation alters lithium reactivity by modifying electronic structure, ion diffusion, and reaction energetics. Vacancies in lithium metal affect lithium-ion migration and reactivity, while dislocations and grain boundaries in cathodes adjust local electronic environments, improving ORR/OER kinetics [16]. Additionally, defects induce localized strain in electrode materials, altering electronic density and promoting efficient electron transfer. While defect engineering enhances battery performance, excessive defect density can trigger side reactions, material degradation, and uneven current distribution. Long-term stability remains a challenge, as defect evolution over cycling influences cycle life and efficiency. Therefore, precise control over defect creation and evolution is essential for achieving durable, high-performance LABs.

20.3 Interface Engineering for Precision Oxygen Reaction

In LABs, interface engineering is pivotal for optimizing ORR and OER efficiency. The surface modification of cathodes enhances catalytic activity and ensures uniform product distribution [17]. Effective strategies include catalyst deposition with transition metal oxides (TMOs) (e.g., MnO_2 and Co_3O_4) and noble metals (e.g. Pt and Ir) to lower reaction barriers [18]. Carbon-based modifiers such as graphene, CNTs, and porous carbons improve conductivity and provide active sites for oxygen reactions. Spinel materials ($NiCo_2O_4$, $FeCo_2O_4$, $MnCo_2O_4$, or ZnS_2Co_4) facilitate oxygen diffusion and create voids for Li_2O_2 storage [19]. Hybrid composite electrodes synergize multiple materials to boost catalytic activity, oxygen transport, and mechanical resilience, while metal oxide-carbon combinations optimize conductivity and efficiency [20].

A stable solid electrolyte interphase (SEI) layer is critical to suppress lithium dendrites and protect the electrolyte. Strategies include electrolyte additives like lithium nitrate ($LiNO_3$) and lithium bis(trifluoromethanesulfonyl)imide (LiTFSI) [4], and controlled cycling for in situ SEI formation, enhancing interfacial stability [21]. Artificial interfacial coatings, such as atomic layer deposition (ALD) and sol–gel processes, mitigate parasitic reactions and improve electrolyte durability, extending the lifecycle of battery.

Electrolyte engineering also modulates interfacial chemistry, with hybrid electrolytes balancing ionic conductivity and stability. Ionic liquid electrolytes offer superior chemical and thermal resilience, improving LAB performance.

Ongoing research focuses on adaptive interfaces that dynamically respond to electrochemical conditions, multifunctional coatings integrating protection and ion transport, and molecular-level modifications to enhance efficiency. These advances are crucial for commercializing Li-air batteries.

20.4 Defect-Induced Stability Enhancements

The long-term stability of LABs depends on suppressing parasitic reactions that degrade electrodes, decompose electrolytes, and cause capacity fading. Li_2O_2 formation and decomposition contribute to structural instability and passivation, exacerbating these challenges. A critical issue in nonaqueous LABs is the decomposition of carbonate-based electrolytes, such as $LiPF_6$ in PC/EC-DMC, which react with oxygen to form irreversible lithium alkylcarbonates and Li_2CO_3 instead of the desired Li_2O_2 [22]. The accumulation of Li_2CO_3 is particularly detrimental due to its high decomposition potential (~4.2 V) compared to Li_2O (~3.3 V) and Li_2O_2 (~3.4 V), leading to severe polarization and capacity loss.

Defect engineering addresses these issues by regulating reaction pathways, enhancing catalytic activity, and improving cyclability. Introducing controlled structural and electronic defects in cathodes, solid-state electrolytes, and lithium anodes mitigates interfacial instabilities and optimizes reaction kinetics. This approach stabilizes Li_2O_2 formation and decomposition while suppressing detrimental byproducts, advancing the practical viability of LABs [23, 24]. This section explores defect-driven enhancements in ORR/OER kinetics, lithium-ion diffusion, and interfacial charge transport.

Cathode defect engineering plays a vital role in improving LAB stability, efficiency, and cycle life. The reversible formation and decomposition of Li_2O_2 are hindered by high charge overpotentials, structural instability, and parasitic side reactions. Introducing controlled defects optimizes catalytic performance by regulating vacancy concentrations, dislocation densities, heteroatom substitutions, and electronic structure modifications. Oxygen vacancies promote surface-mediated Li_2O_2 formation, lowering charge overpotentials and improving reversibility [25]. Similarly, doping with nitrogen, sulfur, or phosphorus enhances electronic conductivity and modulates adsorption energies, facilitating efficient ORR/OER pathways [26].

Hybridizing carbonaceous materials with transition metal derivatives such as oxides, sulfides, and nitrides, further increases defect density while stabilizing intermediates, preventing the formation of insulating byproducts like Li_2CO_3, which compromise long-term performance [27]. Advanced synthesis techniques, including chemical vapor deposition, exfoliation, and pyrolysis, allow precise control over defect distribution and structural integrity, optimizing cathode reactivity [28]. Additionally, density functional theory (DFT) calculations provide insight into defect-induced changes in electronic structure, projected density of states (PDOS), and adsorption energies, guiding a rational material design [29, 30]. By integrating defect engineering with interfacial stabilization strategies, LABs can mitigate charge polarization, electrolyte degradation, and cathode passivation, significantly extending operational lifespan. Cathode optimization strategies focus on enhancing electrocatalysis, oxygen diffusion, structural integrity, and defect-induced performance improvements. The key engineering approaches include (i) bifunctional catalyst integration, (ii) structural engineering, (iii) heteroatom doping, (iv) defect generation, (v) hybrid material design, and (vi) interfacial stabilization strategies [29], as summarized in Figure 20.2.

Defects are crucial in cathode stability by modulating charge transfer, tuning oxygen adsorption energies, and modifying the local chemical environment, enabling efficient Li_2O_2 formation and decomposition [25]. These defect-driven mechanisms enhance LAB performance by facilitating oxygen adsorption and activation,

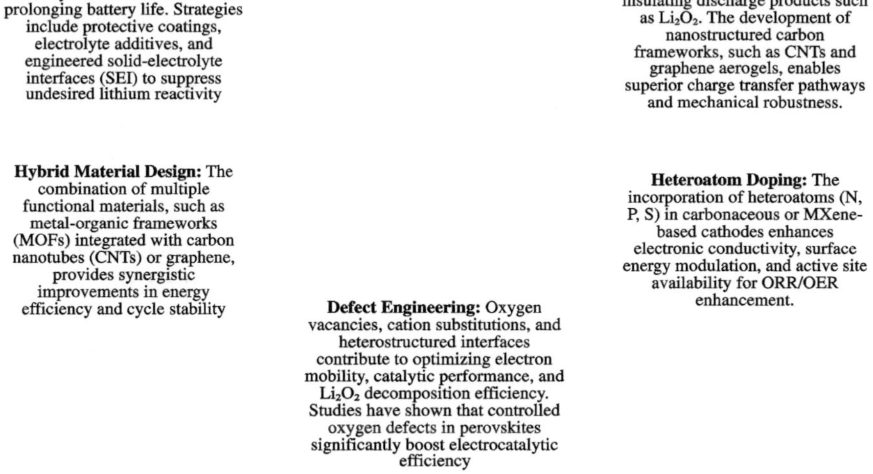

Figure 20.2 Synergistic interactions in defect and interface engineering for LABs.

improving charge transport, suppressing secondary reactions, and controlling the Li_2O_2 morphology to prevent thick deposits that cause irreversible electrode passivation [31].

A key approach in cathode defect engineering is the controlled introduction of oxygen vacancies in TMOs such as Co_3O_4, MnO_2, TiO_2, and perovskites like $La_{1-x}Sr_xCoO_3$ [26]. Oxygen vacancies redistribute charge, altering metal cation oxidation states to enhance electronic conductivity and optimize oxygen adsorption [31]. In Co_3O_{4-x}, these vacancies increase the Co^{3+}/Co^{2+} ratio, boosting oxygen activation and reducing charge overpotentials. Additionally, oxygen-deficient structures promote surface-based Li_2O_2 growth, enhancing system reversibility [32]. Oxygen vacancies can be engineered using plasma treatments, mild hydrogenation (H_2/Ar), and electrochemical activation via cyclic voltammetry, allowing precise defect control [31].

Heteroatom doping further optimizes ORR/OER kinetics by replacing oxygen or carbon atoms with elements such as nitrogen, sulfur, phosphorus, boron, or fluorine. Nitrogen doping in CNTs and graphene introduces pyridinic and graphitic sites, reducing O_2 activation energy and improving Li_2O_2 formation [26, 27]. Co-doping with S and P in carbon frameworks modulates charge distribution, minimizing Li_2CO_3 accumulation and enhancing Li_2O_2 decomposition. These doped materials are synthesized via ammonia-assisted pyrolysis, solvothermal methods with heteroatomic precursors, and electrochemical deposition for precise defect incorporation [26, 31].

Defective carbon materials also play a crucial role in cathode engineering due to their high surface area, conductivity, and tunable reactivity [27]. Atomic vacancies in graphene and CNTs disrupt π-conjugation, creating localized active sites that enhance ORR activity [28]. Stone–Wales defects in graphene facilitate optimal Li_2O_2 nucleation, preventing electrode passivation [25]. Additionally, amorphous carbon and disordered nanomaterials with high-defect densities improve Li_2O_2 solubility and charge transport [33]. These materials are synthesized through ball milling for edge defects, plasma-assisted etching for disorder creation, and thermal activation for microporous structures [27].

MXene-based electrocatalysts are promising for defect engineering in LAB cathodes due to their high conductivity, tunable surface chemistry, and abundant active sites [31]. These two-dimensional transition metal carbides/nitrides ($M_{n+1}X_nT_x$), where M is a transition metal (e.g., Ti, Mo, and V) and T_x are surface terminations (e.g., –OH, –O, and –F), exhibit enhanced catalytic activity when engineered with defects [26]. Oxygen vacancies in MXenes optimize electronic structure and oxygen adsorption, improving ORR/OER kinetics. DFT studies confirm that oxygen vacancies in $Ti_3C_2T_x$ MXenes reduce Li_2O_2 decomposition barriers, lowering overpotentials and increasing charge efficiency. Heteroatom doping in MXenes, such as nitrogen-doped $Ti_3C_2T_x$, enhances oxygen adsorption via charge redistribution, stabilizing intermediates during ORR/OER cycles. Boron, sulfur, and phosphorus doping further improve catalytic efficiency by modulating electronic states near the Fermi level. Electrochemical impedance spectroscopy (EIS) and X-ray photoelectron spectroscopy (XPS) confirm that doped MXenes facilitate faster charge transfer and reduce interfacial resistance [25–27].

MXenes also enable hybrid architectures by combining carbonaceous and TMO materials, leveraging defect engineering from both classes. MXene–graphene composites synergize graphene conductivity with MXenes tunable defect chemistry, improving ORR/OER kinetics and Li_2O_2 solubility. Hybrid MXene/MnO_2 nanosheets demonstrate superior catalytic activity, abundant oxygen vacancies, and enhanced oxygen diffusion, minimizing charge polarization [28, 31]. Furthermore, surface-functionalized MXenes with controlled –OH and –O terminations modulate electrolyte interactions, reducing Li_2CO_3 formation and stabilizing Li_2O_2 intermediates, thereby extending LAB cycle life. Plasma treatment, hydrothermal synthesis, and electrochemical activation are employed to engineer MXene surface defects, ensuring precise catalytic and interfacial properties. Theoretical modeling further elucidates defect-driven mechanisms in MXene-based electrocatalysts. DFT calculations reveal that defect-rich MXenes exhibit lower Gibbs free energy barriers for oxygen desorption, enhancing OER efficiency. Machine learning (ML) is now integrated with computational models to predict optimal defect configurations for next-generation MXene cathodes [27, 29, 31].

Defect engineering in MXene-based electrocatalysts is a frontier strategy for enhancing LAB stability, reversibility, and efficiency. Synergistic effects from oxygen vacancies, heteroatom doping, hybridization with carbon materials, and surface functionalization collectively improve charge–discharge efficiency and cycle life [25]. Future advancements in in situ X-ray absorption near-edge structure (XANES) and environmental transmission electron microscopy (ETEM) will be critical for real-time defect evolution analysis in MXenes [31].

High-entropy oxides (HEOs) have also emerged as promising candidates for next-generation energy storage due to their unique structural and electrochemical properties. Characterized by multicomponent cationic compositions, HEOs exhibit high configurational entropy, stabilizing crystal phases and improving electrochemical performance. Their integration into LABs enhances ORR/OER activity, lowers overpotentials, and stabilizes cathodes over extended cycling. The percolative network created by multiple cations enhances charge transfer, improving reaction kinetics. Furthermore, controlled HEO surface chemistry prevents excessive Li_2CO_3 accumulation, improving electrochemical reversibility and cycle life [34, 35].

The development of high-performance solid-state electrolytes (SSEs) for LABS is crucial for improving ionic conductivity, electrochemical stability, and mechanical robustness. Conventional liquid electrolytes suffer from volatility, lithium dendrite growth, and degradation due to oxygen and CO_2 contamination, limiting cycle life and efficiency. Defect engineering, involving the intentional manipulation of structural and electronic defects in electrolyte materials, addresses these challenges by enhancing lithium-ion transport, stabilizing electrode–electrolyte interfaces, and suppressing parasitic reactions [36].

The ionic conductivity of SSEs is influenced by intrinsic and extrinsic defects. Intrinsic defects, such as Frenkel and Schottky defects, arise from thermodynamic constraints and impact lithium-ion mobility. In garnet-type $Li_7La_3Zr_2O_{12}$ (LLZO) electrolytes, lithium vacancies and interstitials form primary lithium-ion migration pathways [37]. Extrinsic defect engineering, through aliovalent doping (e.g., Al^{3+}, Ga^{3+}, and Ta^{5+}), stabilizes the cubic phase of LLZO, which has superior ionic conductivity compared to the tetragonal phase due to enhanced lithium diffusion channels [21]. In sulfide-based electrolytes, such as $Li_{10}GeP_2S_{12}$ (LGPS), halogen anions (e.g., Cl^-, Br^-, and I^-) reduce grain boundary resistance and enhance lithium-ion percolation pathways, achieving ionic conductivities surpassing many liquid electrolytes [37]. These modifications optimize the defect environment by modulating electrostatic potential and reducing migration barriers.

Interfacial stability is another challenge in SSEs, as uncontrolled defects can increase interfacial impedance, cause electronic leakage, and lead to the formation of insulating interphases like Li_2CO_3. Strategies to mitigate these issues include artificial interlayers, where $LiNbO_3$, $LiAlO_2$, or Li_3PO_4 thin films are deposited at the interface between SSEs and lithium metal anodes to suppress undesirable side reactions while maintaining high ionic conductivity [37]. Point defect tuning, such as oxygen stoichiometry adjustment in oxide-based electrolytes like lithium phosphorus oxynitride (LiPON), optimizes both electrochemical stability and lithium-ion mobility. Additionally, surface functionalization using coatings like LiI, LiF, or Li_3N has been shown to reduce interfacial impedance and enhance lithium wettability, promoting uniform lithium deposition. Polymeric electrolytes offer flexibility and low interfacial resistance but suffer from low ionic conductivity and electrochemical stability. Defect engineering in solid polymer electrolytes (SPEs) like polyethylene oxide (PEO) focuses on tuning segmental motion to improve lithium-ion diffusion. The incorporation of nanofillers (e.g. SiO_2, Al_2O_3, and TiO_2) disrupts polymer crystallinity, increasing amorphous regions that facilitate ion transport [37]. Ionic liquid additives further modify defect density, improving lithium salt dissociation and enhancing ionic conductivity.

Gel polymer electrolytes (GPEs) integrate the benefits of liquid and solid electrolytes, providing high ionic conductivity with mechanical flexibility. Defect engineering in GPEs involves optimizing polymer cross-linking densities and incorporating redox mediators to stabilize Li_2O_2 deposition and mitigate parasitic reactions [21]. Hybrid polymer-ceramic electrolytes, combining polymer matrices (e.g. polyacrylonitrile (PAN), polyvinylidene fluoride (PVDF)) with ceramic SSEs like LLZO or LGPS, enhance both mechanical strength and ionic conductivity, addressing interfacial and bulk transport limitations [37]. Advanced synthesis techniques enable precise defect introduction in SSEs. Sol–gel synthesis facilitates uniform dopant distribution in LLZO, while spark plasma sintering (SPS) densifies SSEs at lower temperatures, preserving defect structures that enhance ionic transport [21]. Pulsed laser deposition (PLD) fabricates thin films with controlled stoichiometry and defect concentrations, making it particularly valuable for interface engineering.

High-entropy oxides (HEOs) are emerging as promising candidates for SSEs due to their unique configurational entropy, which stabilizes structural phases and enhances lithium-ion transport. HEOs create a percolative network for lithium diffusion, minimizing degradation over cycling and preventing excessive Li_2CO_3 formation, thereby extending cycle life. Recent studies on multicomponent garnet-type and NASICON-based HEOs demonstrate significant electrochemical improvements, positioning them as strong contenders for next-generation SSE development [37].

The strategic integration of defect engineering, interface stabilization, and advanced synthesis techniques is essential for optimizing SSEs in LABs. Future research should focus on synergistic defect engineering strategies, such as co-doping with aliovalent cations and anion substitution, to achieve further improvements in SSE properties. Computational methods like DFT and ML-based simulations will help predict optimal defect configurations. Expanding the application of HEOs and refining in situ and operando spectroscopic techniques will provide deeper insights into defect evolution under real operating conditions. By leveraging precise defect control, interface engineering, and novel material designs, future advancements in SSE technology will pave the way for safer, more efficient, and commercially viable LABs.

The performance of LABs depends critically on the optimization of the lithium metal anode, which, despite its high theoretical capacity and low electrochemical potential, faces significant challenges such as dendritic growth, chemical instability, and volumetric expansion during cycling. Defect engineering is a powerful strategy to address these limitations by improving lithium-ion transport, electrochemical stability, and overall battery durability through atomic- and nanoscale modifications [38].

Dendritic lithium growth is one of the most pressing issues in LABs as it leads to short circuits and rapid capacity loss. Defect engineering strategies focus on controlling lithium nucleation and deposition to achieve uniform lithium plating. The introduction of lithiophilic defects via heteroatom doping in carbon hosts, such as N-doped graphene frameworks and S-doped carbon nanofibers, enhances lithium affinity, promoting even lithium deposition and improved cycling stability. Similarly, oxygen vacancies in TMOs facilitate homogeneous lithium-ion flux, reducing current density variations that drive dendrite growth [39].

Another critical challenge is the stability of the SEI as its continuous degradation depletes lithium and electrolyte, limiting cycle life. Defect engineering enables the development of robust artificial SEI layers, such as those enriched in lithium fluoride (LiF), which improve interfacial stability and suppress parasitic side reactions. Surface defects in 2D materials like h-BN and MXenes act as lithium passivation layers, mitigating SEI decomposition and enhancing Coulombic efficiency [31].

The substantial volume changes during lithium plating and stripping generate mechanical stress, risking anode failure and loss of electrical contact. Defect-rich porous architectures, such as hierarchical 3D porous carbon structures with engineered vacancy defects, allow lithium accommodation, mitigating stress and enhancing long-term cycling stability. Studies on defective graphene aerogels and mesoporous carbon scaffolds have demonstrated over 70% capacity retention after 300 cycles, showcasing their effectiveness in reducing volume expansion-induced degradation.

Ionic transport properties within lithium anodes are also influenced by defect engineering. The introduction of oxygen vacancies in ceramic coatings (e.g. Li_3PO_4 and $LiAlO_2$) enhances lithium-ion diffusivity, reducing overpotentials and improving rate capability. In situ X-ray absorption spectroscopy (XAS) studies confirm that oxygen-deficient perovskite coatings improve ionic conductivity, facilitating faster charge transfer and minimizing electrode polarization. Additionally, polymeric protective layers with defect-modified structures (e.g. PEO-based lithium-ion conductors) enhance lithium interface compatibility, reducing charge transfer resistance and improving cycling efficiency [37].

A promising avenue is the hybrid composite anode strategy, which integrates multiple defect engineering approaches for synergistic effects. Combining defective carbon hosts with nanostructured metal oxides (e.g. Co_3O_4) improves lithium deposition kinetics and reduces interfacial resistance. ALD techniques further enable the precise control of defect-engineered coatings on the lithium metal, allowing tunable surface properties and enhanced stability under high current densities.

Advanced characterization techniques and computational modeling play a crucial role in optimizing defect engineering strategies. In situ transmission electron microscopy (TEM) provides atomic-scale insights into lithium deposition dynamics and defect evolution. DFT calculations predict the influence of vacancy concentrations, dopant distributions, and charge transfer pathways in defect-modified anodes. ML algorithms combined with high-throughput screening have accelerated the discovery of novel defect-engineered materials for LAB anodes.

Defect engineering offers a transformative approach to overcoming intrinsic challenges in lithium metal anodes. By tailoring nucleation sites, stabilizing interfaces, accommodating volume changes, and enhancing ionic conductivity, it is possible to significantly improve battery performance and durability. The integration of multi-scale defect engineering strategies, combined with advanced material synthesis and characterization techniques, will be essential for unlocking the full potential of LABs for next-generation energy storage applications.

20.5 Interfaces and Long-Term Cyclability in Li-Air Systems

Achieving long-term cyclability in LABs remains challenging due to unstable electrode–electrolyte interfaces. These regions govern electrochemical stability, reaction kinetics, and degradation pathways. Unlike lithium-ion batteries, LABs involve solid–gas–liquid interactions, discharge product formation, and side reactions affecting both electrodes. Advanced interface engineering is key to minimizing capacity fade, suppressing side reactions, and extending battery life [6, 8].

Direct contact between lithium metal and the electrolyte causes dendritic growth, electrolyte depletion, and interfacial resistance buildup. Defect-engineered SEI layers, incorporating LiF, Li_3N, or Li_2O, improve chemical and mechanical stability, reducing lithium corrosion and excessive SEI growth [2, 17, 22]. Hybrid polymer-ceramic interlayers (e.g. PEO-Li_3PO_4 composites) regulate lithium-ion flux and suppress high-surface-area lithium formation, enhancing cycle life [21, 37]. ALD and molecular layer deposition (MLD) techniques create ultra-thin, defect-controlled coatings that enhance ion transport and suppress dendrite growth, improving stability under high currents. Additionally, lithiophilic nanoparticles (Ag, Au, ZnO) incorporated into interfacial layers promote uniform lithium nucleation, reducing overpotentials and ensuring reversible lithium plating/stripping.

Cathode efficiency and stability depend on ORR and OER dynamics. Large Li_2O_2 aggregates cause passivation and high overpotentials, limiting cycle life. Defect-engineered transition metal oxides (e.g. Co_3O_{4-x} and MnO_{2-x}) modulate oxygen adsorption energies, promoting more uniform Li_2O_2 deposition and preventing electrode clogging [12, 19, 25]. Heteroatom-doped carbons (N-doped graphene, S-doped CNTs) accelerate ORR/OER kinetics, improving Li_2O_2 reversibility [13, 15, 26]. Hybrid cathodes integrating MXene-supported catalysts enhance charge transfer and structural resilience under cycling [27, 28]. Electrolyte modification with redox mediators (e.g. TEMPO and LiI) enables solution-mediated Li_2O_2 dissolution, reducing passivation and voltage hysteresis [21, 31].

Electrolyte decomposition from reactive oxygen species limits nonaqueous LAB performance. Fluorinated ethers and ionic liquid-based electrolytes resist oxidative degradation while maintaining high ionic conductivity [36, 37]. Furthermore, hybrid SSEs combining polymeric lithium-ion conductors with ceramics (LLZO, LATP) suppress lithium dendrite formation and stabilize interfaces, preventing electrolyte decomposition and enhancing oxygen diffusion [21, 31]. Progress in interface engineering is crucial for advancing LAB technology. Integrated approaches combining defect-engineered anodes, catalytically enhanced cathodes, electrolyte modifications, and SSEs will be key to achieving high-efficiency, long-cycle batteries capable of meeting next-generation energy storage demands.

20.6 Future Perspectives in Defect and Interface Engineering for Li-Air Batteries

The evolution of LABs requires a shift from empirical material design to precision-driven atomic-scale modifications. Future advancements will integrate novel materials, in situ characterization, and AI-assisted predictive modeling to accelerate the development of next-generation LAB architectures with high energy density, stability, and cyclability [6, 8, 32].

AI and ML are revolutionizing defect engineering. High-throughput screening and generative adversarial networks (GANs) rapidly identify optimal defect configurations, dopant strategies, and interfacial stabilizers. Automated DFT simulations coupled with reinforcement learning predict defect-induced changes in ionic transport, reaction kinetics, and material stability [30]. AI-driven material databases (e.g., The Materials Project) enable defect formation energy predictions, phase stability assessments, and ion mobility calculations, accelerating material discovery with minimal experimental trial-and-error testing [31].

Electrolyte decomposition remains a critical barrier to LAB commercialization. Hybrid solid-state electrolytes (HSSEs) combining polymer–ceramic composites, perovskite conductors, and HEOs can stabilize interfaces, facilitate uniform oxygen transport, and prevent side reactions [21, 37]. Defect-modulated electrolyte additives (fluorinated/phosphorus-based ionic liquids) enhance chemical durability and interfacial passivation [21, 31]. Single-ion conductors, such as sulfonated polyether ether ketone (sPEEK) and lithium-doped polyimides, improve mechanical stability and minimize interfacial impedance, extending cycle life [22, 37]. Self-healing hybrid electrolytes integrate adaptive polymer networks capable of repairing microfractures and regenerating Li-ion transport pathways, mitigating electrolyte breakdown and enhancing reversibility [38].

Research in defect-engineered electrode architectures is advancing toward self-healing and adaptive materials that dynamically respond to stress, degradation, and phase changes. Self-assembled nanostructures with vacancy-engineered MXenes, doped graphene aerogels, and porous transition metal oxides are crucial for reversible oxygen electrocatalysis and cathode stability [12, 15, 27]. Similarly, adaptive lithium anodes with defect-rich polymer matrices and tunable SEI layers enhance lithium metal stability under extreme cycling. A key challenge in LAB cathodes is uncontrolled Li_2O_2 deposition, which passivates active sites and limits charge transfer. Future designs will incorporate hierarchical nanostructures with defect-rich interfaces to regulate Li_2O_2 nucleation and decomposition. Mesoporous carbon frameworks decorated with oxygen vacancy-rich metal oxides will ensure high catalytic activity and durability under extended cycling [26, 29].

Bio-inspired electrode architectures, mimicking alveolar structures, improve oxygen diffusion and charge–discharge efficiency, preventing electrode clogging and extending battery life [24]. Experimental studies are shifting from ex-situ characterization to operando techniques, capturing real-time defect formation, ion transport, and interfacial evolution. Advanced synchrotron X-ray methods, cryo-electron microscopy (Cryo-EM), and neutron scattering provide insights into defect-driven stability in Li-air systems [38, 39]. Multi-scale modeling, integrating quantum-level simulations with continuum-scale electrochemical modeling, links atomic-scale defect engineering to macroscale battery performance [30]. The development of real-time spectroscopic probes, combining electrochemical strain microscopy (ESM) with soft XAS, enables the tracking of phase transformations at sub-nanometer resolution, optimizing defect configurations for higher efficiency and longer cycle life [18, 21].

20.7 Conclusion

Defect and interface engineering enable enhanced electrochemical performance in LABs by improving cyclability, energy efficiency, and stability. This chapter has explored how controlled defects influence ORR/OER kinetics, lithium-ion transport, and interfacial charge transfer. Heteroatom doping, vacancy engineering, and interface modifications have proven effective in stabilizing reaction intermediates, lowering overpotentials, and reducing side reactions. Despite significant progress, key challenges remain, including defect evolution over cycling, SEI layer degradation, and cathode–electrolyte instability. Advanced operando techniques, AI-driven material discovery, and scalable synthesis methods will be critical in addressing these limitations. Emerging materials, such as HEOs, vacancy-rich TMOs, and MXenes, hold promise for improving electronic conductivity, catalytic activity, and interfacial stability. The integration of adaptive electrode architectures and multifunctional protective coatings will play a pivotal role in extending battery lifespan. Defect and interface engineering will remain fundamental to advancing LABs and accelerating the transition toward high-energy-density, sustainable energy storage technologies.

References

1 Abraham KM, Jiang Z. A Polymer electrolyte-based rechargeable lithium/oxygen battery. *J. Electrochem. Soc.* 1996;143(1):1. doi:https://doi.org/10.1149/1.1836378.

2 Chen K, Yang DY, Huang G, Zhang XB. Lithium-air batteries: air-electrochemistry and anode stabilization. *Acc. Chem. Res.* 2021;54(3):632–641. doi:https://doi.org/10.1021/acs.accounts.0c00772.

3 Chalasani D, Lucht BL. Reactivity of electrolytes for lithium-oxygen batteries with Li_2O_2. *ECS Electrochem. Lett.* 2012;1(2):A38. doi:https://doi.org/10.1149/2.010202eel.

4 Tan P, Kong W, Shao Z, Liu M, Ni M. Advances in modeling and simulation of Li–air batteries. *Prog. Energy Combust. Sci.* 2017;62:155–189. doi:https://doi.org/10.1016/j.pecs.2017.06.001.

5 Grande L, Paillard E, Hassoun J, Park JB, Lee YJ, Sun YK, Passerini S, Scrosati B. The lithium/air battery: still an emerging system or a practical reality? *Adv. Mater.* 2015;27(5):784–800. doi:https://doi.org/10.1002/adma.201403064.

6 Christensen J, Albertus P, Sanchez-Carrera RS, Lohmann T, Kozinsky B, Liedtke R, Ahmed J, Kojic A. A critical review of Li/air batteries. *J. Electrochem. Soc.* 2011;159(1):R1-R30. doi:https://doi.org/10.1149/2.086202jes.

7 Tan P, Shyy W, Wei ZH, An L, Zhao TS. A carbon powder-nanotube composite cathode for non-aqueous lithium-air batteries. *Electrochim. Acta* 2014;147:1–8. doi:https://doi.org/10.1016/j.electacta.2014.09.074.

8 Li J, Hou L, Yu M, Li Q, Zhang T, Sun H. Review and recent advances of oxygen transfer in Li-air batteries. *ChemElectroChem* 2021;8(17):3588–3603. doi:https://doi.org/10.1002/celc.202100560.

9 Lu J, Jung Lee Y, Luo X, Chun Lau K, Asadi M, Wang HH, Brombosz S, Wen J, Zhai D, Chen Z, Miller DJ, Sub Jeong Y, Park JB, Zak Fang Z, Kumar B, Salehi-Khojin A, Sun YK, Curtiss LA, Amine K. A lithium–oxygen battery based on lithium superoxide. *Nature* 2016;529:377–382. doi:https://doi.org/10.1038/nature16484.

10 Tian F, Radin MD, Siegel DJ. Enhanced charge transport in amorphous Li_2O_2. *Chem. Mater.* 2014;26(9):2952–2959. doi:https://doi.org/10.1021/cm5007372.

11 Zhang Z, Xiao X, Zhu X, Tan P. *Addressing Transport Issues in Non-Aqueous Li–air Batteries to Achieving High Electrochemical Performance*. Springer Nature Singapore; 2023. doi:https://doi.org/10.1007/s41918-022-00157-3.

12 Zhang Q, Wang C, Xie Z, Zhou Z. Defective/doped graphene-based materials as cathodes for metal–air batteries. *Energy Environ. Mater.* 2022;5(4):1103–1116. doi:https://doi.org/10.1002/eem2.12293.

13 Yue G, Hong Z, Xia Y, Yang T, Wu Y. Bifunctional electrocatalysts materials for non-aqueous Li–air batteries. *Coatings* 2022;12(8):1227. doi:https://doi.org/10.3390/coatings12081227.

14 Liu T, Zhao S, Xiong Q, Yu J, Wang J, Huang G, Ni M, Zhang X. Reversible discharge products in Li–air batteries. *Adv. Mater.* 2023;35(11):208925. doi:https://doi.org/10.1002/adma.202208925.

15 Li W, Han C, Zhang K, Chou S, Dou S. Strategies for boosting carbon electrocatalysts for the oxygen reduction reaction in non-aqueous metal-air battery systems. *J. Mater. Chem. A* 2021;9(12):6671–6693. doi:https://doi.org/10.1039/d1ta00203a.

16 Booth SG, Nedoma AJ, Anthonisamy NN, Baker PJ, Boston R, Bronstein H, et al. Perspectives for next generation lithium-ion battery cathode materials. *APL Mater.* 2021;9:056106. doi:https://doi.org/10.1063/5.0051092.

17 Guo Y, Niu Q, Pei F, Wang Q, Zhang Y, Du L, Zhang Y, Zhang Y, Zhang Y, Fan L, Zhang Q, Yuan L, Huang Y. Interface engineering toward stable lithium–sulfur batteries. *Energy Environ. Sci.* 2024;17(5):1330–1367. doi:https://doi.org/10.1039/D3EE04183B.

18 Liu MZ, Li C, Wang R, Xiao S, Guo Z, Wang Y. Noble metal catalysts for metal-air batteries: from nano-level to atom-level. *Next Mater.* 2024;2:100126. doi:https://doi.org/10.1016/j.nxmate.2024.100126.

19 Li X, Liu J, Zhang J, Wu S, Zhang D, Shui J. Interface engineering of $NiCo_2O_4$/BCN nanotube for performance enhancement of lithium-oxygen battery. *Chem. Eng. J.* 2021;411:128403. doi:https://doi.org/10.1016/j.cej.2021.128403.

20 Kang JH, Lee J, Jung JW, Park J, Jang T, Kim HS, Nam JS, Lim H, Yoon KR, Ryu WH, Kim ID, Byon HR. Lithium-air batteries: air-breathing challenges and perspective. *ACS Nano* 2020;14(11):14549–14578. doi:https://doi.org/10.1021/acsnano.0c07907.

21 Balaish M, Kraytsberg A, Ein-Eli Y. A critical review on lithium-air battery electrolytes. *Phys. Chem. Chem. Phys.* 2014;16:2801–2822. doi:https://doi.org/10.1039/c3cp54165g.

22 Fang C, Tran TN, Zhao Y, Liu G. Electrolyte decomposition and solid electrolyte interphase revealed by mass spectrometry. *Electrochim. Acta* 2021;399:139362. doi:https://doi.org/10.1016/j.electacta.2021.139362.

23 Barile, C.J. and Gewirth, A.A. (2013). *J. Electrochem. Soc.* 160: A549–A552.

24 Ma Z, Yuan X, Li L, Ma ZF, Wilkinson DP, Zhang L, et al. A review of cathode materials and structures for rechargeable lithium–air batteries. *Energy Environ. Sci.* 2015;8(7):2144–2198. doi:https://doi.org/10.1039/C5EE00838G.

25 Schroeder MA, Pearse AJ, Kozen AC, Chen X, Gregorczyk K, Han X, et al. Investigation of the cathode–catalyst–electrolyte interface in aprotic Li–O_2 batteries. *Chem. Mater.* 2015;27(15):5305–5313. doi:https://doi.org/10.1021/acs.chemmater.5b01605.

26 Chao M, Zeng K, Lu C, Shi Z, Guo J, Chen X, et al. Synergized N and P co-doped $Ti_3C_2T_x$ MXene enabling high-performance Li-air batteries. *J. Colloid Interface Sci.* 2024;657:46–53. doi:https://doi.org/10.1016/j.jcis.2023.11.101.

27 Woo H, Kang J, Kim J, Kim C, Nam S, Park B. Development of carbon-based cathodes for Li-air batteries: present and future. *Electron. Mater. Lett.* 2016;12(4):551–567. doi:https://doi.org/10.1007/s13391-016-6183-1.

28 Suryatna A, Raya I, Thangavelu L, Alhachami FR, Kadhim MM, Altimari US, et al. A review of high-energy density lithium-air battery technology: investigating the effect of oxides and nanocatalysts. *J. Chem.* 2022; 2762647. doi:https://doi.org/10.1155/2022/2762647.

29 Karunarathne S, Malaarachchi CK, Abdelkader AM, Kamali AR. Advances in bifunctional electrocatalysts towards high-performing Li-air batteries. *J. Power Sources* 2024;607:234553. doi:https://doi.org/10.1016/j.jpowsour.2024.234553.

30 Jain A, Shin Y, Persson KA. Computational predictions of energy materials using density functional theory. *Nat. Rev. Mater.* 2016;1(1):1–13. doi:https://doi.org/10.1038/natrevmats.2015.4.

31 Rao Y, Yang J, Chu S, Guo S, Zhou H. Solid-state Li-air batteries: fundamentals, challenges, and strategies. *SmartMat* 2023;4:e1205. doi:https://doi.org/10.1002/smm2.1205.

32 Liu Z, Zhao Z, Zhang W, Huang Y, Liu Y, Wu D, et al. Toward high-performance lithium-oxygen batteries with cobalt-based transition metal oxide catalysts: advanced strategies and mechanical insights. *InfoMat* 2022;4:e12260. doi:https://doi.org/10.1002/inf2.12260.

33 Guo T, Hu P, Li L, Wang Z, Guo L. Amorphous materials emerging as prospective electrodes for electrochemical energy storage and conversion. *Chem* 2023;9:1080–1093. doi:https://doi.org/10.1016/j.chempr.2023.03.032.

34 Liu ZY, Liu Y, Xu Y, Zhang H, Shao Z, Wang Z, et al. Novel high-entropy oxides for energy storage and conversion: from fundamentals to practical applications. *Green Energy Environ.* 2023;8(5):1341–1357. doi:https://doi.org/10.1016/j.gee.2023.04.007.

35 Pikalova EY, Kalinina EG, Pikalova NS, Filonova EA. High-entropy materials in SOFC technology: theoretical foundations for their creation, features of synthesis, and recent achievements. *Materials* 2022;15(24):8783. doi:https://doi.org/10.3390/ma15248783.

36 Tian H, Gu X, Wu Q, Ma T, Cai Y, Yao X, et al. Defect engineering builds organic-inorganic composite electrolytes with a stable electrochemical window and interface for all-solid-state lithium-ion batteries. *Electrochim. Acta* 2024;492:144383. doi:https://doi.org/10.1016/j.electacta.2024.144383.

37 Zhu Q, Ma J, Li S, Mao D. Solid-state electrolyte for lithium-air batteries: a review. *Polymers* 2023;15(11):2469. doi:https://doi.org/10.3390/polym15112469.

38 Zheng J, Zhao Q, Tang T, Yin J, Quilty CD, Renderos GD, et al. Reversible epitaxial electrodeposition of metals in battery anodes. *Science* 2019;366: 645–648. doi:https://doi.org/10.1126/science.aax6873.

39 Zhang R, Chen XR, Chen X, Cheng XB, Zhang XQ, Yan C, et al. Lithiophilic sites in doped graphene guide uniform lithium nucleation for dendrite-free lithium metal anodes. *Angew. Chem. Int. Ed.* 2017;56(27):7764–7768. doi:https://doi.org/10.1002/anie.201702099.

21

Defect and Interface Engineering in Zinc-Air Batteries

Alejandro Arredondo-Espínola[1] *and Noé Arjona*[2]

[1] *Centro de Investigación y Desarrollo Tecnológico en Electroquímica, Posgraduate Deparment, Parque Tecnológico Querétaro St., Pedro Escobedo, Querétaro, C.P. 76703, México*
[2] *Centro de Investigación en Materiales Avanzados S.C., Department of Materials Physics, Miguel de Cervantes St., Complejo Industrial Chihuahua, Chihuahua, Chih., C.P. 31136, México*

21.1 Introduction to Zinc-Air Batteries

ZABs are electrochemical galvanic cells, involving four main components: the anode, the cathode, an electrolyte, and an electrical conductor. During discharge, the Zn oxidation and the oxygen reduction reaction (ORR) take place, while during charging, the reduction of zincate ions (acting as the cathode) and the OER occur (Figure 21.1a) [1]. Because the polarization shifts, it is preferable to call the electrodes as the Zn and air (or bifunctional) electrodes rather than anode(s) and cathode(s). Because of its high conductivity and excellent ZnO dissolution, 6 M KOH is the preferred electrolyte in ZABs, but if faces major challenges such as water evaporation, promotion of the hydrogen evolution reaction (HER) at the Zn anode, and the formation of insoluble carbonates, which block active sites, decreasing the battery rechargeability [2].

Bifunctional electrocatalysts presents several challenges: if hydrophilicity is favored (ORR favored), the discharge processes are impacted by air electrode flooding, leading to poor ZAB performance due to inadequate oxygen diffusion and blockage of active sites. Conversely, if hydrophobicity dominates (OER favored), electrode degradation is reduced, but charging processes are compromised due to weak electrode–electrolyte interaction [3].

The modifications introduced through interface and defect engineering also impact the performance of other components. For instance, a high overpotential required for charging processes in ZABs promotes the HER-induced corrosion of the zinc electrode and dendritic growth (Figure 21.1b). Another major issue is the formation of carbonates, which can be attributed to two causes: the first being the interaction of CO_2 from the air with KOH, and the second being the corrosion of carbon in the air electrode. Carbon corrosion not only reduces the ionic conductivity of the electrolyte but also passivates the electrode surfaces, decreasing the catalyst's active sites due to metal leaching. In addition, it renders

Surface and Interfacial Defects in Nanomaterials for Sustainable Energy Production and Storage, First Edition. Edited by Noé Arjona, Lorena Álvarez Contreras, and Minerva Guerra Balcázar.

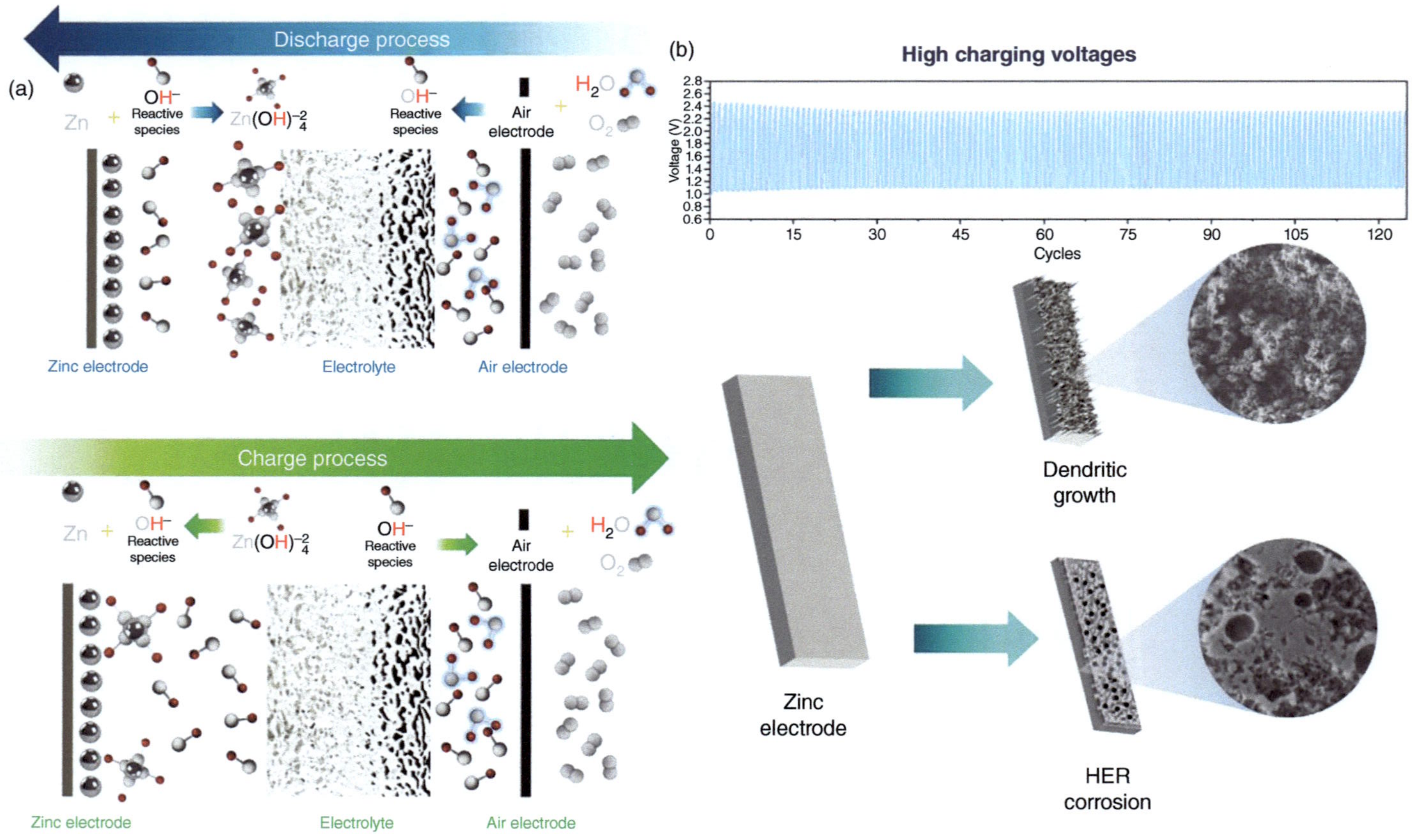

Figure 21.1 (a) Main reactions and processes in a ZAB during the charging/discharging process. (b) Main issues in the zinc electrode at high charging voltages.

the air electrode more hydrophilic, promoting electrolyte depletion and electrode flooding. In most cases, these issues become more significant at charging potentials above 2.0 V [4]. However, depending on the degree of carbon graphitization and the electrode–catalyst synergy, these potentials may not necessarily induce air electrode corrosion [5].

21.2 Types of Bifunctional Electrocatalyst for ZABs

Currently, interface and defect engineering has been extensively studied across various types of electrocatalysts. To address this section clearly and concisely, only three types of bifunctional electrocatalysts will be discussed: noble metals, transition metal, and metal-free carbon-based materials (Figure 21.2).

21.2.1 Noble Metals

Noble metal combinations, especially Pt-IrO_2 and Pt-RuO_2, are known for their favorable thermodynamic potential to drive both ORR and OER, resulting in higher

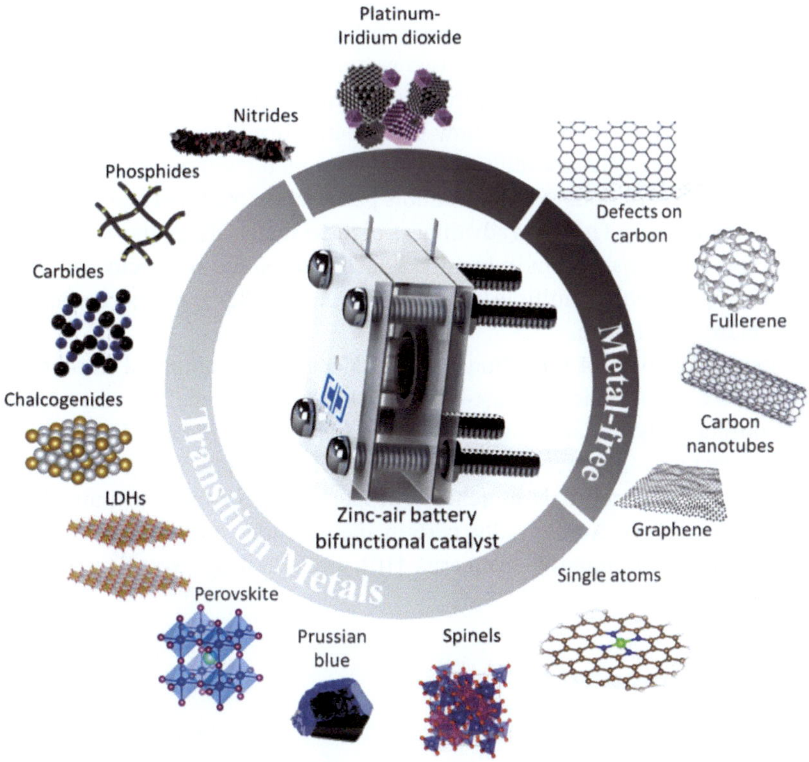

Figure 21.2 Schematic representation of different bifunctional electrocatalysts employed for oxygen reactions in rechargeable ZABs.

power densities and improved efficiency for ZABs. Different strategies have been developed to enhance the efficiency and durability, while reducing the amount of the required noble metals [6]. One approach consists of using Pt and IrO_2 nanoparticles in a 1 : 1 ratio supported on carbon materials to decrease the metal loading (typically is around 20 wt.%). Recently, the development of single-atom catalysts (SACs) to streamline the atomic utilization of metals has gained relevance in the field of the ORR. Liu et al. achieved the super-utilization of Pt for the ORR using a carbon-supported divacancy-anchored Pt SACs [7].

21.2.2 Transition Metals

The most used Pt-group metals (PGMs)–free transition metals (TMs) as electrocatalysts for ORR/OER in ZABs include Ni, Cu, Co, Mn, and Fe. These TMs are utilized in ZABs in the form of transition metal oxides (TMOs), transition metal hydroxides (TMOHs), transition metal chalcogenides (TMCh), transition metal carbides (TMCs), transition metal nitrides (TMNs), and transition metal phosphides (TMPs). Given the wide variety of TMOs, these are typically categorized into single metal oxides, metal oxides in a spinel structure, rutile oxides, and perovskite oxides (POs). The mixtures of transition metal oxides (MTMOs) in a spinel structure have the formulas A_3O_4 (A = Mn, Co, etc.) and AB_2O_4 (A = Ni, Co, Mn, Zn, Li, etc.; B = Mn, Fe, Co, Ni, etc.). The rutile phase of metal oxides is not distinguished by a formula, but by its crystal structure, which consists of two sets of edge-sharing octahedra with the transition metal cations occupying the octahedral centers (e.g. MnO_2, MnO_3, and β-MnO_2). The perovskite oxides follow the general formula ABO_3, where A is a metal with a 12-fold oxygen coordination, and B is a metal with a sixfold oxygen coordination, though variations exist (e.g. double perovskites, triple perovskites, and brownmillerite oxides) [8]. The main advantages of these materials are the versatility in tuning composition, easiness in inducing surface defects, the simplicity in their synthesis, their high stability, multiple valence states, and ability to protect carbonaceous supports from corrosion.

The electric conductivity of transition metal chalcogenides compared to that of metal oxides makes them efficient, cost-effective, and, thus, competitive materials for ZABs. Consequently, because of the semiconducting nature of MTMOs and TMOHs, the defect and interface modifications have been applied to increase the metal-behavior of these materials to achieve properties comparable to transition metal-based materials [9]. Additionally, certain metal oxides pose severe toxicity and environmental contamination concerns [10], necessitating extra care when selecting them for ZABs.

21.2.3 Carbonaceous Materials

Carbon-based materials offer a promising, cost-effective option for developing efficient bifunctional electrocatalysts. The development of metal-free catalysts faces several challenges, including low cyclability, high overpotentials (particularly for OER), and catalyst poisoning due to poor kinetics associated with the two-electron ORR pathway. Therefore, it is essential to optimize, stabilize, and enhance the efficiency of carbon-based electrocatalysts [11]. Among the different carbon allotropes, the defect engineering has been focused on carbon nanotubes and graphene because of their higher surface areas in contrast to carbon black.

21.3 Defect and Interface Engineering Applied to Electrocatalysts

In the early 2000s, electrocatalysts for oxygen reactions were obtained from a nanometric viewpoint, where aspects such as nanometric size and shape together with composition were relevant at that time. Currently, the modifications follow an atomic viewpoint (in the Armstrong scale), where the fine-tuning at atomistic levels is the base of the defect and interface engineering. The incorporation of imperfections into the crystalline lattice through the introduction of defects offers a significant advantage in improving the efficiency of electrocatalysts. The controlled introduction of defects alters the electronic configuration of the material, significantly impacting the interactions between surface atoms and adsorbates. This directly affects the overpotential required for the charge and discharge processes of ZABs. Various types of defects have been studied (Figure 21.3), while the most studied are the heteroatomic doping, edge sites, strains, and the introduction of metal or oxygen vacancies (Ov) [13].

21.3.1 Heteroatomic Doping

It is well established that the activity of a catalyst is primarily determined by its ability to adsorb and desorb reaction intermediates. The doping of heteroatoms in/around the catalytic site alters the coordination environment and electronic cloud due to differences in the electronegativity and atomic radius of the heteroatoms. Whether the dopant directly influences the structure of the active site or that of the support, it can enhance the catalytic activity. This enhancement occurs because dopants can indirectly affect the central atom through delocalized interactions, achieving a regulatory effect similar to direct coordination with the central atom [14]. Depending on the nature of the material to be doped, the procedure can be carried out using various methods, among which co-precipitation synthesis, sol–gel, combustion, and impregnation stand out. For ORR, the most used nonmetallic and metallic heteroatoms as dopants are N, S, P, Fe, and Mn. However, for OER, metallic heteroatoms like Co and Ni are typically introduced to a carbonaceous material to improve electronic conductivity, while providing active sites to adsorb OH species.

21.3.2 Edge Sites

Edge atoms typically exhibit higher electrocatalytic activity than the atoms located at the bulk or at the surface because of the higher unsaturation of actives sites.

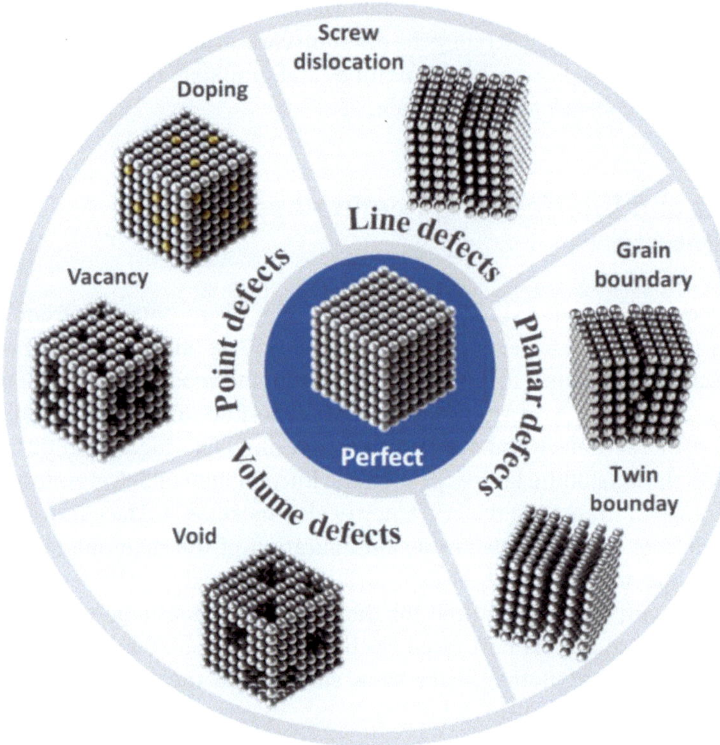

Figure 21.3 Schematic representation of various defects promoted in materials to improve the activity as bifunctional electrocatalysts for ZABs. Source: Adapted from [12].

Consequently, an increased content of edge sites enhances the efficiency of the electrocatalyst by means of an improvement of the inherent reactivity of the active site, while the other pathway involves an increase in the number of active sites. Various strategies exist to increase edge sites, including the creation of vacancies, the formation of ultra-thin nanosheets, and the growth of vertical structures [15].

21.3.3 Strain Defects

When designing structures with strain defects of deformation/dislocation, the atomic configuration can be adjusted to form surface regions with high catalytic activity, thereby improving performance through changes in gas dissociation energies, which promotes the conversion and generation of new species. This also impacts the electronic density near the Fermi level, enhancing conductivity, activity, and catalytic stability. To promote the formation of deformation/dislocation defect-rich structures, specialized equipment is typically used due to the high temperatures and pressures involved, as metals tend to form structurally stable crystals during nucleation and growth. Surface-defect-rich structures will lead to an increase in surface-free energy [16].

21.3.4 Vacancies

One of the most important defects in the development of bifunctional electrocatalysts is Ov, given their ability to change the localized pH, improve the breakage of oxygen bindings, and increase the number of active sites. These defects are widely present in TMOs and greatly affect the material's properties by altering the electronic structure, conductivity, adsorption of species, and can generate edge sites. The main methodologies to promote Ov include thermal treatment, reduction processing, cation/anion doping, plasma treatment, and other technologies such as laser, flame, exfoliation, and template strategy [17].

21.4 Interface and Defect Engineering Applied to Different Rechargeable Zinc-air Batteries

21.4.1 Aqueous/Flow Zinc-Air Batteries

ZABs operated with aqueous 6 M KOH in active (flow battery) or passive flow (traditional rechargeable battery mode) are the most studied system since 6 M KOH displays high ionic conductivity, achieving high battery performance. In this manner, the defect and interface modifications through interface/defect engineering have been reported for a series of bifunctional materials such as those mentioned in Figure 21.2, while in Figure 21.3, a summary of the main defects is displayed. For instance, a defect-engineered Pd material as a family of the PGMs with a Fe_3O_4 skin was enriched with strain/dislocation defects and used as the cathode catalyst in an aqueous ZAB (Figure 21.4a) [18]. This material achieved a battery voltage of 1.48 V and a power density near 250 mW/cm^2, which was 2.5 higher than that achieved by Pt/C. In addition, the rechargeability was twice than that achieved by a mixed Pt/C + IrO_2 bifunctional catalyst. Single metal oxides, such as Co_3O_4, and mixtures of metal oxides/hydroxides (MMOs/MMHs) have been proposed as cost-effective alternatives to PGM catalysts. For instance, Ov were induced to a Co_3O_4 hollow material via Kirkendall effect, as shown in Figure 21.4b [19]. This Ov-defected material was tested in a gel electrolyte ZAB using polyvinyl alcohol (PVA) dissolved in 6 M KOH, resulting in a battery voltage of 1.459 V. Its power density surpasses that of Pt/C + RuO_2, enabling 90 mW/cm^2, while the rechargeability test confirmed higher stability than the benchmarked material. Another interesting single metal oxide, MnO_2, presents different crystalline phases (such as α, β, and γ), and it is well-known for exhibiting good activity toward ORR. The introduction of defects like Ov and metallic heteroatoms like Co, Ni, or Fe yields a bifunctional electrocatalyst. Zhong et al. found that a Co-MnO_2 catalyst exhibited a battery voltage of 1.44 V in an aqueous ZAB with a power density of 167 mW/cm^2 [25]. The rechargeability was tested via galvanic charge–discharge (GCD) curves, finding a voltage difference between the discharge and the charge voltage of 0.75 V, which represents a round-trip efficiency of 63%.

The introduction of Ov has been extended to MMOs, such as those found in a spinel structure ($A^{2+}B^{3+}{}_2O_4$), including $NiCo_2O_4$, $NiMn_2O_4$, $CoMn_2O_4$, and

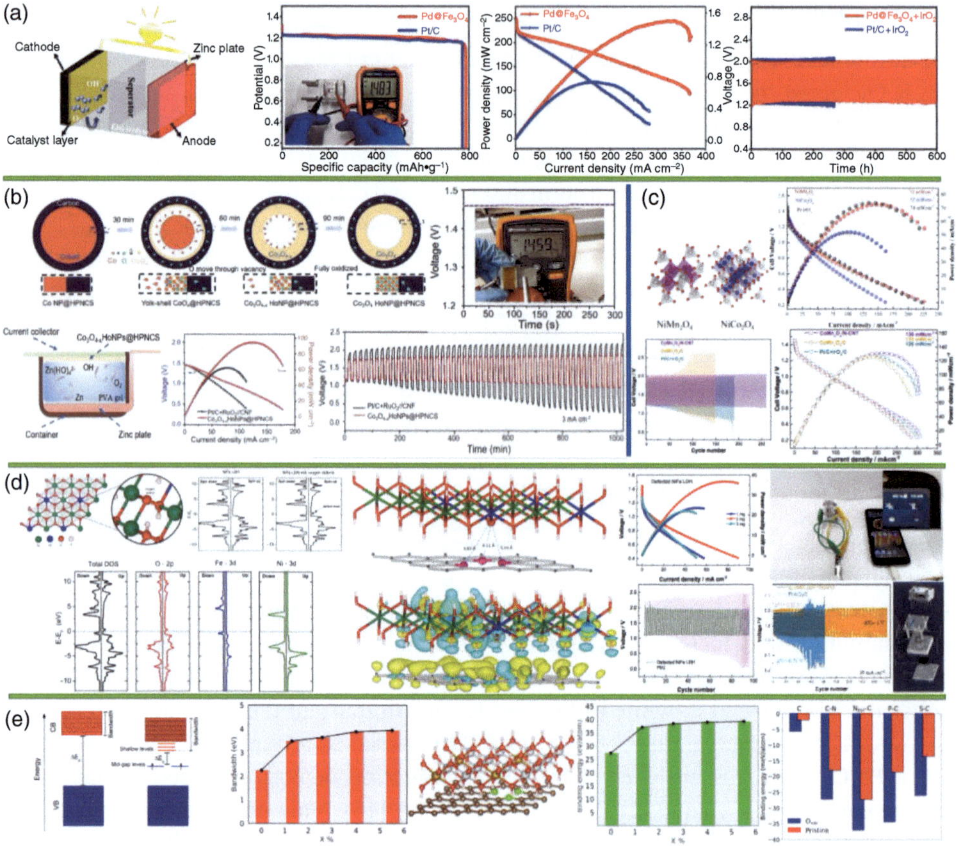

Figure 21.4 (a) Scheme of an aqueous Zn-air battery, specific capacity, and galvanic charge–discharge curves using a defect-engineered Pd PGM with a Fe_3O_4 skin. Source: Reproduced from [18] / with permission of John Wiley & Sons. (b) Scheme of the methodology to produce Co_3O_4 nanoparticles with surface defects, scheme of an aqueous ZAB operated with a gel electrolyte, photography of the battery voltage, and performance/GCD curves. Source: Reproduced from [19] / with permission of John Wiley & Sons. (c) Scheme of the crystalline structure of the $NiMn_2O_4$ and $NiCo_2O_4$ spinels and power density curve. Source: Reproduced with permission from [20]. The power density curve and GCD curve for a $CoMn_2O_4$ spinel with interface engineering by employing an N-doped carbon nanotube are also presented. Source: Reproduced with permission from [21]. (d) Density of states for pristine and oxygen-defected NiFe-layered double hydroxides, and the charge-density (depletion and accumulation) difference for the interface-engineered Ov NiFe LDH/N-doped carbon nanotubes are also presented, as well as the power density and GCD curves in the aqueous ZAB and its application to charge a cell phone. Source: Reproduced with permission from [22]. For comparison purposes, the GCD curve and the aqueous ZAB scheme of a NiMn LDH with oxygen vacancies engineered with a mixed support (N-doped carbon nanotubes + N-doped graphene). Source: Reproduced with permission from [23]. (e) Scheme of the modification of midgap and shallow levels by introducing oxygen defects to a NiFe LDH and calculated bandwidths, and the atomic structure for an interface-engineered NiFe LDH are presented along with the binding energy changes by Ov content. Also, the binding energy as a function of the nonmetallic heteroatom (N, P, or S) is presented for pristine and Ov NiFe. Source: Ramos-Castillo et al. [24] / with permission of American Chemical Society.

$CoFe_2O_4$ [26]. Different surface defects were induced to $NiCo_2O_4$ and $NiMn_2O_4$ spinels (Figure 21.4c) such as metal and Ov, oriented atoms, lattice dislocations, and Frenkel defects [20]. In addition, interface engineering was applied to a $CoMn_2O_4$ spinel by replacing the typical carbon black support with a N-doped carbon nanotube support [21]. The introduction of surface defects to $NiMn_2O_4$ enabled it to exhibit a similar power density to a Pt-IrO_2 catalyst in the aqueous ZAB (72 vs. 74 mW/cm^2, Figure 21.4c). However, the interface engineering using N-doped carbon as a co-catalyst during the ORR enabled $CoMn_2O_4$/N-CNT to display higher power density (130 mW/cm^2), and superior rechargeability than the $CoMn_2O_4$ spinel with the carbon black support (200 cycles more). In this regard, it is clearly observed that the synergistic effect between the introduction of surface defects and the promotion of interfacial effects is a key factor influencing the battery performance.

The interface engineering has been extended to a NiFe-layered double hydroxide (LDH) as an MMH catalyst, in which oxygen defects were induced to the LDH, while N doping was introduced to a carbon nanotube support [22]. The density of states was analyzed for the NiFe LDH with and without Ov (Figure 21.4d), where both materials display ferromagnetic behavior, while Ov induces defect energy levels, changing the electron distribution, while a dangling bond was found in the oxygen atom, enabling the charge transfer during the electrocatalytic process. The incorporation of an N-doped CNT support affects the electron density (depletion and accumulation) as observed in the figure, indicating that a moderated charge is transferred from the LDH to the support. Consequently, the NiFe LDH/N-doped CNT displayed bifunctionality, which is essential for the construction of a rechargeable ZAB. The power density was ~35 mW/cm^2, while during the GCD cycles, this material achieved higher stability than a Pt/C catalyst.

The interface engineering by inducing Ov to LDHs was also evaluated with a Ov-defected NiMn LDH, which was combined with a mixture of carbon nanotubes + graphene both containing N-doped heteroatoms [23]. The mixed support enabled higher stability, and as a consequence, the rechargeability increased from 100 cycles for the NiFe/N-doped CNT to almost 200 cycles for the NiMn/N-doped CNTG. Both materials were tested in an aqueous homemade ZAB, as shown in Figure 21.4d, while the authors show the feasibility of these ZABs by charging a cell phone. Interestingly, Ramos–Castillo et al. studied the impact of the Ov content (from 0 to 5.5%) on a NiFe LDH through DFT [24]. The authors analyzed an increase in the midgap and shallow levels due to the effects of the Ov content (Figure 21.4e). The midgap was found to be dependent on the content, as observed in the figure, suggesting that Ov tailors the electronic properties of the material. Then, the interface engineering was evaluated by adding a graphene layer doped with N, P, or S heteroatoms. For N-doped graphene layer, the binding energy increases in the presence of Ov compared to that of pristine NiFe LDH. The increase in the Ov valued up to 4% did not show significant improvement in bandwidth/binding energies as observed in Figure 21.4e. The binding energy for P and S heteroatoms is also strongly affected by the presence of Ov, with nitrogen showing the highest binding energy per atom. Thus, it is concluded that the presence of Ov improves the metal–support interaction, enabling more stable electrocatalysts for highly durable ZABs. There are other MMOs employed in ZABs such as those with a perovskite structure ($A^{2+}B^{4+}[(X^{2-})_3]$ [27] and metal chalcogenides [28], but for space reasons, they are not herein discussed.

PGMs, metal oxides, and MMOs are nanoparticles, which when contacted with the KOH electrolyte under an electrical field, tend to destabilize the nanoparticles undergoing phenomena such as particle dissolution and the Ostwald ripening. In addition, there is a tendency to develop catalysts with low or ultra-low metal loadings aiming to decrease costs. With these motivations, two approaches arise: the first consists of developing catalysts with an atomic rationalization like the SACs, which are a type of atomically dispersed catalysts (ADCs), where they often present an M–N–C (metal heteroatom co-doped with a non-metallic heteroatom bonded to carbon) structure. The second consists of avoiding the use of metals, such as in metal-free heteroatom-doped carbon allotropes.

SACs are considered as doping defects, and in contrast to ADCs, they are characterized by their atomic accessibility. Among SACs for ZABs, the latest tendency consists of adding two metals into the carbon matrix, in which the typical configuration is depicted in Fig. 21.5a, which is an atomic M–N–C structure being the metal bonded to nitrogen heteroatoms (with $M–N_x$ sites such as MN_4 and Mn_2), which are into the carbon structure [34]. Luo et al. found that a dual-atom Fe/Ni-SAC@NC presented higher activity than their single atom counterparts. In the aqueous ZAB, a cell voltage of 1.54 V was achieved with a power density of 260 mW/cm^2, being 2.7 times higher than that obtained by the PGM catalyst [29]. The rechargeability was tested at two current densities: 5 and 20 mA/cm^2, displaying functionality until 140 and 100 h, respectively. This is particularly important, since the search for higher power densities is not only mandatory but also to achieve bigger rechargeability at high current density demands. Atomically dispersed materials (ADMs) are another interesting class of metal-doped defected catalysts, which can be found in the form of SACs (isolated atoms) or in small groups of atoms (like fully exposed cluster catalysts). The term "atomically dispersed" is a better description of the metal atom distribution on the carbon surface. An atomically dispersed Fe in a M–N–C structure was developed by Shang et al. [30], as shown in the TEM micrograph of Figure 21.5b. The extended X-ray absorption fine structure spectroscopy (EXAFS) presented in the figure demonstrated that this material was found as a single ADM due to the lack of a Fe–Fe bonding peak. This aqueous ZAB presented a cell voltage of 1.45 V and a power density of 140 mW/cm^2. The rechargeability of this ZAB was high, presenting durability after 240 h. There are different carbon allotropes classified as 0D, 1D, 2D, and 3D, as shown in Figure 21.5c [31]. Also, 1D and 2D carbon materials have been greatly utilized in ZABs for the development of metal-free carbon catalysts. For instance, Liu et al. developed graphene-like carbon materials enriched with N doping, and surface defects, such as edge defects, and topological defects (Figure 21.5d) such as steps, forks, cracks, edges, and pores [32]. This material was used in a flow ZAB (which is also called Zn-air fuel cell [35]), displaying a battery voltage of 1.38 V and a maximum power density near 70 mW/cm^2. Because of the electrolyte recirculation, the rechargeability dramatically increased, presenting stability after 500 h.

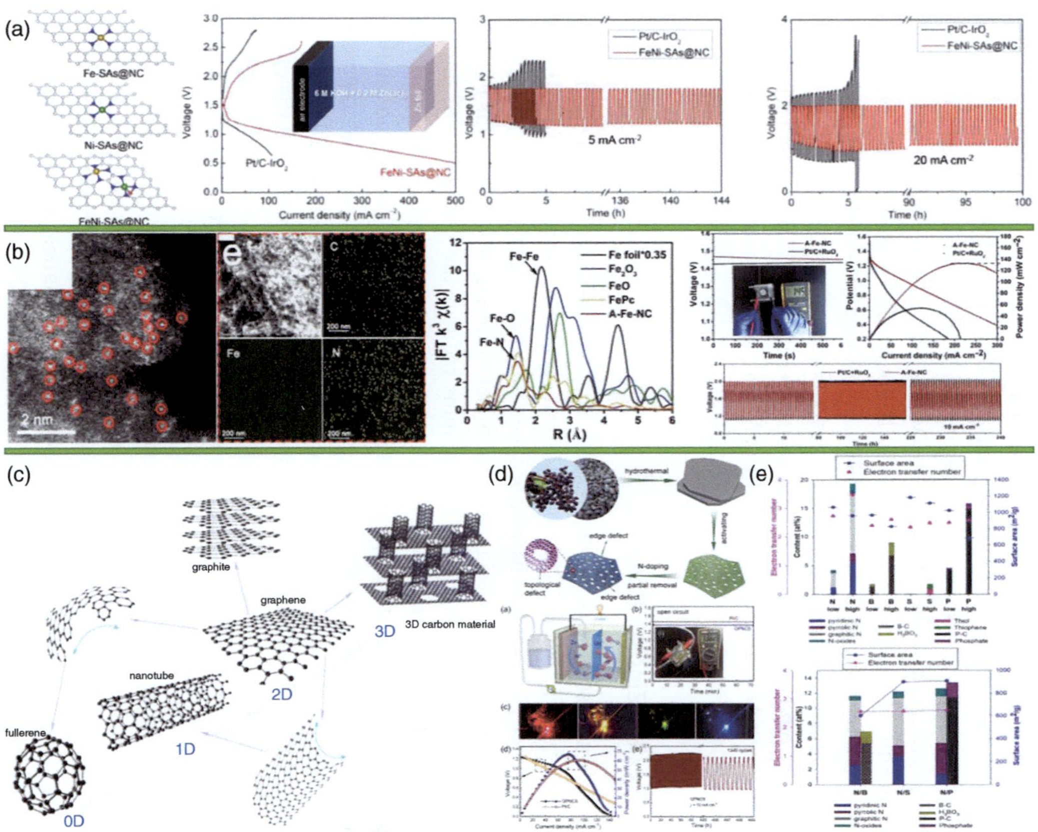

Figure 21.5 (a) Schematic representation of the Fe, Ni, and Fe/Ni dual atom SAC, and battery tests in an aqueous ZAB. Source: Luo et al. [29] / with permission of Elsevier. (b) TEM and TEM-EDS analyses of Fe ADC, EXAFS, and battery tests. Source: Shang et al. [30] / with permission of Elsevier. (c) carbon materials from 0D to 3D. Source: Dai et al. [31] / with permission of American Chemical Society. (d) Defected metal-free carbon material as the catalyst in tests in a flow ZAB. Source: Liu et al. [32] / with permission of American Chemical Society. (e) Correlation plots between heteroatoms and their contents, surface areas, and electron transfer numbers. Source: Preuss et al. [33] / with permission of John Wiley & Sons.

This could be related to carbonates and Zn oxides being removed from the solution, thus reducing the formation of dendrites, passive films, and shape changes.

Finally, the doping defects in carbon materials (metal-free) for ZABs have been a hot topic for the past decade. Preuss et al. investigated the effect of single nonmetallic heteroatoms (N, B, S, and P) and mixtures of heteroatoms (N/B, N/S, and N/P) doped on carbon in the ORR (Figure 21.5e). For single heteroatoms, the authors found that nitrogen doping is easier than other heteroatom doping, and as observed in the figure, S was the most difficult doping material. In addition, nitrogen was the only heteroatom to present real activity toward ORR [33], while B, P, and S act as defect promoters in the carbon matrix. Bejar et al. evaluated the synthesis method, the nitrogen source, and the effect of a surface defect promoter on the activity and performance in an aqueous ZAB [36]. The N-doped CNTs obtained by a thermal treatment method presented higher activity than those obtained by the solvothermal method, and melamine was a better N precursor than urea. The use of hydrogen peroxide promoted surface defects, and consequently, the ZAB operated with this material displayed a power density of 101 mW/cm^2, which was higher to that tested with Pt/C (82 mW/cm^2).

21.4.2 Flexible Zinc-Air Batteries

The use of quasi-solid electrolytes (QSEs) enables the development of flexible zinc-air batteries (FZABs). However, the limited electrolyte volume (6 M KOH) and the presence of a physical barrier between the electrodes (separator) introduce new considerations for the design and optimization of electrocatalysts for FZABs. The limited electrolyte volume in QSE systems reduces the wettability of the air electrode compared to aqueous systems, leading to a decrease in the number of active sites involved in the three-phase interface. Therefore, the proper modulation of the catalytic material's hydrophobicity and hydrophilicity is essential to enable interaction with more active sites without compromising oxygen diffusion within the system. On the other hand, the low ionic conductivity and barrier effects of the QSE result in reduced power density compared to aqueous systems. This necessitates the development of electrocatalysts with a higher density of active sites to ensure the viability of FZABs.

One of the commonly used materials is cobalt (Co), which, when doped with manganese (Mn), increases grain boundaries and the interplanar spacing of Co (111), creating more active sites for electrocatalysis (Figure 21.6a). Mn not only introduces defects into the crystalline structure of Co but also enhances the electrocatalyst interface, promoting oxygen adsorption on its surface 6.28 and 5.63 times that of Pt and Co, respectively. Although this material performs worse than Pt in aqueous electrolytes, it exhibits superior performance in QSE, achieving 1.4 times the current density of Pt and an open-circuit potential (OCP) of 1.573 V (65 mV higher than Pt) [37]. Another approach to improving the hydrophobicity of the electrode through interface and defect engineering has been studied using nitrogen-doped carbon supports (Figure 21.6b). This strategy increases the number of defects that facilitate the growth of TMO nanoparticles (e.g. $MnCo_2O_4$), providing more

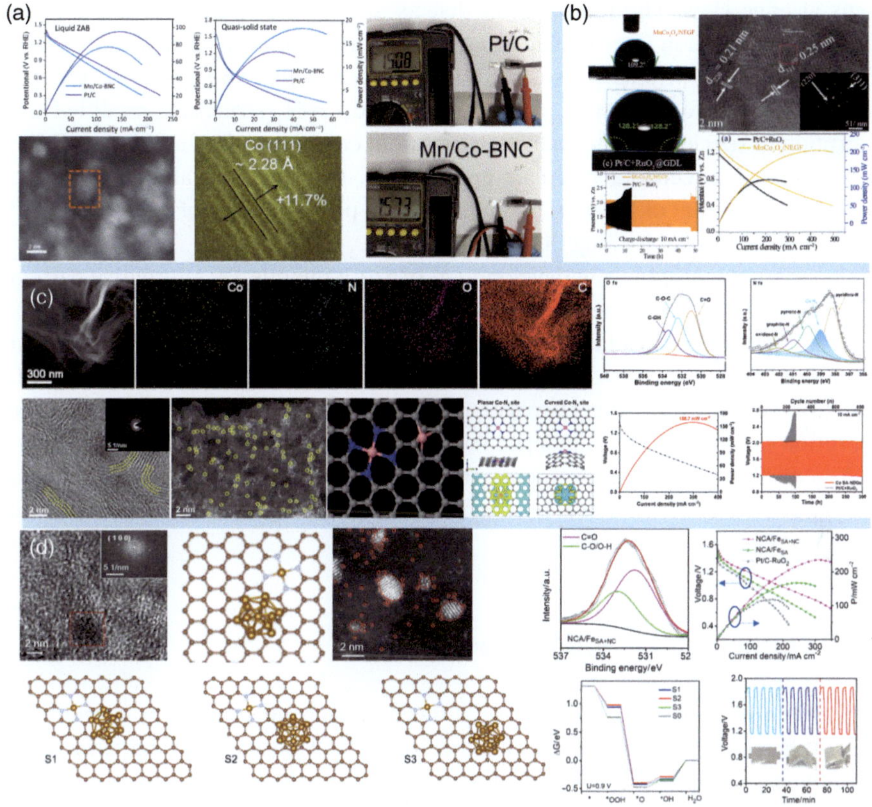

Figure 21.6 (a) Formation of defects and the modulation of the three-phase interface of Co through Mn doping, comparative analysis of the electrocatalyst in aqueous ZABs and with QSE. Source: Wang et al. [37] / with permission of Springer Nature. (b) Effects of hydrophobicity/hydrophilicity modulation through the nitrogen doping of carbon supports. Source: Manna et al. [38] / with permission of American Chemical Society. (c) Co SAC supported on nitrogen-doped graphene, and effects of defect engineering to improve efficiency and the three-phase interface. Source: Wang et al. [39] / Springer Nature / CC BY 4.0. (d) Synergistic effect of an SA and Fe NP electrocatalyst supported nitrogen-doped carbon aerogels. Source: Reproduced from [40] / with permission of Elsevier.

active sites and modulating the catalyst's hydrophobicity. The material becomes sufficiently hydrophilic to enhance active site interactions (nitrogen-doped carbon and TMOs) without compromising the diffusion of the reactive gas (O_2). As a result of the optimized three-phase interaction and the defects in the carbon matrix, a current density of 200 mW cm^{-2} was achieved (81% higher than with Pt-RuO_2) along with a cyclability of 48 h at 5 mA cm^{-2} [38].

A novel alternative, due to its low metal loading, high stability, proper hydrophobicity modulation, and high operating currents in ZABs, is the SACs. For example, a Co SAC supported on nitrogen-doped graphene (Figure 21.6c) exhibits a significant number of defects, which, in conjunction with the porous matrix, promotes the formation of pyridinic nitrogen. This is a result of the positioning of nitrogen atoms at edge sites. The induction of defects altered the energy barriers for the adsorption of reaction intermediates, optimizing the hydrophilicity/hydrophobicity of the three-phase interface associated with C=O, C—OH, and C—O—C groups. Additionally, the formation of planar and curved Co-N_4 structures modified the adsorption of species and the charge distribution, enhancing the material's activity. As a result of defect and interface engineering, this electrocatalyst exhibited a cyclability twice that of Pt-RuO_2 and a maximum power density of 158.7 mW cm^{-2} with only 0.14% metal loading [39].

Other elements used in the development of SACs include Fe, alternative carbon supports (nitrogen-doped aerogels), and catalysts of SACs with NPs (Figure 21.6d). These configurations, in addition to the effects of hydrophilicity/hydrophobicity and the defects induced by nitrogen-doping and NP nucleation previously mentioned, also exhibit a synergistic effect between the positioning of the SA and the NPs in the absorption of reaction intermediates, thereby enhancing the electrocatalyst's activity. As a result of these improvements, the SA and NP electrocatalyst demonstrated a power density of 236 mW cm^{-2}, 66 and 117 mW cm^{-2} higher than with the use of only SA or Pt-RuO_2, respectively [40].

21.5 Conclusions and Perspectives

Defect and interface engineering have allowed to decrease the energy barriers for ORR/OER taking place in the air electrode of ZABs. Consequently, batteries with higher power densities have been obtained. In addition, the rechargeability has been also boosted. However, there are still two aspects in which these modifications need to impact: (i) rechargeability at high current densities to obtain fast-charging batteries, and (ii) better oxygen utilization to improve the battery performance for moving applications such as in drones or vehicles.

References

1 Audeves-Audeves, Y., Arredondo-Espínola, A., Nava, O. et al. (2025). High activity of cobalt-atomically dispersed catalyst on mesoporous carbon for rechargeable Zn-air batteries via effective removal of the hard template. *Microporous Mesoporous Mater.* 381: 113359.

2 Huang, C.L., Wang, P.Y., and Li, Y.Y. (2020). Fabrication of electrospun CO_2 adsorption membrane for zinc-air battery application. *Chem. Eng. J.* 395: 125031.

3 Lu, Q., Zou, X., Bu, Y., and Shao, Z. (2023). Structural design of supported electrocatalysts for rechargeable Zn–air batteries. *Energy Storage Mater.* 55: 166–192.

4 Zhao, S., Liu, T., Wang, J. et al. (2022). Anti-CO_2 strategies for extending zinc-air batteries' lifetime: a review. *Chem. Eng. J.* 450: 138207.

5 Filimonenkov, I.S., Bouillet, C., Kéranguéven, G. et al. (2019). Carbon materials as additives to the OER catalysts: RRDE study of carbon corrosion at high anodic potentials. *Electrochim. Acta* 321: 134657.

6 Kumar, D.B., Nie, W., Jiang, Z. et al. (2023). Recent progress in transition metal carbides and nitrides-based composites as bifunctional oxygen electrocatalyst for zinc air batteries. *J. Alloys Compd.* 960: 170828.

7 Liu, J., Jiao, M., Mei, B. et al. (2019). Carbon-supported divacancy-anchored platinum single-atom electrocatalysts with superhigh Pt utilization for the oxygen reduction reaction. *Angew. Chem.* 131: 1175–1179.

8 Davari, E. and Ivey, D.G. (2017). Bifunctional electrocatalysts for Zn–air batteries. *Sustain. Energy Fuels* 2: 39–67.

9 Tharani, S., Durgalakshmi, D., Kunhiraman, A.K. et al. (2023). Hybrid zinc-air battery (ZAB) with transition metal-based electrocatalysts – a step toward next-generation electrochemical energy storage. *Wiley Interdiscip. Rev.: Energy Environ.* 12: e472.

10 Wei, L., Ang, E.H., Yang, Y. et al. (2020). Recent advances of transition metal based bifunctional electrocatalysts for rechargeable zinc-air batteries. *J. Power Sources* 477: 228696.

11 Arafat, Y., Azhar, M.R., Zhong, Y. et al. (2021). Metal-free carbon based air electrodes for Zn-air batteries: recent advances and perspective. *Mater. Res. Bull.* 140: 111315.

12 Bai, S., Zhang, N., Gao, C., and Xiong, Y. (2018). Defect engineering in photocatalytic materials. *Nano Energy* 53: 296–336.

13 Perivoliotis, D.K., Ekspong, J., Zhao, X. et al. (2023). Recent progress on defect-rich electrocatalysts for hydrogen and oxygen evolution reactions. *Nano Today* 50: 101883.

14 Zhang, Y., Zhu, S., Wang, X. et al. (2023). Heteroatom-doped MNC catalysts for oxygen reduction reactions: doping strategies and active site regulation. *J. Electroanal. Chem.* 943: 117506.

15 Ni, B., Wang, X., Ni, B., and Wang, X. (2015). Face the edges: catalytic active sites of nanomaterials. *Adv. Sci.* 2: 1500085.

16 Shi, J., Qi, D., Zhang, H. et al. (2024). Large-scale atomic strain defects on palladium surfaces for enhanced oxygen reduction and zinc–air batteries. *Small* 20: 2406448.

17 Zhu, K., Shi, F., Zhu, X., and Yang, W. (2020). The roles of oxygen vacancies in electrocatalytic oxygen evolution reaction. *Nano Energy* 73: 104761.

18 Shi, J., Qi, D., Zhang, H. et al. (2024). Large-scale atomic strain defects on palladium surfaces for enhanced oxygen reduction and zinc–air batteries. *Small* 20: 2406448.

19 Ji, D., Fan, L., Tao, L. et al. (2019). The Kirkendall effect for engineering oxygen vacancy of hollow CO_3O_4 nanoparticles toward high-performance portable zinc–air batteries. *Angew. Chem.* 131: 13978–13982.

20 Béjar, J., Delgado, A.D., Espinosa-Magaña, F. et al. (2022). Electrodeposition of small sized NiM_0O_4 spinels (M: Co, Mn) as bifunctional nanomaterials for rechargeable zinc–air batteries. *J. Alloys Compd.* 929: 167266.

21 Béjar, J., Espinosa-Magaña, F., Valdiviezo-Godina, N. et al. (2023). $CoMn_2O_4$ nanoparticles supported on defect-rich N-doped carbon nanotubes as air electrode in rechargeable zinc-air batteries. *J. Electroanal. Chem.* 947: 117754.

22 Ambriz-Peláez, O., Béjar, J., Ramos-Castillo, C.M. et al. (2022). Defected NiFe layered double hydroxides on N-doped carbon nanotubes as efficient bifunctional electrocatalyst for rechargeable zinc–air batteries. *Appl. Surf. Sci.* 601: 154253.

23 Ambriz-Peláez, O., Béjar, J., Delgado, A.D. et al. (2024). NiMn layered double hydroxides with promoted surface defects as bifunctional electrocatalysts for rechargeable zinc–air batteries. *FlatChem* 45: 100664.

24 Ramos-Castillo, C.M., Álvarez-Contreras, L., Arjona, N., and Guerra-Balcázar, M. (2024). Defect engineering of oxygen vacancies in ultrathin NiFe-layered double hydroxides: insights from density functional theory. *J. Phys. Chem. C* 128: 4161–4170.

25 Zhong, Y., Dai, J., Xu, X. et al. (2020). Facilitating oxygen redox on manganese oxide nanosheets by tuning active species and oxygen defects for zinc-air batteries. *ChemElectroChem* 7: 4949–4955.

26 Tatarchuk, T. (2024). Studying the defects in spinel compounds: discovery, formation mechanisms, classification, and influence on catalytic properties. *Nanomaterials* 14: 1640.

27 Dai, Y., Yu, J., Cheng, C. et al. (2020). Mini-review of perovskite oxides as oxygen electrocatalysts for rechargeable zinc–air batteries. *Chem. Eng. J.* 397: 125516.

28 de Souza Dias, G., Costa, J.M., and de Almeida Neto, A.F. (2023). Transition metal chalcogenides carbon-based as bifunctional cathode electrocatalysts for rechargeable zinc-air battery: an updated review. *Adv. Colloid Interface Sci.* 315: 102891.

29 Luo, F., Zhu, J., Ma, S. et al. (2021). Regulated coordination environment of Ni single atom catalyst toward high-efficiency oxygen electrocatalysis for rechargeable Zinc-air batteries. *Energy Storage Mater.* 35: 723–730.

30 Shang, N., Wang, C., Zhang, X. et al. (2021). Atomically dispersed iron on nitrogen-decorated carbon for high-performance oxygen reduction and zinc-air batteries. *Chem. Eng. J.* 426: 127345.

31 Dai, L., Xue, Y., Qu, L. et al. (2015). Metal-free catalysts for oxygen reduction reaction. *Chem. Rev.* 115: 4823–4892.

32 Liu, Y., Sun, K., Cui, X. et al. (2020). Defect-rich, Graphenelike carbon sheets derived from biomass as efficient electrocatalysts for rechargeable zinc-air batteries. *ACS Sustainable Chem. Eng.* 8: 2981–2989.

33 Preuss, K., Siwoniku, A.M., Bucur, C.I., and Titirici, M.M. (2019). The influence of heteroatom dopants nitrogen, boron, sulfur, and phosphorus on carbon electrocatalysts for the oxygen reduction reaction. *Chempluschem* 84: 457–464.

34 Zhang, Q. and Guan, J. (2021). Applications of atomically dispersed oxygen reduction catalysts in fuel cells and zinc–air batteries. *Energy Environ. Mater.* 4: 307–335.

35 Ortiz-Ortega, E., Díaz-Patiño, L., Bejar, J. et al. (2020). A flow-through membraneless microfluidic zinc-air cell. *ACS Appl. Mater. Interfaces* 12: 41185–41199.

36 Béjar, J., Espinosa-Magaña, F., Avelar, J. et al. (2023). Rational design of nitrogen–doped carbon nanotubes by defect engineering for Zn–air batteries with high performance. *Carbon N Y* 204: 411–426.

37 Wang, T., Huang, J., Sang, W. et al. (2024). Correlative Mn-Co catalyst excels Pt in oxygen reduction reaction of quasi-solid-state zinc-air batteries. *Nano Res.* 17: 4118–4124.

38 Manna, N., Singh, S.K., Kurian, M.., Torris, A., and Kurungot, S., Air-cathode interface-engineered electrocatalyst for solid-state rechargeable zinc-air batteries, *ACS Appl. Energy Mater.* 5 (2022) 8756–8768. https://doi.org/10.1021/ACSAEM.2C01266/ASSET/IMAGES/LARGE/AE2C01266_0006.JPEG

39 Wang, Q., Feng, Q., Lei, Y. et al. (2022). Quasi-solid-state Zn-air batteries with an atomically dispersed cobalt electrocatalyst and organohydrogel electrolyte. *Nat. Commun.* 13: 1–10.

40 Chen, Y., He, T., Liu, Q. et al. (2023). Highly durable iron single-atom catalysts for low-temperature zinc-air batteries by electronic regulation of adjacent iron nanoclusters. *Appl. Catal., B* 323: 122163.

22

Addressing Surface and Interfacial Defects in Lithium–Sulfur Batteries

Alexander Suárez-Barajas[1] and Noé Arjona[2]

[1]*Centro de Investigación y Desarrollo Tecnológico en Electroquímica, Departamento de posgrado, Parque Tecnológico Querétaro S/N, SanFandila, Pedro Escobedo, Querétaro, C.P. 76703, México*
[2]*Centro de Investigación en Materiales Avanzados S.C., Departament of Materials Physics, Complejo Industrial Chihuahua, Chihuahua, Chihuahua, C.P. 31136, México*

22.1 Introduction

Over the past decades, lithium-ion batteries (LIBs) have transformed energy storage technologies, offering a balance between high energy density, long cycle life, and fast-charging capability in a compact and lightweight design. These attributes have positioned LIBs as the dominant energy storage solution across a wide spectrum of applications, ranging from portable electronics, such as smartphones and laptops, to electric vehicles (EVs) and large-scale grid energy storage systems. Their pivotal role in advancing e-mobility and powering next-generation consumer electronics underscores their technological significance [1].

As global energy consumption increasingly shifts toward electrification, the demand for sustainable and efficient energy storage systems has grown exponentially. LIBs have become a cornerstone in this transition, owing to their ability to store significant amounts of energy efficiently while adapting to diverse technological applications [2a]. However, despite their widespread adoption, LIBs are approaching their theoretical energy density limit of approximately 387 Wh/kg, with a nominal operating voltage of 3.8 V [2b]. Current commercial LIBs exhibit energy densities in the range of 250–300 Wh/kg, which restricts their potential for advanced applications, such as EVs with driving ranges exceeding 500 km [2].

To address these limitations, the search for alternative battery chemistries has intensified, with lithium–sulfur (Li–S) batteries (LSBs) emerging as one of the most promising candidates. These batteries offer a theoretical energy density of 2600 Wh/kg, nearly seven times that of LIBs, leveraging the unique properties of sulfur as an abundant, low-cost, and environmentally friendly material [3]. However, the commercialization of LSBs remains hindered by significant technical challenges.

This chapter provides an in-depth exploration of the challenges and opportunities associated with LSBs, focusing on the role of surface and interfacial engineering in

Surface and Interfacial Defects in Nanomaterials for Sustainable Energy Production and Storage, First Edition. Edited by Noé Arjona, Lorena Álvarez Contreras, and Minerva Guerra Balcázar.

overcoming their intrinsic limitations. It delves into atomistic-level strategies, such as single-atom catalysts (SACs), heterostructured materials, and defect engineering to enhance the performance and stability of Li-S systems.

22.2 Lithium–Sulfur Batteries: Benefits and Mechanisms

22.2.1 Fundamental Advantages

LSBs have emerged as a frontrunner due to their exceptional theoretical energy density of 2567 Wh/kg and high specific capacity of 1675 mAh/g, far surpassing those of LIBs [4a]. This superior performance is attributed to their unique chemistry, which involves complete conversion reactions and eight-electron transfers. Sulfur, as the active cathode material, is both abundant and inexpensive, with an estimated cost of $100–150 per ton, significantly cheaper than cobalt used in LIBs ($24,000–26,000 per ton). This combination of high capacity and cost-efficiency positions LSBs as a sustainable alternative for applications requiring high energy outputs [4b]. The reaction mechanisms underpinning LSBs can be summarized as follows [4c]:

Phase I: Formation of Soluble Polysulfides

$$S_8 + 2Li^+ + 2e^- \leftrightarrow Li_2S_8$$

$$Li_2S_8 + 2Li^+ + 2e^- \leftrightarrow 2Li_2S_4$$

Phase II: Conversion to Solid Polysulfides (Li_2S_2)

$$Li_2S_4 + 2Li^+ + 2e^- \leftrightarrow 2Li_2S_2$$

Phase III: Formation of Lithium Sulfide (Li_2S)

$$Li_2S_2 + 2Li^+ + 2e^- \leftrightarrow 2Li_2S$$

This stepwise reduction, involving soluble and solid polysulfide intermediates, contributes to the high specific capacity of LSBs.

22.3 Challenges in Lithium–Sulfur Batteries

Despite these advantages, the commercialization of LSBs is hindered by several intrinsic challenges. The high volumetric expansion of sulfur during lithiation, the dissolution of polysulfides into the electrolyte, and the resulting shuttle effect leads to capacity fading and reduced cycle life. Additionally, the use of lithium metal anodes introduces safety concerns due to dendrite formation and the potential for short circuits [5]. Figure 22.1 illustrates a schematic representation of these challenges, highlighting the fundamental obstacles faced in the commercialization of LSBs [5a].

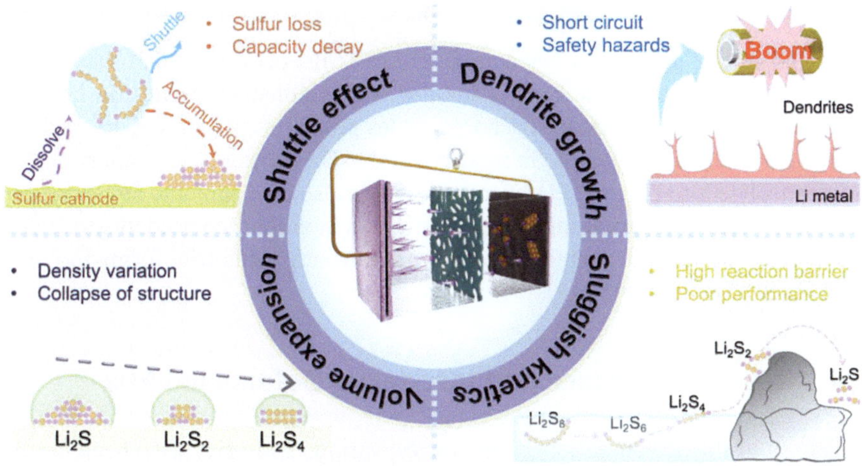

Figure 22.1 Key challenges to overcome in Li–S batteries. Source: Lu et al. [5a] / with permission of Royal Society of Chemistry.

22.3.1 Volumetric Expansion of Sulfur

One of the primary challenges in LSBs is the significant volumetric expansion of sulfur during lithiation. Sulfur molecules transform into lithium sulfide (Li_2S) during discharge, resulting in an 80% increase in volume. This expansion introduces substantial mechanical stress, leading to cracks in the electrode structure and the loss of electrical contact between the active material and the conductive matrix. Over successive cycles, this mechanical degradation causes significant capacity fading and reduces the long-term stability of the battery [5b].

22.3.2 Low Conductivity of Sulfur

Sulfur and its final discharge product, lithium sulfide, exhibit inherently low electrical conductivities ($\sim 5 \times 10^{-30}$ S/cm for sulfur and $\sim 1 \times 10^{-13}$ S/cm for Li_2S). This hinders electron transport and diminishes reaction kinetics. The integration of conductive materials, such as graphene, carbon nanotubes, and doped polymers into cathodes, addresses this issue, improving electronic pathways and facilitating efficient charge transfer [5b].

22.3.3 Polysulfide Dissolution and the Shuttle Effect

The intermediate lithium polysulfides (Li_2S_x, $4 \leq x \leq 8$) formed during cycling are highly soluble in conventional organic electrolytes. These dissolved polysulfides migrate between the cathode and anode, creating a "shuttle effect." The shuttle effect results in parasitic reactions at the lithium anode, leading to the continuous consumption of active materials and electrolytes. This process causes severe capacity decay and low Coulombic efficiency, significantly limiting the practical application of LSBs [5a].

22.3.4 Lithium Metal Anode Challenges

The lithium metal anode is a key component of LSBs, providing high theoretical capacity. However, its application is fraught with challenges. Dendritic lithium growth during repeated charge–discharge cycles poses severe safety risks, including internal short circuits and thermal runaway. The unstable formation of the solid–electrolyte interphase (SEI) layer at the anode also contributes to low Coulombic efficiency and cycle instability. These challenges highlight the need for strategies that stabilize the lithium metal surface while preventing dendrite formation and improving SEI durability [6].

22.4 Impact of Surface and Interfacial Defects in LSBs

In response to the aforementioned challenges, various strategies have been developed to enhance the performance and stability of LSBs. Recent advancements involve a combination of approaches targeting cathode structure optimization, anode design refinement, and the development of innovative materials for separators and electrolytes.

To address the low conductivity of sulfur, researchers have utilized porous carbon matrices, such as graphene and carbon nanotubes, to host sulfur within the cathode. These matrices provide a conductive network that enhances electron transport and facilitates the electrochemical conversion of polysulfides [5b].

The development of separators coated with conductive or catalytic materials has significantly mitigated the shuttle effect. These functional separators are capable of adsorbing soluble polysulfides and catalyzing their conversion to Li_2S, thereby preventing their migration to the anode and improving the overall efficiency of the battery [5c].

Advances in nanomaterial-based anodes have enabled controlled lithium deposition and reduced dendrite formation. Anodes coated with protective layers, such as nitrogen-doped carbon, or incorporating nanoporous structures that promote uniform lithium deposition, have enhanced both long-term cycling stability and the safety of LSBs [5a].

Surface and interfacial engineering have emerged as a critical approach. Innovations include the introduction of defect-engineered materials, such as sulfur hosts with oxygen vacancies or heterostructures combining catalytic and adsorptive properties [7a]. Dual-defect strategies, where vacancies and dopants are simultaneously incorporated, have been shown to significantly improve the adsorption and catalytic conversion of polysulfides. For example, oxygen vacancy-enriched materials accelerate Li^+ diffusion and enhance polysulfide redox kinetics, thereby reducing shuttle effects and improving sulfur utilization [7b–d].

Furthermore, the application of heteroatom-doped materials, such as N- or P-doped carbons, has demonstrated enhanced conductivity and catalytic activity, creating additional active sites for sulfur redox reactions. The integration of these advanced materials into LSB components, such as cathodes, separators, and anodes, represents a promising direction for achieving stable, high-performance LSBs [8].

22.5 The Effect on Sulfur Cathodes in Li–S Batteries

In LSBs, surface and interfacial defects in the sulfur cathode are critical for enhancing electrochemical performance by addressing challenges such as the poor conductivity of sulfur and the polysulfide shuttle effect. Structural defects, including vacancies and heterostructures, introduce additional active sites that improve the adsorption and conversion of lithium polysulfides (LiPS), accelerating redox kinetics [7d, 9a]. For instance, sulfur vacancies in materials like VS_2 facilitate lithium-ion diffusion and improve cathode stability during prolonged charge–discharge cycles [9b].

Heteroatom doping, such as nitrogen or transition metals, modifies electronic properties and enhances cathode polarity, enabling stronger interactions with LiPS. Similarly, heterostructures combining materials with high adsorption capacity and catalytic activity achieve a balanced conversion of LiPS, mitigating capacity losses linked to the shuttle effect [7d, 8b].

Additionally, atomistic designs using single-atom catalysts (SACs) or dual-atom catalysts (DACs) improve atom utilization and catalytic activity, enabling efficient polysulfide conversion with minimal metal loading. These developments highlight the importance of defect engineering in maximizing cathode efficiency, ensuring long-term stability and facilitating the commercialization of LSBs [10].

22.5.1 Heterostructures: Enhancing Redox Kinetics

Heterostructures, which combine distinct materials with complementary properties, have shown remarkable potential in improving the catalytic performance of sulfur cathodes. For instance, ZnO–ZnS heterostructures integrated with reduced graphene oxide (rGO) balance strong polysulfide adsorption and high catalytic activity. ZnO chemically binds LiPS, while ZnS accelerates their conversion to Li_2S, effectively suppressing the shuttle effect. The conductive rGO framework further enhances electron transport, resulting in improved sulfur utilization and long-term cycling stability [7d, 9a].

Another significant development involves $Co_{5.47}N$-Fe_3N heterostructures supported by nitrogen-doped carbon nanotube–graphene hybrids. The synthesis of these heterostructures, achieved through a one-step in situ method, is depicted in Figure 22.2. These heterostructures exploit high catalytic activity at heterointerfaces to enhance the adsorption and conversion of LiPS. Density functional theory (DFT) calculations reveal that heterostructures increase the density of states at the Fermi level, lowering energy barriers for redox reactions and boosting overall performance [7d].

22.5.2 Vacancies: Unlocking Active Sites

Surface vacancies, such as sulfur and oxygen vacancies, create highly reactive sites that enhance polysulfide adsorption and conversion. Sulfur vacancies in vanadium disulfide (VS_2) serve as lithium-ion reservoirs, promoting ion diffusion

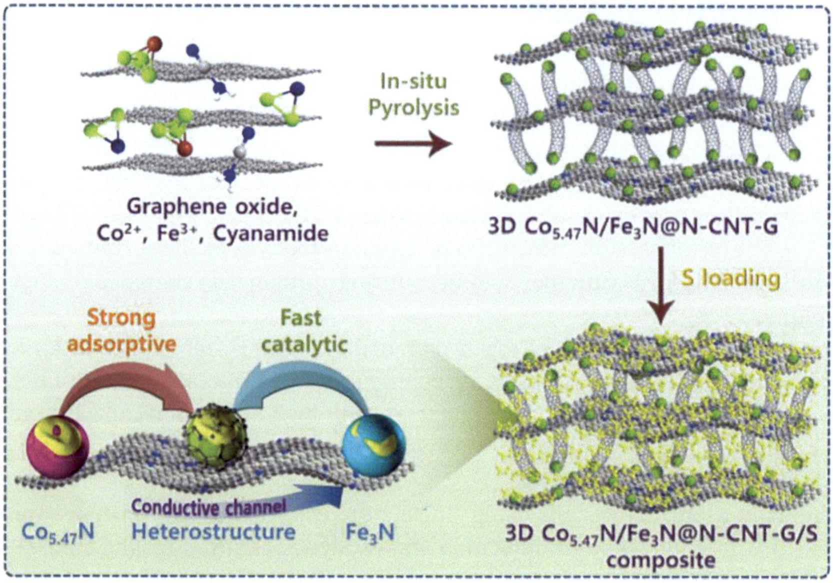

Figure 22.2 In-situ synthesis of 3D $Co_{5.47N}/Fe_3N$@N-CNT-G/S composites for Li–S applications. Source: Nguyen et al. [7d] / with permission of Elsevier.

and accelerating sulfur redox reactions. These vacancies also enhance lithiation behavior during discharge, thus improving the catalytic activity and stability of the cathode [9b].

Oxygen vacancies in materials like Nb_2O_5 increase active site exposure and improve the adsorption and conversion of LiPS. Dual-defect engineering, which combines oxygen vacancies with dopants like phosphorus, creates synergistic effects that further enhance catalytic activity, enabling high specific capacities and excellent cycling stability [7a].

22.5.3 Doping: Tailoring Electronic Properties

Heteroatom doping introduces controlled modifications to the electronic structure of cathode materials, enhancing their chemical reactivity and stability. Nitrogen doping in carbon frameworks, for example, increases polarity and conductivity, enabling stronger chemical interactions with LiPS. In ZIF-67-derived sulfur hosts, nitrogen-doped composites demonstrate superior catalytic activity and reversible capacity by promoting LiPS adsorption and conversion [8b].

Transition metal doping is another effective strategy for improving cathode performance. Cobalt-doped MoS_2 introduces sulfur vacancies and promotes a phase transition from semiconducting to metallic states, significantly enhancing electron transfer and catalytic activity. Cathodes with Co-doped MoS_2 exhibit higher rate capabilities and longer cycle life, underscoring the impact of targeted doping strategies [7d, 11].

22.5.4 Single-Atom Catalysts: Redefining Efficiency

Single-atom catalysts (SACs) represent a transformative advancement in the design of materials for LSBs, particularly in addressing the polysulfide shuttle effect and improving catalytic efficiency. SACs feature isolated metal atoms dispersed on a conductive substrate, maximizing catalytic activity and atomic utilization. Their well-defined active sites, coupled with their ability to manipulate local electronic environments, make SACs highly effective for facilitating polysulfide conversion and stabilizing sulfur redox processes Figure 22.3 illustrates the structure of Fe-N_4 embedded in Fe-N/MHCS nanoreactors, which exhibit exceptional performance in terms of specific energy and lifespan for LSBs [10b].

The atomic dispersion of metal atoms in SACs ensures that nearly all metal atoms are accessible as active sites, unlike conventional catalysts where only surface atoms participate in reactions. This high utilization dramatically enhances catalytic efficiency. For instance, single-nickel atoms embedded in nitrogen-doped porous carbon frameworks have been shown to effectively catalyze the conversion of LiPS to insoluble Li_2S. The nitrogen-doping introduces strong metal-support interactions, stabilizing the nickel sites and promoting their catalytic performance [10a].

By modifying the electronic structure of the substrate and the surrounding environment, SACs significantly lower the energy barriers for sulfur redox reactions. For example, SACs featuring cobalt atoms embedded in a nitrogen-doped graphene matrix demonstrate accelerated reaction kinetics by creating electron-rich catalytic centers. DFT calculations confirm that these centers increase the density of states near the Fermi level, enhancing electron transfer and stabilizing polysulfide intermediates during cycling [10c].

The recent evolution of SACs into dual-atom catalysts (DACs) further amplifies their catalytic efficiency. DACs leverage the synergistic effects between two different metal atoms, each contributing distinct catalytic properties. For example,

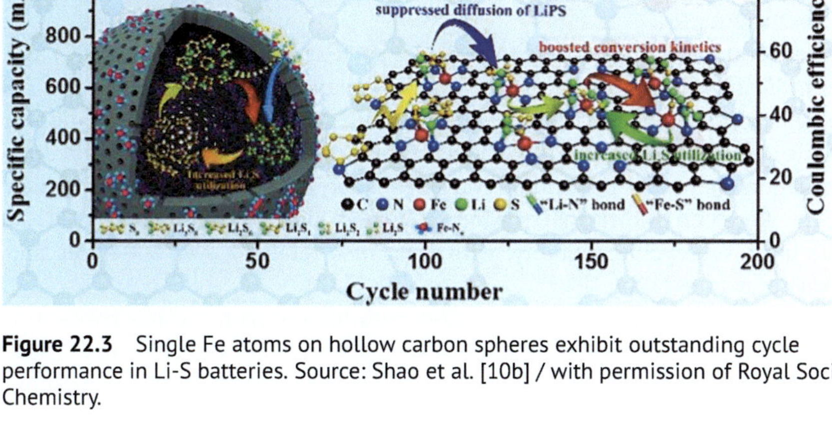

Figure 22.3 Single Fe atoms on hollow carbon spheres exhibit outstanding cycle performance in Li-S batteries. Source: Shao et al. [10b] / with permission of Royal Society of Chemistry.

nickel-cobalt DACs demonstrate improved polysulfide adsorption and faster redox kinetics due to complementary catalytic functionalities. These systems lower the energy barriers for polysulfide transformation and promote uniform sulfur deposition, enhancing the battery's cycling stability [10a].

22.6 Effects of Surface Defects on Separators and Their Role in Addressing Li–S Battery Challenges

The introduction of surface and interfacial defects enhances the functionality of separators by creating active sites that improve polysulfide adsorption, catalyze their conversion, and facilitate ion transport. The shuttle effect, caused by the migration of soluble lithium polysulfides LiPS between the cathode and anode, is mitigated through the incorporation of materials with tailored defects, which trap LiPS and accelerate their transformation into insoluble Li_2S [12].

Defects improve separator performance by modifying the chemical and electronic properties of the separator coatings [13a]. They create strong interactions with LiPS, enabling higher sulfur utilization and more stable cycling performance. Furthermore, advanced designs such as SACs provide atomically precise active sites that enhance polysulfide conversion kinetics and suppress unwanted side reactions. These strategies collectively enable the design of high-performance LSBs that address the inherent challenges of this technology. Batteries incorporating SACs have exhibited high Coulombic efficiency and stable cycling performance over extended periods [13b].

These innovations underscore the pivotal role of defect engineering in separator design, offering a comprehensive solution to the fundamental limitations of LSBs and paving the way for their commercialization.

22.6.1 Engineering Vacancies in Separator Materials

Vacancies, the intentional removal of atoms from a material's structure, serve as active sites that enhance the interaction between separators and polysulfides. These defects significantly improve polysulfide adsorption, catalytic conversion, and redox reaction efficiency.

Oxygen vacancies in transition metal oxides such as tungsten oxides (WO_x) and titanium dioxide (TiO_2) play a pivotal role in enhancing separator performance. These defects provide strong binding sites for LiPS, preventing their migration to the anode and mitigating the shuttle effect. Furthermore, oxygen vacancies reduce the activation energy for polysulfide conversion, accelerating their transformation into Li_2S. Research shows that separators coated with oxygen-deficient TiO_{2-x} achieve improved cycling stability, with capacities retained over 500 cycles and minimal decay rates [12b].

Similarly, tungsten oxide-based separators with high oxygen vacancy concentrations demonstrate improved polysulfide adsorption and catalytic efficiency. These materials achieve high initial capacities of 1169 mAh/g at 0.2C while maintaining stable cycling performance [8a].

Sulfur and selenium vacancies in transition metal dichalcogenides create additional active sites for LiPS interaction. Sulfur-deficient vanadium disulfide (SV-VS_2) enhances lithium-ion storage capacity and improves redox kinetics by facilitating faster reactions. Batteries with SV-VS_2-modified separators exhibit enhanced cycling stability, with capacity decay rates as low as 0.043% per cycle over 880 cycles [9b].

Selenium vacancies, particularly in Co_9Se_8-based systems, further enhance separator performance. These vacancies improve the catalytic and adsorptive properties of separators, enabling effective trapping and conversion of LiPS, which stabilizes the anode and extends battery life [12a].

22.6.2 Doping to Enhance Separator Functionality

Doping, the introduction of heteroatoms into a material's structure, modifies its electronic properties, improving its conductivity and interaction with polysulfides. This strategy has shown significant promise in optimizing separator performance.

Nitrogen doping in carbon-based frameworks significantly enhances polysulfide adsorption and conversion. When integrated with oxygen-deficient TiO_2, nitrogen-doped carbon materials improve electrolyte wettability and ionic conductivity, facilitating more efficient polysulfide trapping and transformation. This results in enhanced cycling stability and reduced shuttle effects [8a].

Dual-metal doping in transition metal oxides, such as Ni and Co in molybdenum oxides (NiCo-MoOx), creates synergistic effects that optimize separator functionality. NiCo-MoOx separators provide dual-active sites: Ni binds polysulfides strongly, while Co accelerates their redox conversion. These separators achieve high discharge capacities and prolonged lifespans, with a decay rate of only 0.049% per cycle [13a]

22.6.3 Heterostructures for Synergistic Performance

Heterostructures, which combine materials with complementary properties, offer a balanced approach to addressing the shuttle effect and improving polysulfide conversion. Advanced materials such as ZnO–ZnS heterostructures incorporated into separators provide selective ion transport, blocking polysulfides while allowing lithium ions to pass through. These separators significantly reduce self-discharge rates and capacity fading, improving overall efficiency and longevity. Batteries using ZnO–ZnS/rGO separators exhibit exceptional performance, delivering initial capacities of 1186 mAh/g and maintaining 671 mAh/g over 500 cycles with a high sulfur loading of 7.6 mg/cm^2 [9a].

Hollow MoS_2-Co_9S_8 structures leverage their high conductivity and abundant active sites for polysulfide adsorption and catalytic conversion. These separators achieve reversible specific capacities of 930.4 mAh/g at 0.5C and retain 72% of their capacity after 500 cycles. This combination of properties enhances sulfur utilization and cycle stability [14].

Bi_2S_3-MoS_2 heterostructures supported on MXene nanosheets optimize electronic redistribution at their interfaces, creating a built-in electric field that enhances polysulfide conversion and mitigates the shuttle effect. Batteries incorporating these separators deliver remarkable capacities of 1306 mAh/g at 0.2C, with minimal degradation over extended cycles [15].

22.6.4 Single-Atom Catalysts (SACs) for Precision Catalysis

Single-atom catalysts (SACs) represent a cutting-edge solution for separator modification, offering atomically dispersed active sites for precise catalysis of polysulfide conversion.

As illustrated in Figure 22.4, SACs anchored on graphene foam separators, such as Fe-SACs, stabilize intermediate polysulfides and promote their catalytic conversion. This design allows high-capacity retention and exceptional cycling stability even under high sulfur loading and low electrolyte-to-sulfur (E/S) ratios. For instance, Fe-SAC-modified separators achieve a discharge capacity of 891.6 mAh/g after 750 cycles at 0.5C, with a retention rate exceeding 83% [13b]. The unique structure

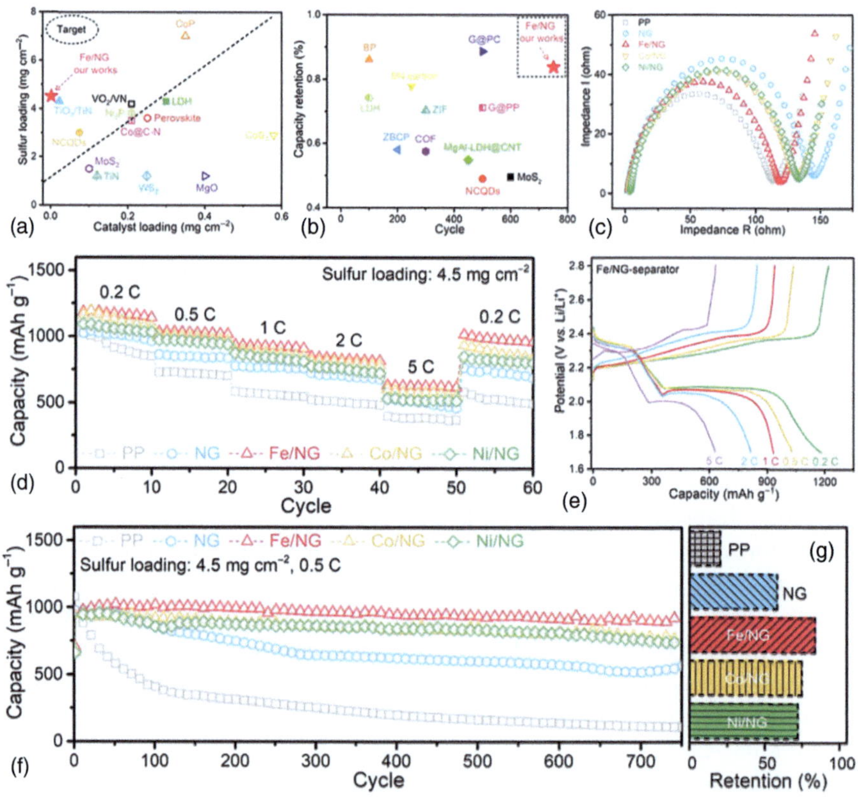

Figure 22.4 Performance evaluation of Li–S batteries with SAC-enhanced separators: (a) Catalyst and sulfur loading, (b) Cycling stability and capacity retention in various Li–S cells. (c) Electrochemical impedance spectroscopy (EIS), (d) Rate capability, (e) Charge–discharge profiles, (f) Long-term cycling performance and (g) Capacity retention after 750 cycles; at various rates using Fe1/NG-modified separators. Source: Zhang et al. [13b] / with permission of American Chemical Society.

of SACs ensures that almost all the embedded metal atoms actively participate in catalytic reactions, unlike traditional catalysts where surface atoms dominate reactivity. Nickel-based SACs supported on nitrogen-doped carbon demonstrate superior catalytic activity compared with cobalt and iron SACs. These catalysts facilitate complete polysulfide conversion, reducing parasitic reactions and improving cycle stability. Ni-SAC-modified separators achieve sulfur utilization with capacities exceeding 1280 mAh/g over 800 cycles [10a].

Innovations in SAC design include dual single-atom systems combining Ni and Co. These systems leverage the complementary properties of each atom, enhancing both polysulfide adsorption and conversion capabilities. This dual-SAC approach reduces the energy barrier for LiPS transformation, suppresses the shuttle effect, and improves overall battery performance [10a].

22.7 Surface and Interfacial Defects in Lithium Metal Anodes for Li–S Batteries

The lithium metal anode is integral to achieving the high energy density of LSBs, but its practical application faces numerous challenges. These include dendritic growth during cycling, unstable SEI formation, and parasitic reactions between lithium and the electrolyte. The inclusion of surface and interfacial defects – achieved through chemical doping, engineered vacancies, and catalytic surface modifications – has emerged as a transformative strategy to overcome these limitations. These defects provide active sites that enhance ionic transport, regulate lithium deposition, and stabilize the SEI, significantly improving the safety, cycling stability, and overall performance of LSBs [6a, 7b, 16].

22.7.1 Protective Coatings and Interlayers

Protective coatings and interlayers incorporating engineered defects have proven to be highly effective in addressing the challenges associated with lithium metal anodes. These materials act as both physical barriers and chemical modifiers for reducing parasitic reactions and promoting uniform lithium deposition.

Oxygen vacancies in protective coatings, such as those in P-Mn_3O_{4-x}, create lithiophilic sites that enhance the interaction between lithium ions and the protective layer, as illustrated in Figure 22.5 [7b]. These vacancies not only reduce overpotential but also guide uniform lithium nucleation for mitigating dendritic growth. Experimental studies demonstrate that batteries using oxygen vacancy-enriched protection layers achieve cycling lifespans exceeding 5000 hours with negligible capacity decay. This improvement is attributed to the defect sites acting as high-energy adsorption centers that redistribute lithium-ion flux during deposition [7b, 10b].

Graphene and nitrogen-doped graphene coatings with engineered vacancies provide both structural stability and enhanced electrochemical performance. These materials act as flexible interlayers that suppress dendritic penetration and ensure

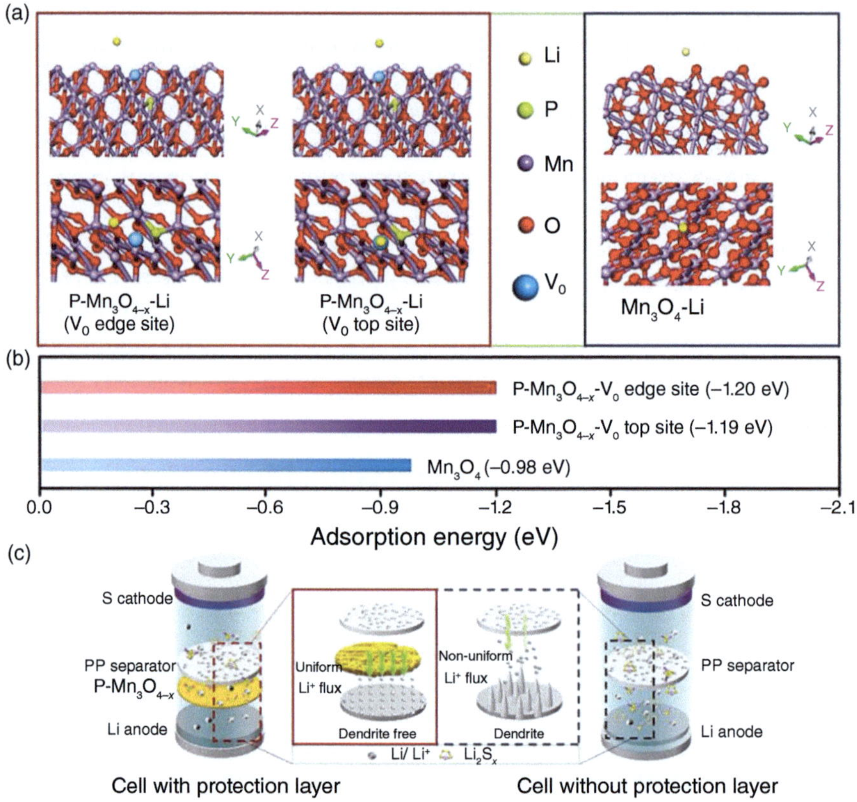

Figure 22.5 Adsorption energies and lithium deposition behaviors using protective P-Mn_3O_{4-x} layers for enhanced lithium–sulfur battery performance. (a) Li binding energies on P-Mn_3O_{4-x} and Mn_3O_4. (b) Comparison of adsorption energies. (c) Schematic of Li deposition on protected vs. bare anode. Source: Liu et al. [7b] / with permission of Elsevier.

stable cycling by homogenizing the electric field and lithium-ion flux. Additionally, these coatings reduce direct contact between the lithium anode and the electrolyte, thereby limiting the extent of parasitic reactions that degrade the SEI layer [6a, 10a].

Interlayers with self-healing properties, such as poly(ether-urethane)-based solid-state electrolytes, further enhance lithium metal anode stability. These polymers, designed with dynamic cross-linking bonds, enable the layer to recover from mechanical stress caused by volume changes during cycling. The inclusion of engineered defects in these interlayers enhances ionic conductivity and reduces resistance at the lithium interface, thereby supporting extended cycling performance at high current densities [6b, 16].

22.7.2 Stable SEI Formation

The formation of a stable SEI is critical for the long-term performance and safety of LSBs. The SEI serves as a passivation layer, preventing direct contact between the electrolyte and the lithium metal while facilitating ion transport. However, traditional SEI layers are often uneven, mechanically fragile, and chemically unstable, leading to poor cycling stability and parasitic reactions. Surface and interfacial defects provide a pathway to address these issues by catalyzing the formation of robust and uniform SEI layers.

Defects such as oxygen vacancies and sulfur vacancies act as catalytic sites for SEI formation. For example, oxygen vacancies in TiO_2-coated separators accelerate the decomposition of electrolyte components to form a uniform SEI. These catalytically active sites promote the creation of a chemically stable and ionically conductive SEI layer, reducing interfacial resistance and improving Coulombic efficiency [7b].

Surface functionalization of lithium anodes with dopants, such as nitrogen or phosphorus, further stabilizes the SEI. These dopants alter the electronic structure of the anode surface, enhancing the chemical interaction between lithium and SEI components. Nitrogen-doped carbon layers, for instance, encourage the formation of a chemically robust SEI while minimizing side reactions with the electrolyte. This results in significantly improved cycling stability and extended battery lifespan [10a].

22.7.3 Non-Dendritic Lithiation and Uniform Lithium Deposition

Dendritic growth is one of the most critical issues in lithium metal anodes, as it poses safety risks and reduces battery efficiency. The inclusion of surface and interfacial defects plays a pivotal role in enabling non-dendritic lithiation and ensuring uniform lithium deposition.

Engineered vacancies in interlayers and coatings create localized electric fields that guide lithium-ion fluxes during deposition. Oxygen vacancies in materials like Mn_3O_{4-x} and TiO_2 not only improve lithium-ion adsorption but also ensure even distribution of lithium ions across the anode surface. This prevents the formation of high-concentration regions that can lead to dendritic growth. As a result, these materials enable uniform and non-dendritic lithium plating, enhancing both safety and cycling performance [7b, 8a, 10b].

SACs, such as nickel or iron atoms embedded in nitrogen-doped carbon matrices, offer precise control over lithium-ion deposition. These single atoms serve as active sites that catalyze lithium nucleation and regulate growth kinetics. SACs also suppress the formation of high-energy dendritic structures by redistributing lithium-ion fluxes, resulting in a smooth and dendrite-free surface. Studies show that SAC-modified anodes achieve high Coulombic efficiency and stable cycling over extended periods [10].

Dual-defect strategies that combine vacancies with dopants have demonstrated synergistic effects in suppressing dendritic growth. For example, materials with oxygen vacancies and nitrogen doping exhibit enhanced ionic diffusion and controlled lithium deposition, ensuring uniform growth patterns. This approach significantly reduces the risk of short circuits and improves overall battery safety [10c].

22.8 Conclusions and Future Perspectives

The strategic integration of structural and chemical defect engineering in LSBs has proven to be instrumental in addressing the fundamental challenges associated with this energy storage technology. Techniques such as oxygen vacancy engineering, heterostructure development, and SAC incorporation have demonstrated significant improvements in polysulfide adsorption, redox kinetics, and overall cycling stability. Structural modifications, particularly oxygen vacancies, enhance the catalytic activity of interlayers by lowering the energy barriers for sulfur species transformation. These modifications ensure efficient lithium-ion transport while stabilizing lithium deposition, thereby mitigating dendritic growth and enhancing safety. SACs, including Fe-N_4 and Ni-based systems, exhibit remarkable catalytic precision, leveraging their well-dispersed atomic sites to promote polysulfide conversion while minimizing electrolyte decomposition.

Looking forward, the scalability of defect engineering techniques must be prioritized to enable large-scale production without compromising material integrity or performance. Advanced synthesis methods, such as in situ atomic layer deposition, hold promise for precise defect modulation at a commercially viable scale. Furthermore, the integration of computational tools like DFT will remain critical for tailoring material properties to achieve synergistic effects between defects, dopants, and substrates.

Future research should also emphasize the development of environmentally benign and cost-effective materials, particularly for separators and interlayers. Multifunctional separators combining high ionic conductivity with catalytic and adsorptive capabilities could further suppress the polysulfide shuttle effect and enhance Coulombic efficiency. Additionally, exploring alternative chemistries and interfacial modifications to stabilize the SEI under prolonged cycling conditions will be essential. The pursuit of these advancements will bridge the gap between laboratory-scale innovations and commercial deployment, thus ensuring that LSBs achieve their full potential as sustainable and high-capacity energy storage systems. This progress will ultimately support the transition toward greener energy solutions in transportation and grid applications.

References

1 (a) Koech, A.K., Mwandila, G., Mulolani, F., and Mwaanga, P. (2024). Lithium-ion battery fundamentals and exploration of cathode materials: a review. *S. Afr. J. Chem. Eng.* (Elsevier B.V.) 50: 321–339. (b) Zhang, J., Huang, H., Zhang, G. et al. (2024). Cycle life studies of lithium-ion power batteries for electric vehicles: a review. *J. Energy Storage.* (Elsevier Ltd) 50: 1–21, 112231. (c) Suganya, R., Joseph, L.M.I.L., and Kollem, S. (2024). Understanding lithium-ion battery management systems in electric vehicles: environmental and health impacts, comparative study, and future trends: a review. results in engineering. *Results Eng.* (Elsevier B.V.) 24: 1–18, 103047.

2 (a) Chen, T., Lai, X., Chen, F. et al. (2024). Empowering lithium-ion battery manufacturing with big data: current status, challenges, and future. *J. Power Sources*. (Elsevier B.V.) 623: 1–13, 235400. (b) Xu, J., Cai, X., Cai, S. et al. (2023). High-energy lithium-ion batteries: recent progress and a promising future in applications. *Energy Environ. Mater.* (John Wiley and Sons Inc) 6: 1–26, e12450. (c) Huang, S., Wang, Z., Von Lim, Y. et al. (2021). Recent advances in heterostructure engineering for lithium–sulfur batteries. *Adv. Energy Mater.* (John Wiley and Sons Inc) 11: 1–27, 2003689.

3 (a) Zheng, S., Khan, N., Worku, B.E., and Wang, B. (2024). Review and prospect on low-temperature lithium-sulfur battery. *Chem. Eng. J.* (Elsevier B.V.) 484: 1–14, 149610. (b) Ma, T., Yue, H., Xiao, Y. et al. (2025). A review of organic sulfur applications in lithium-sulfur batteries. *J. Power Sources*. (Elsevier B.V.) 625: 1–17, 235717. (c) Bandyopadhyay, S. and Nandan, B. (2023). A review on design of cathode, anode and solid electrolyte for true all-solid-state lithium sulfur batteries. *Mater. Today Energy*. (Elsevier Ltd) 31: 1–26, 101201.

4 (a) Han, S., Lee, J.H., Kim, J., and Lee, J. (2024). Recent advances in Li metal anode protection for high performance lithium-sulfur batteries. *Discov. Chem. Eng.* (Springer Nature) 4, 9: 1–46. (b) Wu, B., Chen, C., Danilov, D.L. et al. (2023). All-solid-state thin film Li-ion batteries: new challenges, new materials, and new designs. *Batteries*. (MDPI) 9: 1–49. 9. (c) Patel, M., Mishra, K., Banerjee, R. et al. (2023). Fundamentals, recent developments and prospects of lithium and non-lithium electrochemical rechargeable battery systems. *J. Energy Chem.* (Elsevier B.V.) 81: 221–259.

5 (a) Lu, M., Ding, Y., Xue, Z. et al. (2024). Insight into demand-driven preparation of single-atomic mediators for lithium–sulfur batteries. *J. Energy Chem.* (Elsevier B.V.) 81: 205–219. (b) Cha, E., Patel, M., Bhoyate, S. et al. (2020). Nanoengineering to achieve high efficiency practical lithium-sulfur batteries. *Nanoscale Horiz.* (Royal Society of Chemistry) 5: 808–831. (c) Nakamura, N., Ahn, S., Momma, T., and Osaka, T. (2023). Future potential for lithium-sulfur batteries. *J. Power Sources*. (Elsevier B.V.) 558: 1–15, 232566.

6 (a) Xiao, R., Chen, K., Zhang, X. et al. (2021). Single-atom catalysts for metal-sulfur batteries: current progress and future perspectives. *J. Energy Chem.* (Elsevier B.V.) 54: 452–466. (b) Liang, Z., Shen, J., Xu, X. et al. (2022). Advances in the development of single-atom catalysts for high-energy-density lithium–sulfur batteries. *Adv. Mater.* (Wiley-VCH) 34 (30): 1–12, 2200102.

7 (a) Qu, M., Bai, Y., Yang, W. et al. (2023). Simultaneous dual-defect engineering as a booster for polysulfide adsorption and conversion in Li-S battery. *Chem. Eng. J.* (Elsevier B.V.) 460: 1–10, 141723. (b) Liu, T., Li, J., Cui, H. et al. (2023). Tuning Li nucleation and growth via oxygen vacancy-enriched 3D flexible self-supporting protection layer of P-Mn3O4−x for advanced lithium-sulfur batteries. *J. Energy Chem.* (Elsevier B.V.) 76: 339–348. (c) Zhao, B.S., Zhou, C., Chen, P., and Gao, X.P. (2024). Synergistic interfacial optimization for high-sulfur-content all-solid-state lithium–sulfur batteries. *ACS Appl. Mater. Interfaces* 16 (4): 4679–4688. (d) Nguyen, T.T., Balamurugan, J., Go, H.W. et al. (2022). Dual-functional Co5.47N/Fe3N heterostructure interconnected

3D N-doped carbon nanotube-graphene hybrids for accelerating polysulfide conversion in Li-S batteries. *Chem. Eng. J.* (Elsevier B.V.) 427: 1–12, 131774.

8 (a) Zhou, J., Sun, S., Zhou, X. et al. (2024). Defect engineering enables an advanced separator modification for high-performance lithium-sulfur batteries. *Chem. Eng. J.* (Elsevier B.V.) 487: 1–8, 150574. (b) He, Q., Chen, W., Fan, B. et al. (2024). Heteroatom-doped ZIF-67 for anchoring and catalyzing polysulfides in lithium–sulfur batteries. *Chem. Eng. J.* (Elsevier B.V.) 496: 1–10, 153813.

9 (a) Zhang, W., Hong, D., Su, Z. et al. (2022). Tailored ZnO-ZnS heterostructure enables a rational balancing of strong adsorption and high catalytic activity of polysulfides for Li-S batteries. *Energy Storage Mater.* (Elsevier B.V.) 53: 404–414. (b) He, L., Zhang, X., Yang, D. et al. (2023). Defect-engineered VS_2 electrocatalysts for lithium-sulfur batteries. *Nano Lett.* (American Chemical Society [ACS]) 23 (16): 7411–7418.

10 (a) Chen, L., Sun, Y., Chang, Z. et al. (2024). Single-atomic nickel supported on nitrogen-doped porous carbon to boost polysulfide conversion in lithium-sulfur batteries. *Sci. China Mater.* (Springer Nature) 67 (6): 1938–1946. (b) Shao, Q., Xu, L., Guo, D. et al. (2020). Atomic level design of single iron atom embedded mesoporous hollow carbon spheres as multi-effect nanoreactors for advanced lithium-sulfur batteries. *J. Mater. Chem. A.* (Royal Society of Chemistry) 8 (45): 23772–23783. (c) Liu, Y., Qin, G., Zhang, H. et al. (2024). Atomically dispersed Co/Mn immobilized on O, N dual doped hollow carbon spheres as sulfur host for lithium sulfur batteries. *J. Alloys Compd.* (Elsevier) 1002, 1–12, 175487.

11 Liu, W., Luo, C., Zhang, S. et al. (2021). Cobalt-doping of molybdenum disulfide for enhanced catalytic polysulfide conversion in lithium-sulfur batteries. *ACS Nano.* (American Chemical Society) 15 (4): 7491–7499.

12 (a) Kang, X., Jin, Z., Peng, H. et al. (2023). The role of selenium vacancies functionalized mediator of bimetal (Co, Fe) selenide for high-energy–density lithium-sulfur batteries. *J. Colloid Interface Sci.* (Elsevier) 637: 161–172. (b) Song, Y., Li, H., Li, J. et al. (2023). Rationalizing the impact of oxygen vacancy on polysulfide conversion kinetics for highly efficient lithium-sulfur batteries. *J. Energy Chem.* (Elsevier) 87: 51–60.

13 (a) Chen, J., Lei, J., Zhou, J. et al. (2024). Polysulfides adsorption and catalysis dual-sites on metal-doped molybdenum oxide nanoclusters for Li-S batteries with wide operating temperature. *Nano Res.* (Springer Nature) 17 (4): 9651–9661. (b) Zhang, K., Chen, Z., Ning, R. et al. (2019). Single-atom coated separator for robust lithium-sulfur batteries. *ACS Appl. Mater. Interfaces* 11 (28): 25147–25154.

14 Jin, L., Huang, B., Qian, X. et al. (2022). Hollow bimetallic sulfide composite coated separator for hindering the shuttle of polysulfides in Li-S batteries. *J. Alloys Compd.* (Elsevier) 926: 1–10, 166515.

15 Li, Y., Wang, Z., Wei, C. et al. (2024). Boosting polysulfides conversion kinetics through heterostructure optimization and electrons redistribution for robust lithium-sulfur batteries. *Chem. Eng. J.* (Elsevier) 497: 1–9, 154658.

16 Pei, F., Wu, L., Zhang, Y. et al. (2024). Interfacial self-healing polymer electrolytes for long-cycle solid-state lithium-sulfur batteries. *Nat. Commun.* 15 (1): 1–10, 351.

23

Engineering Defects in Advanced Battery Systems

María Fernanda Bósquez-Cáceres[1,2], Juan P. Tafur[3], and Vivian Morera Córdova[2]

[1] *Yachay Tech University, School of Physical Sciences and Nanotechnology, Hacienda San José Urcuquí, 100115, Ecuador*
[2] *Yachay Tech University, Medical Chemistry Research Group. School of Chemical Sciences & Engineering, Hacienda San José, Urcuquí, 100115, Ecuador*
[3] *Universidad Politécnica de Madrid, G-Aqua Research Group, Department of Mechanical, Chemical and Industrial Design Engineering, ETSIDI, Ronda de Valencia 3, Madrid 28012, Spain*

23.1 Introduction to Advanced Battery Technologies

23.1.1 Overview of Current Battery Technologies

Technologies developed for batteries have been evolving over the last decades. Each offers advantages and limitations for different applications. Many of the limitations are related to the type of material used and its performance, i.e. limited charge and discharge rates [1]. As a starting point on the current landscape of energy storage systems, a comparison of the key metrics of the different battery technologies is presented (Table 23.1). It is worth mentioning that the values reported vary from author to author, and the table presented is intended to give a general overview of the metrics of each type of battery. Lead–acid batteries, known as one of the first rechargeable battery models, are now widely used because of their low cost. They are particularly used in automobile applications despite their low energy density [2]. Li-ion batteries, on the other hand, dominate electric vehicles and portable electronic devices due to their efficiency and high energy density. An intermediate solution is nickel-based batteries or nickel–metal hydride batteries (NiCd and NiMH). Since NiMH batteries offer an improvement in terms of energy density over NiCd batteries, as they dispense the toxic materials of the latter [3]. The main advantage of sodium-sulfur batteries is the abundance of the chemical element sodium as opposed to lithium, and they are used especially in batteries for energy storage in power grids [4]. Flow batteries, represented by vanadium oxide-reduction (VRFB) and zinc–bromine (Zn-Br) systems, offer advantages in terms of scalability by decoupling energy and power aspects, thus allowing operations with ranges of flexibility. Finally, metal-air batteries (Zn–air system in Table 23.1) offer high energy densities; however, this type of battery

Surface and Interfacial Defects in Nanomaterials for Sustainable Energy Production and Storage, First Edition. Edited by Noé Arjona, Lorena Álvarez Contreras, and Minerva Guerra Balcázar.

Table 23.1 Battery systems and general performance metrics.

Performance metrics	Pb-acid	Li-ion	NiCd	NiMH	NaS	VRFB	Zn-Br	Zn–air
Cell voltage (V)	2–2.1	2.5–5	1.2–1.3	1.2–1.35	1.8–2.71	1.2–1.4	1.8	1.45–1.65
Specific energy (Wh/kg)	25–50	80–250	30–80	40–110	150–240	10–130	50–80	350–500
Specific power (W/kg)	150–400	200–2000	80–300	200–300	90–230	50–150	50–150	70–100
Maximum discharge depth (%)	80	100	80	100	100	100	100	100
Efficiency (%)	63–90	75–97	60–90	50–80	75–90	75–90	65–75	50–65
Life cycles (cycles)	250–2000	100–10,000	1000–5000	300–1800	2500–4000	10,000–16,000	2000–3000	300–1000
Lifetime (years)	2–15	5–15	10–20	2–15	10–15	5–15	10–20	3–5
Cost of energy (€/kWh)	40–170	500–2100	680–1300	170–640	250–420	130–850	150–250	100–250
Cost of power (€/kW)	250–500	1000–3400	420–1300	200–470	850–2500	500–1300	700–2500	1000–2000
Operating temperature (°C)	18–45	20–65	−40–50	−30–70	300–350	5–45	20–50	−60–80

does not reach large-scale practical applications due to its low stability and because of cathode degradation problems [5]. In particular, solid-state batteries, still under development, promise improved safety and energy density through the use of solid electrolytes, opening the way as a future replacement for liquid-state systems [6].

23.1.2 Current Limitations in Battery Technologies

Modern battery technologies have certain limitations, mainly relating to the reduction of battery lifetime and overall performance. These limitations are essentially due to the low structural stability of the materials used to manufacture the electrodes and to grow the dendrites on the electrodes. Low structural stability is related to the volumetric changes that occur during charge–discharge cycles, which cause mechanical stress and potential failures. For example, with materials based on the use of silicon in the anode in Li-ion batteries, volumetric expansions exceeding 300% often occur. This results in loss of electrical contact and particulate electrode sputtering and thus low coulombic efficiency, pronounced capacity fading, and poor performance. In turn, the growth of dendrites in batteries based on the use of metals such as zinc and lithium is quite common. Dendrites develop as needle-like structures that form on the surface of the anode during the charging process, and over time, they can perforate the separator, causing short circuits, which reduce the lifetime of the battery or cause failures that threaten the safety of the environment.

Degradation of electrode and electrolyte materials poses challenges to the chemistry underlying the development of these components. In particular, complex interfacial reactions often lead to the formation of layers of a solid–electrolyte interface (SEI) and decomposition products that gradually consume the active battery materials and increase the internal resistance. Over the operating time, these interfaces often become thicker or unstable, thereby reducing ionic transport. On the other hand, battery charge–discharge rates are often limited by the kinetics of ion diffusion to and within the electrode, the electron transport capacity, and the conductivity of the material used as electrolyte. These factors limit the rate at which energy can be stored and released without compromising stability, thus decreasing battery performance in applications that demand fast charging or high-power outputs. Properly understanding and addressing these limitations through the use of advanced material engineering is crucial to developing next-generation battery technologies with improved performance and reliability.

23.1.3 Opportunities Offered by Defect Engineering

Material defects are irregularities or deviations from the ideal structure or composition of a material, which can affect its properties and performance. Defect engineering detects, measures, and analyzes defects in different types of materials such as metals, ceramics, polymers, and composites. It is through it that battery performance metrics can be improved by strategically manipulating material imperfections at the atomic and nanoscale levels. These defects, when strategically controlled, can act as facilitators of ionic transport, consequently improving the mobility of ions through the materials used as electrodes and electrolytes, increasing efficiency by impacting capacity rates and through more efficient charge transfer processes. On the other hand, defect engineering can increase electrode stability by creating more resilient structures that do not exhibit such large volume changes during battery cycling, mitigating the mechanical stress on the battery from repetitive contractions and expansions during cycling, thus reducing the rate of degradation over time. In addition, the cycling capability provides a positive effect through defect control, increasing the durability and lifetime of the battery. For the end user of electronic devices and large-scale energy storage systems, it is imperative that batteries maintain their structural integrity over hundreds or thousands of cycles. This is how the use of nanomaterials offers a powerful approach to controlling defect structures, given their high surface area and properties that can be shaped to improve the performance of the energy storage system in question. The next section of the chapter explores in detail the different types of defects, mechanisms of formation, and how they can be modified to optimize the performance of advanced battery systems.

23.2 Fundamentals of Defect Engineering in Batteries

23.2.1 Types of Defects in Battery Materials and Their Impact on Electrochemical Properties

Materials are not perfect. They often have irregularities that can affect their mechanical, physical, or chemical properties. Defects in materials are defined as deviations observed by experimental methods with reference to the structure described above. They break the periodicity and symmetry of the reticular structures. Depending on the dimension of the space occupied by the defect, they can be classified as point, linear, surface, or volumetric defects [7].

A point defect is any defect involving a single particle and sometimes a very small set of points. Point defects have atomic dimensions. They do not extend in space in any dimension. They can be said to be localized disturbances at the level of the position of individual atoms. These defects can be generated in the material by the movement of the atoms as they gain energy by heating. This type of defect, in the materials used in batteries, plays a crucial role in their final performance. These imperfections at the atomic level include the following types: vacancies, interstices, and substitutions (Table 23.2).

Depending on what causes them, point defects can be either intrinsic or extrinsic defects. Intrinsic defects are the result of energy fluctuations from atomic thermal vibrations. Known Schottky defects may occur because the atom moves out of the original lattice site and causes a vacancy, or Frenkel defects may occur as a result of the atom in the normal lattice position occupying the adjacent interstitial site (Figure 23.1). Extrinsic defects are caused by doping of heteroatoms. They are also called impurity defects and do not change the original matrix lattice but cause distortion.

Table 23.2 Type of point defects and characteristics.

Type of point defect	**Vacancies**	**Interstices**	**Substitutions**
Description	Simpler point defect. Occurs when one or more atoms are missing from their regular position in a crystal lattice.	Occur when additional atoms occupy spaces between lattice sites that are normally unoccupied.	Arise through the replacement of host atoms by different elements.
Impact on electrochemical properties	Create pathways through which ions move more easily, and, as a result, improve the ionic conductivity of the material.	Block or facilitate ion transport depending on the nature of the atom.	Alter the stability of the material and modify its electrochemical properties.
Graphical representation	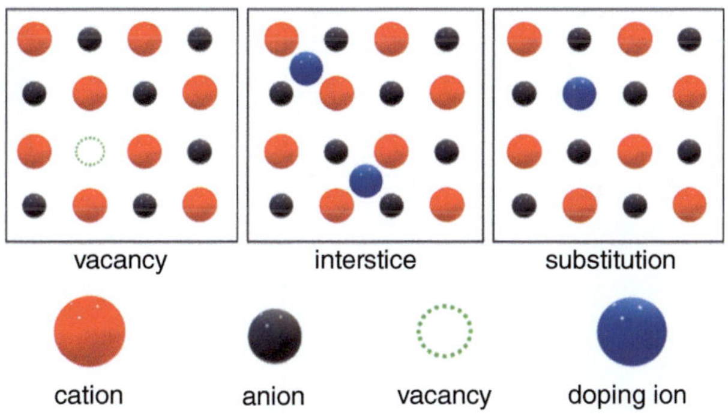		

Point effects can be observed, for example, when lithium vacancies are present in cathode materials, causing routes with enhanced ionic diffusion. Another possibility is to introduce controlled substitutions during synthesis that help stabilize the crystal structure of the material during cycling. However, the incorporation of uncontrolled point defects can generate detrimental effects, such as local structural distortions, affecting phase transitions or causing structural instability.

Line defects, such as dislocations, disclinations, or disconnections, are planar imperfections in the crystalline structure of the material that significantly affects its mechanical behavior during cycling processes (Figure 23.1). They are one-dimensional defects that constitute stress concentration points, resulting in localized deformation when the electrodes undergo volumetric changes. This can be positive because they provide routes for enhanced ionic diffusion. However, they contribute to material fatigue, cracking, and loss of particle contact and ultimately lead to a decrease in battery capacity.

Planar defects, on the other hand, are two-dimensional defects that largely influence ion transport at solid–solid and solid–liquid interfaces within battery systems. They include grain boundaries and stacking faults (Figure 23.1). Grain boundaries

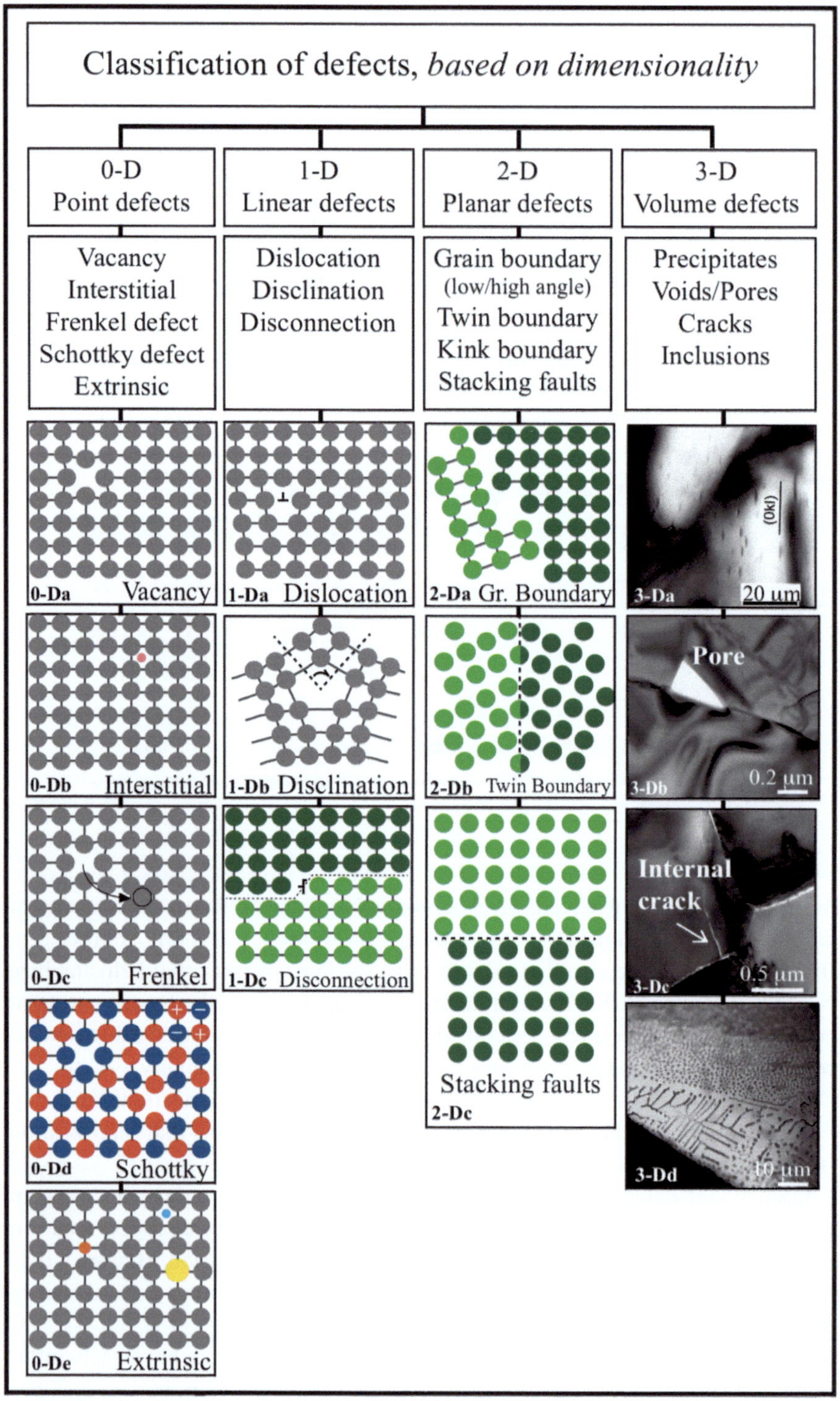

Figure 23.1 Classification of defects based on dimensionality. Source: Demouchy [7] / Copernicus Publications / CC BY 4.0.

occur when two crystallites of different orientations meet in a polycrystalline material, representing regions of atomic mismatch and disorder between adjacent crystalline grains. These defects act as high diffusivity pathways for ions but can also act as barriers to electron transport. Stacking faults occur when there is a shift in the regular stacking pattern of atomic layers within a crystal structure, and the normal arrangement of atoms is affected. Stacking faults have a particular effect on layered electrodes and alter the local electronic structure with the consequent affectation of the kinetics of the ion insertion and extraction processes.

Volume defects are large areas of discontinuity and represent three-dimensional imperfections. They primarily include voids or precipitates that are observed in the material (Figure 23.1). These defects impact battery performance by creating or eliminating sites available for ionic insertion. Additionally, they have an effect on structural stability because they cause localized stress concentrations. Voids or pores within the material cause degradation of the material and can reduce mechanical strength. However, the distribution and size of voids help to improve ionic diffusion by providing additional transport channels. Precipitates are formed by the segregation of atoms and can either enhance or hinder battery performance, since they can act as ion traps or facilitators, depending on their size, distribution, and the characteristics of their interface with the host material.

23.2.2 Defect Formation Mechanisms and Methodologies of Insertion in Battery Materials

Defects can be created intentionally or unintentionally in the material structure and can be introduced during the processing steps as well as during battery operation.

During synthesis and processing, there are several methods to consider for introducing specific defects to improve battery performance. Doping involves the substitution of atoms in the crystal lattice, which alters the electronic or ionic properties of the material. It can result in vacancies or interstitials that facilitate ionic mobility and improve the conductivity and capacity of the battery. Doping can be achieved through solid-state reactions, sol–gel methods, or co-precipitation techniques. Microstructure control is a critical aspect, where porosity, grain size, and surface area are manipulated to optimize defect concentration and distribution. These aspects can be achieved through proper control of synthesis parameters such as temperature, pressure, and reaction time.

Thermal and mechanical treatments are widely employed to induce specific defects. The annealing process can induce vacancy formation or atomic rearrangement, while rapid cooling can freeze defects formed at high temperatures. On the other hand, mechanical milling or pressing can introduce dislocations or stresses in the material, which affect the grain boundary density. Certain advanced techniques such as ion implantation or electron beam irradiation can create precise defect profiles in battery materials.

During battery operation, defect formation is a continuous process influenced by electrochemical cycling and device aging. Repetitive cycling induces material fatigue, which causes the formation of cracks, dislocations, and other structural defects that degrade battery performance over time. Another more common defect formation process during battery operation is the growth of dendrites, which are needle-like structures that form on the anode surface and can penetrate the separator, thus leading to short circuits and widespread failures. To control or mitigate the development of these defects, it is necessary to understand their formation mechanisms in order to improve the efficiency and longevity of these energy storage systems.

23.2.3 Characterization Techniques for Defect Analysis

A comprehensive analysis of the impact of defects requires a multifaceted approach and the use of advanced characterization techniques at different scales. This integral analysis is essentially based on the use of microscopic, spectroscopic, and diffraction techniques.

At the nanometric level, microscopic techniques are crucial for observing defects. Transmission electron microscopy (TEM) uses high-energy electrons that pass through the sample and provide high-resolution images at the atomic level. TEM allows visualization of point defects, dislocations, and grain boundaries. This technique has a very high resolution (less than 0.1 nm). However, it requires very thin samples of the material that are difficult to prepare. Scanning electron microscopy (SEM) uses electrons to scan the material at the surface level, allowing to evaluate the surface morphology and the distribution of defects over larger areas. It has lower resolution than TEM and does not allow observing defects at the atomic scale. Atomic force microscopy (AFM) makes use of a sharp probe that scans the surface of the material and measures the interaction forces at the atomic level. It allows the detection of surface irregularities, topographical defects, and local mechanical characteristics. This technique complements the above techniques by allowing the evaluation of three-dimensional surface topography, in addition to nanoscale mechanical properties, crucial for understanding changes in surface-induced defects.

Spectroscopic techniques are usually employed to investigate the chemical composition and electronic states associated with defects. X-ray photoelectron spectroscopy (XPS) provides information on elemental composition and the resulting chemical bonds, which makes it very useful for identifying oxidation states and surface contaminants that result from defect formation. RAMAN spectroscopy is particularly effective in characterizing vibrational modes in materials, helping to detect structural disorders and subtle structural changes generated in crystal lattices caused by defects. Nuclear magnetic resonance (NMR) is a very powerful technique for defect characterization and analysis of the internal structure of materials. This technique has been developed and successfully applied to the study of solid materials to investigate their defects along with physical and chemical properties. NMR identifies atomic vacancies, dislocations, impurities, and pores in solid materials. NMR can also analyze molecular dynamics and provide information on local electronic environments and ion mobility. This is key to understanding how defects affect the diffusion ability of atoms in materials.

On the other hand, diffraction techniques are widely used to analyze the crystalline structure. X-ray diffraction (XRD), by means of lattice splitting and peak broadening, makes it possible to determine the crystalline structure and phase composition and analyze deformations, all characteristics that can be affected by the presence of defects. When greater sensitivity to light elements such as lithium and the study of magnetic structures is required, XRD results are complemented by neutron diffraction. This technique can provide detailed information on atomic positions and occupancies, crucial for understanding defect distributions in battery materials. It has the advantage of being a nondestructive, fast, accurate, and versatile technique.

23.2.4 Defect Modeling

Today, computational methods are indispensable tools for understanding and predicting defect behavior at the atomic level in battery materials. Density functional theory-related calculations allow to calculate the electronic structure, migration barriers, and defect-related energies and to predict how vacancies, interstitials, and substitutional atoms influence ionic diffusion and conductivity. This is achieved by simulating possible defect configurations, identifying the most favorable conditions for improving battery performance. Atomistic simulations, such as molecular dynamics and Monte Carlo methods, allow studying the evolution of defects and their interactions at longer scales and measurement times, thus allowing the analysis of defect migration and their interactions over the battery operation time. These techniques can also be used to evaluate the defect clustering phenomena and their influence on ion diffusion pathways.

The impact of defects on macroscopic properties must also be considered. Multiscale modeling approaches help to bridge the gap in correlating atomic-scale defects and device performance. For example, phase-field models can simulate dendrite growth or crack propagation evolution, which are often affected by defect structures close to these phenomena. On the other hand, continuum models based on atomistic simulations are useful to predict how defect concentration affects ionic conductivity, charge–discharge rate, and material stability. Therefore, computational predictions are essential to guide experimental efforts in this area, as well as to aid in the interpretation of complex characterization data.

23.3 Case Studies: Enhancing the Performance of Advanced Battery Systems

23.3.1 Defect Engineering in Anode Materials

In an attempt to improve storage capacity and reduce dendrite growth, defect engineering has explored carbon-based anode materials such as graphite, porous carbon, and carbon nanotubes. This has been achieved through two approaches. First, the pairing of carbonaceous materials with coatings and nanoparticles has been used in such a way as to limit undesirable reactions at interfaces. An example is graphite; this material is composed of carbon layers bonded by weak van der Waals forces and allows the intercalation of metal atoms between their planes. However, the formation of a solid–electrolyte interface (SEI) layer during cycling affects performance. This limitation has been addressed through surface modification, applying the coating by atomic layer deposition, passivating the graphite surface, and preventing the unwanted side reactions that lead to SEI formation [8].

On the other hand, the insertion of dopants to perturb the nanostructure of the carbonaceous material creates defect-rich zones. In this way, easier ion storage and transport occur. This is the case of porous carbon, which itself has a high surface area and an interconnected pore structure, leading to effective channels for ion diffusion. To improve the properties of this type of anode, the introduction of impurities such as boron and nitrogen is reported [9]. With the insertion of these atoms, defect-rich zones are created that facilitate ion intercalation, contributing to storage capacity and stability. The use of nitrogen is considered effective for doping carbon nanotubes because it provides additional sites for the intercalation process and boosts the specific capacity and performance at high loading rates [10]. Moreover, plasma functionalization techniques produce extrinsic defects in carbon structures through the inclusion of heterogeneous defects. Finally, fast charging capability through heterogeneous doping is an area of opportunity for future research. For example, three-dimensional nitrogen-doped graphene achieves enhanced fast charge characteristics for electrode application [11]. Additionally, fast charging can be a tool to improve battery lifetime by providing heat to the system, a condition that allows annealing processes to remove dendrites formed during cycling [12]. This is an area of opportunity for future research to determine whether dendrite self-healing can become a viable strategy.

Perovskite-based anode materials, known for their highly tuneable properties, exceptional electrical properties, and defect tolerance, offer several opportunities for optimization of lattice parameters and electrode electronic structure. Among the major advances are the modification of perovskite unit cell dimensions to improve ion storage and the incorporation of heteroatoms to increase electrical conductivity. Perovskite-type oxides (PTOs) and metal halide perovskites (MHPs) are among the most extensively studied materials [13]. In the case of PTOs, the introduction of oxygen vacancies and modifications in the sintering temperature for $LaCoO_3$ and $CeMnO_3$ is highlighted, obtaining significant discharge capacities and improved cyclic stability. For MHPs, such as $CH_3NH_3PbI_3$, the presentation of hybrid intercalation and lithium conversion mechanisms has been achieved; however, side reactions are unavoidable. Doping with lithium atoms in perovskites such as $CaMnO_3$ results in the reduction of crystal size and increases oxygen vacancies, giving rise to new pathways for lithium intercalation [14].

Anode materials based on metal oxides, such as TiO_2, Nb_2O_5, and SnO_2, typically offer high theoretical capacities. However, these present inherent challenges such as low conductivity and significant volumetric changes during charge and discharge cycles. To mitigate these drawbacks, oxygen vacancy insertion, nanostructuring, and doping with cations or anions are considered [15]. The creation of oxygen vacancies in TiO_2 improves electronic conductivity and diffusion rates. In turn, cation doping in SnO_2 with elements such as Al or Zn improves structural stability and suppresses aggregation during cycling, decreasing sputtering effects [16]. Furthermore, in the case of nanostructuring of metal oxides, through the creation of hollow or porous architectures, it provides a better arrangement for volumetric expansion, influencing cycling stability. The combination of the discussed methods with nanometric structures and composite materials, such as graphene, enhances the final materials to obtain high-performance electrodes [17].

23.3.2 Defect Engineering in Cathode Materials

Manganese oxide and vanadium oxide-based cathode materials exhibit high theoretical capabilities and structural diversity. However, both present challenges such as structural degradation during cycling and poor ionic conduction, which limit their long-term performance. The introduction of oxygen vacancies and cation deficiency-type defects can reduce the Gibbs-free energy for the adsorption of metal ions such as Zn^{2+}, facilitating their insertion and extraction [18]. In the case of MnO_2, theoretical studies have confirmed that oxygen vacancy-type defects facilitate H+ insertion into the oxide structure and insertion/extraction kinetics, increasing the number of active sites and decreasing diffusion barriers [19]. In the case of polymorphs of the compound such as α-MnO_2 with tunneling structure and layered δ-MnO_2, when designed with defects, they exhibit improved electrochemical performance by increasing the conductivity and reducing the formation of irreversible by-products during cycling [20]. Finally, these modifications improve stability by mitigating structural collapse and dissolution of the active material.

Similarly, oxygen vacancies improve the performance of vanadium oxide-based materials by generating a larger layered or tunnel structure, which improves the reaction kinetics and storage capacity of metal cations in the battery. In addition, the introduction of defects stabilizes the crystal structure during repeated insertion and removal of cations, which prevents fading of the storage capacity of the system. With proper adjustment of the density and distribution of defects in these metal oxides, higher energy densities and longer life cycles are achieved [20].

Cathode materials based on transition metal dichalcogenides (TMD), such as MoS_2, TiS_2, and WS_2, exhibit exceptional electrochemical properties due to their layered structure and high surface area. However, their application in batteries is limited due to problems such as low intrinsic conductivity and aggregation. One strategy to prevent the re-aggregation of TMD layers is the expansion of the interlamellar space by introducing foreign species, using ammonium ions or organic compounds. In this way, a high active surface area is maintained during the charging–discharging cycles [21]. This expansion is also beneficial in facilitating ion insertion and removal and improves the capacity and cyclic stability of the system. In addition, the creation of sulfur vacancies in MoS_2 and WS_2 TMDs improves ionic and electronic conductivity, because these vacancies act as active sites that facilitate the intercalation of metals such as Li and Na [22]. On the other hand, sulfur vacancies also have an effect on amplifying the interlayer spacing, reduce the energy barrier required for ion diffusion, and improve the charge–discharge rate. For the controlled production of these vacancies, methods of heat treatment in controlled atmosphere of hydrogen or oxygen and modulation of solvothermal processes are employed [23].

For TMD cathodes, doping with heteroatoms such as nitrogen and phosphorus has also been addressed. Through this doping, the mechanical resistance against possible volumetric expansions during cycling is increased, additional electronic pathways are created within the TMD structure, and conductivity is increased [24]. Additionally, the phase control engineering of the semiconductor-to-metallic TMDs increases not only their electrochemical performance for application in batteries but also in terms of structural stability, critical for the operation of the device. The 1T (metallic) phase provides more conductive pathways than the 2H (semiconducting) phase [25]. This control can be achieved through solvothermal synthesis, exfoliation, and chemical vapor deposition (CVD) [26]. Also, the presence of ammonium, carbon, and graphene ions hinders the formation of the 2H phase and leads to the 1T phase.

Carbon-based cathode materials, such as graphene, carbon nanotubes, and porous carbon composites, also play a role as cathodes, especially in lithium–sulfur (Li–S) batteries. They feature high electrical conductivity, an adjustable and flexible structure. Defect engineering comes into play to resolve inherent drawbacks of the Li-S system, such as the insulating nature of sulfur and lithium sulfide [27]. The insertion-type defects of nitrogen or oxygen functional groups facilitate strong interactions with sulfur species, mitigating the dissolution and displacement of lithium polysulfides, which is usually one of the main causes of capacity loss in this type of battery. On the other hand, the improvement in surface chemistry resulting from defect engineering allows this type of cathode to better retain sulfur, increasing cycle stability and system capacity [28].

23.3.3 Defect Engineering in Solid Electrolytes

Solid metal halide electrolytes such as A_3MX_6 ($A = Li^+, Na^+$; M = trivalent metal; and X = halide) are considered as a promising material to improve ionic conductivity and electrochemical stability in solid-state batteries. The compositional flexibility they present allows to control ionic migration pathways, since the structure can be tuned to reduce the migration barrier and allow the development of vacancies that enhance ion mobility. Aliovalent substitution, such as the incorporation of Zr^{4+} or Hf^{4+} at the trivalent cation site, creates mobile vacancies that allow ion migration and enhance ionic conductivity [29] (Figure 23.2). In addition, these substitutions have effects on the crystal structure by adjusting migration paths and reducing activation barriers [30].

Isovalent substitution plays a crucial role in increasing the moisture resistance and structural robustness of the electrolyte. This is achieved by adjusting the volume of the migration paths and the polarizability of the bonding environment, leading to reduced migration barriers and improved conductivity. For example, replacing the

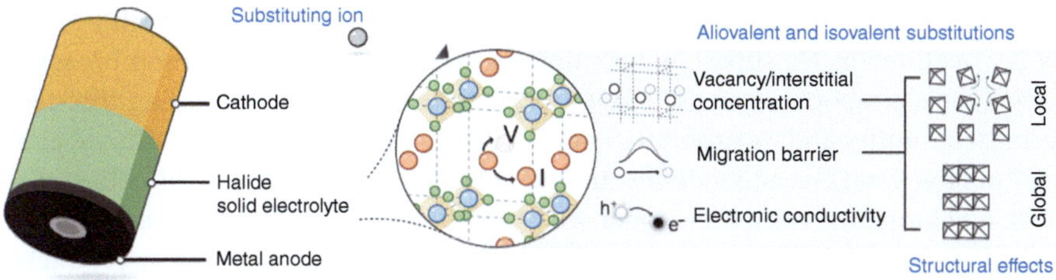

Figure 23.2 Aliovalent and isovalent substitutions in solid electrolytes affect the mobile ion vacancy and interstitial concentrations, migration barrier of the diffusing mobile ion, and electronic conductivity. Substitutions induce local, and, in some cases, global structural changes that affect mobile ion defect concentrations and migration barriers. Source: Combs et al. [30] / with permission of IOP Publishing.

halides present with larger ions, such as iodide, leads to an expansion of the unit cell volume, facilitating ion diffusion [30].

Solid electrolytes of complex hydrides $NaAlH_4$ and Na_3AlH_6 for sodium batteries have shown potential as solid electrolytes due to their high Na^+ ion transfer number. However, the most important limitation they present at room temperature is the low ionic conductivity. To overcome this drawback, chemical modification of the anions, mixing of anions, and reduction of the crystal size are proposed. In the case of the introduction of large anions, $B_{12}H_{12}{}^{2-}$ and $B_{10}H_{10}{}^{2-}$ are reported, which increase the ionic conductivity (>0.1 S/cm) since they present a structure rich in cationic vacancies. These transitions occur at temperatures higher than those practical for solid-state batteries [31]. On the other hand, the substitution of hydrogens by halogens in anions of the $Na_2B_{12}X_{12}$ type (X = Cl, Br, I) modifies the static cation–anion interactions and causes a reduction in the phase transition temperature, which optimizes ionic conductivity at lower temperatures. Furthermore, the mixture of different anions introduces geometric frustrations into the system that cause the elimination of the phase transition, in cases such as $Na_2(B_{12}H_{12})_{0.5}(B_{10}H_{10})_{0.5}$, a compound with high ionic conductivity without evident phase transitions [32].

Organic polymeric solid electrolytes have excellent flexibility and interfacial contact with the electrodes, but their low ionic conductivity and low ion transfer number constitute important limitations. The most addressed strategies are based on increasing the amorphous regions in the polymer matrix and producing the movement of the segmental chains more easily and, consequently, greater ion migration. This can be achieved through copolymerization, mixing, cross-linking, and addition of plasticizers. The use of organic nanofillers such as SiO_2 and TiO_2 can increase amorphous regions and improve ionic conductivity [33].

23.3.4 Interface Engineering for Improved Battery Performance

Cathode–electrolyte interface (CEI) engineering is essential to improve the performance of battery systems, especially in solid-state batteries with chemical compatibility and interfacial contact issues. The introduction of controlled defects, such as cationic vacancies and doping with aliovalent cations, optimizes the chemical stability and ionic conductivity of the electrolyte. For example, doping with Mg^{2+} and Sc^{3+} cations in Na superionic conducting materials (NASICON) can broaden positive ion transport channels [32]. CEI can benefit from defect tuning through coating techniques and incorporation of layers of conductive materials with the aim of optimizing ionic flux and reducing interfacial resistance. The methods include the addition of inorganic compounds such as NASICON or Na_3SbS_4, which act by forming a stable passivation layer and reducing the formation of decomposition products at the cathode. On the other hand, the in situ preparation of compounds during the formation of the solid electrolyte and the active material of the cathode influences the obtaining of an intimate and uniform contact, which reduces the interfacial resistance. This is achieved through solution deposition techniques, which results in uniform coverage of the electrolyte on the cathode surface and minimizes charge transfer resistance [34]. This reduction in interfacial resistance can also be obtained through the addition of liquid solvents, such as ionic liquids [35]. The use of plastic crystal electrolytes is another effective strategy, since these materials are capable of reversibly deforming with volumetric changes in the cathode, relieving the mechanical stresses generated during cycling [36].

Anode–electrolyte interface (AEI) engineering faces the challenge of the formation of unstable or reactive passivation layers at the interface that affects cycling efficiency and increases interfacial resistance. These effects can be mitigated through defect adjustment and the addition of buffer layers between the electrolyte and the anode. To this end, materials, such as HfO_2, ZrO_2, and Sc_2O_3, have been used as buffer layers to stabilize the interface and reduce unwanted chemical reactions at the anode. Additionally, these layers act as chemical and physical barriers, decreasing the formation of dendrites [37]. On the other hand, the formation of controlled passivation layers at the anode–electrolyte interface can prevent unwanted reactions. Finally, as with CEI, the in situ preparation of solid electrolyte composites and anode active materials ensures uniform and intimate contact, significantly reducing interfacial resistance [38].

23.4 Challenges and Future Perspectives in Defect Engineering

23.4.1 Current Limitations in the Implementation of Defect Engineering for Batteries

The advantages of defect engineering are obvious. However, there are some limitations that hinder its application on a large scale. One of the main problems is the scalability of the techniques used in defect engineering. Although several of the processes for introducing and controlling defects, such as doping and ion substitution, are very effective at laboratory scale; at large scale, they are difficult to reproduce due to their complexity and the precision required to achieve beneficial effects. Uniformity within materials is crucial to ensure optimal device performance and lifetime. Keeping precise control of defect formation in large batches is an obstacle that affects consistency and reliability in final battery production.

Moreover, in terms of cost and sustainability, the materials and processes applied in defect engineering are often costly and have environmental impacts. Thus, the production of advanced materials and the use of sophisticated techniques increase the costs of battery material production considerably. In addition, to reduce the environmental impact associated with battery production, nontoxic materials must be used, and manufacturing processes must be sustainable.

Finally, the long-term stability of defects is a major challenge. Ensuring that inserted defects remain stable during cycling and aging is crucial in the long term since defects can migrate or change over time, degrading battery performance. These stability issues put into perspective the need for further research into ways to achieve long-term defect retention and the development of methods to ensure defect stability throughout the lifetime of a battery.

23.4.2 Future Perspectives

The future of defect engineering in materials with potential use in batteries looks exciting. Advances in synthesis and simulation techniques are setting precedents on the road to high-precision defect control and prediction. New synthesis methodologies, such as atomic layer deposition and plasma-assisted synthesis, allow for proper defect control at the atomic level. Also, advances in computational and molecular dynamics simulation methods will facilitate the prediction of defects and their impact on the electrochemical properties of new materials. This, in turn, will lead to greater rationality and efficiency in the design and synthesis of new materials.

Research focused on two-dimensional (2D) and nanoscale materials offers new opportunities. Graphene, MXene, and other low-dimensional composites have large surface areas and tuneable electronic properties, allowing finer control of defects and significantly improved electrochemical properties [39]. Thus, 2D materials should increasingly be able to be integrated into electrode construction and electrolyte preparation to improve overall battery performance.

Interface engineering is another key area with great potential. The combination of defect engineering at solid–solid and solid–liquid interfaces can further optimize battery performance. Thus, the creation of defect-controlled interfaces can improve ion transfer and reduce interfacial resistance, resulting in more efficient and longer-lasting batteries. Finally, the insertion of defects in new battery architectures, such as solid-state and metal–air batteries, is a promising avenue in this field. Achieving the ability to control and stabilize potential defects inserted in these new configurations should lead to significant advances in terms of energy density, lifetime, and battery safety.

References

1 Kumar Samanta, B., Hiranmay, S., Kr. Deb, N., and Hiranmay, S. (2020). Battery storage technologies: a review. In: *Intelligent Electrical Systems: A Step* *towards Smarter Earth* (ed. S. Chakrabarti, A. Kumar Panja, A. Mukherjee, and A. Kr. Bar), 257–268. CRC Press.

2 Yanamandra, K., Pinisetty, D., and Gupta, N. (2023). Impact of carbon additives on lead-acid battery electrodes: a review. *Renew. Sust. Energ. Rev.* 173: 113078.

3 Koohi-Fayegh, S. and Rosen, M.A. (2020). A review of energy storage types, applications and recent developments. *J. Energy Storage* 27: 101047.

4 Wang, Y., Zhang, Y., Cheng, H. et al. (2021). Research progress toward room temperature sodium sulfur batteries: a review. *Molecules* 26 (6): 1535.

5 Leong, K.W., Wang, Y., Ni, M. et al. (2022). Rechargeable Zn-air batteries: recent trends and future perspectives. *Renew. Sust. Energ. Rev.* 154: 111771.

6 Šimić, Z., Knežević, G., Topić, D., and Pelin, D. (2021). Battery energy storage technologies overview. *Int. J. Electr. Comput. Eng. Syst.* 12 (1): 53–65.

7 Demouchy, S. (2021). Defects in olivine. *Eur. J. Mineral.* 33 (3): 249–282.

8 Fedorov, R.G., Maletti, S., Heubner, C. et al. (2021). Molecular engineering approaches to fabricate artificial solid-electrolyte interphases on anodes for Li-Ion batteries: a critical review. *Adv. Energy Mater.* 11 (26): 2101173.

9 Wang, Q., Xie, Z., Liang, Y. et al. (2019). Facile synthesis of boron-doped porous carbon as anode for lithium–ion batteries with excellent electrochemical performance. *Ionics* 25 (5): 2111–2119.

10 Zhong, S., Liu, H., Wei, D. et al. (2020). Long-aspect-ratio N-rich carbon nanotubes as anode material for sodium and lithium ion batteries. *Chem. Eng. J.* 395: 125054.

11 Mo, R., Li, F., Tan, X. et al. (2019). High-quality mesoporous graphene particles as high-energy and fast-charging anodes for lithium-ion batteries. *Nat. Commun.* 10 (1): 1–10.

12 Weiss, M., Ruess, R., Kasnatscheew, J. et al. (2021). Fast charging of lithium-ion batteries: a review of materials aspects. *Adv. Energy Mater.* 11 (33): 2101126.

13 Verduzco, J.C., Bettes, B., Hu, Q., and Marinero, E.E. (2023). A review on defect engineering of anode materials for solid-state battery applications. *Ionics* 29 (2): 439–453.

14 Fang, M., Yao, X., Li, W. et al. (2018). The investigation of lithium doping perovskite oxide $LiMnO_3$ as possible LIB anode material. *Ceram. Int.* 44 (7): 8223–8231.

15 Wang, J., Zhou, Y., Li, J. et al. (2024). Synthesis, characterizations, and applications of vacancies-containing materials for energy storage systems. *DeCarbon* 3: 100037.

16 Dou, P., Cao, Z., Wang, C. et al. (2017). Multilayer Zn-doped SnO_2 hollow nanospheres encapsulated in covalently interconnected three-dimensional graphene foams for high performance lithium-ion batteries. *Chem. Eng. J.* 320: 405–415.

17 Dong, C., Dong, W., Lin, X. et al. (2020). Recent progress and perspectives of defective oxide anode materials for advanced lithium ion battery. *EnergyChem* 2 (6): 100045.

18 Yadav, N., Qin, J., Maiyalagan, T. et al. (2024). Optimizing the defect engineering of $MnCo_2O_4$ spinel structure through Zn-substitution as a cathode for high-performance Zinc-ion batteries. *Electrochim. Acta* 500: 144776.

19 Han, M., Huang, J., Liang, S. et al. (2020). Oxygen defects in β-MnO_2 enabling high-performance rechargeable aqueous zinc/manganese dioxide battery. *iScience* 23 (1): 100797.

20 Er Pam, M., Yan, D., Yu, J. et al. (2021). Microstructural engineering of cathode materials for advanced Zinc-Ion aqueous batteries. *Adv. Sci.* 8 (1): 2002722.

21 Chen, B., Chao, D., Liu, E. et al. (2020). Transition metal dichalcogenides for alkali metal ion batteries: engineering strategies at the atomic level. *Energy Environ. Sci.* 13 (4): 1096–1131.

22 Askari, M.B., Salarizadeh, P., Veisi, P. et al. (2023). Transition-metal dichalcogenides in electrochemical batteries and solar cells. *Micromachines.* 14 (3): 691.

23 Ullah, N., Guziejewski, D., Yuan, A., and Shah, S.A. (2023). Recent advancement and structural engineering in transition metal dichalcogenides for alkali metal ions batteries. *Materials* 16 (7): 2559.

24 Prabhu, P., Jose, V., and Lee, J.M. (2020). Design strategies for development of TMD-based heterostructures in electrochemical energy systems. *Matter* 2 (3): 526–553.

25 Yin, X., Tang, C.S., Zheng, Y. et al. (2021). Recent developments in 2D transition metal dichalcogenides: phase transition and applications of the (quasi-)metallic phases. *Chem. Soc. Rev.* 50 (18): 10087–10115.

26 Li, P., Yang, Y., Gong, S. et al. (2019). Co-doped 1T-MoS_2 nanosheets embedded in N, S-doped carbon nanobowls for high-rate and ultra-stable sodium-ion batteries. *Nano Res.* 12 (9): 2218–2223.

27 Shi, Z., Li, M., Sun, J., and Chen, Z. (2021). Defect engineering for expediting Li–S chemistry: strategies, mechanisms, and perspectives. *Adv. Energy Mater.* 11 (23): 2100332.

28 Liu, H., Jia, G., Zhu, S. et al. (2022). Functionalized carbon-based composite materials for cathode application of lithium-sulfur batteries. *Acta Chim. Sin.* 80 (1): 89–99.

29 Park, J., Han, D., Kwak, H. et al. (2021). Heat treatment protocol for modulating ionic conductivity via structural evolution of $Li_{3-x}Yb_{1-x}M_xCl_6$ (M = Hf^{4+}, Zr^{4+}) new halide superionic conductors for all-solid-state batteries. *Chem. Eng. J.* 425: 130630.

30 Combs, S.R., Todd, P.K., Gorai, P., and Maughan, A.E. (2022). Editors' choice—review—designing defects and diffusion through substitutions in metal halide solid electrolytes. *J. Electrochem. Soc.* 169 (4): 040551.

31 Udovic, T.J., Matsuo, M., Unemoto, A. et al. (2014). Sodium superionic conduction in $Na_2B_{12}H_{12}$. *Chem. Commun.* 50 (28): 3750–3752.

32 Lu, Y., Li, L., Zhang, Q. et al. (2018). Electrolyte and interface engineering for solid-state sodium batteries. *Joule* 2 (9): 1747–1770.

33 Bósquez-Cáceres, M.F., Hidalgo-Bonilla, S., Morera Córdova, V. et al. (2021). Nanocomposite polymer electrolytes for zinc and magnesium batteries: from synthetic to biopolymers. *Polymers (Basel).* 13 (24): 4284.

34 Yang, H. and Wu, N. (2022). Ionic conductivity and ion transport mechanisms of solid-state lithium-ion battery electrolytes: a review. *Energy Sci. Eng.* 10 (5): 1643–1671.

35 Wu, F., Chen, Z., Fang, S. et al. (2023). The role of ionic liquids in resolving the interfacial chemistry for (quasi-) solid-state batteries. *Energy Storage Mater.* 63: 103062.

36 Meng Q, Lou S, Shen B, Wan X, Xiao X, Ma Y, et al. Reevaluating flexible lithium-ion batteries from the insights of mechanics and electrochemistry. *Electrochem. Energy Rev.* 2022 52. 2022;5(2):1–31.

37 Tang, H., Deng, Z., Lin, Z. et al. (2018). Probing solid-solid interfacial reactions in all-solid-state sodium-ion batteries with first-principles calculations. *Chem. Mater.* 30 (1): 163–173.

38 Zhou, Q., Xiong, X., Peng, J. et al. (2024). Tailored engineering on the interface between lithium metal anode and solid-state electrolytes. *Energy Environ. Mater.* 0: e12831.

39 Dong, Y., Yan, C., Zhao, H., and Lei, Y. (2022). Recent advances in 2D heterostructures as advanced electrode materials for potassium-ion batteries. *Small Struct.* 3 (3): 2100221.

24

Defect and Interface Engineering in Electrochemical Pseudocapacitors Based on Carbon

Zhipeng Sun and Xiaoyan Shi

Guangdong University of Technology, School of Materials and Energy, No. 100, Outer Ring West Road, Guangzhou Higher Education Mega Center, Panyu District, Guangzhou, 510006, Guangdong, P. R. China

24.1 Introduction

Carbon, with its distinctive physical, chemical, and electronic properties, has been widely explored as a functional material for electrochemical energy storage and conversion systems. Defect engineering, which involves the deliberate modification of carbon structures to introduce controlled defects, has emerged as a critical strategy to enhance the performance of carbon-based materials. Compared to pure carbon, defect-rich carbon matrices exhibit improved electronic properties, increased exposed active sites, and enhanced electrochemical redox reactions, ultimately improving electrochemical performance [1–6].

This chapter reviews recent advances in defect and surface engineering in carbon materials, focusing on their applications in electrochemical pseudocapacitor systems. We discuss various defect types, construction strategies, and characterization techniques while emphasizing the importance of defect–performance relationships. By integrating experimental findings and theoretical insights, this chapter aims to provide guidance for the large-scale and industrial applications of defect-rich carbon materials.

24.2 Defect Engineering in Carbon Materials: Insights and Applications

24.2.1 Classification of Defects in Carbon Materials

Defects in carbon materials can be categorized into intrinsic defects (e.g. vacancies, edge defects, topological defects) and extrinsic defects (e.g. heteroatom doping with N, B, S, P) [2, 7].

Surface and Interfacial Defects in Nanomaterials for Sustainable Energy Production and Storage, First Edition. Edited by Noé Arjona, Lorena Álvarez Contreras, and Minerva Guerra Balcázar.

24.2.1.1 Intrinsic Defects

Intrinsic defects arise naturally within the carbon matrix, influencing electronic properties and electrochemical activity, while extrinsic defects are introduced through doping or modification techniques to enhance material performance.

Early studies on the modification of carbon materials mainly emphasized surface functionalization and heteroatom doping. However, intrinsic carbon defects were often overlooked. Recent research has demonstrated that intrinsic defects possess superior electrochemical activity compared to heteroatom doping, making them highly promising for energy storage applications.

(i) Types of Intrinsic Defects

Intrinsic defects in carbon materials manifest as topological distortions, vacancy defects, and edge defects, each affecting the material's electronic structure and electrochemical behavior in distinct ways (Figure 24.1).

- **Edge Defects**: The presence of unpaired π-electrons at edge sites enhances local charge redistribution, making them highly active for redox reactions and electrocatalysis. The zigzag and armchair edge configurations influence the spin density and chemical reactivity, affecting the efficiency of carbon-based electrodes. Graphene nanoribbons (GNRs) and graphene quantum dots (GQDs) leverage these edge defects to improve electrochemical response.
- **Vacancy Defects**: Missing carbon atoms create vacancies in the hexagonal lattice, which can be classified into monovacancy (MV), divacancy (DV), and multivacancy (where three or more atoms are removed). These vacancies disrupt local charge distribution, resulting in enhanced electrochemical activity. Theoretical calculations have shown that defect-induced bond length variations (e.g. from ~1.42 Å in perfect hexagonal lattices to 1.30–1.81 Å in defective lattices) significantly modify the electronic density of states, impacting charge storage capabilities.
- **Topological Defects**: These result from the distortion of the hexagonal carbon lattice, leading to the formation of unconventional polygonal structures such as pentagons (C5) and Stone–Wales (SW) defects (C5-7-5-7). The redistribution of asymmetric electrons in these defects alters the electronic properties, influencing charge transport and electrocatalytic activity.

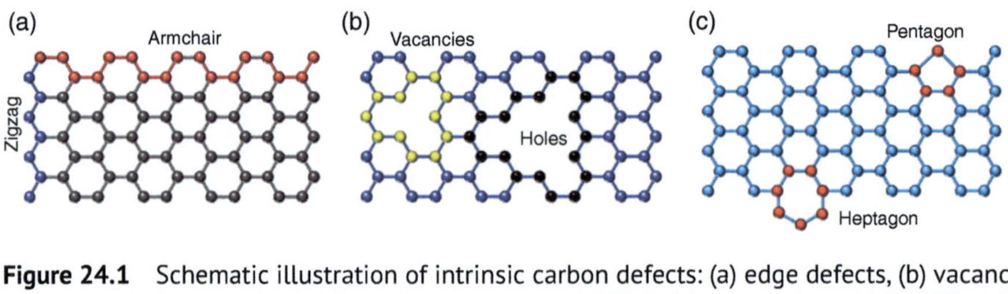

Figure 24.1 Schematic illustration of intrinsic carbon defects: (a) edge defects, (b) vacancy defects, and (c) topological defects. Source: Luo et al. [7] / with permission of John Wiley & Sons.

(ii) Influence of Intrinsic Defects on Electrochemical Performance

Intrinsic defects significantly impact the electrochemical properties of carbon materials by modulating mechanical strength, electrical conductivity, and catalytic behavior:

- **Mechanical Strength**: Molecular dynamics simulations suggest that topological defects contribute to the fracture toughness of graphene sheets, while high defect densities may reduce structural integrity, particularly in carbon nanotubes (CNTs).
- **Electronic Conductivity**: The energy gap between the highest occupied molecular orbital (HOMO) and the lowest unoccupied molecular orbital (LUMO) decreases in defective carbon lattices, enhancing electron transfer efficiency. Density functional theory (DFT) studies confirm that defects such as Stone–Wales and divacancy structures enhance redox reactivity, making them advantageous for battery and supercapacitor electrodes.

24.2.1.2 Extrinsic Defects

Intrinsic carbon defect sites possess a high level of compatibility and can function as anchor sites for heteroatoms and clusters, leading to the formation of coordination active centers. Exotic atoms with varying electronegativities and electronic structures can replace carbon atoms in the skeleton, be incorporated into the intrinsic carbon defects, or be firmly adsorbed onto the carbon surface. These interactions can induce intramolecular charge transfer or spin redistribution, ultimately enhancing the electrochemical activity of the carbon material. Research aimed at improving the activity of carbon materials through metal-free heteroatom doping has developed rapidly and matured over time, receiving significant attention from the scientific community. Studies have documented various aspects of this field, emphasizing the concept of metal-free heteroatom-doped defects, which has garnered substantial support among researchers. Additionally, metal atoms – ranging from noble metals to transition metals – demonstrate high dispersion and atomic utilization. More importantly, these metals can be anchored onto carbon materials, allowing for the regulation of electronic distribution and the chemical environment.

24.2.1.3 Metal-Free Heteroatom-Doped Defects

Doping with metal-free heteroatoms, such as nitrogen (N), phosphorus (P), sulfur (S), and halogens (F, Cl, Br, I), can effectively regulate the structure and electron density of carbon materials, improving their physical and chemical properties. The incorporation of heteroatoms creates new bonds with carbon in the lattice, leading to alterations in local chemical reactivity. Numerous experimental and theoretical studies on metal-free heteroatom doping of carbon materials have been published, highlighting the benefits of such modifications. The doping process enhances the charge and spin density of surrounding carbon atoms, creating active sites that contribute to improved electrochemical reactivity (Figure 24.2).

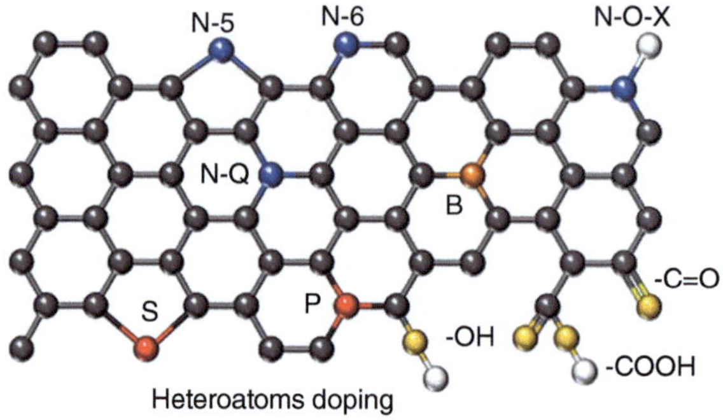

Figure 24.2 Schematic illustration of extrinsic carbon defects: heteroatom doping defects. Source: Luo et al. [7] / with permission of John Wiley & Sons.

24.2.1.4 Charge-Dominated Mechanisms: Nitrogen Doping

The effects of doping can be categorized into charge-dominated, electron-spin-dominated, and charge-spin coupling mechanisms, depending on the electronegativity differences between the doped atoms and carbon. Nitrogen doping is a classic example of the charge-dominant mechanism. Due to its higher electronegativity compared to carbon, nitrogen increases the positive charge density around its doped site, which in turn elevates the electron spin density of neighboring carbon atoms. This enhancement is attributed to the proximity of nitrogen's valence electrons and empty orbitals to the Fermi level, facilitating increased electronic transitions and resulting in improved electron transfer capabilities and electrochemical reactivity [8].

Nitrogen in nitrogen-doped carbon materials typically exists in several forms: graphitic-N (quaternary N), pyridinic-N, and pyrrolic-N, each associated with distinct intrinsic defect structures. Graphitic-N can donate electrons to the p-conjugated systems (n-type doping), thereby enhancing the nucleophilicity of the adjacent carbon rings. The N–C bond length for graphitic-N measures 1.3927 Å. Pyridinic-N, often found in mono-vacancy (MV), di-vacancy (DV), and SW defects, is the most stable configuration near MV with a bond length of 1.33 Å. Pyrrolic-N, on the other hand, occurs within the pentagonal structure of the SW defect, showing an N–C bond length of 1.372 Å.

24.2.1.5 Halogen Doping: A Versatile Strategy

Doping with halogen atoms is another effective method to introduce active sites within carbon structures. Halogens, with their high electronegativities, can form C–X bonds (X = F, Cl, Br, or I) and act as electron acceptors. The incorporation of halogen atoms disrupts the p bonds between carbon atoms, and their larger atomic radii make doping at the edges more stable. In this process, halogen atoms replace hydrogen atoms in sp^2 carbon, altering the hybridization state of carbon atoms to sp^3, which subsequently influences the electronic properties and local structure of the carbon material. Although the C—F bond length is slightly shorter than that of C—C in pristine graphene, the distances between other halogen atoms and adjacent carbon atoms are generally larger than 1.425 Å.

The substitution of halogen atoms significantly impacts the electron distribution and structural framework of carbon materials. While halogen doping increases repulsive interactions within the carbon lattice, it retains the hexagonal symmetry characteristic of graphene. This substitution leads to a reduction in the p-conjugated electron clouds above and below the graphene plane. The degree of halogenation plays a crucial role in controlling the electronic properties of the material. Moderate levels of C—X bonds can enhance the polarization of the highly electronegative halogen functional groups, improving electronic conductivity and transport properties. However, excessive halogenation can render the material insulating; for instance, graphene fully substituted with fluorine exhibits a charge carrier mobility three orders of magnitude lower than that of pristine graphene.

24.2.1.6 Silicon, Boron, and Phosphorus Doping

Silicon (Si) shares similarities with carbon in terms of structure and electronic configuration due to their positioning in the same group of the periodic table. While Si's electronegativity is considerably lower than that of carbon, Si doping can still be understood through a charge-dominant mechanism. The reactivity of Si-doped carbon materials has been well established.

Boron (B) and phosphorus (P) doping, classified as charge–spin coupling mechanisms, also exhibit notable effects. B, which has an atomic size comparable to that of carbon, integrates easily into the carbon framework. However, the longer B–B bond leads to strain on the C–C bonds, which enhances reactivity. Although boron's electronegativity is lower than that of carbon, it often requires electrons to fill empty orbitals, functioning as an electron acceptor near C–B bonds. The incorporation of boron results in a positive charge density around adjacent carbon atoms, facilitating charge transfer and improving electrical conductivity. Furthermore, unpaired pz orbitals in boron create vacancies, inducing relatively high spin densities.

Phosphorus, with slightly lower electronegativity than carbon, can donate electrons to form P—C bonds, contributing to positive charge density. However, due to its larger atomic radius, phosphorus protrudes more prominently in the carbon plane, resulting in longer C—P bonds and complicating the doping process. The structural deformation induced by the larger phosphorus atoms also leads to lower spin density compared to other dopants.

24.2.1.7 Sulfur Doping and Electron Spin Contribution

Sulfur is another common dopant for modifying sp2 carbon lattices. However, due to the similarity in electronegativity between sulfur and carbon, the effects of sulfur doping on charge distribution are minimal. The active sites in sulfur-doped carbon materials are primarily attributed to the electron spin contribution mechanism. Theoretical calculations indicate that edge-selectively sulfurized graphene demonstrates higher activity compared to both pristine graphene and surface sulfur-doped graphene despite similar charge densities. This enhanced reactivity is particularly pronounced in edge sulfur oxide-doped structures.

24.3 Strategies for Defect Engineering

Defect regulation involves the intentional introduction of defects in carbon materials to enhance their electrochemical properties. Strategies for introducing defects include mechanical ball-milling, plasma-assisted treatment, chemical oxidation, and chemical vapor deposition (CVD). "Defect engineering" aims to control the type, concentration, configuration, and spatial distribution of defects to tailor the electrochemical properties of carbon materials.

24.3.1 Carbon Strategies for Introducing Intrinsic Defects in Carbon Materials

Intrinsic carbon defects can be classified into various types, including edge defects and vacancy defects. Edge defects occur at the edges of carbon structures, where carbon atoms possess unpaired p electrons that can effectively modify the local charge state and facilitate electron transfer during electrochemical reactions. Edge carbon atoms typically exhibit a higher electron density of states compared to basal carbon atoms, making them favorable as active sites for electrochemical processes. Vacancy or hole defects arise from the removal of several carbon atoms from the lattice, promoting changes in the bonding geometry of adjacent carbon atoms to achieve a lower energy state [7, 9, 10].

(i) **Types of Topological Defects**

Topological defects, such as pentagon, heptagon, and octagon carbons, are also common in carbon lattices. SW defects (SW5775 or SW7557), for instance, can be generated by the rotation of C–C bonds, forming non-hexagonal carbon rings without loss of carbon atoms.

(ii) **Mechanical Ball-Milling**

Mechanical ball-milling is an effective method for producing various types of defects, including translation, delamination, warping, curvature, and edge defects. This technique can create various carbon nanostructures such as closed shells, tubes, and nanoarches. For example, after high-energy ball-milling, the crystallite size of expanded graphite (EG) decreased from 15.4 to 11.3 nm, resulting in a rich presence of warped graphene planes, stacking faults, carbon nanoarches, and in-plane defects. However, the low efficiency and high energy consumption of this method hinder its large-scale applications.

(iii) **Plasma-Assisted Treatment**

Plasma-assisted treatment is another effective strategy for producing defects by controlling the gas source, power level, and treatment duration. This method can generate multiple vacancy defects and facilitate heteroatom doping within the carbon lattice through plasma etching. For instance, graphene nanoribbons (GNRs) with rich zigzag edges can be obtained by Ar plasma etching, where CNTs are partially encapsulated in a polymer and then exposed to etching.

(iv) **Chemical Oxidation**

Chemical oxidation is a classic method for creating defects, where a precursor is mixed with oxidizing agents to produce a porous structure rich in edge and topological defects. For example, CNTs can be produced through the lengthwise cutting and unraveling of multi-walled carbon nanotubes (MWCNTs) in an oxidative solution, with $KMnO_4$ acting as the oxidizing agent. While chemical oxidation can yield carbons with high surface areas, the subsequent washing process to remove residuals is time consuming and costly.

(v) **Chemical Vapor Deposition (CVD)**

Chemical vapor deposition (CVD) is a straightforward method for obtaining graphene or defect graphene, useful for introducing intrinsic defects or heteroatom doping into the carbon lattice. Using CH_4 as a carbon precursor, controllable growth of three-dimensional graphene with high-density sharp edge sites can be achieved under atmospheric pressure CVD. However, the low efficiency and high cost of this method limit its large-scale application.

(vi) **Template-Assisted Treatment**

Template-assisted treatment is a strategy for constructing carbon defects by removing or treating a template. For instance, defect-rich carbons can be synthesized by cutting fullerene (C60) into fragments through in situ alkaline etching. However, this method is also tedious and expensive, restricting its large-scale adaptability.

(vii) **Nitrogen Removal Strategy**

The nitrogen removal strategy is an innovative method for constructing intrinsic carbon defects, particularly topological defects. Simulation results have shown that the initial carbon structure is crucial for obtaining specific nitrogen configurations. Controlled removal of nitrogen can yield various defect structures, such as divacancies and separated pentagons.

24.3.2 Carbon Strategies for Introducing Extrinsic Defects in Carbon Materials

Extrinsic carbon defects can be introduced through the use of carbon precursors containing heteroatoms or via post-treatment processes. This section explores methods for obtaining doped carbons and highlights self-doping in biomass-derived carbons [11].

(i) **Self-Doping from Biomass**

Biomass-derived carbons can naturally yield self-doped carbons with inherent heteroatoms such as O, N, P, and S. For example, *Artocarpus heterophyllus* seeds can be transformed into N-doped porous carbon through chemical activation. Other biomass sources, such as pinecones and various plant materials, can also be converted into N/S/O self-doped carbons using one-step carbonization combined with activation.

(ii) **Artificial Doping Methods**

Artificially doped carbons involve the introduction of heteroatoms during preparation or post-treatment processes. Various sources, such as $(NH_4)_2S_2O_8$, H_2O_2, and HNO_3, can be utilized to introduce oxygen functional groups. Additionally, numerous nitrogen sources such as polypyrrole, melamine, and ammonia solutions have been employed to produce heteroatom-doped carbons.

24.3.3 Challenges and Future Directions

Despite the various defect regulation strategies discussed, precisely introducing a single type of defect into carbon materials remains a challenge. The current methods often result in random defects rather than targeted modifications. Therefore, the development of more controllable and effective preparation strategies is essential for understanding the specific influence of a single type of defect on different electrochemical reactions.

24.4 Defect Characterization in Carbon Materials

Characterizing defects in carbon materials is a critical aspect of understanding their electrochemical performance. While acquiring correlations between defects and electrochemical behavior remains challenging, several advanced techniques have been developed to study these defects. These techniques can be classified into direct visualization and indirect observation methods [7].

24.4.1 Direct Visualization of Carbon Defects

Direct visualization refers to the use of high-resolution microscopy techniques to observe the microscopic and atomic structure of carbon materials. These methods allow the researchers to examine defects with exceptional spatial resolution.

(i) **Electron Microscopy Techniques**

Electron microscopy is an essential tool for visualizing defects at the nanoscale, with spatial resolution that can extend to atomic scales. Various types of electron microscopy are used, including:

- **Scanning Electron Microscopy (SEM)**: Provides surface images and can identify large-scale defects.
- **Transmission Electron Microscopy (TEM)**: Offers high-resolution imaging of internal structures, capable of visualizing edges and some large holes but not fine defects like vacancies.
- **Scanning Transmission Electron Microscopy (STEM)**: Combines the capabilities of SEM and TEM to enable atomic-resolution imaging.
- **Atomic Force Microscopy (AFM)**: Offers ultrahigh spatial resolution at the atomic scale, allowing the direct observation of surface structures.
- **Scanning Tunneling Microscopy (STM)**: Provides atomic-level resolution of surface structures and defects, particularly useful for observing nanographene formation.

(ii) **High-Resolution Transmission Electron Microscopy (HRTEM)**

HRTEM is commonly used to visualize the morphology of carbon materials. With the addition of an aberration corrector (AC-HRTEM), spatial resolution can reach sub-angstrom levels, enabling the observation of finer defects like SW defects. AC-HRTEM can directly visualize atomic defects such as pentagons, heptagons, and octagons, providing valuable insight into the defect structures.

(iii) **Aberration-Corrected STEM (AC-HAADF-STEM)**

High-angle annular dark-field scanning transmission electron microscopy (AC-HAADF-STEM) is the most commonly used technique for observing defects in carbon materials. By collecting high-angle scattered electrons, AC-HAADF-STEM avoids the complications of diffraction contrast, providing high-resolution images that reflect finer structural details.

24.4.2 Indirect Observation of Carbon Defects

In addition to direct visualization, several spectral characterization techniques offer indirect methods for analyzing the microstructure and defects of carbon materials. These methods provide chemical, electronic, and structural information [7].

(i) **X-Ray Photoelectron Spectroscopy (XPS)**

XPS is a surface-sensitive technique used to measure the chemical and electronic states of elements in carbon materials. The typical C1s peak in XPS can be deconvoluted into various subpeaks representing sp^2/sp^3-hybridized carbon, C-O, C=O, -COOH, and shake-up satellite features. This method allows researchers to determine the content and type of defects, such as oxygen-containing functional groups.

(ii) **Electron Energy Loss Spectroscopy (EELS)**

EELS is a technique used for studying the defects in carbon materials by measuring the energy loss of electrons after interacting with the material. It can provide detailed information on the bonding environment and electronic structure of carbon.

(iii) **X-Ray Absorption Spectroscopy (XAS)**

XAS, including X-ray absorption near-edge structure (XANES) and extended X-ray absorption fine structure (EXAFS), offers valuable insights into the local structure around absorption atoms. XANES can reveal the oxidation state and coordination environment of atoms, while EXAFS provides information on atomic distances and disorder in the material's structure. For example, C1s XANES spectra can show peaks corresponding to sp^2 bonding and structural changes due to defects.

(iv) **Raman Spectroscopy**

Raman spectroscopy is a nondestructive technique widely used for characterizing carbon materials. It provides information about the degree of disorder in the material. The key peaks in the Raman spectrum are the D-band (around $1350\,cm^{-1}$) and

G-band (around 1580–1600 cm^{-1}), corresponding to the disordered graphitic lattice and the ideal graphitic lattice of sp^2-hybridized carbon, respectively. The intensity ratio between the D and G bands (I_D/I_G) is often used to assess the level of disorder and defects in carbon materials.

(v) **X-Ray Diffraction (XRD)**

XRD can provide information about the crystal structure of carbon materials. The broadening of diffraction peaks indicates a decrease in crystallinity and an increase in the defect degree. Amorphous carbon, for example, typically shows broad peaks at ~25° and ~43°, corresponding to the (002) and (100) planes of graphitic carbon, respectively.

(vi) **Electron Paramagnetic Resonance (EPR) Spectroscopy**

EPR spectroscopy is highly sensitive to materials with unpaired electrons, which are often present at defect sites such as vacancies. This technique can help identify paramagnetic defects in carbon materials.

24.4.3 Comprehensive Characterization of Carbon Defects

To fully understand the impact of defects on the properties of carbon materials, a combination of direct and indirect characterization techniques is necessary. While AC-HAADF-STEM and XAS are particularly effective at providing detailed structural information, other techniques such as HRTEM, Raman spectroscopy, and XPS are also indispensable for a comprehensive understanding of carbon defects.

24.5 Applications in Electrochemical Pseudocapacitor Systems

The presence of defects in carbon materials can disrupt the structural integrity and alter the electronic configuration, leading to changes in charge and spin redistribution. These changes significantly influence the electrochemical performance of carbon-based materials. Consequently, the following sections explore the impact of defects in carbon materials on their applications in supercapacitors (SCs), rechargeable batteries, and electrocatalysis [12].

24.5.1 Supercapacitors

Carbon materials containing defects and heteroatom doping (e.g. N, B, S, P) exhibit improved capacitive performance, primarily due to an increase in active sites and enhanced electron/ion transport. Common carbon materials used as electrode materials in SCs include activated carbons (AC), graphene, CNTs, and carbon aerogels. Despite their widespread use, the relationship between the structural features of these materials and their capacitive performance remains unclear due to the complexity of their microstructures. This section specifically focuses on the role of intrinsic defects and heteroatom-doped carbon in improving SC performance [7, 13, 14].

(i) Intrinsic Defects on Carbons and their Application in SCs

The total capacitance (C_{total}) of carbon materials typically consists of the electrical double-layer (EDL) capacitance (C_{dl}) in series with the quantum capacitance (C_Q), represented by the equation: $C_{total} = C_{dl} + C_Q$. The EDL capacitance depends on the surface properties of the electrode materials and the effective thickness of the EDL. In contrast, quantum capacitance is linked to the electronic density of states (DOS), which can be described by the formula: $C_Q = e^2 \times DOS(E_F)$, where e is the elementary electronic charge, and DOS(E_F) refers to the electronic density of states near the Fermi level (E_F). For example, single-layer graphene with a specific surface area of around 2675 m^2/g could theoretically yield a specific capacity of approximately 21 $\mu F/cm^2$ (or 550 F/g). However, the experimental-specific capacitance of most carbon materials typically ranges between 100 and 250 F/g due to factors such as low electrical conductivity and limited ion-accessible surface area. This gap between theoretical and experimental capacitance can primarily be attributed to low quantum capacitance arising from the low electronic DOS near the Fermi level, and limited EDL capacitance caused by the restricted ion-accessible surface area for charge storage.

Defect engineering has emerged as a potential method to enhance the capacitive performance of carbon-based electrode materials. The introduction of defects disrupts the carbon atom arrangement, influences the band structure, and leads to space charge redistribution around the defects. Additionally, defects may generate new states at the Fermi level, thus elevating quantum capacitance. For instance, defect-enriched graphene blocks (DGB), created through ball milling, demonstrated an ultrahigh specific areal capacitance of 7.91 F/m^2, with the "self-doping" defects contributing 4.81 F/m^2 – significantly higher than graphene's theoretical capacitance of 0.21 F/m^2. Despite having a lower specific surface area (29.7 m^2/g), DGB still achieved a higher gravimetric capacitance of 235 F/g. The defect contribution, in this case, was estimated at 114 F/g at 1 A/g, which indicates the effect of local charge redistribution in carbon networks, improving binding affinity and electrochemical reactivity. Moreover, carbons with abundant pentagon defects exhibited much higher EDL capacitance compared with pure hexagonal carbons.

(ii) Use of Heteroatom-Doped Carbons in SCs

Heteroatom doping plays a crucial role in modifying the electrochemical properties of carbon materials. By introducing heteroatoms, the surface heterogeneity of carbon structures is altered, which directly influences their electrochemical performance. Enhanced capacitive behavior in heteroatom-doped carbons is typically attributed to the creation of additional active sites, improved wettability, and contributions from pseudocapacitance [14].

A fundamental distinction in ion adsorption capacity lies between in-plane sp^2-hybridized carbon atoms and edge-connected sp^3-hybridized carbon atoms. For example, graphene's edge surfaces possess numerous electrochemically active sites, resulting in a capacitance approximately four times higher than that of its basal plane, alongside a significantly faster electron transfer rate. Quantitative analysis further reveals that the estimated capacitance contributions of the basal, edge, and defect surfaces in a graphitic layer are approximately 0.04, 1.65, and 7.95 μF/cm^2, respectively. This data underscores the dominant role of defect sites in enhancing overall capacitance. Additionally, the presence of pentagon defects within the carbon lattice can induce a contraction in the bandgap and a reorganization of local charge distribution, further amplifying electrochemical reactivity and binding affinity. Consequently, carbons rich in pentagon defects exhibit significantly higher electrical double-layer (EDL) capacitance compared to those composed purely of hexagonal carbon configurations.

Among various heteroatoms, oxygen is particularly effective in improving the capacitive performance of carbon materials, primarily through the introduction of surface oxygen functional groups (OFGs), such as hydroxyl, carboxyl, epoxy, ether, and carbonyl groups. These functional groups not only enhance surface wettability but also introduce additional pseudocapacitance, thereby boosting overall energy storage capacity. A striking example of this effect is observed in oxygen plasma-treated vertical graphene nanosheets (VGN), where the formation of superhydrophilic surfaces significantly enhances electrochemical performance. Plasma-treated VGN, enriched with C—O bonding (including carbonyl, ether, epoxy, and hydroxyl groups), exhibits increased wettability and higher capacitance compared to untreated samples. In particular, in situ-treated VGN demonstrates a 10fold increase in capacitance, attributed to a greater ion-accessible surface area and enhanced ion diffusion. However, while oxygen groups contribute positively to capacitance in aqueous electrolytes, their instability in organic electrolytes can be detrimental. To mitigate this issue, hydrogen thermal reduction has been employed to remove unstable oxygen species, thereby enhancing the capacitance of porous carbon materials in organic systems [15].

Nitrogen doping is another powerful strategy for enhancing the capacitive performance of carbon materials. The introduction of nitrogen atoms reduces the bandgap between carbon and nitrogen, facilitating higher electron transfer rates. With an electronegativity of 3.04 (compared to carbon's 2.55), nitrogen alters the electronic structure of sp^2-hybridized carbon, improving charge storage efficiency. Various nitrogen configurations exist within carbon materials, including pyrrolic-N, pyridinic-N, quaternary-N, and pyridinic-N-oxide. Among these, pyridinic-N is particularly beneficial for electron transfer due to its lower energy band, while pyrrolic-N – often found in pentagon ring structures – functions as a defect site but is relatively unstable. Quaternary-N, where nitrogen is bonded to three carbon atoms, enhances electrical conductivity, whereas pyridinic-N-oxide, akin to oxygen functional groups, improves both wettability and pseudocapacitance. Experimental evidence underscores the effectiveness of nitrogen doping; for instance, N-doped graphene exhibits a specific capacitance of approximately 280 F/g, significantly surpassing the ~69 F/g recorded for undoped graphene. Moreover, N-doped graphene nanosheets with high nitrogen content (10.13 at.%) achieve an even greater capacitance of 326 F/g in 6 M KOH, outperforming reduced graphene oxide (203 F/g). The enhanced performance is attributed to the combined effects of pseudocapacitance from pyrrolic-N and pyridinic-N, along with the improved conductivity introduced by quaternary-N. Additionally, these materials demonstrate excellent cycle stability (~100%) and Coulombic efficiency (~99.58%) after 2000 cycles.

Beyond oxygen and nitrogen, other heteroatoms – such as boron (B), sulfur (S), fluorine (F), and phosphorus (P) – can also significantly alter the surface properties of carbon materials, further enhancing their supercapacitive performance. Fluorine, with its high electronegativity, forms polarized C–F bonds that can exist in ionic, semi-ionic, or covalent forms, depending on the electronegativity difference between fluorine and carbon. This unique bonding character makes fluorinated carbons particularly promising for supercapacitor applications.

In many cases, co-doping with multiple heteroatom species yields synergistic effects that further enhance capacitance. Combinations such as N/O, N/B, N/F, N/P, N/S, and N/O/P have been explored to optimize electrochemical properties. For example, nitrogen and fluorine co-doped carbon microspheres (CM-NF) exhibit an ultrahigh volumetric capacitance of 521 F/cm^3, maintaining excellent cyclic stability with no capacitance loss after 10,000 cycles in both basic and acidic electrolytes at a current density of 5 A/g. The presence of fluorine plays a crucial role in redistributing nitrogen's charge, reducing the bandgap, and lowering charge transfer resistance, thereby doubling the volumetric capacitance compared to nonfluorinated counterparts [13, 16, 17].

24.5.2 Hybrid Capacitors

Carbon materials play a crucial role in hybrid capacitors like zinc-ion hybrid capacitors (ZIHCs) and potassium-ion hybrid capacitors (KIHCs), where defective carbon matrices enhance energy density and cycling stability through improved ion storage capabilities and optimized ion transport channels.

24.5.2.1 Lithium-Ion Hybrid Capacitors (LIHCs)

One innovative approach in lithium-ion hybrid capacitors involves the development of a cost-effective and sustainable carbon anode (DNC) derived from biowaste sepia. The DNC, produced via direct pyrolysis, features a defect-rich and nitrogen-doped structure that significantly contributes to its high specific capacity of 580.3 mAh/g at 0.05 A/g, excellent rate performance, and impressive cycle life (1000 cycles). When paired with a polyimide-derived NPCM-A cathode in the LIC device, it achieves an energy density of 101.7 Wh/kg, alongside a remarkable rate capability of 56.3 Wh/kg at 12.5 kW/kg. Furthermore, the device maintains 82.2% of its capacity after 3000 cycles, underscoring the potential of biowaste-derived materials in high-performance energy storage applications [18, 19].

24.5.2.2 Sodium-Ion Hybrid Capacitors (SIHCs)

In the realm of sodium-ion hybrid capacitors, challenges persist in achieving high capacity, fast charge/discharge rates, and long cycle life, especially with regard to carbon anodes. A promising solution lies in the self-induced internal electric field (SIEF), which enhances Na^+adsorption and overall performance. To optimize the anode material, sulfur–nitrogen co-doped hollow carbon nanofibers (SNHCF) are synthesized using a template-assisted electrospinning method, followed by low-temperature pyrolysis. These nanofibers, rich in micropores and chemisorption sites, promote the exposure of active sites and reduce volume expansion while also shortening Na+ diffusion distances. The co-doping enhances electronic interactions, forming an SIEF that aids Na+ adsorption. As a result, the SNHCF anode delivers an exceptional specific capacity of 588.2 mAh/g at 0.05 A/g, with an impressive rate performance of 264.3 mAh/g at 20 A/g. The full-carbon sodium-ion capacitor demonstrates an energy density of 119 Wh/kg and maintains 80.06% of its capacity after 12,000 cycles, showcasing outstanding long-term stability [20].

24.5.2.3 Potassium-Ion Hybrid Capacitors (PIHCs)

For potassium-ion hybrid capacitors (PIHCs), enhancing K+ storage capacity is critical. A key strategy involves designing hierarchical porous carbon anodes with defects and abundant active sites. This study introduces phosphorus-nitrogen double-doped foamed porous carbon (PN-FPC), which features meso-microporous structures that promote long-range mass transfer. The ultrathin carbon bubble walls within these structures maximize the availability of chemisorption sites. DFT calculations indicate that the synergistic effect of P and N co-doping enhances the reversible adsorption–desorption of K+. The PIHCs employing PN-FPC anodes, coupled with P, N double-doped activated porous carbon cathodes, exhibit impressive performance metrics: ultra-high energy density of 155.6 Wh/kg, power density of 17,000 W/kg, and exceptional cycling stability, with 90.9% capacity retention after 10,000 cycles [21].

Amorphous carbon anodes, typically challenged by unsatisfactory capacities at low temperatures, can be significantly enhanced through heteroatom-induced defect engineering. A novel preparation method for nitrogen/sulfur co-doped lignin-derived porous carbon (NSLPC) introduces an ultrahigh nitrogen doping level (21.6 at.% N) and sulfur content (0.8 at.% S) through a supramolecule-mediated pyrolysis strategy. This method, which combines biomolecular precursors with nitrogen-doping agents, results in high heteroatom doping levels, improving potassium storage performance. The NSLPC anode benefits from its defect-rich structure, abundant edge defects, and enhanced K+ adsorption, offering a promising route for high-performance potassium-ion storage [22].

Further advancements are achieved with phosphorus and nitrogen dual-doped hollow carbon spheres (PNHCS), synthesized using a nanoscale-confined in situ polymerization process followed by conventional carbonization. The PNHCS anode demonstrates superior sodium and potassium ion storage capabilities due to its hierarchical porous structure and dual doping. DFT calculations and characterizations such as Raman spectroscopy and XPS confirm that P/N doping enhances the electronic structure and ion adsorption, resulting in excellent electrochemical performance. Sodium and potassium-ion hybrid capacitors (SIHCs and PIHCs) utilizing PNHCS anodes deliver high energy densities of 178.8 Wh/kg and show remarkable cycling stability, offering new strategies for efficient and rapid ion storage in sodium and potassium-ion batteries [1].

24.5.2.4 Zinc-Ion Hybrid Capacitors (ZIHCs)

The development of advanced cathode materials for aqueous zinc-ion hybrid capacitors (ZIHCs) has been a major focus, as researchers aim to enhance their storage capacity, cycling stability, and power density. Despite the promising potential of these materials, challenges such as limited active sites and slow ion diffusion remain. Defect engineering strategies have been proposed to address these issues and improve electrochemical performance [23–25].

One approach utilizes a combination of self-assembled supermolecules and defect engineering techniques. By introducing edge-nitrogen doping and grafting small electronegative molecules, this method significantly enhances ion adsorption, resulting in a high capacity of 218 mAh/g and outstanding electrochemical performance for ZIHCs. Another method incorporates nitrogen and sulfur co-doping, along with vacancy defects in the porous carbon structure. This combination improves the material's ability to store zinc and sulfate ions, with the resulting HC-0.2 cathode achieving a high capacity of 245.8 mAh/g, 82.3% cycling stability after 10,000 cycles, and remarkable energy and power densities of 164.1 Wh/kg and 30.1 kW/kg, respectively [26, 27].

A third strategy explores the use of multichannel hollow carbon nanofibers (MCHCNF), where a pore construction-pore expansion strategy creates a multichannel structure with a graphene-like shell. This innovative approach improves ion diffusion, increases active site availability, and leads to a high capacity of 191.4 mAh/g, impressive cycling stability (93.0% retention over 15,000 cycles), and an energy density of 133.1 Wh/kg. The exceptional performance of these materials has been attributed to the combination of high-density defects, optimal pore sizes, and heteroatom-induced defects that create abundant active sites for electrochemical reactions. In another study, B, N co-doped porous carbon (BN-C) was synthesized using a salt template strategy. This cathode material enhanced the physical adsorption of anions and chemical absorption of Zn^{2+}, resulting in a high capacity of 190.2 mAh/g and an energy density of 105.1 Wh/kg. The B/N dual-doping significantly improved ion/electron transport and Zn^{2+} absorption at the electrolyte-carbon interface, making the material highly effective for energy storage. Moreover, a quasi-solid-state pouch-type battery utilizing this cathode demonstrated excellent durability and flexibility, highlighting its potential for flexible power sources [28, 29].

Further advancements in zinc-ion hybrid capacitors (ZICs) were made by developing a supermolecule-mediated pyrolysis strategy to convert sodium lignosulfonate (LS) into heteroatom-doped porous carbons with large mesopores. These carbons, doped with 14.9 at.% nitrogen and 4.7 at.% oxygen, exhibited a high specific capacitance of 266 F/g and excellent rate capability, owing to enhanced pseudocapacitance. First-principle calculations indicated that nitrogen and oxygen dopants facilitated the reversible adsorption of Zn^{2+} ions and protons, while pores smaller than 6.8 Å negatively impacted performance. This study emphasizes the importance of optimizing pore size and heteroatom doping for efficient Zn^{2+} ion storage and better electrochemical performance. In another notable study, heteroatom-doped carbon materials derived from biomass waste were explored as an alternative approach. Using a simple pyrolysis method, N, P co-doped foamy-like carbons (NPFC) were produced from osmanthus flowers. These materials exhibited uniform doping, porous structures, and high conductivity, providing abundant active sites for ion adsorption, enhancing ion transfer kinetics, and facilitating electron transport. When used as a cathode for zinc-ion hybrid capacitors, NPFC delivered a high energy density of 85.7 Wh/kg and long cycle life (97.4% retention after 20,000 cycles). The ZHC assembled with NPFC demonstrated excellent capacity (163.6 mAh/g) and power density (35.9 kW/kg), making it an attractive candidate for high-performance energy storage applications [11].

Finally, a novel approach to ZICs involved using a dual-ion adsorption mechanism in the cathode to achieve exceptional energy and power densities. By employing a templating/activating co-assisted carbonization process, a protein-rich biomass was converted into defect-rich carbon with a high surface area of 3657.5 m^2/g and 8.0 at.% heteroatom content. This material enabled the reversible adsorption of both cations and anions at heteroatom sites and defects, achieving a capacity of 257 mAh/g, a power density of 61,700 W/kg, and an energy density of 168 Wh/kg. The ZICs, which showed excellent performance in flexible solid-state pouch and cable-type configurations, hold promise for use in wearable devices and other advanced energy storage applications.

24.6 Surface/Interface Engineering

Addressing challenges in carbon cathodes, Liu et al. utilized a surface functionalization strategy through solution chemistry to enhance the specific capacitance of commercial AC powder. The improvements were attributed to pseudocapacitive redox reactions at C=O sites, which play a key role in fabricating high-performance carbon cathodes for lithium-ion capacitors (LICs). Further enhancement of electron conductivity and capacitive performance was achieved through heteroatom doping, which modifies the electron distribution in carbon materials. This doping not only improves the capacity but also accelerates the reaction kinetics in carbon cathodes by facilitating pseudocapacitive reactions [30–32].

Building on this, Xia et al. designed B and N dual-doped 3D porous carbon nanofibers (BNC), which created abundant active sites that contribute significantly to surface capacity. The BNC//BNC LIC device demonstrated impressive performance, achieving a high energy density of 220 Wh/kg, a power density of 22.5 kW/kg, and excellent cycling stability with 81% capacitance retention after 5000 cycles. These results emphasize the potential of carbon-based nanocomposites, which, through their synergistic effects, offer significant promise for energy storage applications [19].

Further advancements in the field involve interface engineering of carbon-based nanocomposites. Through methods like structural design, synthesis, and characterization of interfaces between carbon and noncarbon materials, electrochemical properties can be significantly improved. This review systematically explores research on the preparation of these nanocomposites via both noncovalent and covalent approaches. It also examines their effects on the electrochemical performance of supercapacitors, lithium-ion batteries, and lithium-sulfur batteries, while highlighting the challenges and future directions for the development of advanced carbon-based nanocomposite electrode materials [33].

24.6.1 Functionalization

Surface modifications, such as introducing oxygen functional groups and cyclic ether groups, have been explored to enhance electrochemical performance. These modifications are crucial in improving the overall energy storage capabilities while minimizing any adverse effects on conductivity.

24.6.2 Interface Engineering

The integration of CNTs, graphene, and metal oxides into nanocomposites has proven to be a highly effective strategy. This combination enhances electron and ion transport within the material, thereby improving the capacitive performance and overall efficiency of energy storage devices.

24.6.3 Characterization

To fully understand and optimize these materials, several characterization techniques are employed:

- **Direct Visualization:** SEM, TEM, and AFM are commonly used to directly observe defects in the material's structure. These tools provide detailed images that help in understanding the material's morphology and surface features.
- **Indirect Observation:** Raman spectroscopy, XPS, and XAS are used to gather insights into the chemical environment and electronic structure of defective carbons. These techniques allow for the analysis of defects at a molecular level and the evaluation of how they affect the material's electrochemical properties.

24.6.4 Applications in Electrochemical Pseudocapacitor Systems

Among various carbon materials, carbon nanostructures like 0D carbon nanospheres, 1D CNTs, 2D graphene, and 3D porous carbons are commonly used. These materials exhibit desirable physical and chemical properties that enable fast charge/discharge rates and high-power density. However, in their pure form, these materials typically have limited energy storage capacity and low energy density. This limitation has driven the development of hybrid electrodes that combine carbon with pseudocapacitive materials. The interactions between these components at the interface can enhance charge storage capacity and cycling stability, providing a more efficient and robust solution.

Notably, carbon/carbon nanocomposites such as CNT/mesoporous carbon, CNT/graphene, and graphene/porous carbon have been explored in depth. These composites show significant improvements in capacitance and rate performance, largely due to the optimized interfaces between the components. For instance, graphene and CNT composites benefit from enhanced interactions, including covalent bonding or π–π interactions, which contribute to improved stability and higher capacitance. In some cases, CNTs are grown between graphene layers to form a 3D structure, which further improves conductivity and overall electrochemical performance.

In addition to carbon-based materials, surface functionalization plays a key role in enhancing the performance of supercapacitors. The introduction of oxygen-functional groups (OFGs) on carbon surfaces can influence EDLC performance. While common OFGs may reduce conductivity and ion migration, cyclic ether groups (CEGs) have shown promise in enhancing pseudocapacitance without disrupting the carbon network. Surface functionalization with CEGs helps preserve the π-conjugation system of carbon, minimizing the degradation of the material's conductive properties while improving both capacitance and rate performance.

In summary, advances in carbon-based nanocomposites, interface engineering, and surface functionalization are critical to improving the performance of supercapacitors. These developments not only enhance energy storage capacity but also ensure high power density and excellent cycling stability, opening up new possibilities for practical energy storage applications.

One notable study addresses the challenge of increasing the mass loading of active materials in energy storage devices while maintaining high performance. A nitrogen-doped carbon interface with high conductivity, porosity, and electrolyte affinity was engineered on a flexible cellulose electrode, enabling the accommodation of 1D active materials. This design improves electron transport and ion migration, resulting in an ultrahigh capacitance of 9501 mF cm^{-2} for the flexible anode (with CNTs) and 6212 mF cm^{-2} for the flexible cathode (with polypyrrole nanotubes). The flexible quasi-solid-state asymmetric supercapacitor demonstrates a maximum energy density of 1.42 mWh cm^{-2} (2.2 V, 2105 mF cm^{-2}), the highest among reported flexible supercapacitors. The design approach involves creating a nitrogen-doped carbon interface on cellulose textile, which serves as a porous framework to house 1D nanomaterials. This structure enhances electron transport while preventing agglomeration, enabling high mass loading and capacitance without sacrificing performance. The assembled flexible ASC device exhibits excellent energy density and performance, offering a promising solution for wearable electronics and advancing energy storage beyond conventional supercapacitors.

Another significant advancement presents a novel approach to constructing an atomic-scale, seamless interface between various dimensional carbon nanostructures to enhance the transport of masses, electrons, ions, and energy in energy storage applications. Inspired by the structure of the *Catalpa bungei* tree, which efficiently transports nutrients through its thorns, a fractal-tree structure of nitrogen-doped graphene/CNTs (N-G/CNTs) composite was designed. By using NiMgAl-layered double hydroxides (LDHs) as a CVD substrate, CNTs and graphene were grown and covalently connected, reducing interface and thermal resistance while maintaining excellent mechanical and electrical properties. CO2 etching was applied to create defects in the matrix, enabling the anchoring of metal nanoparticles and graphene oxide sheets. The resulting N-G/CNTs composite exhibited a high specific surface area of 190.66 m^2/g, a narrow pore size distribution of 4–6 nm, and an excellent conductivity, leading to outstanding capacitor performance. A quasi-solid-state supercapacitor made from this composite delivered a high specific capacitance of 198.1 F/g at 0.2 A/g and demonstrated excellent cycle stability, with 91.1% capacitance retention after 2000 cycles. Furthermore, the supercapacitor successfully powered LED lamps, showcasing its practical potential for energy storage applications. This approach offers a versatile strategy for developing high-performance carbon-based materials for advanced energy storage solutions [34].

24.7 Future Perspectives

Challenges: The precise control of defect types and concentrations remains a significant challenge in the development of advanced energy storage systems and catalytic applications. Defects, including vacancies, edge sites, and oxygen-containing functional groups, play a crucial role in modifying the electronic and chemical properties of carbon materials. However, controlling these defects with high precision, particularly at the atomic level, continues to be difficult. Additionally, the evolution of these defects during electrochemical reactions such as charge/discharge cycles or catalytic processes, is not yet fully understood. To address these challenges, advanced in situ characterization techniques such as STM, XPS, and AFM are required to monitor defect formation and transformation in real time. These tools could provide valuable insights into how defects impact the long-term stability and efficiency of carbon-based electrodes and catalysts during operation, guiding the design of more robust materials with superior performance.

Opportunities: Despite these challenges, the development of defect-rich carbon materials with tunable properties offers immense potential for next-generation energy storage devices and catalytic applications. When strategically introduced, defects can enhance the electrochemical reactivity, surface area, and charge storage capabilities of carbon materials. For instance, defects in the carbon lattice can serve as active sites for ion adsorption, greatly improving the capacitive performance of supercapacitors. Defect engineering can also increase conductivity, reduce charge transfer resistance, and provide more accessible pathways for ion movement. Beyond energy storage, defect-engineered carbons are also attracting attention in catalytic processes, where they can act as highly active sites for reactions such as oxygen reduction, hydrogen evolution, and carbon dioxide reduction. The ability to tune the density and nature of defects enables the design of materials with specific properties tailored to enhance performance in these diverse applications. Therefore, defect-rich carbon materials hold the potential to revolutionize not only energy storage systems but also catalysis, positioning them as key candidates for sustainable technologies in the future.

24.8 Conclusion

Defect engineering in carbon materials is a powerful approach to tailoring their electrochemical properties for advanced energy storage and catalytic applications. By strategically introducing intrinsic and extrinsic defects, such as vacancies, edge sites, and heteroatoms, electron transport, ion diffusion, and overall electrochemical performance can be significantly enhanced. This engineering allows for the design of materials with optimized properties capable of meeting the demands of next-generation energy devices. This document provides a comprehensive overview of how defects and heteroatom doping contribute to improving the performance of carbon materials in energy storage, catalysis, and electrochemical applications. It emphasizes the importance of defect engineering, surface/interface modifications, and advanced characterization techniques in developing next-generation carbon-based materials, which are crucial for driving sustainable energy technologies forward. Future research should focus on achieving precise defect control, advancing large-scale synthesis techniques, and successfully integrating defect-rich carbons into cutting-edge energy devices.

References

1. Gao, J., Wang, G., Wang, W. et al. (2022). Engineering electronic transfer dynamics and ion adsorption capability in dual-doped carbon for high-energy potassium ion hybrid capacitors. *ACS Nano* 16: 6255–6265.
2. Sahu, R., Ponnusami, S.A., Weimer, C., and Harursampath, D. (2024). Interface engineering of carbon fiber composites using CNT: a review. *Polym. Compos.* 45: 9–42.
3. Liu, Z., Bol, A.A., and Haensch, W. (2011). Large-scale graphene transistors with enhanced performance and reliability based on interface engineering by phenylsilane self-assembled monolayers. *Nano Lett.* 11: 523–528.
4. Yang, F., Zhang, L., Zuzuarregui, A. et al. (2015). Functionalization of defect sites in graphene with RuO_2 for high capacitive performance. *ACS Appl. Mater. Interfaces* 7: 20513–20519.
5. Chen, J., Han, Y., Kong, X. et al. (2016). The origin of improved electrical double-layer capacitance by inclusion of topological defects and dopants in graphene for supercapacitors. *Angew. Chem.* 55: 13822–13827.
6. Bandaru, P.R., Yamada, H., Narayanan, R., and Hoefer, M. (2017). The role of defects and dimensionality in influencing thecharge, capacitance, and energy storage of graphene and 2D materials. *Nanotechnol. Rev.* 5: 421–433.
7. Luo, X., Zheng, H., Lai, W. et al. (2023). Defect engineering of carbons for energy conversion and storage applications. *Energy Environ. Mater.* 6: e12402.
8. Zhao, Z., Chen, H., Zhang, W. et al. (2023). Defect engineering in carbon materials for electrochemical energy storage and catalytic conversion. *Mater. Adv.* 4: 835–867.

9 Zhao, Y., Yang, B., Liu, Y. et al. (2022). Capturing carriers and driving depolarization by defect engineering for dielectric energy storage. *ACS Appl. Mater. Interfaces* 14: 6547–6559.

10 Ma, R., Luo, W., Yan, L. et al. (2023). Constructing the quinonyl groups and structural defects in carbon for supercapacitor and capacitive deionization applications. *J. Colloid Interface Sci.* 645: 685–693.

11 Wei, F., Wei, Y., Wang, J. et al. (2022). N, P dual doped foamy-like carbons with abundant defect sites for zinc ion hybrid capacitors. *Chem. Eng. J.* 450: 137919.

12 Yuan, R., Dong, Y., Zhang, S. et al. (2021). Efficient utilization of the active sites in defective graphene blocks through functionalization synergy for compact capacitive energy storage. *ACS Appl. Mater. Interfaces* 13: 57092–57099.

13 Jin, H., Feng, X., Li, J. et al. (2019). Heteroatom-doped porous carbon materials with unprecedented high volumetric capacitive performance. *Angew. Chem.* 58: 2397–2401.

14 Mofokeng, T.P., Tetana, Z.N., and Ozoemena, K.I. (2020). Defective 3D nitrogen-doped carbon nanotube-carbon fibre networks for high-performance supercapacitor: transformative role of nitrogen-doping from surface-confined to diffusive kinetics. *Carbon* 169: 312–326.

15 Polaki, S.R., Sahoo, G., Anees, P. et al. (2021). Engineering the edge-terminations and defect-density to enhance the electrochemical capacitance performance of vertical graphene nanosheets. *Appl. Surf. Sci.* 545: 149045.

16 Ranjith, K.S., Lee, S.-Y., Ghoreishian, S.M. et al. (2023). Defect and interface engineering of MXene-tagged N,F-doped carbon-$CoSe_2$ heterostructure forsuperior hydrogen evolution reactions and supercapacitors. *Carbon* 206: 246–259.

17 Ji, Z., Tang, G., Chen, L. et al. (2024). Defect engineering of hollow porous N, S Co-doped carbon spheres-derived materials for high-performance hybrid supercapacitors. *Chem. Eng. J.* 480: 148213.

18 Jiang, J., Zhang, Y., Li, Z. et al. (2020). Defect-rich and *N*-doped hard carbon as a sustainable anode for high-energy lithium-ion capacitors. *J. Colloid Interface Sci.* 567: 75–83.

19 Zou, K., Cai, P., Cao, X. et al. (2020). Carbon materials for high-performance lithium-ion capacitor. *Curr. Opin. Electrochem.* 21: 31–39.

20 Peng, H., Hou, W., Ma, L. et al. (2024). Hollow carbon nanofibers with self-induced internal electric field for high-performance full-carbon sodium ion capacitors. *Chem. Eng. J.* 500: 157145.

21 Liu, Z., Peng, H., Xie, X. et al. (2024). Coupling of reasonable micro-defect structure and multiple chemisorption sites for boosting the K^+ hybrid capacitors. *Energy Storage Mater.* 65: 103100.

22 Zhao, L., Sun, S., Lin, J. et al. (2023). Defect engineering of disordered carbon anodes with ultra-high heteroatom doping through a supermolecule-mediated strategy for potassium-ion hybrid capacitors. *Nano Lett.* 15: 41.

23 Zhang, W., Yin, J., Jian, W. et al. (2022). Supermolecule-mediated defect engineering of porous carbons for zinc-ion hybrid capacitors. *Nano Energy* 103: 107827.

24 Yuan, R., Wang, H., Shang, L. et al. (2023). Revealing the self-doping defects in carbon materials for the compact capacitive energy storage of Zn-Ion capacitors. *ACS Appl. Mater. Interfaces* 15: 3006–3016.

25 Fan, W., Ding, J., Ding, J. et al. (2021). Identifying heteroatomic and defective sites in carbon with dual-ion adsorption capability for high energy and power zinc ion capacitor. *Nano-Micro Lett.* 13: 59.

26 Li, H.-X., Shi, W.-J., Liu, L.-Y. et al. (2024). Coupling effect of vacancy defects and multi-adsorption sites in porous carbon cathode for high-performance aqueous zinc-ion hybrid capacitors. *Chem. Eng. J.* 487: 150630.

27 Zhang, J., Qin, F., Zuo, P., and Shen, W. (2025). Supermolecule-opitimized defect engineering of rich nitrogen doped porous carbons for advanced zinc-ion hybrid capacitors. *ChemSusChem* 18: e202401311.

28 Zhao, L., Sun, S., Lin, J. et al. (2023). Defect engineering of disordered carbon anodes with ultra-high heteroatom doping through a supermolecule-mediated strategy for potassium-ion hybrid capacitors. *Nano Lett.* 15: 41.

29 Zhang, Y., Zhu, C., Xiong, Y. et al. (2023). Multi-channel hollow carbon nanofibers with graphene-like shell-structure and ultrahigh surface area for high-performance zn-ion hybrid capacitors. *Small Methods* 7: e2300714.

30 Liang, Y., Zhang, W., Wu, D. et al. (2018). Interface engineering of carbon-based nanocomposites for advanced electrochemical energy storage. *Adv. Mater. Interfaces* 5, 1800430.

31 Chen, R., Ling, H., Huang, Q. et al. (2022). Interface engineering on cellulose-based flexible electrode enables high mass loading wearable supercapacitor with ultrahigh capacitance and energy density. *Small* 18: e2106356.

32 Han, C., Li, Y.-H., Qi, M.-Y. et al. (2020). Surface/interface engineering of carbon-based materials for constructing multidimensional functional hybrids. *Sol. RRL* 4, 1900577.

33 Zhang, Q., Wang, C., Zhang, H. et al. (2020). Designing ultrahard nanostructured diamond through internal defects and interface engineering at different length scales. *Carbon* 170: 394–402.

34 Ye, J., Hu, B., Jin, Y. et al. (2020). Interface engineering integrates fractal-tree structured nitrogen-doped graphene/carbon nanotubes for supercapacitors. *Electrochim. Acta* 349: 136372.

25

Metal Oxide-Based Electrochemical Supercapacitors: Performance Enhancement by Defects and Interface Engineering

Poovitha Ganesan[1], Yuvashree Jayavelu[2], D. Paul Joseph[2], V. Ganesh[1], Rathika Rajendran[3], and Kovendhan Manavalan[1]

[1] *SRM Institute of Science and Technology, Department of Physics and Nanotechnology, Kattankulathur, Chengalpattu, Tamil Nadu, 603203, India*
[2] *National Institute of Technology, Department of Physics, Hanamkonda, Warangal, Telangana, 506004, India*
[3] *St. Theresa's Arts and Science College for Women, Department of Physics, Mayiladuthurai District, Sathangudi, Tharangambadi, Tamil Nadu, 609313, India*

25.1 Introduction

Energy storage devices play a crucial role in balancing energy supply and demand in modern society. At the earlier stages, electricity was stored from the nonrenewable energy sources such as coal, oil, and natural gas, but it created a lot of problems in our daily life. Because burning of fossil fuels emits carbon dioxide and other greenhouse gases into the atmosphere, causing climate change and respiratory problems for humans. To address these challenges, one can obtain an energy from renewable energy sources such as solar energy, wind energy, nuclear energy, biomass energy, hydroelectric energy, geothermal energy, tidal energy, and so on. The renewable energy sources offer many advantages, although there are still major challenges that make it difficult to use them effectively in various situations. For example, solar cell produces electricity from sunlight only during daytime because of the accessibility of sunlight, and the mechanism of energy storage is lacking. Seasonal energy sources such as wind and tidal energy also have the same difficulties. The intermittent nature of these sources causes challenges in continuous energy supply with demand. Some renewable energies such as nuclear energy, biomass energy, hydroelectric energy, and geothermal energy produce continuous power. However, due to their exclusive requirements, these power stations need specific locations. For example, a nuclear power station needs a vast amount of water for the purpose of cooling the system, which causes the coastline construction. These types of renewable energy sources are not used when the people need it most, because large amount of potential energy is wasted [1]. Therefore, electrochemical energy storage systems such as batteries, supercapacitors, and fuel cells were developed, and these offer high energy storage capabilities, rapid charge, and discharge rate. The Ragone plot (Figure 25.1) shows the comparison between different electrochemical energy

Surface and Interfacial Defects in Nanomaterials for Sustainable Energy Production and Storage, First Edition. Edited by Noé Arjona, Lorena Álvarez Contreras, and Minerva Guerra Balcázar.

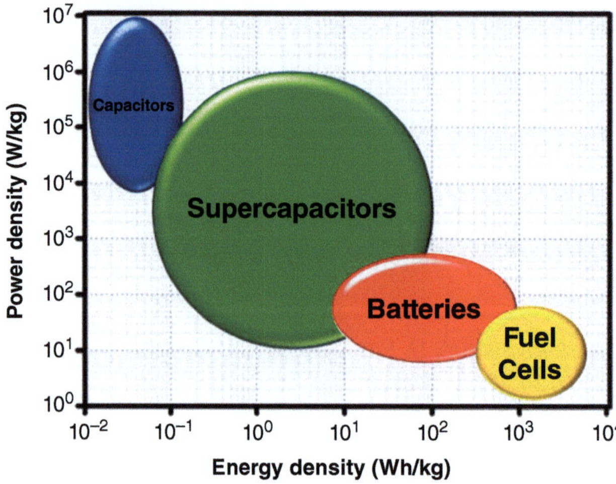

Figure 25.1 Ragone plot for different energy storage devices. Source: Ref: Shivasharma, K., Upadhyayet, N., Deshmukhal, T.B. et al. (2023). Exploring vacuum-assisted thin films toward supercapacitor applications: present status and future prospects. *ACS Omega* 8(41): 37685–37 719 / American Chemical Society / CC BY-NC-ND 4.0.

storage devices. Batteries and fuel cells have high energy density and low power density, which means they can store high energy; however, their charge–discharge rate is low. They have limitations such as low power density, high internal resistance, short life span, and high weight, due to this reason they have a limited application. Capacitors have high power density and low energy density, which means they can store low energy and their charge/discharge rate is high. There are limitations in capacitors such as low energy density, leakage current, and cost; due to this reason, they have limited applications [2]. Supercapacitors, also known as ultracapacitor or electrochemical capacitors, have a moderate power and energy density, which means they can store low energy, but they have fast charge–discharge rate. Due to this reason, they can be used for energy storage applications, including power backup systems, vehicles, flashlights, and wind turbines. Supercapacitors bridge the gap between conventional capacitors and batteries. The profitability of supercapacitors includes the extended lifespan (charge/discharge rate is greater than 50,000 times), including applicability in wide temperature range, minimal charge loss at the time of charging and discharging, and minor equivalent series resistance (ESR). Due to the high input and output current, the supercapacitor can charge and discharge rapidly.

A supercapacitor consists of two electrodes (anode and cathode), an electrolyte (aqueous or non-aqueous) and a separator (Figure 25.3 (a)). Based on energy storage mechanism, the supercapacitors can be classified into three types: (i) electric double-layer capacitors (EDLCs), (ii) pseudocapacitors (PCs), and (iii) hybrid capacitors (HCs). EDLCs store charge via electrode–electrolyte interface. Electrodes used in EDLCs are carbon-based materials because they have high surface area, low cost, and availability. Examples of carbon-based electrode materials are activated carbon, carbon aerogels, carbon nanotubes, graphene, etc. PCs store charge via redox Faradaic reaction. Electrode materials used in PCs include metal oxides and conducting polymers. The metal oxides have high specific capacitance and low resistance. Examples of metal oxides-based electrode materials include NiO, RuO_2, MnO_2, and IrO_2. Conducting polymers has high conductivity, capacitance, and equivalent series resistance. Examples of conducting polymer-based electrode materials include polyaniline, polypyrrole, and polyacetylene. The energy storage mechanism of hybrid capacitor is a combination of both electric double-layer capacitors and PCs. The materials used in the hybrid supercapacitors are carbon-based materials, metal oxides, and conducting polymers, which are based on the type of hybrid supercapacitors used. Although these materials have their own disadvantages, such as metal oxides having a high capacitance but a lows conductivity, carbon-based materials are conductive but have a low capacitance, and conductive polymers are unstable despite having a high capacitance and conductivity [3]. Defect and interface engineering have emerged as promising approaches to overcome these limitations. By introducing defect states, such as oxygen vacancies and optimizing the interfaces between active materials and conductive substrates, substantial improvements can be made in conductivity, ion transport, and structural stability. This chapter delves into the role of defect and interface engineering in enhancing the electrochemical performance of supercapacitors.

25.1.1 Background of Various Energy Storage Device

25.1.1.1 Fuel Cell

The first fuel cell was successfully developed in 1932. Fuel cell is a electrochemical device that converts chemical energy into electrical energy, and it is not an energy storage device [4]. The components used in the fuel cells include a positive electrode (or cathode), a negative electrode (or anode), separator, and electrolyte. Fuel cells store energy and generate power continuously as long as fuels (e.g. propane, natural gas, diesel, hydrogen) and oxygen are supplied. The working mechanism of fuel cells includes supplying fuel to anode to initiate oxidation reaction and releasing electrons. These electrons flow through an external circuit, generating electricity. The electrolyte conducts ions between the electrodes in order to complete the circuit. At the cathode, the oxidant gains electrons and combines with the ions from the electrolyte and produce water or other byproduct. Transportation, stationary power generation, and portable electronics are some of the applications of the fuel cells. The cost for hydrogen production and the need for continuous supply of fuels remain as a big challenge for developing fuel cells.

25.1.1.2 Capacitor

Capacitor, first discovered using Leyden jar in 1740s, is an energy storage device that stores energy in the form of electric charge. It consists of two conductors (plates) separated by an insulating material, also known as dielectric. When a voltage is applied across a capacitor's plates, electrons move from the negative terminal of the source to one plate, giving it a negative charge. At the same time, electrons leave the other plate toward the positive terminal, making the other plate as positively charged. This setup generates an electric field between the plates, and the energy from the voltage source is then stored. The most common applications of the capacitors include energy storage, electronic noise filtering, and remote sensing. The capacitors encounter some drawbacks such as low energy density, leakage current, and voltage limitations.

The energy stored (E) in a capacitor is given by:

$$E = \frac{1}{2}CV^2 \tag{25.1}$$

25.1.1.3 Battery

Battery was first invented by Alessandro Volta, whereas rechargeable battery was invented by Gaston Plante. The components in the batteries include the cathode (negative side), the anode (positive side), and the electrolyte. Batteries are mainly classified into two types: primary batteries (nonrechargeable battery) and secondary batteries (rechargeable battery). Primary batteries usually consist of one or more electrochemical cells that transform the chemical energy into electrical energy. When the chemical reactants are exhausted or used up, the battery loses its functionality, become inactive and can no longer be utilized. Common examples of primary batteries are alkaline batteries, zinc–carbon batteries, and lithium batteries. These are commonly used in applications such as portable electronic devices, toys, and flashlights. Secondary batteries are also known as rechargeable batteries; they can store and release electrical energy repeatedly and can be used for long term. They can be recharged by applying an external electrical current that reverses the chemical reactions within the batteries. This allows the battery to restore its energy capacity and can be reused. Examples of secondary batteries are lead–acid batteries, lithium-ion batteries, and nickel–metal hydride batteries. Secondary batteries are widely used in applications such as electric vehicles, portable electronics, renewable energy systems, and grid storage, where the ability to recharge and reuse the battery is essential for sustainability and cost-effectiveness.

25.2 Classification of Supercapacitor

25.2.1 Electric Double-Layer Capacitor (EDLC)

EDLCs having charge storage mechanism is similar to conventional capacitors. They consist of two carbon-based electrode materials, an electrolyte and a separator. Carbon-based materials include graphene, xerogels, CNFs, AC, carbide-derived carbon, and CNTs. The EDLCs exhibit outstanding conductivity, stability, and larger specific surface area. Benefits of these materials include having high power density, large cycle life, extraordinary rate performance, and high cyclic efficiency [5]. In Figure 25.3 (b) EDLCs store charge via electrostatically or non-Faradic process, which involves no transfer of charge between electrode and the electrolyte [6]. Energy storage in EDLCs is facilitated by the formation of an electric double layer between the electrode and the electrolyte. They cannot undergo any sort of redox reactions. When voltage is applied to the setup, the device starts storing charges at the pore surface of the electrode material through adsorption and de-adsorption of ions across the separator. Simultaneously, oppositely charged ions move from the electrolyte to electrodes in order to maintain neutrality. The EDLC device is highly reversible and exceedingly stable, with no volumetric change due to the electrostatic charge storage. Since no charge is transferred between electrode and electrolyte under this operation, the concentration of the electrolyte is never changed [7]. The specific capacitance of the EDLCs is calculated by using the equation:

$$C = \frac{A\varepsilon_0\varepsilon_\mathrm{r}}{d} \tag{25.2}$$

The higher charge storage capacity of EDLCs is due to two main factors, namely, (i) the distance between the charges on the electrode surface and the opposite ions in the electrolyte is extremely small, resulting in a greater capacitance; and (ii) porous carbon-based materials such as activated carbon, carbon nanotubes, and graphene are commonly used because they offer a large surface area, which enables a greater accumulation of charge at the electrode–electrolyte boundary, improving capacitance and energy storage. As a result, porous carbon materials are highly suitable for EDLCs, providing a large surface area and efficient charge storage at the interface. The limitations in EDLCs include lower electrochemical capacitance and less energy density. These limitations may restrict their effectiveness in energy storage applications.

25.2.2 Pseudocapacitor

Pseudocapacitors significantly enhance the energy density of supercapacitors as they store energy through rapid surface or near surface redox reactions occurring at the electrode material, resulting in higher capacitance compared with traditional double-layer capacitors. Pseudocapacitance materials having similar electrochemical mechanism to those of EDLCs are further enhanced by reversible redox reactivity that improves the electrochemical properties of electrode materials. When an external voltage is applied to the pseudocapacitor electrode, fast and reverse reduction–oxidation process happens at the current collector and electrolyte interface, resulting in high energy density and a short cyclic lifespan (Figure 25.3 (c)). Pseudocapacitors can be categorized into three types based on electrochemical perspectives: underpotential deposition, redox pseudocapacitance, and intercalation pseudocapacitance (Figure 25.2). Underpotential deposition mainly occurs when metal ions form an adsorption monolayer from a solution on a different metal surface above the redox potential (Figure 25.2 (a)). Redox pseudocapacitance occurs when the adsorption of metal ions from an electrolytic solution takes place onto the surface of the active material. Charge is stored by the process of Faradic redox reaction (Figure 25.2 (b)). Intercalation pseudocapacitance occurs when ions from an electrolytic solution intercalate into the pore materials and then undergo redox reactions with the surrounding atoms to transfer electrons (Figure 25.2 (c)). The remarkable difference between battery and pseudocapacitive materials is the charging and discharging mechanisms of pseudocapacitive

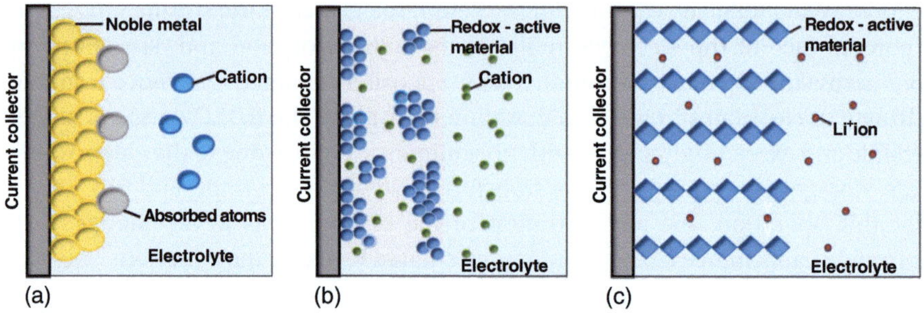

Figure 25.2 Different types of pseudocapacitors: (a) underpotential deposition, (b) redox pseudocapacitance, and (c) intercalation pseudocapacitance.

materials that are in the order of seconds and minutes. Pseudocapacitance materials have both high energy and high power density. Due to this reason, the researchers get interest in studying and developing pseudocapacitive materials.

Transition metal oxides (TMOs) are the most commonly used electrode material in pseudocapacitors because of their very high conductivity. Further, they also exhibit high specific capacitance and low resistance. RuO_2 is the most commonly used metal oxides for various applications in the earlier stage of the development of pseudocapacitors.

Although RuO_2 has low effective series resistance (ESR) and very high specific capacitance it is very costly compared to other TMOs. In transition metals, *d* orbitals are in the partially filled state; they belong to Groups 3–11. These elements have oxidation states in which low oxidation states tend to form ionic bonds and high oxidation states tend to form covalent bonds. Electronegativities increase with rise in oxidation states. Metallic bonding occurs when delocalized *d* electron enhances cohesion by increasing the number of shared electrons. Transition metals usually have high densities and high melting and boiling points. Various electrochemical investigations on these metal oxides have been explored by varying morphology and porosity. Metal oxides allow facile intercalation and de-intercalation of protons, thereby facilitating oxidation–reduction reaction [8]. Usually in the pseudocapacitors, theoretical capacitance was calculated by using the equation:

$$C = \frac{n \times F}{M \times V} \tag{25.3}$$

where C is the theoretical capacitance, n is the number of electrons transferred in the Faradaic process, F is a Faraday constant, M is a molar mass of active material, and V is a voltage window. Furthermore, contradictory to TMOs, conducting polymers also shows more specific capacitance. Conducting polymers such as polyaniline (PANI), polypyrrole (PPy), polythiophene (PT), and polyacetylene (PA) are other types of pseudocapacitive electrode materials. The pseudocapacitors have capacitance of 10–100 times greater than EDLCs. The challenges in pseudocapacitors when compared to EDLCs have lower power density. The reason for this is generally, the Faradaic processes are slower than non-Faradaic processes. Moreover, in pseudocapacitors, the electrodes store energy by surface redox reaction, where the ions in the electrolyte interact with the surface of the electrode material. The electrode materials often undergo structural or volume change since the ions in the electrolytes are absorbed or deliver during the charging and discharging cycles [9].

25.2.3 Hybrid Supercapacitors

Hybrid supercapacitor (HC) is a type of energy storage capacitor, and EDLCs offer high power density and good cyclic stability but have lower capacitance. Pseudocapacitors offer high specific capacitance at a lower power density, and the drawback is poor cyclic stability performance. Hybrid supercapacitors overcome the limitations, including energy density of EDLC, which is lower than midrange batteries. Pseudocapacitor has a low power density. The important role of HCs is to increase energy density. This can be accomplished by the incorporation of high-performance electrodes and electrolyte materials or by the incorporation of hybrid supercapacitors. In Figure 25.3 (d) hybrid supercapacitors, non-Faradaic electrode from EDLCs acts as a negative electrode and the Faradaic electrode from PCs acts as a positive electrode, and so it provides a better electrochemical performance and has the ability to store charge through electrostatic and electrochemical mechanisms [7]. HCs integrate with two or more types of electrode materials, and its energy storage mechanism is a combination of both EDLCs and pseudocapacitors. HCs have a high potential window and a high specific capacitance [10]. Hybrid electrodes can be classified into three types based on the electrode material: composite, asymmetric, and battery type. The composite electrode is composed of carbon-based materials with either metal oxides or conducting polymers. The asymmetric electrode is composed of both Faradaic and non-Faradaic process by coupling EDLCs and pseudocapacitors. In these type of supercapacitors, carbon-based materials are used as negative electrode, while metal oxides or conducting polymers are used as positive electrode. The battery-type hybrid supercapacitor contains battery-type electrode material and supercapacitor electrode [11]. The supercapacitors with overall output is interconnected to the resistance, which consists of internal resistance, interfacial resistance, and diffusion resistance, and it is known as the equivalent series resistance (ESR).

25.2.3.1 Symmetric and Asymmetric Supercapacitors

The hybrid supercapacitor provides high specific energy and power. The advantages of HCs include high cyclability, reversible non-Faradaic cathode, and an extraordinary high capacitance of anode. A symmetric hybrid device is composed of both electrodes (anode and cathode) made of similar materials; for instance: both the anode and the cathode are made of –carbon-based materials. Further, an asymmetric hybrid device is made of carbon-based materials with either metal oxides or conducting polymers, paring a negative and positive electrode [12]. TMOs and conducting polymers are the promising electrode material for pseudocapacitive behavior of the asymmetric hybrid supercapacitors. These pseudocapacitive electrodes show poor performance, and these electrodes integrated into asymmetric hybrid system enhance the properties such as cyclic stability, efficiency, energy, and power density. The fundamental difference between symmetric and asymmetric hybrid devices is mainly the electrode material used. Using two similar carbon-based electrode materials, the symmetric supercapacitor performance is investigated in both aqueous and nonaqueous electrolyte [13, 14].

25.2.3.2 Battery-Type Supercapacitors

A battery-type hybrid system is similar to asymmetric-type hybrid system. The battery-type hybrid system consists of two different electrodes, one is battery-type electrode and another one is supercapacitor electrode. The combination of these two electrodes is known as battery-type hybrid supercapacitor. The battery-type electrodes have the characteristics of increasing energy, and the supercapacitor electrodes have the quality of rising power. During charging and discharging, the battery-type electrode experiences a Faradaic redox reaction, whereas the supercapacitor electrode experiences an electric double-type charging and discharging. This combination provides good cyclic stability, power density, and energy density. Carbon-based materials, metal oxides, and conducting polymers are used as supercapacitor electrode, whereas lithium ion is used as battery-type electrode [9].

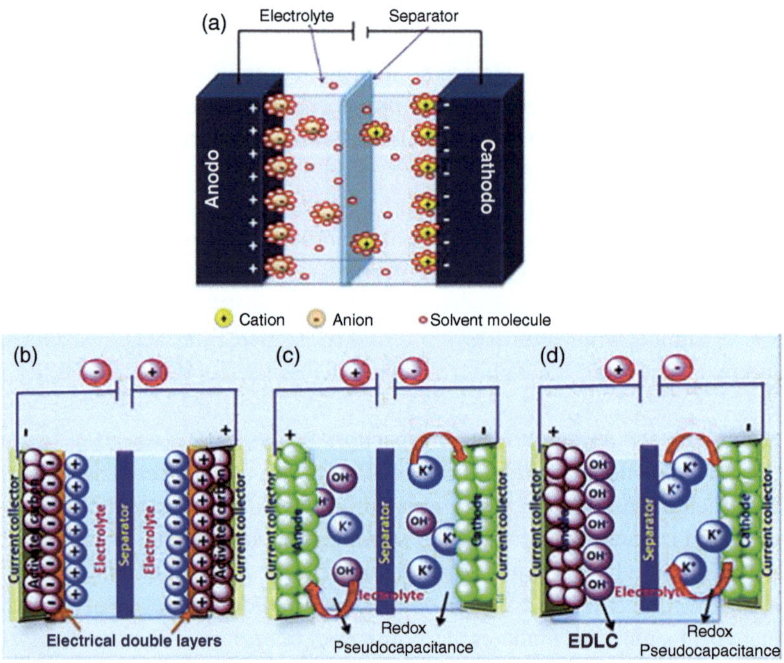

Figure 25.3 (a) Schematic representation of the structure of supercapacitor. Types of supercapacitors: (b) EDLC, (c) pseudocapacitor, and (d) hybrid supercapacitor. Source: Ref: Chaluvachar, P., Sudhakar, Y.N., Mahesha, G.T. et al. (2024). Emerging role of graphitic carbon nitride in advanced supercapacitors: a comprehensive review. *J. Energy Chem.* 103: 498–524/Elsevier / CC BY 4.0.

25.3 Supercapacitor Components

A supercapacitor is a type of electrochemical capacitor that stores energy through Faradaic and non-Faradaic processes. The supercapacitors efficiency is determined by the selection of active electrode material, electrolyte, separator, and current collector. Researchers aim to improve the energy density, power density, cycle life, and voltage using the various components of the electrode and electrolyte.

25.3.1 Electrode

In EDLC, the electrode material used is carbon-based materials. The excellent conductivity, inertness, a very high surface area, and a tunable pore size are the features of EDLCs. The drawback is it has a lower energy density [15].

The basement of the pseudocapacitors is made of the metal oxide-based electrode materials that has good conductivity, high-temperature stability, long-term chemical stability, and high corrosion resistance. Metal oxides such as RuO_2, V_2O_2, MnO_2, Co_3O_4, Fe_2O_3, and NiO are the most commonly used electrode materials. When a voltage is applied, ions from the electrolyte are adsorbed onto the electrode surface. It triggers a chemical transformation where electrons are exchanged between the electrode and these absorbed ions. The process in which the electron exchange leads to a chemical reaction is known as Faradaic process. The challenges in supercapacitor metal oxides are

- Some metal oxides can be expensive or difficult to synthesize
- Some metal oxides may exhibit poor conductivity or low-rate capability
- Metal oxide electrodes may suffer from capacity fading over time.

Despite these challenges, metal oxide electrodes remain a promising material for pseudocapacitors due to their high performance and versatility.

The electrode materials used in the hybrid supercapacitors are carbon-based materials, metal oxides, metal hydroxides, and battery-type electrodes. These types of materials increase the energy and power density of the supercapacitors.

25.3.2 Electrolyte

Due to the higher conductivity, low cost, nontoxicity, and environmental friendliness of the aqueous electrolyte, it has gained more popularity in recent years. Based on the pH value, the aqueous electrolyte can be classified into three types: acidic, alkaline, and neutral electrolyte [16, 17]. H_2SO_4 and HCl electrolytes belong to acidic electrolytes. KOH, NaOH, and LiOH belong to alkaline electrolytes, and Na_2SO_4, LiCl, and NaCl belong to neutral electrolytes. The potential window of the electrolyte plays a crucial role for the development of supercapacitors. To improve the supercapacitors energy density, one of the most important factors is to improve the potential window of the electrolyte. In neutral electrolyte, the ionic conductivity is lower when compared with acidic and alkaline electrolytes [18, 19]. Neutral electrolytes potential window is 1.4–2 V; it has the higher potential window compared with acidic and alkaline electrolytes. In neutral electrolytes, pseudocapacitive electrode material operates at high voltage compared to other types of supercapacitors. Acidic and alkaline aqueous electrolytes have high ionic conductivity and ionic mobility, and these electrolytes are widely used in supercapacitors. These electrolytes provide high charge storage capacity and better rate performance. Potential window of acidic electrolyte is 1.2 V. Further, the widely used acidic electrolyte is H_2SO_4, especially for carbon-based electrode materials as it has a lower potential window. Other electrolytes including $HClO_4$, HClO, and HNO_3 are the rarely used electrolytes because they have high strong oxidation properties. Nowadays, researchers focus on preparing the other types of electrolytes, which may improve the potential window and overall performance of the supercapacitors. In alkaline electrolytes, the metal oxides, hydroxides, and sulfides have high pseudocapacitance. The carbon-based symmetric supercapacitors have a voltage window below 1.2 V. A asymmetric supercapacitor is composed of metal hydroxides and carbon-based materials for providing the high electrochemical performance and has voltage window of 1.6 V [20].

25.3.3 Separator

The separator acts as the barrier between the two same or different electrodes. It is mainly used to avoid short circuiting between the two electrodes while allowing the electrolyte ions to pass through. The separator material must have the following criteria:

- high ionic conductivity, mechanical strength, thermal stability, and low cost.
- minimum intrinsic resistance with electrolyte ion permeability.

Glass, paper, and ceramics are also used as separator material. Polymer-based separators have low cost and porous nature. Further, the polymer-based separator is chosen from polypropylene, polypropylene-carbonate, polyethylene, polyamide, and polyvinylidene difluoride. Among these polymers, polypropylene is mostly used due to its ease of wettability [21, 22]. Nowadays, researchers develop a flexible, economical, and durable paper-based separator, because it has flame retardant and good mechanical stability.

25.3.4 Current Collectors

The main function of the current collector is to transport electrons to external circuits while also helping to dissipate heat generated inside the cell [23]. The most commonly used current collector materials in the supercapacitors include carbon-based materials (carbon black, carbon nanotubes), various polymers, and metal current collectors (foil, porous metal). The choice of current collectors depends on the type of electrode material and electrolyte used. The limiting factor of metallic current collectors is corrosion of metal current collectors in a water-based electrolyte. The supercapacitors internal resistance is affected by the conductivity of the electrode material and the current collector, as well as the resistivity of the electrolyte and the thickness of the separator. The desirable solution for the corrosion of metal current collectors and the diminishing electronic resistance at the interface between the current collector and the active electrode layer is rectified by physicochemical variation of the current collector surface and the deposition of an additional intermediate layer (binder) between the active electrode layer and the current collector. Binders are used for better adhesion of the active material with current collectors; a polymer such as polyvinylidene fluoride (PVDF) and polyacrylonitrile (PAN) is used as a binding material. The conducting polymer is used as a binder. The current collector should have the characterization of high electrical conductivity, low contact resistance with the active electrode material, high specific surface area, high electrochemical stability, and cost features, which must be considered for choosing the current collectors.

25.4 Synthesis Strategies for Electrode Materials

25.4.1 Hydrothermal Method

Hydrothermal method is one of the most important methods used for the synthesis of metal oxides in microscale and nanoscale. The term hydrothermal indicates that "hydro" means water and "thermal" means heat. In this method, water is used as the solvent, and the basic principle of hydrothermal synthesis is to dissolve the materials in water; the precursors are taken in the autoclave, and the temperature in the range of higher than 100 °C is required, and also high pressure is required to dissolve the materials [24]. By controlling reaction parameters such as reaction temperature and pressure and concentration of the reactants and reaction time, it is possible to control the chemical composition, particle size, and morphology of the prepared electrode material. The synthesis is performed for 6–48 hours to achieve a high crystalline material. The advantages of hydrothermal method are simple operation and cost saving, and the size and morphology of the materials are controllable.

25.4.2 Solvothermal Method

The solvothermal method is similar to the hydrothermal method. In the solvothermal method, either an organic or a mixed solvent is used, but in the hydrothermal process, water is used as the solvent, and further temperature is based on the solvent used, and usually temperature used in this method is higher than the boiling point of the solvent used [24]. The pressure (1–10,000 atm) and temperature (100–1000 °C) are carried out for the synthesis of the nanoparticles. The advantages of this method are shape, size, and surface area, and the morphology of the synthesized material is also controlled by changing the experimental conditions, such as the reaction time and temperature. Some disadvantages of this method include the need of long reaction times, high-pressure equipment (autoclaves), safety concerns due to high temperatures and pressures, toxicity of solvents used, and the possibility of producing unwanted byproducts depending on the reaction condition.

25.4.3 Sol–Gel Method

The sol–gel method has emerged as a promising synthesis technique for synthesizing nanoparticles. In this method, the synthesis is performed at a low temperature (less than 100 °C) in the liquid state, and the final product is in the solid state via hydrolysis and polymerization reactions. The precursor solution contains metal alkoxides such as metal chlorides or nitrates, and this solution is prepared by dissolving metal alkoxides in a solvent. This method is simple and cheap, and it produces high-purity materials. The main disadvantage of this method is that they form dense films without using template, so there is a challenge in producing porous films.

25.4.4 Chemical Precipitation Method

The chemical precipitation method is a widely used synthesizing technique that produces solid materials from a solution through a chemical reaction, known as precipitation. If the concentration of one solid is above the solubility of the host material, then the precipitation will occur. In this method, precipitation occurs fast, and the difficulty in this method is to control the morphology of the material due to the fast precipitation.

25.4.5 Electrochemical Deposition Method

Electrochemical deposition is a simple and widely used synthetic technique for fabricating highly ordered and uniform nanomaterials. In this method, the metal ions in the electrolyte move toward the cathode. The metal ions at the cathode gain electrons, and they are reduced to their elemental form. These metal atoms are deposited on the surface of the cathode, forming a solid layer.

25.5 Defect and Interface Engineering in Pseudocapacitors

Based on dimensionality, defects can be classified into four types: (i) vacancies and dopants are zero dimensionality defects, and these are known as point defects; (ii) screw dislocations and edge dislocations are one dimensionality defects and these are known as line defects; (iii) grain boundaries and twin boundaries are two dimensionality defects and are known as planar defects; and (iv) voids and lattice disorders are three dimensionality defects and are known as there is no changes volume defects [25]. Point defects are deviations from the orderly arrangement of atoms and molecules around a point. It also changes the property of the material. These defects disturb the regular arrangement of atoms and molecules and change the material's property. Vacancies are simply defined as missing atoms at a lattice point. Another type of point defects are heteroatoms doping, where atoms of different elements (heteroatoms such as N, S, B, and P) are inserted into the material's structure. The pivotal role of dopant is to increase the electrical conductivity and decrease the resistance and also enhance the energy storage capacity, power density, and cycle life.

25.5.1 Oxygen Vacancy Defects

Oxygen vacancy is defined as the vacancies in metal oxides or other compounds containing an oxygen atom, and it is created by the separation of an oxygen atom in the crystal structure. TMOs are a promising electrode material for energy storage applications, such as supercapacitors; this type of material belongs to pseudocapacitors, and it stores charge through Faradaic process. It has high specific capacity and enhances the supercapacitors energy density, but the TMOs have low electrical conductivity that leads to a decrease in the performance of the supercapacitors. To address this problem, researchers introduced an oxygen vacancy defect to improve the performance of the electrode material in supercapacitors [26]. Gopi et al. synthesized reduced graphene oxide–Lathanum nickelate (rGO-LNO) nanocomposite by a facile template-free synthesis method, and it belongs to pseudocapacitve electrode. The XRD results confirmed that the electrode material belongs to a hexagonal rhombohedral structure, and the space group of LNO and composite rGO-LNO is R3C. SEM and HR-TEM confirmed the morphology of rGO-LNO, and they showed the uniform distribution of LNO particles onto the rGO. Oxygen deficiency characteristics of rGO-LNO nanocomposite are analyzed using Raman spectroscopy. XPS studies for rGO-LNO nanocomposites exhibit result with faster oxygen ion diffusion due to higher oxygen vacancy concentrations. The electrochemical analysis shows that the nanocomposites have a specific capacitance of $920\,Fg^{-1}$ at $1\,Ag^{-1}$ current density in a 6 M KOH electrolyte, and it revealed that the electrode material has pseudocapacitive behavior. The coulombic efficiency after 10,000 charge–discharge cycles at $10\,Ag^{-1}$ current density in two-electrode and three-electrode setups shows 99% and 95% of the efficiency, respectively [27]. Liu et al. synthesized $ZnMoO_4$ (ZMO) electrodes by the hydrothermal method for supercapacitor applications. Oxygen vacancy defects are introduced in the ZMO electrodes, and the presence of oxygen vacancy defects is analyzed using XRD and XPS. Conductivity of the electrode material and the active sites of ZMO nanomaterials are increased, which may be due to the introduction of oxygen vacancies in the ZMO; nanosheet such as ZMO electrode was obtained at 140 °C, and it shows the best electrochemical performance. ZMO model and OV-ZMO models having electronic and adsorption properties are compared using DFT analysis. The intrinsic conductivity of the ZMO nanomaterial is changed due to the introduction of oxygen vacancy defects, and it also diminishes the adsorption energy of hydroxide, and redox reduction is stimulated between the ZMO material and the anion in electrolyte. As a result, oxygen vacancy defect enhances the conductivity, minimizes the OH adsorption energy, and improves the electrochemical properties of ZMO electrode [26] (Table 25.1).

25.5.2 Heteroatom Doping

Heteroatom doping refers to the intended introduction of foreign atoms (different from the host material) into a material's crystal structure. This process significantly modifies the physical, chemical, and electronic properties of the material, so as to improve their performance in applications such as catalysis, energy storage, and sensors. Doping of heteroatoms can be classified into two types, namely, nonmetal

Table 25.1 Electrochemical performance of oxygen vacancies-based metal oxide. materials.

Material	Method of preparation	Capacitance	Cyclic stability	References
$NiCo_2O_4$	Hydrothermal method	997.50 F g^{-1} at 1.5 A g^{-1}	100% retention after 3000 cycles	[28]
MoO_3/Ce	Hydrothermal method	1446.3 at 5Ag^{-1}	94.8% after 5000 cycles	[29]
MnO_2/MnO_2	Molten-salt method	452.4 at 1 A g^{-1}	79.2% after 1000 cycles	[30]
Ni-Sn oxide/rGO	Hydrothermal method	1800 F g^{-1} at 2 Ag^{-1}	80% after 10,000 cycles	[31]

doping and metal doping. Nonmetal doping (N, S, P, and B) and metal doping (Co, Ni, Fe, and Na), formed by doping the heteroatom into the electrode material, result in high electrical conductivity and increase in capacitance and cycling stability. Guo et al. synthesized a sulfur-doped zinc–nickel–cobalt oxide (S-ZNCO) nanorod materials for supercapacitors by hydrothermal and ion exchange method. TMOs such as ZNCO have a poor electrical conductivity; hence, heteroatom doping with TMO may increase the conductivity of metal oxide materials. Comparison was made between S-ZNCO and ZNCO, which reveals the S-ZNCO has lower crystallinity than ZNCO, and this may be due to the introduction of oxygen vacancies and S-doped heteroatom defects in the ZNCO electrode material. The electrode S-ZNCO shows a specific capacitance of 2919.6 Fg^{-1} at 1 A g^{-1}. Moreover, the capacitance retention after 5000 charge–discharge cycles is 92.65%. The prepared electrode material S-ZNCO-NF/AC achieves an energy density of 72.97 WhKg^{-1} at a power density of 825 WKg^{-1} [32]. Nitrogen-doped graphene@Co_3O_4 nanocomposite was synthesized using hydrothermal method by Liu et al. Co_3O_4 nanoparticles, sized between 80 and 100 nm, were evenly distributed on graphene nanosheets, resulting in a composite with a specific capacitance of 315.6 Fg^{-1} at a current density of 1 Ag^{-1}. After 1000 cycles, the material retained 84.5% of its capacitance. This performance is significantly better than the needle-like Co_3O_4@graphene structure developed by Guan et al., which had a lower specific capacitance of 155.7 Fg^{-1} at 0.1 Ag^{-1}. The improvement is largely due to the coordination between the high nitrogen content (11.7%) in the nitrogen-doped graphene and the uniformly dispersed Co_3O_4 nanoparticles, which enhances the electrochemical properties of the composite electrode [33]. Arun Kumar et al. developed a hybrid full-cell device by combining Co_3O_4 nanowires on nickel foam (NF) as the positive electrode material and nitrogen-doped g-C_3N_4 nanosheets as the negative electrode material for supercapacitor applications in an aqueous electrolyte. The device achieves an energy density of 22.26 WhKg^{-1} and a power density of 4000 WKg^{-1} while maintaining 82.4% cycling stability. These impressive results are attributed to the hybridization of Co_3O_4 nanowires and nitrogen-doped g-C_3N_4 nanosheets, which provide ample space for the electrochemical intercalation and de-intercalation of

Table 25.2 Electrochemical performance of heteroatom doping-based metal oxide material.

Material	Method of preparation	Capacitance	Cyclic stability	References
S-doped $MnMoO_4$	Hydrothermal method	2526.7 at 1 Ag^{-1}	95.56% after 2000 cycles	[36]
B-$NiCo_2O_4$/CF	Hydrothermal method	1452.7 at 1 Ag^{-1}	94.7% after 5000 cycles	[37]
S-Co_3O_4	Hydrothermal method	638.6 at 0.5 Ag^{-1}	87.5% after 5000 cycles	[38]
P-$NiMoO_4$-N_2	Hydrothermal method	2180 at 1 Ag^{-1}	96.1% after 5000 cycles	[39]

K^+ ions during the energy storage process. This hybrid supercapacitor devices demonstrate promising potential for various applications [34]. Asaithambi et al. synthesized pure SnO_2 and transition metal (Fe, Cu, Zn)-doped SnO_2 spherical nanoparticles by chemical co-precipitation method. To maintain the local charge balance, the oxygen vacancies would be formed. Cation vacancies can be engineered for charge balance; when the high valance state dopants replace the host low valance state metal ion, these enhance the electrochemical performance of the prepared electrode material. Compared to Cu-SnO_2 and Zn-SnO_2, the Fe-SnO_2 electrode material demonstrates high specific capacitance of 270 F/g at a current density of 0.5 A/g in 2 M KOH electrolyte. After 2000 charging–discharging, Fe-SnO_2 has an excellent cyclic stability of 98% [35] (Table 25.2).

25.6 Characterization Techniques for Defects and Interface Analysis

25.6.1 Physiochemical Characterization Technique

The physiochemical properties of the prepared electrodes were characterized by X-ray diffraction analysis (XRD), scanning electron microscope (SEM), transmission electron microscope (TEM), and X-ray photoelectron spectroscopy (XPS). XRD is used to determine the crystal structure, crystallite size, dislocation density, and microstrain of the material. In XRD, we can identify oxygen vacancies by observing the changes in the peak intensity or position due to the lattice distortion produced by oxygen vacancies. Fourier transform infrared (FT-IR) spectroscopy is used for the purpose of identifying functional groups and chemical bonds present in the materials. The presence of oxygen vacancies and the creation of new chemical bonds due to heteroatom doping were detected using FT-IR analysis. SEM provides the high-resolution images of the surface morphology and microstructure. It is used to visualize the particle size and shape of the material and is also used to identify the defects and porosity. TEM is used for the detailed analysis of the internal structure and also has higher magnification than SEM. TEM is also used to study the atomic structure and identify defects such as dislocations and grain boundaries. XPS is used to determine the elemental composition and oxidation state of the surface, and it provides information about the presence and concentration of oxygen vacancies and heteroatom doping. Due to these reasons, the chemical shift can occur in the spectra and thereby indicate the changes in the bonding environment of atoms.

25.6.2 Electrochemical Characterization Techniques

25.6.2.1 Cyclic Voltammetry

Cyclic voltammetry (CV) is a electrochemical technique that is commonly used to investigate the reduction and oxidation process of electrode materials. As a result of CV measurement a graph is plotted against current vs applied voltage. This is a suitable method for characterization of electric-double-layer capacitor, pseudocapacitor, and hybrid capacitor behavior of supercapacitors. CV uses a three-electrode system, consisting of a working electrode, a reference electrode, and a counter electrode. The potential is measured between the working electrode and the reference electrode, while the current flows between the working electrode and the counter electrode. In Figure 25.4 (a–d), the CV curve for EDLCs shows a rectangular shape, and in pseudocapacitors, the CV curve is no longer in the rectangular shape due to oxidation and reduction reaction. Thus, in a hybrid supercapacitor, the CV performance is nonrectangular shape [40].

25.6.2.2 Galvanostatic Charge–Discharge Analysis

The capacitance of the electrode material is analyzed by the GCD test and is evaluated by the following equation:

$$C = \frac{I \times t}{\Delta V} \tag{25.4}$$

In Eq. (25.4), C is the differential capacitance (Fg^{-1}), I is a current density (Ag^{-1}), ΔV is a working potential window, and t is a charging/discharging time (s) of the electrode. In EDLCs, there are no limitations on the potential window range due to the independent nature of charge storage. However, for intercalation-type pseudocapacitors or standard Faradaic pseudocapacitors, the potential window range is fixed. This means that while determining the capacitance of hybrid supercapacitors, there is no need to set a specific potential window for the EDLC portion. In contrast, the potential window for the pseudocapacitive part must be predetermined, based on the type of pseudocapacitive electrode used. The total capacitance of the hybrid system is then calculated by summing the individual capacitances of each component.

25.6.2.3 Electrochemical Impedance Spectroscopy

Electrochemical impedance spectroscopy (EIS) is an electrochemical characterization technique used to study the electrochemical behavior of electrode

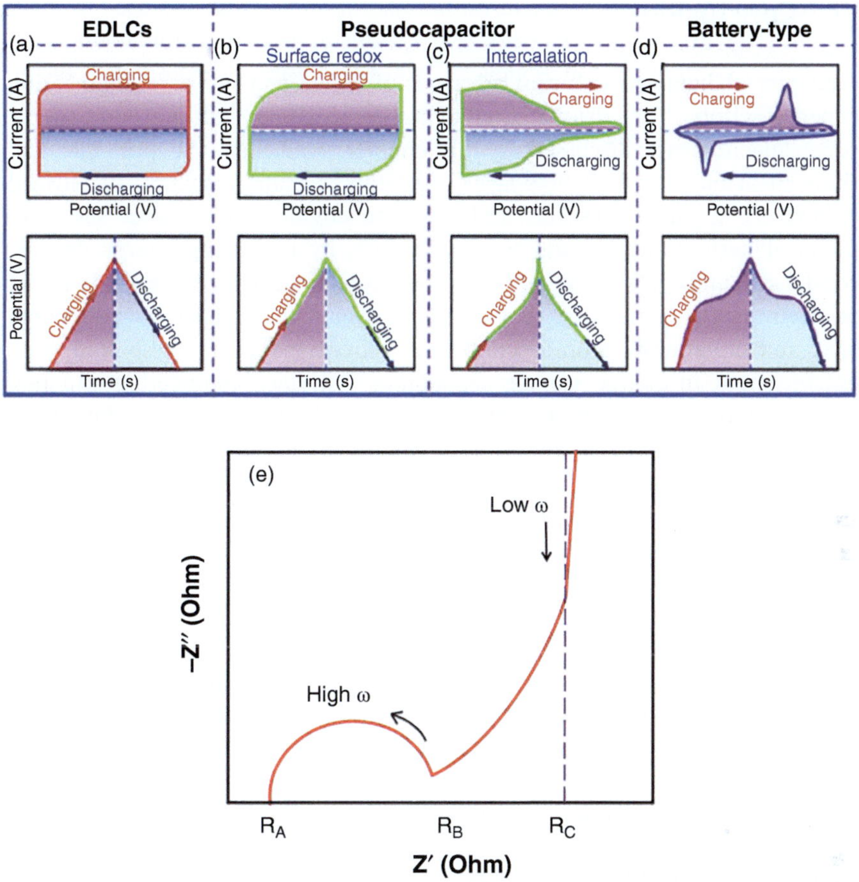

Figure 25.4 CV and GCD curves for (a) EDLCs, (b–c) pseudocapacitor, and (d) hybrid supercapacitor. Source: Ref: Wu, Qianghong, et al. (2021). Cyclic stability of supercapacitors: materials, energy storage mechanism, test methods, and device. *J. Mater. Chem. A* 9(43): 24094–24147. (e) Schematic of a typical Nyquist plot for supercapacitors / with permission of Royal Society of Chemistry.

materials. It involves applying a small AC voltage to a system and measuring the resulting current. The impedance, which is the complex ratio of voltage to current, provides insights into the resistance, capacitance, and other electrochemical properties of the system. In Figure 25.4 (e), a Nyquist plot is a graphical representation of the real part (resistance) and imaginary part (reactance) of the impedance measured over a range of frequencies. The Nyquist plot consists of three distinct regions. The high-frequency region (above 10^4 Hz) is characterized by a semicircle, which corresponds to interfacial resistance. The second region, spanning from high to medium frequencies (between 10^4 and 10^1 Hz), is represented by a vertical line, indicating the resistance associated with charge transfer. Finally, the low-frequency region (below 1 Hz) shows an extension of the vertical line, illustrating the capacitive behavior of the system.

25.7 Conclusion

This chapter elaborates the recent advances of defect and interface engineering of supercapacitors. The classification of supercapacitors and the components of supercapacitors are discussed in this chapter. And also, the principle and mechanism of the electric-double-layer capacitors, pseudocapacitors, and hybrid capacitors are discussed. Defects and interface engineering enhance the electrochemical performance of the supercapacitors, by improving energy density, power density, and cyclic stability. When the oxygen vacancies make the material more electronically conductive, they are more chemically reactive toward charge storage, and structurally more open for ion movement, all of these contribute to enhancement in charge storage capabilities. Interface engineering and heteroatom doping increased the energy storage capability and cyclic performance. Defect-engineered materials and precise interface modifications are used to achieve supercapacitors with higher efficiency, longer lifespans, and sustainable, cost-effective manufacturing processes.

References

1 Zhang, J., Min, G., and Chen, X. (2023). Supercapacitors for renewable energy applications: a review. *Micro Nano Eng.* 21: 100229.

2 Kularatna, N. and Gunawardane, K. (2021). Capacitors as energy storage devices: simple basics to current commercial families. In: *Energy Storage Devices for Renewable Energy-Based Systems* (ed. N. Kularatna and K. Gunawardane), 181–197. Academic Press.

3 Priyadarshini, P. and Parida, K. (2024). Two-dimensional metal-organic frameworks and their derived materials: properties, synthesis and application in supercapacitors field. *J. Energy Storage* 87: 111379.

4 Sharma, S. and Chand, P. (2023). Supercapacitor and electrochemical techniques: a brief review. *Results Chem.* 5: 100885.

5 Chen, X., Paul, R., and Dai, L. (2017). Carbon-based supercapacitors for efficient energy storage. *Natl. Sci. Rev.* 4 (3): 453–489.

6 Iro, Z.S., Subramani, C., and Dash, S.S. (2016). A brief review on electrode materials for supercapacitor. *Int. J. Electrochem. Sci.* 11 (12): 10628–10643.

7 Temam, A.G., Alshoaibi, A., Getaneh, S.A. et al. (2023). Recent progress on phosphate-based electroactive materials for supercapacitor applications. *Results Surf. Interfaces* 13: 100154.

8 Gaikwad, P., Tiwari, N., Kamat, R. et al. (2024). A comprehensive review on the progress of transition metal oxides materials as a supercapacitor electrode. *Mater. Sci. Eng. B* 307: 117544.

9 Lee, S.-H., Kim, J.H., and Yoon, J.-R. (2018). Laser scribed graphene cathode for next generation of high-performance hybrid supercapacitors. *Sci. Rep.* 8 (1): 8179.

10 Gupta, S.P., Sanap, P.P., Deore, M.K. et al. (2023). Molybdenum chalcogenides for supercapacitor applications: a critical review. *J. Energy Storage* 73: 109065.

11 Muzaffar, A., Basheer Ahamed, M., Deshmukh, K., and Thirumalai, J. (2019). A review on recent advances in hybrid supercapacitors: design, fabrication and applications. *Renew. Sust. Energ. Rev.* 101: 123–145.

12 Chodankar, N.R., Dubal, D.P., Ji, S.-H., and Kim, D.-H. (2019). Highly efficient and stable negative electrode for asymmetric supercapacitors based on graphene/$FeCo_2O_4$ nanocomposite hybrid material. *Electrochim. Acta* 295: 195–203.

13 Sun, J., Chen, W., Sun, X. et al. (2017). Recent progresses in high-energy-density all pseudocapacitive-electrode-materials-based asymmetric supercapacitors. *J. Mater. Chem. A* 5 (20): 9443–9464.

14 Gao, Z., Zhang, L., Chang, J. et al. (2018). $ZnCo_2O_4$-reduced graphene oxide composite with balanced capacitive performance in asymmetric supercapacitors. *Appl. Surf. Sci.* 442: 138–147.

15 Shiraishi, S. (2003). Electric double layer capacitors. In: *Carbon Alloys* (ed. E.-i. Yasuda, M. Inagaki, K. Kaneko, et al.), 447–457. Elsevier Science.

16 Xiang, M., He, L., Qiuyao, S. et al. (2024). Supercapacitor properties of N/S/O co-doped and hydrothermally sculpted porous carbon cloth in pH-universal aqueous electrolytes: mechanism of performance enhancement. *Chem. Eng. J.* 485: 149835.

17 Yang, G., Zhu, Y., Hao, Z. et al. (2024). An aqueous all-quinone-based redox flow battery employing neutral electrolyte. *Adv. Energy Mater.* 26: 2400022.

18 Sánchez-Valdez, A.G., de la Parra-Arciniega, S.M., Sánchez-Cervantes, E.M., and Torres-González, L.C. (2023). Neutral pH Na_2SO_4/glycerol/PVA polymer hydrogel electrolyte prepared at room temperature for activated carbon supercapacitors. *J. Solid State Electrochem.* 27 (11): 2917–2925.

19 He, Z.-H., Gao, J.-F., and Kong, L.-B. (2024). Electrolyte effect on electrochemical behaviors of manganese fluoride material for aqueous asymmetric and symmetric supercapacitors. *Rare Metals* 43 (3): 1048–1061.

20 Lu, J., Zhang, J., Wang, X. et al. (2024). A review of advanced electrolytes for supercapacitors. *J. Energy Storage* 103: 114338.

21 Ahankari, S., Lasrado, D., and Subramaniam, R. (2022). Advances in materials and fabrication of separators in supercapacitors. *Mater. Adv.* 3 (3): 1472–1496.

22 Verma, K.D., Sinha, P., Banerjee, S. et al. (2020). Characteristics of separator materials for supercapacitors. In: *Handbook of Nanocomposite Supercapacitor Materials I: Characteristics* (ed. K.K. Kar), 315–326. Springer Cham.

23 Lei, Z., Christov, N., and Zhao, X.S. (2011). Intercalation of mesoporous carbon spheres between reduced graphene oxide sheets for preparing high-rate supercapacitor electrodes. *Energy Environ. Sci.* 4 (5): 1866–1873.

24 Sun, J., Chunju, X., and Chen, H. (2021). A review on the synthesis of $CuCo_2O_4$-based electrode materials and their applications in supercapacitors. *J. Mater.* 7 (1): 98–126.

25 Lu, W., Yan, L., Ye, W. et al. (2022). Defect engineering of electrode materials towards superior reaction kinetics for high-performance supercapacitors. *J. Mater. Chem. A* 10 (29): 15267–11529.

26 Lu, W., Yan, L., Ye, W. et al. (2022). Defect engineering of electrode materials towards superior reaction kinetics for high-performance supercapacitors. *J. Mater. Chem. A* 10 (29): 15267–15296.

27 Liu, Y., Peiran, X., Zhang, S. et al. (2024). Oxygen vacancies promoting the electrochemical performance of $ZnMoO_4$ electrode: an experimental and theoretical study. *J. Energy Storage* 94: 112361.

28 Gopi, U.V., Bhojane, P., and Smaran, K.S. (2022). Engineering oxygen-deficient nanocomposite comprising $LaNiO_{3-\delta}$ and reduced graphene oxide for high-performance pseudocapacitors. *J. Energy Storage* 54: 105301.

29 Kumar, S., Tahira, A., Bhatti, A.L. et al. (2024). Oxygen vacancies and tailored redox activity encountered with $NiCo_2O_4$ nanostructures for promising applications in supercapacitor and water oxidation. *J. Energy Storage* 77: 109994.

30 Xu, L., Zhou, W., Chao, S. et al. (2022). Advanced oxygen-vacancy Ce-doped MoO_3 ultrathin nanoflakes anode materials used as asymmetric supercapacitors with ultrahigh energy density. *Adv. Energy Mater.* 12 (16): 2200101.

31 Fu, Y., Gao, X., Zha, D. et al. (2018). Yolk–shell-structured MnO_2 microspheres with oxygen vacancies for high-performance supercapacitors. *J. Mater. Chem. A* 6 (4): 1601–1611.

32 Paul, A., Ghosh, S., Kolya, H. et al. (2023). New insight into the effect of oxygen vacancies on electrochemical performance of nickel-tin oxide/reduced graphene oxide composite for asymmetric supercapacitor. *J. Energy Storage* 62: 106922.

33 Guo, Y., Yan Wang, Y., Zhang, Y.Z., and Cai, W. (2022). Functional sulfur-doped zinc-nickel-cobalt oxide nanorods materials with high energy density for asymmetric supercapacitors. *J. Alloys Compd.* 896: 163053.

34 Yadav, S. and Devi, A. (2020). Recent advancements of metal oxides/Nitrogen-doped graphene nanocomposites for supercapacitor electrode materials. *J. Energy Storage* 30: 101486.

35 Kumar, S., Arun, I., Sarasamreen, C. et al. (2023). Elevates the electrochemical stability performance of hydrothermally synthesized Co_3O_4 nanowires/NF for hybrid supercapacitors. *Inorg. Chem. Commun.* 158: 111506.

36 Asaithambi, S., Sakthivel, P., Karuppaiah, M. et al. (2020). Investigation of electrochemical properties of various transition metals doped SnO_2 spherical nanostructures for supercapacitor applications. *Journal of Energy Storage* 31: 101530.

37 Zhang, X., Zhang, X., Guangmeng, Q. et al. (2020). Nickel-cobalt double oxides with rich oxygen vacancies by B-doping for asymmetric supercapacitors with high energy densities. *Appl. Surf. Sci.* 512: 145621.

38 Li, Y., Li, W., Yang, C. et al. (2020). Engineering coordination polymer-derived one-dimensional porous S-doped Co_3O_4 nanorods with rich oxygen vacancies as high-performance electrode materials for hybrid supercapacitors. *Dalton Trans.* 49 (30): 10421–10430.

39 Zhu, Z., Zang, L., Chu, M. et al. (2022). Oxygen-vacancy and phosphorus-doping enriched $NiMoO_4$ nanoarrays for high-energy supercapacitors. *J. Energy Storage* 54: 105314.

40 Moniruzzaman, S.,.M., Yue, C.Y., Ghosh, K., and Jena, R.K. (2016). Review on advances in porous nanostructured nickel oxides and their composite electrodes for high-performance supercapacitors. *J. Power Sources* 308: 121–140.

26

Defect and Interface Engineering in Electrochemical Pseudocapacitors Based on Pseudocapacitive Materials

Próspero Acevedo-Peña

SECIHTI-Instituto Politécnico Nacional. Laboratorio Nacional de Conversión y Almacenamiento de Energía. CICATA unidad Legaria. Clz Legaria S/N, 11500. Mexico City, Mexico

26.1 Introduction

Developing energy storage devices (ESD) that can be fully charged in short periods of time has become a continuously more requisitioned characteristic for portable gadgets and even for electromobility applications. In this arena, supercapacitors have an advantage over batteries since the processes involved in energy storage are mainly confined to the surface of the electrode materials, conferring the characteristic fast kinetics of these devices. In contrast, in batteries, these processes are typically slow due to the bulk transformation of the materials.

In supercapacitors, energy can be stored by two mechanisms: charging–discharging of the double layer and pseudocapacitance [1], schematized in Figure 26.1. Carbon-based materials are conventionally employed in supercapacitors due to their typically high electrical conductivity that facilitates fast electron transport and the high specific surface area that ensures a proportional amount of charge in the double layer formed at the solid–electrolyte interface. However, restraining the charge storage process only at the interface limits the energy that can be provided by the device. In contrast, when pseudocapacitive materials are used, redox reactions confined to the surface of the electrode are responsible for the charge stored. Since the diffusion of charge-compensating ions does not control the energy storage process, a pseudocapacitive material exhibits a fast kinetic-like conventional double-layer charging.

Currently used pseudocapacitive materials commonly store energy by combining the abovementioned mechanisms (double layer and fast kinetic redox reactions). This is due to the processing of materials on the nanometer scale, which confers a high specific surface area, and the electric conductivity of the materials, intrinsic or provided by forming composites. For this reason, defect and surface engineering of the materials represents an attractive path to maximize the performance of pseudocapacitive materials. Below are summarized some of the strategies more recently reported in the specialized literature to improve the electrochemical performance of pseudocapacitive materials through the surface and defect engineering of

Surface and Interfacial Defects in Nanomaterials for Sustainable Energy Production and Storage, First Edition. Edited by Noé Arjona, Lorena Álvarez Contreras, and Minerva Guerra Balcázar.

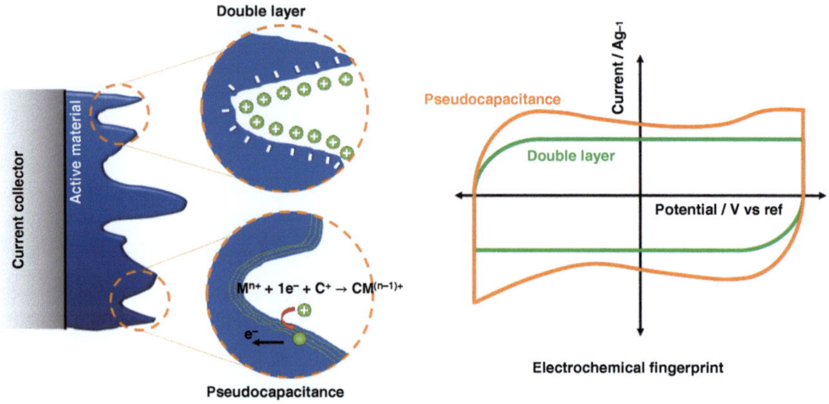

Figure 26.1 Scheme of the charge storage mechanisms involved in supercapacitors: electric double layer, pseudocapacitance (surface-confined redox reactions), and electrochemical fingerprint recorded by cyclic voltammetry.

high-impact materials in energy storage applications, such as MXenes [2], transition metal nitrides (TMN) [3], and conducting polymers (CPs) [4].

26.2 MXenes

The MXenes ($M_{n+1}X_n$, $n = 1$–3) are synthesized from MAX phases (M is a transition metal, A is an A-group element, and X is C and/or N), where layers of $M_{n+1}X_n$ are interleaved with A-element layers. Although MXenes materials can be considered new, their synthesis protocols are well documented in the literature [5]. Generally, a wet method is used to dissolve the A-element layer from the MAX phase selectively. After etching, the surface of the solid is terminated by different groups according to the chemicals employed, such as –O, –OH, –F, –NH_3, or others. These materials exhibit high surface area and conductivity, making them ideal for supercapacitor applications due to the accumulation of significant charge in the double layer. Additionally, the transition metals in the MXenes can experience redox reactions that would contribute to the charge stored. Combined with this, the redox processes in MXenes are typically independent of the diffusion of charge-compensating ions since the metallic centers are mainly exposed (particularly after exfoliation), offering fast kinetics, a suitable characteristic for supercapacitor applications.

The high potential of MXenes for supercapacitor application has contributed to a broad interest in design strategies that allow exploiting its characteristics. Among the strategies employed to tune the electrochemical behavior of MXenes for pseudocapacitor applications, from the point of view of defect and surface engineering, are the functionalization of the terminal groups of the MXenes, doping, creation of vacancies, surface and morphology modification, and incorporation of molecules or other materials to avoid the restacking of the lamellas. Figure 26.2 schematizes the structure and defects commonly formed in MXenes.

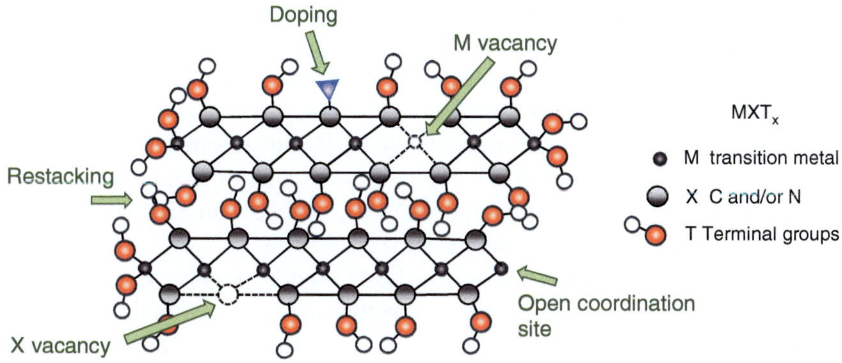

Figure 26.2 Scheme of the structure and defects commonly found in MXene materials.

The pseudocapacitive processes originate mainly from a surface-confined redox reaction that contributes to fast energy storage, similar to a conventional double-layer supercapacitor. Energy storage in MXenes is attributed to a combination of high double-layer capacitance provided by the high surface area of the exfoliated two-dimentional (2D) material and the proton adsorption–desorption paired with the partial reduction and oxidation of the transition metal in the MXene. The charge-transfer processes, which are the origin of the pseudocapacitance, take place within –O and –OH groups that are also involved in the double-layer formation. Bo et al. used a computational approach to show that the protonation/deprotonation of surface groups in MXenes contributes to comparable EDLC and pseudocapacitance [6]. It is worth pointing out that the calculation indicates that only a fraction of the surface groups is involved in reaching the maximum capacitance, leaving the side to a fraction of inert function groups, such as -F. These surface groups can be used to engineer the surface of the MXene further and provide new characteristics that favorably impact its performance.

Zhan et al. passivated the surface of $Ti_2C_3T_x$ with tartaric acid to offer higher chemical stability to the MXene layer [7]. The coordination of tartaric acid to titanium in the MXene, through the carboxylic terminal groups, provided improved chemical stability in water and allowed for the avoidance of the formation of TiO_2. The authors assembled a symmetric supercapacitor with this material and a conducting polymer, displaying a short time constant and high capacitance in acidic electrolytes. The current registered from CV exhibits a linear dependence with the scan rate at values up to 500 $mV\,s^{-1}$, showing the main contribution of fast redox processes to the charge stored. The current registered from CV exhibits a linear dependence with the scan rate at values up to 500 $mV\,s^{-1}$, showing the main contribution of fast redox processes to the charge stored, which is a suited characteristic for pseudocapacitor-type materials. The assembled device delivered a capacitance of 1.149 $mF\,cm^{-2}$ at 120 Hz with a phase angle of 80° in a microcapacitor assembly.

Functional groups in MXenes can be tuned by processing, principally during the etching step of the MAX phase. These groups modulate properties such as conductivity and play an essential role in the pseudocapacitive energy storage in these materials. Recently, Zhu et al. synthesized $Ti_3C_2Br_2$ by etching TiAlC MAX phase using Br_2 gas and then was thermally treated in a P vapor with an H_2 atmosphere to replace –Br with –P partially [8]. P substitution derived in a notorious increase of the capacitance of the solid in acid electrolyte. While $Ti_3C_2T_x$ and $T_3C_2Br_2$ exhibited capacitances of 50 and 100 $mF\,cm^{-2}$, respectively, $Ti_3C_2PBr_x$ reached a capacitance of 360 $mF\,cm^{-2}$ at a scan rate of 20 $mV\,s^{-1}$. Molecular dynamics simulation indicated that H_3O^+ ions aggregate more onto $Ti_3C_2PBr_x$ surface than $Ti_3C_2Br_2$, which favored an increase in the charge stored in acidic electrolyte.

Other elements and molecules have also been used to modify the terminal groups in MXenes. Yang et al. described a solvothermal route to obtain N-doped Ti_3C_2 MXene using a double strategy, diethanolamine as the source of N ions, and isopropyl alcohol in the interlamellar space of the MXene and favor the transport of ions [9]. The modified material reaches a capacitance of 1004.9 $F\,g^{-1}$ at 5 $mV\,s^{-1}$, surpassing the 643.9 $F\,g^{-1}$ recorded for pristine Ti_3C_2. In a similar approach, Sun et al. synthesized N- and S-doped MXene in two steps (combined thermal and solvothermal treatments) [10]. DFT calculations show that N and S sites on MXene promote weaker H_3O^+ adsorption than standard O and F sites, easing the redox processes during energy storage. The doped material exhibits a capacitance of 266 $F\,g^{-1}$ at 0.5 $A\,g^{-1}$, larger than ~200 $F\,g^{-1}$ recorded for the pristine material under the same conditions. Additionally, the N- and S-doped MXene retains up to 75% of this value at a current of 10 $A\,g^{-1}$ due to the fast kinetics of the redox processes.

The standard process to obtain MXenes involves etching under harsh conditions, typically fluoride-containing highly acidic solutions. This step is needed to dissolve the A-element layer in the MAX phase and obtain the 2D material. However, these environments might also generate M (Ti, Nb, Mo) or X (C and/or N) vacancies in the material, which will modulate the electronic structure and performance of the resulting material. Ibragimova et al. studied the formation of titanium and nitrogen/carbon vacancies in MXenes-type Ti_3C_2 and Ti_2N [11]. The results indicate that the functional groups directly determine metal vacancies; while O groups protect the material, OH or bare surfaces favor the generation of vacancies. On the other hand, metal vacancy generation in Ti_3C_2 has higher energy than in Ti_2N. However, X vacancies (C or N) are favored in Ti_3C_2. Additionally, the authors constructed Pourbaix-type diagrams that show the energy for vacancy generation. Particularly interesting is that during the exfoliation process, commonly performed at neutral pH, Ti vacancy generation can be favored due to lowering the energy needed for its formation. These diagrams can qualitatively guide the identification of the experimental conditions required to obtain MXene with the desired type and number of defects. In this regard, Wang's research group evaluated through a computational approach the impact of material vacancies and N-doped sites on the energy storage of MXenes in supercapacitor-type devices [12]. They found that metal vacancy generation might be a more straightforward approach to improve the electrochemical energy storage properties of MXene compared to N-doped sites. The generation of vacancies favors the charge stored in the electric double layer by altering, in a more significant extension, the electronic structure of the solid. Additionally, metal vacancies induce the formation of low-coordinated functional groups, which have more reversibility in the electrochemical protonation/deprotonation of the solid. This explains the higher capacitance values, the low charge-transfer resistances, and the high-rate capability experimentally recorded when these defects are induced in the material.

After MXene exfoliation, a suspension of the solid is typically obtained. The MXene sheets are electrostatically stabilized and can be further processed to assemble the electrodes employed for supercapacitors, such as drop casting and vacuum filtering. However, incorporating cations into the suspension will cause a shrinking in the double layer formed over the MXenes, diminishing the surface charge excess and destabilizing the suspension. This gives origin to the formation of MXenes monoliths with stacked sheets whose porosity and properties can be tuned during the processing, offering an easy strategy to engineer the surface of the electrodes. Recently, Sikdar et al. created nanosized tunable holes in 3D MXene by incorporating tetra-amine zinc (II) complex while forming a hydrogel [13]. The creation of this defect allows to tackle the increased ion diffusion resistance that this structure might experiment with during energy storage, deriving in a material with a capacitance of $359.2\,F\,g^{-1}$ at $10\,mV\,s^{-1}$, with an outstanding high-rate capability, retaining up to 79% of its capacitance at a scan rate of $5\,V\,s^{-1}$. However, this property can be compromised at high material loading. To overcome this issue, the conductivity of the MXene needs to be ensured, and composites with other conductive materials, such as carbon nanostructures, can be a suitable strategy. Zhao et al. obtained monoliths of face-to-face stacked GO/MXene sheets by alkaline-ion-assisted self-assembling [14]. The sheet arrangement during the formation of these monoliths provided more active sites for Na^+ storage, promoting a surface-confined redox process that enhanced the pseudocapacitive contributions, ensuring higher energy storage values than monoliths synthesized without alkaline ions.

Modifying the surface of MXenes with other compounds can also contribute to the redox reaction involved in the amount of energy stored, thus increasing the capacitance of the composite. Hedge et al. treated films of Nb_2CT_x-graphene oxide composites to induce the reduction of the GO, the exfoliation of Nb_2CT_x, and the formation of niobium oxides, improving the electrochemical performance of the materials in solid micro-supercapacitors [15]. The laser exfoliation and GO incorporation increased the interlamellar distance to 1.22 nm (from 1 nm for the pristine solid). The device reached an aerial capacitance of $16.25\,mF\,cm^{-2}$ at $0.1\,mA\,cm^{-2}$ and an energy density of $4.42\,\mu Wh\,cm^{-2}$. Additionally, the device retained up to 80% of its initial capacitance after 10,000 galvanostatic charging–discharging (GCD) cycles. In another approach, Ambade et al. hybridized $Ti_3C_2T_x$ with VS_2 by a hydrothermal treatment to obtain a composite for supercapacitors in neutral electrolytes [16]. The material's hybridization boosted the solids' electrochemical performance, reaching up to $750\,F\,g^{-1}$ and retaining 93% of its initial capacitance after 10,000 GCD cycles.

Another problem to overcome when working with MXenes is the restacking of the layers due to the H-bonds and van der Waals forces between the sheets.

MXene sheet's self-restacking impacts both main processes involved in energy storage, limiting the area of the solid in contact with the electrolyte, hindering the double-layer formation, and reducing the access of the active redox centers to store energy through the electrochemical reactions. In this way, different materials have been added to the interlamellar space of the MXenes to act as spacers, protect the electroactive sites, and increase conductivity.

Tomar et al. modified $Ti_3C_2T_x$ with $LaMnO_3$ to avoid restacking and favor the intercalation of anions and cations to boost the energy storage properties [17]. The composite also increased the capacitive potential window of the raw materials, allowing the assembly of a symmetric supercapacitor device with a voltage of 1.6 V and a specific energy of 34.1 Wh kg^{-1}. The device retained up to 96.2% of its initial capacitance after 10,000 GCD cycles. Similarly, Sun et al. assembled an MXene/NH_2-functionalized graphene composite to form a porous network of interconnected sheets [18]. Additionally, MXene was partially oxidized to obtain TiO_2 to avoid restacking the MXene layers and enhance the lithium-ion storage. The composite exhibited a capacitance of 188 F g^{-1} @ 1 A g^{-1}, retaining 78% of this capacitance at 10 A g^{-1}. The authors formulated a ternary electrolyte EmimTFSI/ACN/LiTFSI to promote the redox reaction in the composite and enhance the capacitance. Similarly, Dang's research group evaluated the incorporation of highly defective MoO_{3-x} to avoid MXene restacking and provided a solid, contributing to faster energy storage kinetics [19]. The composite allows assembling a device with a capacitance of 869.1 F cm^{-3} @ 0.5 A cm^{-3} and an energy density of 77.3 mWh cm^{-3} at a power density of 204.8 mW, retaining 86.4% of its initial capacitance after 5000 GCD cycles.

The modification with graphene materials to avoid the restacking of MXenes has attracted particular attention since these materials can be obtained as thin sheets, facilitating their intercalation to promote contact and synergetic interactions. Ali et al. show thorough DFT calculations that incorporating graphene into MXene significantly improves its thermodynamic stability and mechanical strength, additionally providing high electrical conductivity [20]. Moreover, doping the graphene layers was shown to be more efficient in increasing the quantum capacitance of the composite than inducing defects directly on the MXene. On the other hand, Mahajan et al. incorporated 1D graphene nanoribbons to form a porous network of interconnected MXene layers and favor titanium reduction in the MXene to increase the charge stored [21]. The modification caused an increment of the specific capacitance from 87.1 and 93.9 F g^{-1} in the pristine state of the 1D graphene nanoribbons and MXene, respectively, to a value of up to 239 F g^{-1} for the composite.

MXenes are flexible materials whose processing facilitates surface and defect engineering to improve their electrochemical performance for energy storage applications, particularly in supercapacitors. The different alternatives described here show some of the routes followed by researchers to tune the electrochemical performance of these materials, primarily focused on increasing the amount of charge stored in the solid by increasing the number and accessibility of the sites involved in the double-layer formation and redox reactions. However, there is a considerably lower amount of effort to increase the capacitive potential window of the MXenes, which is also a route to boost the energy stored in the supercapacitors. This can be reached through surface engineering of the solid to provide chemical and electrochemical stability of the solid in a more significant potential window and avoid electrolyte decomposition.

26.3 Transition Metal Nitrides

Transition metal nitrides (TMNs) are chemically inert materials with large hardness and high conductivity that have attracted attention for supercapacitor applications. Compared with carbon materials, TMN has higher density, offering the possibility of reaching high volumetric energy density, making them suitable materials for micro-supercapacitors. Although TMN has been explored for supercapacitor applications since the end of the last century, these materials have not permeated on the same scale as others with more versatile synthetic routes, such as MXenes or CPs. One of the main challenges for TNM is developing synthetic routes that allow their obtention under mild conditions without requiring sophisticated equipment. Regarding the defect and surface engineering of TNM, there is also room to enhance its performance. The primary approaches involve the formation of TMN with a mixed composition, material doping, the creation of interconnected porous structures, and surface modification with other materials to provide electroactive sites and enhance the energy storage process, see Figure 26.3.

Usually, the oxidation of the transition metals is involved in the energy storage process, forming oxides or hydroxides on the surface, limiting the capacitive potential window of the material and causing poor capacitance retention during cycling

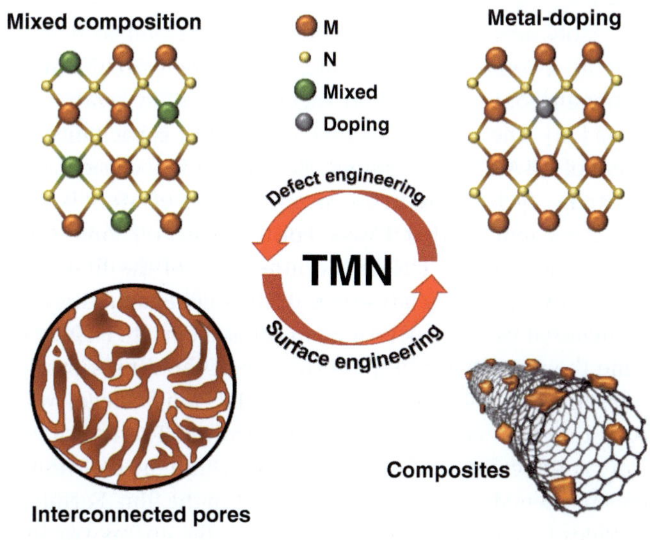

Figure 26.3 Graphical representation of the defect and surface engineering reported for transition metal nitrides.

when these species are not stable in the electrolyte. Li et al. studied the mechanisms involved in the capacitance fading of vanadium nitride electrodes for supercapacitors [22]. When the materials are cycled in a capacitive potential window in which partial oxidation occurs, the nitride irreversibly transforms into oxides and then dissolves into the electrolyte. This phenomenon limits the capacitive potential window of the electrode and limits the cyclability. However, the authors showed that KOH electrolytes containing VO_4^{3-} anions overcome this issue, retaining 120% of their initial values after 1000 cycles. Contrastingly, the formation of oxides distributed in the ruthenium nitrides has been pointed out as responsible for the pseudocapacitive energy storage. Dinh et al. evaluated the impact of deposition variables such as pressure, inert gas ratio, and substrate temperature on the electrochemical performance of RuN films formed by magnetron sputtering [23]. Although RuN exhibits high electrical conductivity, the improvement in the electrochemical performance was related to the formation of RuO_2 (approx. 30%) particles evenly distributed in the film. After tuning all the deposition parameters, a 16-μm-thick RuN electrode with a capacitance of 0.8 $F\,cm^{-2}$ was obtained. The enhanced capacitance of the formed electrode was also related to the film porosity.

The growth of porous structures offers several advantages for energy storage applications such as the shortening of the ion and electrons transport pad, the increase in the electrode–electrolyte contact area, and the exposure of active sites for charge storage. Gao et al. synthesized porous single crystal Nb_4N_5 (conductor) and Ta_3N_5 (semiconductor) and tested their behavior in symmetrical supercapacitors [24]. Both electrodes exhibited excellent stability (no degradation) during prolonged cycling; however, Nb_4N_5 showed considerably larger capacitance values (35.08 $F\,cm^{-2}$) compared to Ta_3N_5 (1.79 $F\,cm^{-2}$), with no degradation during prolonged cycling (10,000). This outstanding performance highlights the potentiality of this kind of material for energy storage applications.

Palraj et al. synthesized MoN crystals through a multistep methodology that involved a heat treatment at a temperature considerably lower (650 °C) than that employed to conventionally obtain transition metal nitrides, to synthesize TMN under mild conditions [25]. The material exhibited a specific capacitance of 383 $F\,g^{-1}$ at a specific current of 0.8 $A\,g^{-1}$ and retained 99% of its initial capacitance after 6000 charging–discharging cycles. The main inconvenience of MoN is the poor electrical conductivity compared to other TMNs. For this reason, Zhang et al. synthesized a bilayered film composed of TiN in the inner part of the film and $Ti_xMo_yN_z$ in the outermost part [26]. Taking advantage of the high conductivity of TiN, an inner layer of this material was first deposited to act as a current collector. Meanwhile, looking to tune the energy storage performance of the film, a mixed nitride ($Ti_xMo_yN_z$), responsible for the pseudocapacitive K^+ ion storage, was deposited in the outer layer that is in contact with the electrolyte. The transition metal's ratio tuned the film's conductivity and capacitance, reaching the best result (4.48 $F\,cm^{-3}$ at 0.01 $mA\,cm^{-2}$) when Mo:Ti was in a ratio of 2 : 1 in the film. Forming mixed transition metal nitrides is a viable strategy to improve the energy storage characteristic of these materials. Ranjan et al. synthesized mixed NiMoN films onto stainless steels to assemble flexible supercapacitors [27]. Ni incorporation into Mo_2N enhanced the areal capacitance from 34.94 mF cm^{-2} for pure Mo_2N to 43.11 mF cm^{-2}. Additionally, the NiMoN film retained its initial capacitance after 10,000 cycles. Although the bare NiN film exhibited poor electrochemical performance, the formation of the mixed nitride increased the currents and the capacitive potential window of the electrode.

Metal element doping is also a well-established strategy for improving the energy storage characteristics of TMN. Zhu et al. explored the impact of Li- and Na-doped GaN on its electrochemical performance as supercapacitors [28]. In both cases, 5% of the doping element was needed to increase the capacitance from 70 to 160 F g^{-1} (Li-doped GaN) and 188 F g^{-1} (Na-doped GaN). The authors attributed the improved behavior to the increment of the specific area, the total pore volume, and the increased affinity to proton hydrated provided by the doping. The authors also show through computational calculations that doping causes the reduction of the bandgap, increases the electron transition during the charge-transfer reaction, and induces an anti-d band center phenomenon that favors the adsorption ability of the material surface. Following a similar strategy, Wei et al. co-sputtered Cr–V nitride films to enhance the energy storage properties of CrN [29]. V incorporation enhances the film's conductivity and increases the capacitance to around 24.5 mF cm^{-2} at 0.5 mA cm^{-2} from around 14 mF cm^{-2} at 0.5 mA cm^{-2} for undoped CrN. Additionally, the film retained 93.9% of its initial capacitance after 20,000 charging–discharging cycles.

A common strategy to modulate interfacial properties is through the generation of heterojunctions, which overcomes individual flaws of the raw materials and provides synergistic interactions to enhance their performance. The formation of heterostructures offers a route to customize the electronic structure, enhance the surface area, and provide an electroactive center to increase the stored charge. Lakshmanan et al. synthesized graphitic carbon nitride/AlN composites through a thermal plasma arc discharge approach and self-assembling [30]. The size and morphology of the AlN can be controlled to some degree through the synthetic route. Meanwhile, the graphitic carbon nitride incorporation increases the area of the composite. The combination of these materials ended up in a composite that surpassed the performance of raw materials, reaching a capacitance of 434.1 F g^{-1} @ 1 A g^{-1}, which is larger than capacitances recorded for bare graphitic carbon nitride and AlN, 106.4 and 57.7 F g^{-1}, respectively. The composite retains 93.2% of its initial capacitance after 5000 cycles. In a similar strategy, Liu et al. formed VN/N-doped porous carbon composites to lessen the structural damage induced by the volume changes experienced by the nitride during energy storage [31]. The composite reached a specific capacitance of 272.3 F g^{-1} and retained 93% of its initial capacitance after 10,000 cycles.

The main challenge in nitrides is the precise control of composition under mild conditions. This will not only open the door to large-scale application of these materials but also boost their electrochemical performance by simplifying the defect and surface engineering of these materials, allowing to control the vacancy generation, the formation of a protective layer on top of the materials, and the intentional incorporation of redox-active centers to the materials.

26.4 Conducting Polymers

Conducting polymers (CPs) constitute another family of materials widely employed for supercapacitor application. The origin of the pseudocapacitive behavior exhibited by CPs is the reversible redox reaction paired with the doping/de-doping processes by ions in the electrolyte, which enables the transport of counterion dopants and ionic species through the polymer matrix. The widespread use of these materials relies on the low processing costs, processability at large-scale, high easy-to-tune electrical conductivity, and high capacitance values. For example, CP conductivity can be controlled by low processing temperatures, the length of the conjugated polymer chain, and the type of dopants. However, the main drawback of CPs is their poor stability during cycling due to volume swelling during ion doping and de-doping and the compromise of mechanical integrity when aging, compared to their inorganic counterparts for energy storage applications. Figure 26.4 illustrates the various strategies employed to address these flaws and enhance the stability and electrochemical performance of CPs.

One common strategy to improve CPs' electrochemical energy storage performance is hybridization. The growth of CPs over high surface area substrates enhances the stability of the polymer to cycling by buffering the volume changes experienced during the redox reactions. Chang et al. covalently linked aniline tetramers to carboxylic acid-functionalized carbon nanotubes [32]. This strategy achieves 80% retention of the initial capacitance after 20,000 cycles, which is four times larger than the 5000 cycles required when the aniline tetramers are not covalently bonded to the CNT. The variation in the performance was related to the detachment and dissolution of the polymer. Analogously, Saborio et al. synthesized a three-component composite (P[nBA-stat-BzMA], rGO, and Lignosulfonate) through emulsion polymerization and physical mixing, obtaining a material with improved electrochemical characteristics [33]. The composite reached a volumetric capacitance of 0.46 F cm^{-3} @ 5 mV s^{-1} in 1 M H_3PO_4. Using a carbon substrate takes advantage of its high surface area and inherent conductivity. Park et al. growth PANI nanorods over activated carbon derived from ZIF-8 [34]. The material

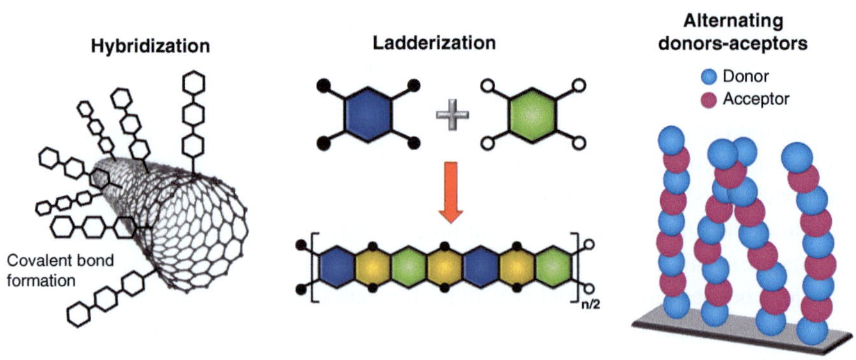

Figure 26.4 Schematic representation of the strategies employed to enhance the capacitance of conducting polymers.

ratio, monomer:oxidant ratio, temperature, anion doping, and growth time were optimized to achieve a material with exceptional energy storage capacitance. The composite reached a specific capacitance of 571.1 $F g^{-1}$ @ 1 $A g^{-1}$ and retained 75% of its initial capacitance after 5000 cycles at 5 $A g^{-1}$. Hernández-cortés et al. incorporated multiwalled carbon nanotubes into free-standing sponges of polypyrrole [35]. In this arrangement, the conducting polymer plays a bifunctional role, acting both as an active material for energy storage and as a current collector. A maximum capacitance of 450 $F g^{-1}$ at 0.5 $A g^{-1}$ was reached by controlling the electrodeposition time.

Hybridization can also be done with other conducting polymers used as current collectors. Lai et al. obtained a high-rate electrode by assembling a PANI crosslinked film over PEDOT:PSS-coated polyacrylonitrile nanofibrous film through electropolymerization [36]. The electrode reached an areal capacitance of 252 $mF cm^{-2}$ @ 0.1 $V s^{-1}$, and the solid symmetric supercapacitor assembled with this electrode retained up to 75% of its initial capacitance after 5000 cycles at 1 $V s^{-1}$. Additionally, the functionalization of the polymer may enhance its interaction with other materials, thereby drastically improving their performance. Poh et al. designed a polymer/MXene composite whose interaction was tuned by functionalizing PPy spheres with CTAB to promote the self-assembling of the composite and enhance the interaction with the MXene layer [37]. The composite reached a capacitance of 619.7 $F g^{-1}$ @ 5 $A g^{-1}$, retaining up to 81.6% of its initial capacitance after 10,000 cycles.

Conjugated ladder polymers are characterized by multiple strands of rings connected by two or more atoms, which offer high stability. Leng et al. synthesized ladder-type polyaniline analogs looking to improve the stability of polyaniline in energy storage applications [38]. The rings fused in these polymers restrict their conformational freedom and control the solubility but may sacrifice the conductivity. The conductivity of the new polymer, "P1a," was optimized to a value of 7 $mS cm^{-1}$. Surprisingly, this new material exhibited long-term stability during cycling, increasing its capacity by 65% with the number of cycles, compared with a loss of approximately 15% of the initial capacity measured for commercial polyaniline after the same number of cycles. Developing ladder polymers inspired by conventional CPs is a promising route for future generations of CPs with enhanced characteristics for energy storage applications.

The alternation of donor (D)–acceptor (A) units in conducting polymers is another recurrent strategy to lower the optical bandgap and increase the charge carriers' mobility, obtaining new materials with tuned characteristics for energy storage. Wavhal et al. synthesized perylenebisimide-based D–A conjugated polymer and incorporated MWCNT to provide higher area, stability, and conductivity [39]. The composite reached a capacitance of 92.9 $F g^{-1}$ at 0.5 $A g^{-1}$ and retained 70% of its initial capacitance after 45,000 cycles. On the other hand, Pati et al. employed an A–A design to synthesize a family of materials derived from benzodithiophenedione and diketopyrrolepyrrole [40]. The insertion of electron-donating thiophene rings derived in polymers improved electrochemical performance, reaching specific capacitances of 126.5 $F g^{-1}$ @ 0.5 $A g^{-1}$ and retaining up to 74.2% of its initial capacitance after 4000 cycles.

CPs main drawback is its poor cycling stability. However, several strategies drive improved behavior. Particularly interesting are the formation of ladder-type CPs and the alternation of D-A units, as both approaches enable the formulation of new CPs with finely tuned characteristics for energy storage applications. On the other hand, the generation of a nanocomposite of conducting polymer with other pseudocapacitive materials can also open up the opportunity to create composites with outstanding performance for supercapacitor applications.

References

1 Panchu, S.J., Raju, K., and Swart, H.C. (2024). Emerging two–dimensional intercalation pseudocapacitive electrodes for supercapacitors. *ChemElectroChem* 11 (15): e202300810.

2 Li, K., Li, J., Zhu, Q., and Xu, B. (2022). Three-dimensional MXenes for supercapacitors: a review. *Small Methods* 6 (4): 2101537.

3 Kadam, S.A., Kate, R.S., Peheliwa, V.M. et al. (2023). A comprehensive review on transition metal nitrides electrode materials for supercapacitor: syntheses, electronic structure engineering, present perspectives and future aspects. *J. Energy Storage* 72: 108229.

4 Suriyakumar, S., Bhardwaj, P., Grace, A.N., and Stephan, A.M. (2021). Role of polymers in enhancing the performance of electrochemical supercapacitors: a review. *Batteries Supercaps* 4 (4): 571–584.

5 Anasori, B., Lukatskaya, M.R., and Gogotsi, Y. (2017). 2D metal carbides and nitrides (MXenes) for energy storage. *Nat. Rev. Mater.* 2 (2).

6 Bo, Z., Chen, Y., Yu, Q. et al. (2024). Unveiling the energy storage mechanism of MXenes under acidic conditions through transitions of surface functionalizations. *J. Phys. Chem. C* 128 (6): 2352–2361.

7 Zhang, M., Héraly, F., Yi, M., and Yuan, J. (2021). Multitasking tartaric-acid-enabled, highly conductive, and stable MXene/conducting polymer composite for ultrafast supercapacitor. *Cell Rep. Phys. Sci.* 2 (6): 100449.

8 Zhu, J., Zhu, S., Cui, Z. et al. (2024). Dual redox reaction sites for pseudocapacitance based on Ti and –P functional groups of $Ti_3C_2PBr_x$ MXene. *Angew. Chem. Int. Ed.* 63 (27): e202403508.

9 Yang, C., Tang, Y., Wei, C., and Que, W. (2022). Improved volumetric pseudocapacitance electrode obtained by solvothermal treatment of diethanolamine and different auxiliary solvents with MXene. *J. Alloys Compd.* 898: 162882.

10 Sun, P., Liu, J., Liu, Q. et al. (2022). Nitrogen and sulfur co-doped MXene ink without additive for high-performance inkjet-printing micro-supercapacitors. *Chem. Eng. J.* 450: 138372.

11 Ibragimova, R., Rinke, P., and Komsa, H.P. (2022). Native vacancy defects in MXenes at etching conditions. *Chem. Mater.* 34 (7): 2896–2906.

12 Chen, G., Xie, Y., Tang, Y. et al. (2024). Unraveling the role of metal vacancy sites and doped nitrogen in enhancing pseudocapacitance performance of defective MXene. *Small* 20 (12): 2307408.

13 Sikdar, A., Héraly, F., Zhang, H. et al. (2024). Hierarchically porous 3D free-standing holey-MXene framework via mild oxidation of self-assembled MXene hydrogel for ultrafast pseudocapacitive energy storage. *ACS Nano* 18 (4): 3707–3719.

14 Zhao, J., Li, Q., Shang, T. et al. (2021). Porous MXene monoliths with locally laminated structure for enhanced pseudo-capacitance and fast sodium-ion storage. *Nano Energy* 86: 106091.

15 Hegde, S., Hakkeem, H., Alagar, S. et al. (2025). Laser-driven engineering of modulated structural defects in Nb2CTx MXene-rGO based flexible microsupercapacitors for energy storage application. *J. Energy Storage* 111: 115330.

16 Ambade, R.B., Ambade, S.B., Veerasubramani, G.K. et al. (2024). Unlocking highly stable and reversible in-situ integration of defect-rich 2D vanadium sulfide with Ti3C2Tx MXene heterostructures: boosting asymmetric supercapacitor performance. *J. Power Sources* 608: 234615.

17 Tomar, A.K., Kshetri, T., Kim, N.H., and Lee, J.H. (2022). Cation and anion (de)intercalation into MXene/Perovskite oxides for high-rate intercalation pseudocapacitance. *Energy Storage Mater.* 50: 86–95.

18 Sun, L., Chen, Y., Zhuo, K. et al. (2023). Nitrogen-doped Ti3C2Tx/TiO2-graphene hybrid materials with 3D interconnected structure compatible with ionic liquid-based electrolyte for high performance supercapacitors. *Electrochim. Acta* 466: 143005.

19 Sun, Y., Li, T., Liu, X. et al. (2024). Modulating oxygen vacancies in MXene/MoO3-x smart fiber by defect engineering for ultrahigh volumetric energy density supercapacitors and wearable SERS sensors. *Chem. Eng. J.* 494: 152911.

20 Ali, M. and Alqahtani, S.M. (2023). Role of defects in graphene-passivated Ti3C2 MXene for energy conversion and storage applications: a first-principles study. *ACS Appl. Energy Mater.* 6 (14): 7535–7544.

21 Mahajan, P., Sardana, S., and Mahajan, A. (2024). 1D graphene nanoribbons-mediated defect engineering in 2D MXene for high-performance supercapacitors. *Appl. Phys. Lett.* 124 (11): 111602.

22 Li, X.-L., Song, H., Zhang, Y.-H. et al. Capacitance decay mechanism of vanadium nitride supercapacitor electrodes in KOH electrolytes. *Rare Metals* 1–11.

23 Dinh, K.H., Whang, G., Huve, M. et al. (2024). High capacitance porous ruthenium nitride films with high rate capability for micro-supercapacitors. *Small* 20 (40): 2402607.

24 Gao, X., Ma, G., Luo, C. et al. (2024). Porous single-crystal nitrides for enhanced pseudocapacitance and stability in energy storage applications. *Adv. Sci.* 12 (1): 2410429.

25 Palraj, J. and Therese, H.A. (2025). Phase-pure 2D molybdenum nitride (Mo2N) symmetric supercapacitor: paving the way for superior energy storage. *J. Energy Storage* 107: 114957.

26 Zhang, B., Wang, M., Sui, J. et al. (2025). Titanium molybdenum nitride/titanium nitride laminated films prepared by a high-throughput method for on-chip microsupercapacitors. *Electrochim. Acta* 514: 145687.

27 Ranjan, B. and Kaur, D. (2024). Pseudocapacitance powered nickel molybdenum nitride nanocomposite reactively cosputtered on stainless-steel mesh toward advanced flexible supercapacitors. *ACS Appl. Energy Mater.* 7 (10): 4513–4527.

28 Zhu, Y., Zhou, K., Liu, W. et al. (2025). Alkali metal ion-doped GaN for ultra-fast electrochemical capacitor: doping mechanism, structural adjustment, and structure–performance relationship. *Chem. Eng. J.* 504: 158761.

29 Wei, B., Wu, J., Hou, Z. et al. (2024). Binder-free V-doped CrN thin film electrode enables high performance symmetric supercapacitor. *J. Alloys Compd.* 977: 173424.

30 Lakshmanan, K., Chidambaram, S., and Gurusamy, S. (2024). High-performance supercapacitor with plasma-assisted AlN and graphitic carbon nitride composite electrode. *ACS Appl. Electron. Mater.* 6 (7): 5101–5116.

31 Liu, L., Zhou, X., Wang, W. et al. (2024). Vanadium nitride/nitrogen-doped porous carbon prepared by one-pot molten salt pyrolysis for all-solid-state asymmetric supercapacitors. *J. Alloys Compd.* 1006: 176274.

32 Chang, X., Yang, Z., Huang, A. et al. (2024). Understanding the degradation mechanisms of conducting polymer supercapacitors. *Macromol. Rapid Commun.* 45 (1): 2300237.

33 Saborio, M.G., Privat, K., Tran, B.N. et al. (2022). Polymer/reduced graphene oxide/lignosulfonate nanocomposite films as pseudocapacitor cathodes. *ACS Appl. Nano Mater.* 5 (3): 3686–3700.

34 Park, J., Kim, D., Kim, S. et al. (2024). Controlled growth of redox-active polymer nanorods on MOF-derived activated nanoporous carbons: 3D reconstruction study and high-performance supercapacitor. *Chem. Eng. J.* 498: 155122.

35 Hernández-Cortés, G., Acevedo-Peña, P., Díaz-Góngora, J.A.I. et al. (2023). Self-standing multiwalled carbon nanotubes/polypyrrole foams for aqueous symmetric supercapacitors. *Synth. Met.* 299: 117479.

36 Lai, H., Bai, C., Wang, Y. et al. (2022). Highly crosslinked conductive polymer nanofibrous films for high-rate solid-state supercapacitors and electromagnetic interference shielding. *Adv. Mater. Interfaces* 9 (9): 2102115.

37 Poh, W.S., Hoo, D.Y., Zheng, S.Q. et al. (2024). Ti3C2TX encapsulation of CTAB-functionalized polypyrrole nanospheres towards pseudocapacitive intensification in supercapacitors. *Chem. Eng. J.* 497: 154440.

38 Leng, M., Koripally, N., Huang, J. et al. (2023). Synthesis and exceptional operational durability of polyaniline-inspired conductive ladder polymers. *Mater. Horiz.* 10 (10): 4354–4364.

39 Wavhal, A.A., Dilwale, S., Puthiyaveetil, P.P. et al. (2024). High-voltage symmetric supercapacitor electrodes via in situ synthesized multiwalled carbon nanotube-doped perylenebisimide-based donor-acceptor conjugate polymer P(PDI2OD-T2). *J. Phys. Chem. C* 128 (31): 12808–12821.

40 Pati, S.K., Patra, D., Muduli, S. et al. (2023). Energy storage application of conducting polymers featuring dual acceptors: exploring conjugation and flexible chain length effects. *Small* 19 (26): 2300689.

Index

Surface and Interfacial Defects in Nanomaterials for Sustainable Energy Production and Storage, First Edition. Edited by Noé Arjona, Lorena Álvarez Contreras, and Minerva Guerra Balcázar.

d

f